Springer INdAM Series

Volume 45

This series will publish textbooks, multi-authors books, thesis and monographs in English language resulting from workshops, conferences, courses, schools, seminars, doctoral thesis, and research activities carried out at INDAM - Istituto Nazionale di Alta Matematica, http://www.altamatematica.it/en. The books in the series will discuss recent results and analyze new trends in mathematics and its applications.
THE SERIES IS INDEXED IN SCOPUS

More information about this series at http://www.springer.com/series/10283

Paolo Ciatti • Alessio Martini
Editors

Geometric Aspects of Harmonic Analysis

Springer

Editors
Paolo Ciatti
Dipartimento di Ingegneria Civile,
Edile e Ambientale - ICEA
University of Padova
Padova, Italy

Alessio Martini
School of Mathematics
University of Birmingham
Birmingham, UK

ISSN 2281-518X ISSN 2281-5198 (electronic)
Springer INdAM Series
ISBN 978-3-030-72057-5 ISBN 978-3-030-72058-2 (eBook)
https://doi.org/10.1007/978-3-030-72058-2

This Springer imprint is published by the registered company Springer Nature Switzerland AG.
The registered company address is: Gewerbestrasse 11, 6330 Cham, Switzerland

To Fulvio Ricci, with gratitude.

Preface

On 25–29 June 2018 the INdAM Meeting "Geometric Aspects of Harmonic Analysis" took place in Cortona. This conference, which saw the participation of over 120 mathematicians from around the world, was organised on the occasion of Fulvio Ricci's 70th birthday.

This short introduction is not meant to discuss the interest and relevance of Fulvio Ricci's mathematical contributions, which are witnessed by his bright career, the quality of his scientific production, the awards he received and the level of the scholars who participated in the conference. Some words in that direction can be found in the letter by Elias Stein included in this volume. Instead, we would like to express our appreciation of Fulvio and our gratitude to him for the humanity, the rigour, the fairness he has always shown in mathematics and life and, last but not least, his great openness to interact and collaborate with mathematicians of all ages and from all over the world.

This volume originated in talks given in Cortona and presents timely syntheses of several major fields of mathematics as well as original research articles contributed by some of the finest mathematicians working in these areas.

It is our pleasure to thank all the organisations that contributed generously to the conference with their financial support: the Istituto Nazionale di Alta Matematica-INdAM, the Clay Mathematics Institute, the US National Science Foundation, the Scuola Normale Superiore di Pisa, the Università degli Studi di Milano Bicocca, and the Università degli Studi di Padova. Special thanks are due to the University of Wisconsin–Madison, which kindly hosted the website of the conference.

On behalf of the entire organising committee of the conference we would like to acknowledge our great appreciation to the director of INdAM, Professor Giorgio Patrizio, and to the former director of SNS, Professor Vincenzo Barone. Their efforts and suggestions helped to make this a most fruitful and enjoyable meeting.

We are also pleased to thank all the speakers for the distinguished and outstanding lectures they gave.

It is our pleasure to thank all people working at the Centro Convegni Sant'Agostino and the Palazzone, which were the meeting's venues, for their friendliness, kindness and effectiveness; special thanks are particularly due to Mrs Rita Santiccioli and Mrs Benedetta Biagiotti.

We owe a debt of gratitude to all the other organisers of the conference: Luigi Ambrosio, Gian Maria Dall'Ara, Bianca Di Blasio, and especially the US organisers Loredana Lanzani, Betsy Stovall and Brian Street, who applied for the NSF funding.

Finally, we would like to take this opportunity to thank all the participants in the conference. We hope that the warm atmosphere of those days in Cortona will be a nice memory for all of them.

Padova, Italy
Birmingham, UK
October 2020

Paolo Ciatti
Alessio Martini

At the Occasion of Fulvio's Conference

Elias M. Stein could not attend the June 2018 conference in Cortona. Instead, he sent a letter, which was read during the conference and is reproduced below.

Dear Fulvio and friends,

I'm sorry that I'm missing this wonderful celebration in your honor, Fulvio—I can only blame my overly cautious doctor for this. But I want to take this opportunity to say a few words of appreciation of your many remarkable achievements, and then indulge in a few reminiscences.

First, we all know and recognize that your work continues to have broad impact and wide influence—indeed your efforts have played a major role in transforming a number of diverse area in analysis. Your constant urge to try to look at things differently, your deep insights, great energy, and your keen appreciation for what is really important, has made all of this possible. In working with others (you've had at least 20 collaborators), your wisdom and warmth have brought out the best in your coworkers, and in many cases made them even better than they thought possible, as I can readily attest.

I will indicate the sweep of your interests and contributions by sketching only a partial list of the main areas of your work.

- Harmonic analysis of singular integrals of Radon-type on nilpotent groups.
- Geometry and analysis of non-symmetric harmonic spaces, and the study of their boundary groups.
- The theory of solvability of invariant differential operators on the Heisenberg group.
- The study of maximal functions and singular integrals associated to polynomial maps.
- Spectral multipliers on the Heisenberg group, their connection with the Hodge-Laplacian, and the origin of flag kernels.
- The general theory of operators with flag kernels on nilpotent groups, and most recently, the theory of singular integrals controlled by multiple norms.

Fulvio—allow me now to come to some personal recollections. I'm not sure when we first met. It might have been before 1980, but we really got to know each other a few years later when you came to the Institute for the whole academic year with Sandra and Alberto. We began working together then, and wrote a nice (but forgettable) paper. However what was important is that we learned to appreciate each other, that we had mathematical empathy, and that we could easily talk together in that common language we both loved.

There followed a series of visits by you in Princeton, and by me in Torino. Besides all the mathematics we did together—which I will always treasure—I remember with nostalgia the hotel Bologna near the train station, the cafes in the elegant Piazza San Carlo, and the pleasant walks to the Politecnico where we worked all day, interrupted only by lunch (not at a mensa!), but with paninis in the nice cafes in the area.

We also had the good fortune to twice spend one-week stays during the summer (with our families and a few friends) at the Villa Ronconi, right on the shore of Lake Como, with its marvelous grounds and stunning views. However, soon thereafter my university, in its wisdom, decided to dispose of this unique holding, and we were thus expelled from our own private paradise. Nevertheless, a few years later we had the lucky chance to spend (again with our families and some good friends) a summer month in Berkeley. While not paradise, Berkeley and its surroundings were the next best thing on earth! It was there that Alex Nagel joined our collaboration, and a few years later we also attracted Steve Wainger to our common effort.

And now, after these few warm recollections of the past, I come to some words about the present and future. Having myself passed this milestone a number of years ago, I can say with some certainty that this is a new beginning—maybe not what one would like as an ideal starting point—but nevertheless bracing, full of interesting challenges to undertake and try to master, and rich in the achievements that can be hoped for, and the joy and satisfaction they entail. So with this in view, I wish you all the best of fortune in your further life and adventures!

Happy birthday!
Eli

Contents

An Extension Problem and Hardy Type Inequalities for the Grushin Operator

Rakesh Balhara, Pradeep Boggarapu, and Sundaram Thangavelu

Dedicated to Professor F. Ricci on his 70th birthday

Abstract In this paper we study the extension problem associated to the Grushin operator $G = -\Delta - |x|^2\partial_w^2$ on $\mathbb{R}^{n+1}$ and use the solutions to prove trace Hardy and Hardy inequalities for fractional powers of G.

Keywords Grushin operator · Extension problem · Hardy and trace Hardy Inequalities

1 Introduction and Main Results

In this article we are interested in proving Hardy type inequalities for fractional powers of the Grushin operator $G = -\Delta - |\xi|^2\partial_w^2$ on $\mathbb{R}^{n+1}$. Recall that in the case of Laplacian Δ on $\mathbb{R}^n$ such inequalities are well known and there is a vast literature on the topic. For $0 < s < 1$, two kinds of Hardy inequalities for the fractional powers $(-\Delta)^{s/2}$ have been studied. The inequality

$$((-\Delta)^{s/2} f, f) \geqslant c_{n,s} \int_{\mathbb{R}^n} \frac{|f(x)|^2}{(1+|x|^2)^s}\, dx$$

R. Balhara · S. Thangavelu (✉)
Department of Mathematics, Indian Institute of Science, Bangalore, India
e-mail: veluma@iisc.ac.in

P. Boggarapu
Department of Mathematics, BITS Pilani, KK Birla Goa Campus, Zuarinagar, Goa, India
e-mail: pradeepb@goa.bits-pilani.ac.in

P. Ciatti, A. Martini (eds.), *Geometric Aspects of Harmonic Analysis*,
Springer INdAM Series 45, https://doi.org/10.1007/978-3-030-72058-2_1

with a sharp explicit constant $c_{n,s}$ is known as Hardy's inequality with non-homogeneous weight function whereas the inequality

$$((-\Delta)^{s/2} f, f) \geqslant C_{n,s} \int_{\mathbb{R}^n} \frac{|f(x)|^2}{|x|^s} dx$$

is the Hardy inequality with homogeneous weight. The constant $C_{n,s}$ is also known to be sharp and explicit. It is of interest to prove such inequalities when Δ is replaced by more general elliptic/subelliptic operator. A particularly interesting case is the one where we have the sublaplacian $\mathcal{L}$ on Heisenberg groups $\mathbb{H}^n$ in place of the Laplacian Δ. In the articles [13] and [14] the authors have established Hardy type inequalities for (conformally invariant) fractional powers of $\mathcal{L}$.

In this work we are mainly interested in proving Hardy type inequalities for fractional powers of G. There are several ways of proving Hardy inequalities for the Laplacian, see [2, 10] and [20]. For the case of sublaplacian $\mathcal{L}$ the authors in [13] have used the method of ground state representation developed by Frank, Lieb and Seiringer [9] in proving a version of Hardy inequality for the sublaplacian with non-homogeneous weight. Later, in [14] the same authors have used a different method in proving analogues of both inequalities making use of solutions of the so called extension problem for the sublaplacian. The extension problem for the Laplacian studied by Caffarelli and Silvestre [4] deals with the initial value problem

$$(\Delta + \partial_\rho^2 + \frac{1-s}{\rho}\partial_\rho)u(x,\rho) = 0, \quad u(x,0) = f(x), x \in \mathbb{R}^n, \rho > 0.$$

The solutions of this problem can be written down explicitly and using them one proves the following inequality known as trace Hardy inequality: for reasonable real valued functions φ from the domain of $(-\Delta)^{s/2}$ one has

$$\int_0^\infty \int_{\mathbb{R}^n} |\nabla_{x,\rho} u(x,\rho)|^2 \rho^{1-s} dx d\rho \geq c_s \int_{\mathbb{R}^n} u(x,0)^2 \frac{(-\Delta)^{s/2}\varphi(x)}{\varphi(x)} dx$$

valid for all real valued functions $f \in C_0^\infty(\mathbb{R}^{n+1})$. When u is a solution of the extension problem with initial condition f, the left hand side of the above reduces to a constant multiple of $((-\Delta)^{s/2} f, f)$. Further, the choice $\varphi(x) = (1+|x|^2)^{-(n-s)/2}$ allows us to simplify the right hand side and we obtain the Hardy inequality

$$((-\Delta)^{s/2} f, f) \geq c_{n,s} \int_{\mathbb{R}^n} \frac{|f(x)|^2}{(1+|x|^2)^s} dx.$$

When $f(x) = (1+|x|^2)^{-(n-s)/2}$ both sides of the above inequality are equal with $c_{n,s} = 2^s \frac{\Gamma(\frac{n+s}{2})}{\Gamma(\frac{n-s}{2})}$.

All of these are well known in the case of the Laplacian on $\mathbb{R}^n$. Recently, in [14] the authors have carried out similar analysis for the sublaplacian $\mathcal{L}$ on the

Heisenberg group $\mathbb{H}^n$. Our aim in this article is to show that the same analysis can be done also for the case of the Grushin operator. Thus we will be studying the extension problem for the Grushin operator and use the solutions to prove trace Hardy and Hardy inequalities for fractional powers of the Grushin operator.

In the Euclidean case, an important role is played by the identity

$$(-\Delta)^{s/2}(1+|x|^2)^{-(n-s)/2} = 2^s \frac{\Gamma(\frac{n+s}{2})}{\Gamma(\frac{n-s}{2})}(1+|x|^2)^{-(n+s)/2}$$

which follows from the transformation property of the Macdonald function. This can be easily proved by taking the Fourier transform: writing

$$\varphi_s(x) = (1+|x|^2)^{-(n+s)/2} = \frac{1}{\Gamma(\frac{n+s}{2})}\int_0^\infty e^{-t(1+|x|^2)} t^{\frac{n+s}{2}-1} dt$$

and taking the Fourier transform we see that

$$\widehat{\varphi_s}(\xi) = \frac{(4\pi)^{-n/2}}{\Gamma(\frac{n+s}{2})}\int_0^\infty e^{-t} e^{-\frac{1}{4t}|\xi|^2} t^{\frac{s}{2}-1} dt.$$

The integral on the right hand side is given in terms of the Macdonald function $K_{-s/2}$, see [12], page 407):

$$K_{-s/2}(|\xi|^2) = 2^{s/2-1}|\xi|^{-s}\int_0^\infty e^{-t} e^{-\frac{1}{4t}|\xi|^2} t^{\frac{s}{2}-1} dt.$$

The change of variables $u = \frac{|\xi|^2}{4t}$ proves that

$$|\xi|^s \widehat{\varphi_{-s}}(\xi) = 2^s \frac{\Gamma(\frac{n+s}{2})}{\Gamma(\frac{n-s}{2})} \widehat{\varphi_s}(\xi)$$

as desired. The corresponding identity used in the case of the sublaplacian $\mathcal{L}$ is the Cowling-Haagerup [6] formula

$$\mathcal{L}_s((1+|z|^2)^2+16t^2)^{-(n+1-s)/2} = 4^{2s}\frac{\Gamma(\frac{n+1+s}{2})^2}{\Gamma(\frac{n+1-s}{2})^2}((1+|z|^2)^2+16t^2)^{-(n+1+s)/2}.$$

This identity is a consequence of certain transformation property of the Kummer's function. In our case we make use of the following relation which is the analogue of the above identity:

$$\tilde{G}_s((1+|\xi|^2)^2+w^2)^{-(n+2-2s)/4} = 2^{2s}\frac{\Gamma(\frac{n+2+2s}{4})^2}{\Gamma(\frac{n+2-2s}{4})^2}((1+|\xi|^2)^2+w^2)^{-(n+2+2s)/4}.$$

Here $\tilde{G}_s$ stands for the conformally invariant fractional power of the Grushin operator and the above identity can be proved by expanding the functions involved in terms of Laguerre functions and making use of an identity proved for Laguerre operators in [5].

2 Preliminaries on the Grushin Operator

By the Grushin operator we mean the degenerate elliptic operator $G = -\Delta - |\xi|^2\partial_w^2$, $\xi \in \mathbb{R}^n$, $w \in \mathbb{R}$ on $\mathbb{R}^{n+1}$. Here Δ stands for the standard Laplacian on $\mathbb{R}^n$. When f is an integrable function on $\mathbb{R}^{n+1}$ let

$$f^\lambda(\xi) = \int_{-\infty}^{\infty} f(\xi, w)e^{i\lambda w}dw$$

stand for the inverse Fourier transform of f in the last variable. Then it follows that $(Gf)^\lambda(\xi) = H(\lambda)f^\lambda(\xi)$ where $H(\lambda) = -\Delta + \lambda^2|x|^2$ is the scaled Hermite operator on $\mathbb{R}^n$. The spectral decomposition of G can be written in terms of Hermite expansions. Let $P_k(\lambda)$ stand for the projections of $L^2(\mathbb{R}^n)$ onto the $k-$th eigenspace of $H(\lambda)$ with eigenvalue $(2k+n)|\lambda|$ so that

$$H(\lambda) = \sum_{k=0}^{\infty}((2k+n)|\lambda|)P_k(\lambda).$$

Then the spectral decomposition of G is given by

$$Gf(\xi, w) = (2\pi)^{-1}\int_{-\infty}^{\infty} e^{-i\lambda w}\Big(\sum_{k=0}^{\infty}((2k+n)|\lambda|)P_k(\lambda)f^\lambda(\xi)\Big)d\lambda.$$

For a bounded function m defined on the spectrum of G, viz. $\mathbb{R}^+$ we can define the operator $m(G)$ by

$$m(G)f(\xi, w) = (2\pi)^{-1}\int_{-\infty}^{\infty} e^{-i\lambda w}\Big(\sum_{k=0}^{\infty}m((2k+n)|\lambda|)P_k(\lambda)f^\lambda(\xi)\Big)d\lambda$$

which is clearly a bounded linear operator on $L^2(\mathbb{R}^{n+1})$. The choice $m(a) = e^{-ta}, t > 0$ leads to the heat semigroup e^{-tG} generated by the Grushin operator. For information on the spectral theory of the Hermite operator we refer to [17].

We make use of the following representation of the Heisenberg group $\mathbb{H}^n$ in order to transfer operators in the Heisenberg setting into the setting of Grushin. On $L^2(\mathbb{R}^{n+1})$ we define the representation π by

$$\pi(z, t)f(\xi, w) = f(\xi - y, w - t - \xi \cdot x + \frac{1}{2}x \cdot y), \quad f \in L^2(\mathbb{R}^{n+1}),\ z = x + iy.$$

It is easy to see that π is a strongly continuous unitary representation of $\mathbb{H}^n$ on $L^2(\mathbb{R}^{n+1})$. More generally, for any $f \in L^p(\mathbb{R}^{n+1})$, $1 \le p \le \infty$, we can check that $\pi(z,t)f$ converges to f in $L^p(\mathbb{R}^{n+1})$ as (z,t) goes to 0. The connection between the sublaplacian $\mathcal{L}$ and the Grushin operator G arises from the following. We can easily check that $\pi(X_j) = -\xi_j \frac{\partial}{\partial w}$ and $\pi(Y_j) = -\frac{\partial}{\partial \xi_j}$ where $X_j = \frac{\partial}{\partial x_j} + \frac{1}{2} y_j \frac{\partial}{\partial t}$ and $Y_j = \frac{\partial}{\partial y_j} - \frac{1}{2} x_j \frac{\partial}{\partial t}$ are the vector fields on $\mathbb{H}^n$ which along with $T = \frac{\partial}{\partial t}$ form a basis for the Heisenberg Lie algebra. Thus we see that $\pi(\mathcal{L}) = G$ and this allows us to express certain functions of G in terms of operators related to $\mathcal{L}$. For example, the heat semigroup e^{-tG} generated by the Grushin operator can be written as

$$e^{-tG} f(\xi, w) = \int_{\mathbb{H}^n} q_t(z,a)\pi(z,a) f(\xi, w)\, dz da \tag{1}$$

where $q_t(z,a)$ stands for the heat kernel associated to $\mathcal{L}$. A simple proof of this goes as follows.

$$\int_{\mathbb{H}^n} q_t(z,a)\pi(z,a)f(\xi,w)\,dzda = \int_{-\infty}^{\infty}\int_{\mathbb{C}^n} q_t(z,a) f(\xi - y, w - a - \xi\cdot x + \frac{1}{2} x\cdot y)\,dzda$$

which by Plancherel theorem for the Euclidean Fourier transform simplifies to

$$\int_{\mathbb{C}^n}\int_{-\infty}^{\infty} e^{-i\lambda w} e^{i\lambda(-x\cdot\xi + \frac{1}{2}x\cdot y)} q_t^\lambda(z) f^\lambda(\xi - y)\, d\lambda dz.$$

Recalling the definition of the Schrödinger representation $\pi_\lambda(z,a)$ of $\mathbb{H}^n$ (see [18]) and using the fact that $q_t(x+iy,a)$ is even in y we get

$$\int_{\mathbb{H}^n} q_t(z,a)\pi(z,a)f(\xi,w)\,dzda = \int_{-\infty}^{\infty}\int_{\mathbb{C}^n} e^{-i\lambda w} q_t^\lambda(z)\pi_\lambda(z,0) f^\lambda(\xi)\, dz d\lambda.$$

But it is well known that

$$\int_{\mathbb{C}^n} q_t^\lambda(z)\pi_\lambda(z,0)\,dz = \int_{\mathbb{H}^n} q_t(z,a)\pi_\lambda(z,a)\,dzda = e^{-tH(\lambda)}$$

(see Sections 2.8, 2.9 in [18]). In view of the spectral resolution of G we obtain the desired representation:

$$\int_{\mathbb{H}^n} q_t(z,a)\pi(z,a)f(\xi,w)\,dzda = e^{-tG}f(\xi,w).$$

The above representation gives us an easy proof of the fact that $e^{-tG}f$ converges to f in $L^p(\mathbb{R}^{n+1})$ for all $1 \le p < \infty$. This can be seen as follows. It is well known that q_t is a Schwartz class function $\mathbb{H}^n$ with $\int_{\mathbb{H}^n} q_t(z,a)\,dzda = 1$ and it satisfies $q_t(z,a) = t^{-n-1} q_1(t^{-1/2}z, t^{-1}a)$. Therefore, making a change of variables in (1) we get

$$e^{-tG}f - f = \int_{\mathbb{H}^n} q_1(z,a)\big(\pi(t^{1/2}z, ta)f - f\big)dzda \tag{2}$$

from which our claim is immediate. More generally, the following is true and we will make use of it later.

Lemma 1 *Suppose $\varphi \in L^1(\mathbb{H}^n)$ with $\int_{\mathbb{H}^n} \varphi(z,a)dzda = c$ and for $t > 0$ define $\varphi_t(z,a) = t^{-n-1}\varphi(t^{-1/2}z, t^{-1}a)$. Then for any $f \in L^p(\mathbb{R}^{n+1})$, $1 \leq p < \infty$, $\pi(\varphi_t)f$ converges to cf in the norm as $t \to 0$.*

3 An Extension Problem for the Grushin Operator

In this section we study the following extension problem for the Grushin operator G. Given $f \in L^p(\mathbb{R}^{n+1})$ we are interested in finding solutions $u(\xi, w, \rho)$ of the equation

$$\Big(-G+\partial_\rho^2+\frac{1-2s}{\rho}\partial_\rho+\frac{1}{4}\rho^2\partial_w^2\Big)u(\xi,w,\rho)=0, \quad u(\xi,w,0)=f(\xi,w). \tag{3}$$

It might appear to be natural to study the extension problem

$$\Big(-G+\partial_\rho^2+\frac{1-2s}{\rho}\partial_\rho\Big)u(\xi,w,\rho)=0, \quad u(\xi,w,0)=f(\xi,w) \tag{4}$$

instead of the above. However, the problem (3) is more suitable for the study of trace Hardy and Hardy inequalities. The solutions of (4) are related to pure powers G^s of the Grushin operator whereas those of (3) are related to the conformally invariant fractional powers G_s (see Section 4 for the definitions of G^s and G_s). A solution of (4) is given by the following formula of Stinga–Torrea [15]:

$$u(\xi,w,\rho)=\frac{\rho^{2s}}{\Gamma(s)}\int_0^\infty e^{-\frac{1}{4t}\rho^2}e^{-tG}f(\xi,w)t^{-s-1}dt \tag{5}$$

where e^{-tG} is the heat semigroup generated by G. Then it is not difficult to see that $u(\xi,w,\rho)$ solves (4) and $u(\xi,w,\rho)$ converges to f in $L^p(\mathbb{R}^{n+1})$ as $\rho \to 0$ for $1 \leq p < \infty$. It is also known, see [15], that $\rho^{1-2s}\partial_\rho u(\xi,w,\rho)$ converges to a constant multiple of $G^s f$ as $\rho \to 0$.

By modifying the Stinga–Torrea formula (5) we can also write down a solution of the extension problem (3). Let $p_{t,s}(\rho, w)$ be the heat kernel associated to the generalised sublaplacian (see [1])

$$\mathcal{L}(s)=\partial_\rho^2+\frac{1+2s}{\rho}\partial_\rho+\frac{1}{4}\rho^2\partial_w^2$$

on $\mathbb{R}^+ \times \mathbb{R}$. Then the solution of the above extension problem can be written down explicitly in terms of the function $e^{-tG}f$. Indeed, we have the following analogue of the Stinga–Torrea formula.

Theorem 2 *For $f \in L^p(\mathbb{R}^{n+1})$, $1 \leq p \leq \infty$ a solution of the extension problem (3) is given by*

$$u(\xi, w, \rho) = \rho^{2s} \int_0^\infty \int_{-\infty}^\infty p_{t,s}(\rho, w') e^{-tG} f(\xi, w - w') dw' dt. \tag{6}$$

As ρ tends to zero, the solution $u(\xi, w, \rho)$ converges to $C_s f$ in $L^p(\mathbb{R}^{n+1})$ for $1 \leq p < \infty$ where $C_s = \frac{1}{4}\Gamma(s)\pi^{-s-1}$.

Proof Applying G to the function u and noting that $e^{-tG}f(\xi, w)$ satisfies the heat equation $-Gu_t(\xi, w) = \partial_t u_t(\xi, w)$ we see that

$$Gu(\xi, w, \rho) = -\rho^{2s} \int_0^\infty \int_{-\infty}^\infty p_{t,s}(\rho, w') \partial_t e^{-tG} f(\xi, w - w') dw' dt.$$

Integrating by parts in the t variable we can transfer the t derivative to $p_{t,s}(\rho, w)$ and since it satisfies the heat equation associated to $\mathcal{L}(s)$ we obtain

$$Gu(\xi, w, \rho) = \rho^{2s} \left(\partial_\rho^2 + \frac{1+2s}{\rho}\partial_\rho + \frac{1}{4}\rho^2\partial_w^2\right) \times \int_0^\infty \int_{-\infty}^\infty p_{t,s}(\rho, w') \partial_t e^{-tG} f(\xi, w - w') dw' dt.$$

A simple calculation shows that

$$\rho^{2s}\left(\partial_\rho^2 + \frac{1+2s}{\rho}\partial_\rho + \frac{1}{4}\rho^2\partial_w^2\right) v(\xi, w, \rho) = \left(\partial_\rho^2 + \frac{1-2s}{\rho}\partial_\rho + \frac{1}{4}\rho^2\partial_w^2\right)(\rho^{2s} v(\xi, w, \rho))$$

for any function $v(\xi, w, \rho)$. This proves that u satisfies the extension problem.

Since the heat semigroup e^{-tG} is contractive on L^p spaces it follows that

$$\|u(\cdot, \rho)\|_p \leq \rho^{2s} \left(\int_0^\infty \int_{-\infty}^\infty p_{t,s}(\rho, w') dw' dt \right) \|f\|_p.$$

We also know that (see [1])

$$\int_{-\infty}^\infty p_{t,s}(\rho, w) e^{i\lambda w} dw = (4\pi)^{-s-1} \left(\frac{\lambda}{\sinh(t\lambda)} \right)^{s+1} e^{-\frac{1}{4}\lambda \coth(t\lambda)\rho^2}.$$

In view of this we obtain

$$\|u(\cdot,\rho)\|_p \leq C\rho^{2s}\Big(\int_0^\infty e^{-\frac{1}{4t}\rho^2}t^{-s-1}dt\Big)\|f\|_p \leq C\|f\|_p.$$

In order to prove that $u(\cdot,\rho)$ converges to f we make use of the fact that $e^{-tG}f$ converges to f in $L^p(\mathbb{R}^{n+1})$ as t tends to zero for $1 \leq p < \infty$. From the explicit form of $p_{t,s}(\rho, w)$ we note that $p_{\rho^2 t,s}(\rho, w) = \rho^{-2s-4}p_{t,s}(1, w/\rho^2)$. Thus the solution u of the extension problem is given by the integral

$$u(\xi, w, \rho) = \rho^{-2}\int_0^\infty\int_{-\infty}^\infty p_{t,s}(1, w'/\rho^2)e^{-t\rho^2 G}f(\xi, w-w')dw'dt.$$

Letting

$$C_s = \rho^{-2}\int_0^\infty\int_{-\infty}^\infty p_{t,s}(1, w'/\rho^2)dw'dt$$

$$= (4\pi)^{-s-1}\int_0^\infty t^{-s-1}e^{-\frac{1}{4t}}dt = \frac{1}{4}\Gamma(s)\pi^{-s-1}$$

we write $u(\xi, w, \rho) - C_s f(\xi, w)$ as the sum of the following two terms:

$$I_1(\xi, w, \rho) = \rho^{-2}\int_0^\infty\int_{-\infty}^\infty p_{t,s}(1, w'/\rho^2)\big(e^{-t\rho^2 G}f(\xi, w-w') - f(\xi, w-w')\big)dw'dt$$

and

$$I_2(\xi, w, \rho) = \rho^{-2}\int_0^\infty\int_{-\infty}^\infty p_{t,s}(1, w'/\rho^2)\big(f(\xi, w-w') - f(\xi, w)\big)dw'dt.$$

Clearly,

$$\|I_1\|_p \leq \rho^{-2}\int_0^\infty\int_{-\infty}^\infty p_{t,s}(1, w'/\rho^2)\|e^{-t\rho^2 G}f - f\|_p dw'dt$$

and hence converges to zero as ρ goes to zero. On the other hand $\|I_2\|_p$ also converges to zero as translation is continuous on L^p and $p_{t,s}(\rho, w)$ satisfies the estimate $p_{t,s}(\rho, w) \leq Ct^{-s-1}e^{-\frac{c}{t}(\rho^2+|w|)}$ for some constants C and c. □

Remark 3 The solution of the extension problem for the Grushin operator given in (6) can be written as $\rho^{2s}\pi(\Phi_{s,\rho})$ for a suitable function $\Phi_{s,\rho}$ on the Heisenberg group. In fact let us define

$$\Phi_{s,\rho}(z, w) = \int_0^\infty\Big(\int_{-\infty}^\infty p_{t,s}(\rho, w')p_t(z, w-w')dw'\Big)dt.$$

Using the fact that $\pi(p_t) = e^{-tG}$ it can be easily shown that

$$\rho^{2s}\pi(\Phi_{s,\rho}) = \rho^{2s}\int_0^\infty\int_{-\infty}^\infty p_{t,s}(\rho, w')\pi(0, w')e^{-tG}dt.$$

Since $\pi(0, w')f(\xi, w) = f(\xi, w - w')$ it follows that

$$\rho^{2s}\pi(\Phi_{s,\rho})f(\xi, w) = \rho^{2s}\int_0^\infty\int_{-\infty}^\infty p_{t,s}(\rho, w')e^{-tG}f(\xi, w - w')dt \tag{7}$$

is the solution defined in (6). Using the homogeneity properties of the heat kernels $p_{t,s}$ and p_t we can check that $\rho^{2s}\Phi_{s,\rho}(z, w) = \rho^{-2n-2}\Phi_{s,1}(\rho^{-1}z, \rho^{-2}w)$ and $\|\Phi_{s,1}\|_{L^1(\mathbb{H}^n)} = \frac{1}{4}\Gamma(s)\pi^{-1-s}$. Thus the solution of the extension problem is given by

$$u(\xi, w, \rho) = \rho^{-2n-2}\int_{\mathbb{H}^n}\Phi_{s,1}(\rho^{-1}z, \rho^{-2}a)\pi(z, a)f(\xi, w)dzda \tag{8}$$

which gives, in view of Lemma 1, another proof that $u(\xi, w, \rho)$ converges to $\frac{1}{4}\Gamma(s)\pi^{-1-s}f(\xi, w)$ as ρ goes to 0.

Remark 4 We can also rewrite the solution in the form

$$u(\xi, w, \rho) = \int_{\mathbb{R}^{n+1}} K_\rho(\xi, y, w - w')f(y, w')dydw'$$

where the kernel K_ρ satisfies the homogeneity condition

$$K_\rho(x, y, w) = \rho^{-n-2}K_1(\rho^{-1}x, \rho^{-1}y, \rho^{-2}w).$$

The kernel K_ρ is expressible in terms of $\Phi_{s,\rho}$. We also remark that the functions $\Phi_{s,\rho}$ are known explicitly (see [13]). By using explicit formulas for the kernels $p_{t,s}$ and p_t we can calculate the above integral obtaining

$$\Phi_{s,\rho}(z, w) = C_1(n, s)\big((\rho^2 + |z|^2)^2 + w^2\big)^{-(n+1+s)/2},$$

where $C_1(n, s) = 2^{n+s-1}\pi^{-n-s-2}\Gamma\left(\frac{n+1+s}{2}\right)^2$.

In the above theorem we have shown that the solution defined by (6) satisfies the uniform estimates $\|u(\cdot, \rho)\|_p \leq C\|f\|_p$. It is therefore natural to ask if all the solutions of (3) satisfying the uniform estimates $\|u(\cdot, \rho)\|_p \leq C, \rho > 0$ are given by the formula (6) for some $f \in L^p(\mathbb{R}^{n+1})$.

Theorem 5 *Assume* $1 \leq p < \infty$ *and let* $u(\xi, w, \rho)$ *be a solution of the extension problem* (3) *which satisfies the uniform estimates* $\|u(\cdot, \rho)\|_p \leq C, \rho > 0$. *Then there exists a unique* $f \in L^p(\mathbb{R}^{n+1})$ *such that* u *can be expressed as in* (6).

Proof Under the hypothesis on u it follows that there is a subsequence ρ_k tending to 0 and an $f \in L^p(\mathbb{R}^{n+1})$ such that $u(\xi, w, \rho_k)$ converges to f weakly. With this f let us define

$$v(\xi, w, \rho) = \rho^{2s} \int_0^\infty \int_{-\infty}^\infty p_{t,s}(\rho, w')e^{-tG} f(\xi, w - w')dw'dt.$$

The theorem will follow once we show that $u = v$. In order to prove this we make use of the uniqueness theorem for solutions of the extension problem for the sublaplacian proved in [14]. This theorem for the sublaplacian was proved as an easy consequence of results from [3] and [7].

We make use of the fact that $\mathcal{L}(\pi(z,t)\varphi, \psi)) = (\pi(z,t)G\varphi, \psi)$ for any two functions φ, ψ on $\mathbb{R}^{n+1}$. Therefore, if $u(\xi, w, \rho)$ is a solution of the extension problem for the Grushin operator with initial value 0 then for any $\varphi \in L^{p'}(\mathbb{R}^{n+1})$

$$\Big(-\mathcal{L} + \partial_\rho^2 + \frac{1-2s}{\rho}\partial_\rho + \frac{1}{4}\rho^2\partial_t^2\Big)\big(\pi(z,t)u(\cdot, \rho), \varphi\big)$$
$$= \big(\pi(z,t)(-G + \partial_\rho^2 + \frac{1-2s}{\rho}\partial_\rho + \frac{1}{4}\rho^2\partial_w^2)u(\cdot, \rho), \varphi\big) = 0.$$

Hence the hypothesis on u shows that $\|(\pi(\cdot,\cdot)u(\cdot, \rho), \varphi)\|_\infty \leq C$ and so by the uniqueness theorem for the sublaplacian (see Theorem 1.1 in [14]) we conclude that $(\pi(z,t)u(\cdot, \rho), \varphi) = 0$ for all φ and hence $u = 0$. □

4 Fractional Powers of the Grushin Operator

Given a bounded function m on the spectrum of G one can define the operator $m(G)$ via spectral theorem by

$$m(G)f(\xi, w) = (2\pi)^{-1} \int_{-\infty}^\infty e^{-i\lambda w} m(H(\lambda)) f^\lambda(x) d\lambda.$$

Thus we can think of $m(G)$ as an operator valued multiplier for the Euclidean Fourier transform on $\mathbb{R}$. Indeed, by identifying $L^2(\mathbb{R}^{n+1})$ with $L^2(\mathbb{R}, X)$ where $X = L^2(\mathbb{R}^n)$ the above can be rewritten as

$$m(G)F(w) = (2\pi)^{-1} \int_{-\infty}^\infty e^{-i\lambda w} M(\lambda)\hat{F}(\lambda)\, d\lambda$$

where $F(w)(\xi) = f(\xi, w)$ and $M(\lambda) = m(H(\lambda))$. Assuming that $m(H(\lambda))$ is a bounded linear operator on $X = L^2(\mathbb{R}^n)$ the above is precisely the definition of operator valued Fourier multipliers studied by L. Weis in [19]. The operator valued function $M(\lambda)$ is known as the multiplier corresponding to $m(G)$. With this terminology, the fractional powers G_s, $0 < s < 1$ are defined via the multiplier

$$M_s(\lambda) = (2|\lambda|)^s \sum_{k=0}^{\infty} \frac{\Gamma(\frac{2k+n+1+s}{2})}{\Gamma(\frac{2k+n+1-s}{2})} P_k(\lambda).$$

More explicitly,

$$G_s f(\xi, w) = (2\pi)^{-1} \int_{-\infty}^{\infty} e^{-i\lambda w} (2|\lambda|)^s \Big(\sum_{k=0}^{\infty} \frac{\Gamma(\frac{2k+n+1+s}{2})}{\Gamma(\frac{2k+n+1-s}{2})} P_k(\lambda) \Big) f^\lambda(\xi) d\lambda.$$

Observe that $G_s f$ is well defined as an L^2 function under the assumption that $M_s(\lambda) f^\lambda(\xi)$ is an L^2 function of (ξ, λ) on $\mathbb{R}^{n+1}$. By Stirling's formula, $\frac{\Gamma(\frac{2k+n+1+s}{2})}{\Gamma(\frac{2k+n+1-s}{2})}$ behaves like $(2k+n)^s$ and hence $G_s f$ will be in $L^2(\mathbb{R}^{n+1})$ if for every λ, $H(\lambda)^s f^\lambda \in L^2(\mathbb{R}^n)$ and

$$\int_{-\infty}^{\infty} \int_{\mathbb{R}^n} |H(\lambda)^s f^\lambda(\xi)|^2 d\xi d\lambda < \infty.$$

The domain of G_s consists precisely of those $f \in L^2(\mathbb{R}^{n+1})$ for which the above condition is satisfied. It is clear that all Schwartz functions are in the domain and therefore G_s is densely defined.

Returning to the solution of the extension problem (3) we can now prove the following result which is the analogue of the result proved in [4] (see equation (3.1)) for the Laplacian on $\mathbb{R}^n$.

Theorem 6 *Assume $0 < s < 1$ and let $f \in L^p \cap L^2(\mathbb{R}^{n+1})$ be such that $G_s f$ also belongs to $L^p \cap L^2(\mathbb{R}^{n+1})$. Let $u(\xi, w, \rho)$ be the solution of the extension problem (3) defined by (6). Then $-\rho^{1-2s}\partial_\rho u(\xi, w, \rho)$ converges to $B_s G_s f$ in $L^p \cap L^2(\mathbb{R}^{n+1})$ as ρ goes to 0, where $B_s = 2^{-1-2s}\pi^{-1-s}\Gamma(1-s)$.*

Proof In order to prove this theorem we make use of the formula (7) for the solution of the extension problem. We claim that there is an explicit constant $C_{n,s}$ such that $-\rho^{1-2s}\partial_\rho\big(\rho^{2s}\pi(\Phi_{s,\rho})\big) = C_{n,s}\pi(\psi_{s,\rho})G_s$ as operators where $\psi_{s,\rho}(z, w) = \rho^{-2n-2}\psi_{s,1}(\rho^{-1}z, \rho^{-2}w)$ with $\|\psi_{s,1}\|_{L^1(\mathbb{H}^n)} = C_2(n, s)$, where

$$C_2(n, s) = 2^{-n+s}\pi^{n+1}\Gamma(1-s)/\Gamma\Big(\frac{n+1-s}{2}\Big)^2.$$

Once we have this claim, it follows from Remark 3 that $-\rho^{1-2s}\partial_\rho u(\xi, w, \rho) = C_{n,s}\pi(\psi_{s,\rho})G_s f(\xi, w)$ and hence the theorem follows from Lemma 1 with $B_s = C_{n,s}C_2(n, s)$. In order to prove the claim, we make use of the formula

$$\mathcal{L}_s\Phi_{-s,\rho}(z, w) = (2\pi)^{2s}\rho^{2s}\Phi_{s,\rho}(z, w) \tag{9}$$

which has been proved in Cowling–Haagerup [6] (see also [5]). Here $\mathcal{L}_s$ is the conformally invariant fractional power of the sublaplacian which is defined by the relation $\widehat{\mathcal{L}_s f}(\lambda) = \hat{f}(\lambda)M_s(\lambda)$ where $M_s(\lambda)$ is the same family of operators used in the definition of G_s and $\hat{f}$ stands for the operator valued group Fourier transform of f on $\mathbb{H}^n$. From (9) we obtain

$$\pi(\Phi_{-s,\rho})G_s f(\xi, w) = (2\pi)^{2s}\rho^{2s}\pi(\Phi_{s,\rho})f(\xi, w) = (2\pi)^{2s}u(\xi, w, \rho). \tag{10}$$

In [14] the authors have calculated that $-\rho^{1-2s}\partial_\rho\varphi_{-s,\rho} = \rho^{-2n-2}\psi_{s,1}(\rho^{-1}z, \rho^{-2}w)$ for an explicit function $\psi_{s,1}$ and constant $C_2(n, s)$, where $\varphi_{s,\rho}(z, w) = ((\rho^2 + |z|^2)^2 + 16w^2)^{-\frac{n+s+1}{2}}$. Thus differentiating both sides of (10) by ρ and multiplying by $-\rho^{1-2s}$, we obtain

$$C_1(n, -s)\pi(\psi_{s,p})G_s f(\xi, w) = -(2\pi)^{2s}\rho^{1-2s}\partial_\rho u(\xi, w, \rho).$$

This proves our claim with $C_{n,s}{=}(2\pi)^{-2s}C_1(n, -s){=}2^{n-3s-1}\pi^{-n-s-2}\Gamma(\frac{n+1-s}{2})^2$. Finally we calculate B_s using the value of $C_2(n, s)$ calculated in [14]:

$$\begin{aligned} B_s &= \left(2^{n-3s-1}\pi^{-n-s-2}\Gamma(\frac{n+1-s}{2})^2\right)\times\left(2^{-n+s}\pi^{n+1}\Gamma(1-s)/\Gamma(\frac{n+1-s}{2})^2\right) \\ &= 2^{-1-2s}\pi^{-1-s}\Gamma(1-s). \end{aligned}$$

The proof is complete. □

5 Trace Hardy and Hardy Inequalities

Consider the vector fields $X_j = \xi_j\frac{\partial}{\partial w}$, $Y_j = \frac{\partial}{\partial \xi_j}$ and $T = \frac{1}{2}\rho\frac{\partial}{\partial w}$ on $\mathbb{R}^{n+1}\times\mathbb{R}^+$. Let X be one of these vector fields. For real valued functions u, v defined on $\mathbb{R}^{n+1}\times\mathbb{R}^+$ consider

$$\int_0^\infty \int_{\mathbb{R}^{n+1}} \left(Xu - \frac{u}{v}Xv\right)^2 \rho^{1-2s}\, d\xi\, dw d\rho.$$

Using integration by parts and assuming that u and v are such that $\frac{u^2}{v}Xv$ vanishes at infinity, we have

$$\int_{\mathbb{R}^{n+1}} \frac{u}{v}Xu\, Xv\, d\xi dw = -\int_{\mathbb{R}^{n+1}} \frac{u}{v}Xu\, Xv\, d\xi dw - \int_{\mathbb{R}^{n+1}} u^2 X(\frac{1}{v}Xv)\, d\xi dw.$$

Simplifying, we get

$$\int_{\mathbb{R}^{n+1}} \frac{u^2}{v^2}(Xv)^2\, d\xi - 2\int_{\mathbb{R}^{n+1}} \frac{u}{v} Xu\, Xv\, d\xi = \int_{\mathbb{R}^{n+1}} \frac{u^2}{v} X^2 v\, d\xi. \tag{11}$$

Consequently,

$$\begin{aligned}
&\int_0^\infty \int_{\mathbb{R}^{n+1}} \left(Xu - \frac{u}{v} Xv\right)^2 \rho^{1-2s}\, d\xi\, dw d\rho \\
&= \int_0^\infty \int_{\mathbb{R}^{n+1}} (Xu)^2 \rho^{1-2s}\, d\xi\, dw d\rho + \int_0^\infty \int_{\mathbb{R}^{n+1}} \frac{u^2}{v}(X^2 v)\rho^{1-2s}\, d\xi\, dw d\rho.
\end{aligned}$$

In a similar way, using integration by parts, we can check that

$$\begin{aligned}
&\int_0^\infty \frac{u^2}{v^2}(\partial_\rho v)^2 \rho^{1-2s}\, d\rho - 2\int_0^\infty \frac{u}{v} \partial_\rho u\, \partial_\rho v\, \rho^{1-2s}\, d\rho \\
&= \int_0^\infty \frac{u^2}{v} \partial_\rho(\rho^{1-2s}\partial_\rho v)\, d\rho + \lim_{\rho\to 0}\left(\frac{u^2}{v}\rho^{1-2s}\partial_\rho v\right)
\end{aligned}$$

which leads to the equation

$$\begin{aligned}
&\int_0^\infty \int_{\mathbb{R}^{n+1}} \left(\partial_\rho u - \frac{u}{v}\partial_\rho v\right)^2 \rho^{1-2s}\, d\xi\, d\rho \\
&= \int_0^\infty \int_{\mathbb{R}^{n+1}} (\partial_\rho u)^2 \rho^{1-2s}\, d\xi\, d\rho + \int_0^\infty \int_{\mathbb{R}^{n+1}} \frac{u^2}{v} \partial_\rho(\rho^{1-2s}\partial_\rho v)\rho^{1-2s}\, d\xi\, d\rho \\
&\qquad + \int_{\mathbb{R}^{n+1}} \frac{u^2(\xi,0)}{v(\xi,0)} \lim_{\rho\to 0}(\rho^{1-2s}\partial_\rho v)(\xi,\rho)\, d\xi.
\end{aligned}$$

Let us now consider the gradient

$$\nabla u = (X_1 u, \cdots, X_n u, Y_1 u, \cdots, Y_n u, \frac{1}{2}\rho\, \partial_w u, \partial_\rho u). \tag{12}$$

Adding the above equations we obtain the identity

$$\begin{aligned}
&\int_0^\infty \int_{\mathbb{R}^{n+1}} |\nabla u - \frac{u}{v}\nabla v|^2 \rho^{1-2s}\, d\xi\, dw d\rho = \int_0^\infty \int_{\mathbb{R}^{n+1}} |\nabla u|^2 \rho^{1-2s}\, d\xi\, dw d\rho \\
&\qquad + \int_{\mathbb{R}^{n+1}} \frac{u^2(\xi,0)}{v(\xi,0)} \lim_{\rho\to 0}(\rho^{1-2s}\partial_\rho v)(\xi,\rho)\, d\xi\, dw d\rho \\
&\qquad + \int_0^\infty \int_{\mathbb{R}^{n+1}} \frac{u^2}{v}(-G + \frac{1}{4}\rho^2\partial_w^2 + \partial_\rho)(\rho^{1-2s}\partial_\rho v) d\xi\, dw d\rho.
\end{aligned}$$

If v solves the extension problem

$$\left(-G+\partial_\rho^2+\frac{1-2s}{\rho}\partial_\rho+\frac{1}{4}\rho^2\partial_w^2\right)v(\xi,\rho)=0,$$

then the last term in the above vanishes. As the left hand side is non-negative, this leads to the following inequality known as trace Hardy inequality in the literature.

Proposition 7 *Let u be a real valued compactly supported continuous function on $\mathbb{R}^{n+1}\times[0,\infty)$ which is smooth for $\rho>0$. Let v be a real valued function which solves the extension problem for the Grushin operator. Then*

$$\int_0^\infty\int_{\mathbb{R}^{n+1}}|\nabla u(\xi,w,\rho)|^2\rho^{1-2s}\,d\xi\,dwd\rho$$
$$\geqslant-\int_{\mathbb{R}^{n+1}}\frac{u^2(\xi,w,0)}{v(\xi,w,0)}\lim_{\rho\to0}(\rho^{1-2s}\partial_\rho v)(\xi,w,\rho)\,d\xi dw.$$

When v is the solution of the extension problem for the Grushin operator G with initial condition $\varphi\in L^2(\mathbb{R}^{n+1})$ defined by (6) then we have proved that $\lim_{\rho\to0}\rho^{1-2s}\partial_\rho v=-B_sG_s\varphi$ where B_s is given in Theorem 4.1. Thus the inequality in the above proposition takes the form

$$\int_0^\infty\int_{\mathbb{R}^{n+1}}|\nabla u(\xi,w,\rho)|^2\rho^{1-2s}\,d\xi\,dwd\rho\geqslant c_s\int_{\mathbb{R}^{n+1}}\frac{u^2(\xi,w,0)}{\varphi(\xi,w)}G_s\varphi(\xi,w,\rho)\,d\xi dw$$

where $c_s=2^{1-2s}\frac{\Gamma(1-s)}{\Gamma(s)}$. In the case of the sublaplacian on H-type groups, there are explicit functions $\varphi_{s,\delta}$ such that $\mathcal{L}_s\varphi_{-s,\delta}=\delta^sC_{n,s}\varphi_{s,\delta}$ with explicit constant $C_{n,s}$ which have allowed the authors in [14] to simplify the quotient $\frac{\mathcal{L}_s\varphi_{-s,\delta}}{\varphi_{-s,\delta}}$ to get a sharp trace Hardy inequality. Unfortunately in our context, though we can find analogues of $\varphi_{s,\delta}$ the quotient $\frac{G_s\varphi_{-s,\delta}}{\varphi_{-s,\delta}}$ does not seem to simplify. But things are not so bad if we slightly modify the definition of the fractional power G_s.

Recall that G_s is defined in terms of the multiplier

$$M_s(\lambda)=(2|\lambda|)^s\sum_{k=0}^\infty\frac{\Gamma(\frac{2k+n+1+s}{2})}{\Gamma(\frac{2k+n+1-s}{2})}P_k(\lambda).$$

Instead of this we use the slightly different multiplier

$$\tilde{M}_s(\lambda)=(2|\lambda|)^s\sum_{k=0}^\infty\frac{\Gamma(\frac{2k+n+2+2s}{4})}{\Gamma(\frac{2k+n+2-2s}{4})}P_k(\lambda)$$

in defining the modified fractional power $\tilde{G}_s$. Note that $\tilde{G}_s$ is nothing but $(\frac{1}{2}G)_s$. The operators G_s and $\tilde{G}_s$ are comparable. For $0<s<1$ let $C_1(s)$ and $C_2(s)$ be defined by

$$C_1(s) = \inf_k \frac{\Gamma(\frac{2k+n+1+s}{2})}{\Gamma(\frac{2k+n+1-s}{2})} \frac{\Gamma(\frac{2k+n+2-2s}{4})}{\Gamma(\frac{2k+n+2+2s}{4})}, \; C_2(s)^{-1} = \inf_k \frac{\Gamma(\frac{2k+n+1-s}{2})}{\Gamma(\frac{2k+n+1+s}{2})} \frac{\Gamma(\frac{2k+n+2+2s}{4})}{\Gamma(\frac{2k+n+2-2s}{4})}.$$

In view of Stirling's formula for the gamma function, these constants are positive and finite. Then it follows that

$$C_1(s)\langle \tilde{G}_s f, f\rangle \leq \langle G_s f, f\rangle \leq C_2(s)\langle \tilde{G}_s f, f\rangle. \tag{13}$$

The operator $\tilde{G}_s$ is better behaved as we can see from the following proposition.

For any $s \in \mathbb{R}$ and $\delta > 0$ let $u_{s,\delta}(x, w) = ((\delta + |\xi|^2)^2 + w^2)^{-\frac{n+2+2s}{4}}$ defined on $\mathbb{R}^{n+1}$. We have an explicit expression for the action of $\tilde{G}_s$ on $u_{-s,\delta}$.

Proposition 8 *For any $\delta > 0$ and $0 < s < 1$, we have $\tilde{G}_s u_{-s,\delta}(\xi, w) = C_{n,s}\delta^s u_{s,\delta}(\xi, w)$, where $C_{n,s} = \frac{2^{2s}\Gamma(\frac{n+2+2s}{4})^2}{\Gamma(\frac{n+2-2s}{4})^2}$.*

Proof Since $u_{-s,\delta}$ is radial in ξ the action of G on $u_{-s,\delta}$ is the same as that of the generalised sublaplacian $\mathcal{L}(n/2-1) = -\partial_r^2 - \frac{(n-1)}{r}\partial_r - r^2\partial_w^2$. Therefore, the result follows from Theorem 3.11 in [5]. This can be proved by expanding the function $u^\lambda_{-s,\delta}(r)$ in terms of Laguerre functions of type $(n/2 - 1)$. We leave the details to the reader. □

Using the above proposition it is easy to prove a Hardy inequality for the modified fractional power $\tilde{G}_s$. If we let T_s to stand for the operator

$$T_s f(\xi, w) = ((\delta + |\xi|^2)^2 + w^2)^{\frac{n+2}{4}} \tilde{G}_s\big(((\delta + |\xi|^2)^2 + w^2)^{-\frac{n+2}{4}} f\big)(\xi, w),$$

then it follows that

$$T_s((\delta + |\xi|^2)^2 + w^2)^{\frac{s}{2}} = C_{n,s}\delta^s((\delta + |\xi|^2)^2 + w^2)^{-\frac{s}{2}}.$$

Therefore, using Schur test we can prove the following inequality (see Section 5.1 in [13]).

Theorem 9 *Let $f \in L^2(\mathbb{R}^{n+1})$ be real valued and assume that $G_s f \in L^2(\mathbb{R}^{n+1})$. Then for any $\delta > 0$ we have the inequality*

$$\langle \tilde{G}_s f, f\rangle \geqslant A_1(n, s)\delta^s \int_{\mathbb{R}^{n+1}} \frac{(f(\xi, w))^2}{((\delta + |\xi|^2)^2 + w^2)^s}\, d\xi\, dw$$

where $A_1(n, s) = 4^s\frac{\Gamma(\frac{n+2+2s}{4})^2}{\Gamma(\frac{n+2-2s}{4})^2}$. The inequality is sharp and equality holds when $f = u_{-s,\delta}$.

Remark 10 In view of (13) we also have the Hardy inequality for G_s, namely

$$\langle G_s f, f\rangle \geqslant C_1(s)A_1(n,s)\delta^s \int_{\mathbb{R}^{n+1}} \frac{(f(\xi,w))^2}{((\delta+|\xi|^2)^2+w^2)^s}\, d\xi\, dw.$$

Finally, using the above Hardy inequality we can prove the following trace Hardy inequality for the Grushin operator G.

Theorem 11 *Let $\delta > 0$ and $0 < s < 1$. For any real valued compactly supported continuous function u on $\mathbb{R}^{n+1} \times [0,\infty)$ which is smooth for $\rho > 0$, we have the inequality*

$$\int_0^\infty \int_{\mathbb{R}^{n+1}} |\nabla u(\xi,w,\rho)|^2 \rho^{1-2s}\, d\xi\, dw d\rho$$

$$\geqslant C_1(s)B_s A_1(n,s)\delta^s \int_{\mathbb{R}^{n+1}} \frac{u(\xi,w,0)^2}{((\delta+|\xi|^2)^2+w^2)^s}\, d\xi\, dw.$$

In view of Hardy's inequality for G_s all we have to do is to prove the following energy estimate for the Grushin operator. The following result is the analogue of Theorem 1.2 in [8] proved in the context of Heisenberg groups.

Theorem 12 *Let $\delta > 0$ and $0 < s < 1$. For any real valued compactly supported continuous function u on $\mathbb{R}^{n+1} \times [0,\infty)$ which is smooth for $\rho > 0$, we have the inequality*

$$\int_0^\infty \int_{\mathbb{R}^{n+1}} |\nabla u(\xi,w,\rho)|^2 \rho^{1-2s}\, d\xi\, dw d\rho \geqslant B_s \langle G_s f, f\rangle$$

where $f(\xi,w) = u(\xi,w,0)$ and $B_s = 2^{-1-2s}\pi^{-1-s}\Gamma(1-s)$.

Proof In proving this theorem we closely follow the proof of Theorem 1.2 in [8]. We therefore give only a sketch of the proof referring to [8] for details. In what follows we assume that u is smooth. The general case can be dealt with using an approximation argument.

Let $H^s(\mathbb{R}^{n+1})$ be the completion of $C_0^\infty(\mathbb{R}^{n+1})$ with respect to the norm $\|f\|_{(s)}^2 = \langle G_s f, f\rangle$. It can be verified that the dual of $H^s(\mathbb{R}^{n+1})$ is $H^{-s}(\mathbb{R}^{n+1})$. If $g \in H^{-s}(\mathbb{R}^{n+1})$, it follows that $h = G_s^{-1} g = G_{-s} g \in H^s(\mathbb{R}^{n+1})$. Let $H(\xi,w,\rho)$ be the solution of extension problem with initial condition h defined as in (6). In view of Theorem 2 and Theorem 6, $H(\xi,w,\rho)$ converges to $C_s h$ with $C_s = \frac{1}{4}\Gamma(s)\pi^{-s-1}$ and $-\rho^{1-2s}\partial_\rho H(\xi,w,\rho)$ converges to $B_s G_s h$ with $B_s = 2^{-1-2s}\pi^{-1-s}\Gamma(1-s)$. If we let $W(\xi,w,q) = H(2^{-1/2}\xi, 2^{-1}w, \rho)$ with $q = \rho^2/2$, then W satisfies the equation

$$2\big(q\partial_q^2 + (1-s)\partial_q + q\partial_w^2 - G\big)W(\xi,w,q) = 0. \tag{14}$$

Moreover, as $q \to 0$, we have $W(\xi, w, q) \to C_s h(\xi, w)$ in $H^s(\mathbb{R}^{n+1})$ and $-q^{1-s}\partial_q W(\xi, w, q) \to 2^{s-1} B_s G_s h(\xi, w)$ in $H^{-s}(\mathbb{R}^{n+1})$. We define $U(\xi, w, q) = u(\xi, w, \rho)$ with $q = \rho^2/2$ and proceed as in the proof of Theorem 1.2 in [8]. We leave the details to the reader. □

6 An Isometry Property of the Solution Operator Associated to the Extension Problem

In this section we will prove an isometry property of the solution operator associated to the extension problem for the Grushin operator. Such a property has been already proved in the context of $\mathbb{R}^n$ and $\mathbb{H}^n$ in [11], see also [14]. Let $u(\xi, w, \rho)$ be the solution of the extension problem given by (6). For $s > 0$, for which G_s makes sense (e.g. $0 < s < (n+1)$), recall that the Sobolev space $H^s(\mathbb{R}^{n+1})$ is defined as the completion of $C_0^\infty(\mathbb{R}^{n+1})$ under the norm

$$\|f\|_{(s)}^2 = \langle G_s f, f\rangle = \|G_s^{1/2} f\|_{L^2(\mathbb{R}^{n+1})}^2.$$

Let Φ_α^λ, $\alpha \in \mathbb{N}^n$ be the scaled Hermite functions which are eigenfunctions of scaled Hermite operator $H(\lambda) = -\Delta + \lambda^2|\xi|^2$. In view of the spectral decomposition of G_s we have that

$$\|f\|_{(s)}^2 = (2\pi)^{-1} \int_{-\infty}^{\infty} (2|\lambda|)^s \left(\sum_{\alpha \in \mathbb{N}^n} \frac{\Gamma(\frac{2|\alpha|+n+1+s}{2})}{\Gamma(\frac{2|\alpha|+n+1-s}{2})} |\langle f^\lambda, \Phi_\alpha^\lambda\rangle|^2 \right) d\lambda.$$

We think of the solution $u(\xi, w, \rho)$ as a function on $\mathbb{R}^{n+3}$ which is radial in the third variable. Thus $U(\xi, w, \zeta) = u(\xi, w, |\zeta|)$ is a function on $\mathbb{R}^{n+3}$. For $(\alpha, \beta) \in \mathbb{N}^n \times \mathbb{N}^2$, let $\Phi_{\alpha,\beta}^\lambda(\xi, \zeta) = \Phi_\alpha^\lambda(\xi)\Phi_\beta^{\lambda/2}(\zeta)$, where $\Phi_\alpha^\lambda(\xi)$ and $\Phi_\beta^{\lambda/2}(\zeta)$ are Hermite functions on $\mathbb{R}^n$ and $\mathbb{R}^2$ respectively. Now we define the Sobolev space $\widetilde{H}^{s+1}(\mathbb{R}^{n+3})$ in terms of $\Phi_{\alpha,\beta}^\lambda(\xi, \zeta)$ as the space of all functions $U \in L^2(\mathbb{R}^{n+3})$ for which the following norm is finite.

$$\|U\|_{\widetilde{H}^{s+1}(\mathbb{R}^{n+3})}^2$$

$$= (2\pi)^{-1} \int_{-\infty}^{\infty} (2|\lambda|)^{s+1} \left(\sum_{(\alpha,\beta) \in \mathbb{N}^n \times \mathbb{N}^2} \frac{\Gamma(\frac{2|\alpha|+|\beta|+n+1+1+s+1}{2})}{\Gamma(\frac{2|\alpha|+|\beta|+n+1+1-s-1}{2})} |\langle U^\lambda, \Phi_{\alpha,\beta}^\lambda\rangle|^2 \right) d\lambda$$

where U^λ is defined as usual by

$$U^\lambda(\xi; \zeta) = \int_{-\infty}^{\infty} U(\xi, w, \zeta) e^{i\lambda w} d\lambda.$$

We begin with the following expansion of the function $U(\xi, w, \zeta)$ in terms of Hermite functions. We let $L(\lambda, a, b)$ be defined by

$$L(\lambda, a, b) = \int_0^\infty e^{-\lambda(2t+1)} t^{a-1}(1+t)^{-b}\, dt, \qquad \lambda > 0, a > 0, b \in \mathbb{C}.$$

Proposition 13 *If $U(\xi, w, \zeta) = u(\xi, w, |\zeta|)$, where $u(\xi, w, \rho)$ is the solution of the extension problem given in (6), then*

$$U^\lambda(\xi; \zeta) = \sum_\alpha a^\lambda_{\alpha,\zeta}(s)\langle f^\lambda, \Phi^\lambda_\alpha\rangle \Phi^\lambda_\alpha(\xi) \tag{15}$$

where the coefficients are given by

$$a^\lambda_{\alpha,\zeta}(s) = (4\pi)^{-s-1}(2|\lambda|)^s|\zeta|^{2s} L\Big(\frac{|\lambda||\zeta|^2}{4}, \frac{2|\alpha|+n+1+s}{2}, \frac{2|\alpha|+n+1-s}{2}\Big).$$

Proof It is easy to see that

$$U^\lambda(\xi, \zeta) = |\zeta|^{2s} \int_0^\infty p^\lambda_{t,s}(|\zeta|) e^{-tH(\lambda)} f^\lambda(\xi)dt \tag{16}$$

which follows from the spectral decomposition of G. We also have

$$e^{-tH(\lambda)} f^\lambda(\xi) = \sum_{\alpha\in\mathbb{N}^n} e^{-t(2|\alpha|+n)|\lambda|}\langle f^\lambda, \Phi^\lambda_\alpha\rangle \Phi^\lambda_\alpha(\xi) \tag{17}$$

and the heat kernel for the generalised sublaplacian is given by

$$p^\lambda_{t,s}(|\zeta|) = (4\pi)^{-s-1}\left(\frac{\lambda}{\sinh(t\lambda)}\right)^{s+1} e^{-\frac{1}{4}\lambda\coth(t\lambda)|\zeta|^2}. \tag{18}$$

We substitute (17) and (18) in (16) and we get

$$U^\lambda(\xi, \zeta) = \sum_\alpha a^\lambda_{\alpha,\zeta}(s)\langle f^\lambda, \Phi^\lambda_\alpha\rangle \Phi^\lambda_\alpha(\xi)$$

where the coefficients are given by the integral

$$a^\lambda_{\alpha,\zeta}(s) = (4\pi)^{-s-1}|\zeta|^{2s} \int_0^\infty \left(\frac{\lambda}{\sinh(t\lambda)}\right)^{s+1} e^{-\frac{1}{4}\lambda\coth(t\lambda)|\zeta|^2} e^{-t(2|\alpha|+n)|\lambda|} dt.$$

We make use of the change of variables $\coth(t|\lambda|) = 2u + 1$ or $u = \frac{1}{e^{2t|\lambda|}-1}$ in the above integral and note that $e^{2t|\lambda|} = 1 + \frac{1}{u}$, $\sinh(t\lambda) = \frac{1}{2}u^{-1/2}(1+u)^{-1/2}$ and $dt = \frac{-1}{2|\lambda|u(1+u)}du$. The above integral then simplifies to

$$a^{\lambda}_{\alpha,\zeta}(s) = (4\pi)^{-s-1}(2|\lambda|)^s|\zeta|^{2s}L\Big(\frac{|\lambda||\zeta|^2}{4}, \frac{2|\alpha|+n+1+s}{2}, \frac{2|\alpha|+n+1-s}{2}\Big). \tag{19}$$

The proof is complete. □

Lemma 14 *The Hermite expansion of* $L\Big(\frac{|\lambda||\zeta|^2}{4}, a, b\Big)$ *in* ζ*-variable is given by*

$$2\pi|\lambda|^{-1}\Gamma(b-a+1)\sum_{\beta\in\mathbb{N}^2}\frac{\Gamma(a+\frac{|\beta|}{2})}{\Gamma(b+\frac{|\beta|}{2}+1)}\Phi^{\lambda/2}_{\beta}(0)\Phi^{\lambda/2}_{\beta}(\zeta).$$

Proof We make use of the following two dimensional Mehler's formula:

$$\sum_{\beta\in\mathbb{N}^2} r^{|\beta|}\Phi^{\lambda/2}_{\beta}(\zeta)\Phi^{\lambda/2}_{\beta}(\zeta') = (2\pi)^{-1}|\lambda|(1-r^2)^{-1}e^{-\frac{|\lambda|}{4}\frac{1+r^2}{1-r^2}(|\zeta|^2+|\zeta'|^2)+\frac{|\lambda|}{1-r^2}\zeta\cdot\zeta'}.$$

If we take $r = \sqrt{\frac{t}{1+t}}$ and $\zeta' = 0$, then the above yields

$$e^{-\frac{|\lambda||\zeta|^2}{4}(2t+1)} = 2\pi|\lambda|^{-1}\sum_{\beta\in\mathbb{N}^2} t^{\frac{|\beta|}{2}}(1+t)^{-(\frac{|\beta|}{2}+1)}\Phi^{\lambda/2}_{\beta}(0)\Phi^{\lambda/2}_{\beta}(\zeta). \tag{20}$$

Recall that

$$L\Big(\frac{|\lambda||\zeta|^2}{4}, a, b\Big) = \int_0^{\infty} e^{-\frac{|\lambda||\zeta|^2}{4}(2t+1)}t^{a-1}(1+t)^{-b}dt.$$

In view of (20), it is easy to see that

$$L\Big(\frac{|\lambda||\zeta|^2}{4}, a, b\Big) = 2\pi|\lambda|^{-1}\sum_{\beta\in\mathbb{N}^2}\Phi^{\lambda/2}_{\beta}(0)\Big(\int_0^{\infty} t^{\frac{|\beta|}{2}+a-1}(1+t)^{-(\frac{|\beta|}{2}+b+1)}dt\Big)$$
$$\times\,\Phi^{\lambda/2}_{\beta}(\zeta).$$

Since $\int_0^{\infty}(1+t)^{-b}t^{a-1}dt = \frac{\Gamma(a)\Gamma(b-a)}{\Gamma(b)}$, we get the required expansion. □

We are now ready to prove the following theorem. Let $P_s f = U$ where $u(\xi, w, \rho)$ is the solution of the extension problem given by (6) and $U(\xi, w, \zeta) = u(\xi, w, |\zeta|)$. The operator P_s has the following isometry property.

Theorem 15 *For $0 < s < n+1$, the operator $P_s : H^s(\mathbb{R}^{n+1}) \to \widetilde{H}^{s+1}(\mathbb{R}^{n+3})$ satisfies*

$$\|U\|^2_{\widetilde{H}^{s+1}(\mathbb{R}^{n+3})} = c_{n,s}\|f\|^2_{(s)}$$

for an explicit constant $c_{n,s} = \frac{s\Gamma(s)^2}{4\pi^{2s+1}}$.

Proof In view of Proposition 13, the function U^λ has the expansion

$$U^\lambda(\xi, \zeta) = \sum_{\alpha\in\mathbb{N}^n} a^\lambda_{\alpha,\zeta}(s)\langle f^\lambda, \Phi^\lambda_\alpha\rangle\Phi^\lambda_\alpha(\xi) \tag{21}$$

where the coefficients are given by (with $a = \frac{2|\alpha|+n+1+s}{2}$ and $b = \frac{2|\alpha|+n+1-s}{2}$)

$$a^\lambda_{\alpha,\zeta}(s) = (4\pi)^{-s-1}(2|\lambda||\zeta|^2)^s L\Big(\frac{|\lambda||\zeta|^2}{4}, a, b\Big). \tag{22}$$

Since the $L-$function has the transformation property (see Cowling–Haagerup [6])

$$L(\lambda, a, b) = \frac{\Gamma(a)}{\Gamma(b)}(2\lambda)^{b-a}L(\lambda, b, a),$$

for all $a, b \in \mathbb{C}$ and $\lambda > 0$, the coefficients $a^\lambda_{\alpha,\zeta}(s)$ can be written as

$$a^\lambda_{\alpha,\zeta}(s) = (4\pi)^{-s-1}2^{2s}\frac{\Gamma(a)}{\Gamma(b)}L\Big(\frac{|\lambda||\zeta|^2}{4}, b, a\Big). \tag{23}$$

We use Lemma 14 to write the coefficients $a^\lambda_{\alpha,\zeta}(s)$ as

$$a^\lambda_{\alpha,\zeta}(s) = d_s|\lambda|^{-1}\frac{\Gamma(a)}{\Gamma(b)}\sum_{\beta\in\mathbb{N}^2}\frac{\Gamma(\frac{2|\alpha|+|\beta|+n+1-s}{2})}{\Gamma(\frac{2|\alpha|+|\beta|+n+3+s}{2})}\Phi^{\lambda/2}_\beta(0)\Phi^{\lambda/2}_\beta(\zeta), \tag{24}$$

where $d_s = 2^{-1}\pi^{-s}\Gamma(s+1)$. From (21) and (24), we have that

$$\langle U^\lambda, \Phi^\lambda_{\alpha,\beta}\rangle = d_s|\lambda|^{-1}\frac{\Gamma(a)}{\Gamma(b)}\frac{\Gamma(\frac{2|\alpha|+|\beta|+n+1-s}{2})}{\Gamma(\frac{2|\alpha|+|\beta|+n+3+s}{2})}\Phi^{\lambda/2}_\beta(0)\langle f^\lambda, \Phi^\lambda_\alpha\rangle.$$

Therefore, recalling the norm in $\widetilde{H}^{s+1}(\mathbb{R}^{n+3})$ we see that

$$\|U\|^2_{\widetilde{H}^{s+1}(\mathbb{R}^{n+3})}$$
$$= \frac{(2d_s)^2}{2\pi}\int_{-\infty}^{\infty}(2|\lambda|)^{s-1}\left(\sum_{(\alpha,\beta)}\frac{\Gamma(a)^2}{\Gamma(b)^2}\frac{\Gamma(\frac{2|\alpha|+|\beta|+n+1-s}{2})}{\Gamma(\frac{2|\alpha|+|\beta|+n+3+s}{2})}\Phi_\beta^{\lambda/2}(0)^2|\langle f^\lambda,\Phi_\alpha^\lambda\rangle|^2\right)d\lambda$$
$$= \frac{(2d_s)^2}{2\pi}\int_{-\infty}^{\infty}(2|\lambda|)^{s-1}\sum_{\alpha}\frac{\Gamma(a)^2}{\Gamma(b)^2}C_\alpha^\lambda(s)|\langle f^\lambda,\Phi_\alpha^\lambda\rangle|^2 d\lambda$$

where

$$C_\alpha^\lambda(s) = \left(\sum_{\beta}\frac{\Gamma(\frac{2|\alpha|+|\beta|+n+1-s}{2})}{\Gamma(\frac{2|\alpha|+|\beta|+n+3+s}{2})}\Phi_\beta^{\lambda/2}(0)^2\right)$$

which can be simplified further.

If h_j's are the one dimensional Hermite functions, then we know that

$$\Phi_\beta^{\lambda/2}(\zeta) = (|\lambda|/2)^{1/2}h_j\Big((|\lambda|/2)^{1/2}\zeta_1\Big)h_l\Big((|\lambda|/2)^{1/2}\zeta_2\Big)$$

for $\zeta = (\zeta_1,\zeta_2)\in\mathbb{R}^2$ and $\beta = (j,l)\in\mathbb{N}^2$. We also know that $\Phi_\beta^{\lambda/2}(0) = 0$ unless $\beta = (2j,2l)$ for $j,l\in\mathbb{N}$ and in this case we have that (see Eq. (1.1.21) in [16])

$$\Phi_\beta^{\lambda/2}(0)^2 = \frac{|\lambda|}{2\pi^2}\frac{\Gamma(j+1/2)\Gamma(l+1/2)}{\Gamma(j+1)\Gamma(l+1)};\quad \beta = (2j,2l).$$

We also make use of the following properties of the hypergeometric function $F(a,b;c;z)$:

$$\sum_{j=0}^{\infty}\frac{\Gamma(a+j)\Gamma(b+j)}{\Gamma(c+j)\Gamma(j+1)}z^j = \frac{\Gamma(a)\Gamma(b)}{\Gamma(c)}F(a,b;c;z)$$

and

$$F(a,b;c;1) = \frac{\Gamma(c)\Gamma(c-a-b)}{\Gamma(c-a)\Gamma(c-b)},\qquad \mathrm{Re}(c-a-b)>0.$$

The above two equations imply

$$\sum_{j=0}^{\infty}\frac{\Gamma(a+j)\Gamma(b+j)}{\Gamma(c+j)\Gamma(j+1)} = \frac{\Gamma(a)\Gamma(b)\Gamma(c-a-b)}{\Gamma(c-a)\Gamma(c-b)} \tag{25}$$

Consider now

$$\sum_{\beta} \frac{\Gamma(\frac{2|\alpha|+|\beta|+n+1-s}{2})}{\Gamma(\frac{2|\alpha|+|\beta|+n+3+s}{2})} \Phi_{\beta}^{\lambda/2}(0)^2$$

$$= \frac{|\lambda|}{2\pi^2} \sum_{j,l=0}^{\infty} \frac{\Gamma(\frac{2|\alpha|+2j+2l+n+1-s}{2})}{\Gamma(\frac{2|\alpha|+2j+2l+n+3+s}{2})} \frac{\Gamma(j+1/2)\Gamma(l+1/2)}{\Gamma(j+1)\Gamma(l+1)}$$

$$= \frac{|\lambda|}{2\pi^2} \sum_{j=0}^{\infty} \left(\sum_{l=0}^{\infty} \frac{\Gamma(\frac{2|\alpha|+2j+n+1-s}{2}+l)\Gamma(1/2+l)}{\Gamma(\frac{2|\alpha|+2j+n+3+s}{2}+l)\Gamma(l+1)} \right) \frac{\Gamma(j+1/2)}{\Gamma(j+1)}$$

$$= \frac{|\lambda|}{2\pi^2} \sum_{j=0}^{\infty} \frac{\Gamma(\frac{2|\alpha|+2j+n+1-s}{2})\Gamma(1/2)\Gamma(s+1/2)}{\Gamma(s+1)\Gamma(\frac{2|\alpha|+2j+n+2+s}{2})} \frac{\Gamma(j+1/2)}{\Gamma(j+1)}$$

$$= \frac{|\lambda|}{2\pi^2} \frac{\Gamma(1/2)\Gamma(s+1/2)}{\Gamma(s+1)} \sum_{j=0}^{\infty} \frac{\Gamma(\frac{2|\alpha|+n+1-s}{2}+j)\Gamma(1/2+j)}{\Gamma(\frac{2|\alpha|+n+2+s}{2}+j)\Gamma(j+1)}$$

$$= \frac{|\lambda|}{2\pi^2} \frac{\Gamma(1/2)\Gamma(s+1/2)}{\Gamma(s+1)} \frac{\Gamma(\frac{2|\alpha|+n+1-s}{2})\Gamma(1/2)\Gamma(s)}{\Gamma(s+1/2)\Gamma(\frac{2|\alpha|+n+1+s}{2})}$$

$$= \frac{|\lambda|}{2\pi s} \frac{\Gamma(\frac{2|\alpha|+n+1-s}{2})}{\Gamma(\frac{2|\alpha|+n+1+s}{2})}.$$

Thus the expression for $C_{\alpha}^{\lambda}(s)$ simplifies to yield

$$C_{\alpha}^{\lambda}(s) = \frac{|\lambda|}{2\pi s} \frac{\Gamma(\frac{2|\alpha|+n+1-s}{2})}{\Gamma(\frac{2|\alpha|+n+1+s}{2})}$$

and consequently we obtain

$$\|U\|_{\widetilde{H}^{s+1}(\mathbb{R}^{n+3})}^2 = (2\pi)^{-1} \frac{d_s^2}{\pi s} \int_{-\infty}^{\infty} (2|\lambda|)^s \sum_{\alpha} \frac{\Gamma(\frac{2|\alpha|+n+1+s}{2})}{\Gamma(\frac{2|\alpha|+n+1-s}{2})} |\langle f^{\lambda}, \Phi_{\alpha}^{\lambda}\rangle|^2 d\lambda$$

$$= \frac{s\Gamma(s)^2}{4\pi^{2s+1}} \|f\|_{(s)}^2.$$

This completes the proof of the theorem. □

7 Hardy–Littlewood–Sobolev Inequality for G_s

In this section we study the $L^p - L^q$ mapping properties of the operator $G_{-s}, 0 < s < 1$. In [13] the authors have shown that the integral kernel of the operator $\mathcal{L}_{-s}$ is given by $k_s(z,t) = c_{n,s}|(z,t)|^{-Q-2s}$ where $Q = 2n+2$ is the homogeneous dimension of the Heisenberg group $\mathbb{H}^n$. In view of the relation $\pi(\mathcal{L}) = G$, it follows that G_{-s} is also an integral operator whose kernel is given in terms of k_s. Indeed, by spectral theorem

$$G_{-s}f(\xi,w) = \pi(\mathcal{L}_{-s})f(\xi,w) = \int_{\mathbb{H}^n} k_s(z,t)\pi(z,t)f(\xi,w)dzdt.$$

A simple calculation shows that

$$G_{-s}f(\xi,w) = \int_{\mathbb{R}^{n+1}} K_s(\xi,\eta,w-w')f(\eta,w')d\eta dw', \quad (\xi,\eta \in \mathbb{R}^n, w, w' \in \mathbb{R})$$

where the kernel K_s is given by the integral

$$K_s(\xi,\eta,w) = c_{n,s}\int_{\mathbb{R}^n}\Big(\big(|x|^2+|\xi-\eta|^2\big)^2+\big(w-\frac{1}{2}x\cdot(\xi+\eta)\big)^2\Big)^{-\frac{n+1-s}{2}}dx,$$

for $\xi,\eta \in \mathbb{R}^n$, $w \in \mathbb{R}$. From the above expression, it follows that the kernel has the homogeneity

$$K_s(\delta\xi,\delta\eta,\delta^2 w) = \delta^{-(n+2)+2s}K_s(\xi,\eta,w), \quad \delta > 0$$

and consequently a necessary condition for the boundedness of G_{-s} from $L^p(\mathbb{R}^{n+1})$ into $L^q(\mathbb{R}^{n+1})$ is that $\frac{1}{q} = \frac{1}{p} - \frac{2s}{n+2}$. In the following theorem we establish the boundedness of G_{-s} from L^p into L^q under the above condition on p and q.

Theorem 16 *Let $1 < p < q < \infty$ be such that $\frac{1}{q} = \frac{1}{p} - \frac{2s}{n+2}$. Then there exists a constant $C_{n,s}(p)$ such that for all $f \in L^p(\mathbb{R}^{n+1})$ the inequality $\|G_{-s}f\|_q \le C_{n,s}(p)\|f\|_p$ holds.*

Proof In order to prove the theorem we split the kernel into two parts as $K_s = K_s^0 + K_s^\infty$ with $K_s^0(\xi,\eta,w) = K_s(\xi,\eta,w)\chi_{|\xi-\eta|\le\mu}(\xi,\eta,w)$ where μ is to be chosen later. We observe that

$$\int_{\mathbb{R}^{n+1}} K_s^0(\xi,\eta,w)d\eta dw$$
$$= \int_{\mathbb{R}^n}\int_{\{|\xi-\eta|\le\mu\}}\int_{-\infty}^{\infty}\Big(\big(|x|^2+|\xi-\eta|^2\big)^2+\big(w-\frac{1}{2}x\cdot(\xi+\eta)\big)^2\Big)^{-\frac{n+1-s}{2}}dwd\eta dx$$

can be evaluated as follows:

$$\int_{\mathbb{R}^n}\int_{\{|\xi-\eta|\le\mu\}}\int_{-\infty}^{\infty}\left(\left(|x|^2+|\xi-\eta|^2\right)^2+\left(w-\frac{1}{2}x\cdot(\xi+\eta)\right)^2\right)^{-\frac{n+1-s}{2}}dwd\eta dx$$
$$=\int_{\mathbb{R}^n}\int_{\{|\xi-\eta|\le\mu\}}\int_{-\infty}^{\infty}\left(\left(|x|^2+|\xi-\eta|^2\right)^2+w^2\right)^{-\frac{n+1-s}{2}}dwd\eta dx$$

which after a change of variables leads to

$$\int_{\mathbb{R}^n}\int_{\{|\xi-\eta|\le\mu\}}\left(|x|^2+|\xi-\eta|^2\right)^{-n+s}d\eta dx$$
$$=\int_{\{|\eta|\le\mu\}}|\eta|^{-2n+2s+n}\left(\int_{\mathbb{R}^n}(1+|x|^2)^{-n+s}dx\right)d\eta=C\mu^{2s}.$$

This estimate on the kernel K_s^0 allows us to conclude that

$$\left\|\int_{\mathbb{R}^{n+1}}K_s^0(\xi,\eta,w-w')f(\eta,w')d\eta dw'\right\|_p\le C\mu^{2s}\|f\|_p.$$

On the other hand, by Hölder's inequality we estimate

$$\left|\int_{\mathbb{R}^{n+1}}K_s^\infty(\xi,\eta,w-w')f(\eta,w')d\eta dw\right|$$
$$\le\left(\int_{\mathbb{R}^{n+1}}|K_s^\infty(\xi,\eta,w-w')|^{p'}d\eta dw'\right)^{1/p'}\|f\|_p.$$

We now claim that

$$\left(\int_{\mathbb{R}^{n+1}}|K_s^\infty(\xi,\eta,w-w')|^{p'}d\eta dw'\right)^{1/p'}\le C\mu^{-\frac{n+2}{q}}$$

where q is as in the theorem. In view of the definition and the Minkowski integral inequality, the above integral is bounded by

$$\int_{\mathbb{R}^n}\left(\int_{\{|\xi-\eta|>\mu\}}\int_{-\infty}^{\infty}\left(\left(|x|^2+|\xi-\eta|^2\right)^2\right.\right.$$
$$\left.\left.+\left(w-w'-\frac{1}{2}x\cdot(\xi+\eta)\right)^2\right)^{-\frac{n+1-s}{2}p'}dw'd\eta\right)^{1/p'}dx$$
$$=\int_{\mathbb{R}^n}\left(\int_{\{|\xi-\eta|>\mu\}}\int_{-\infty}^{\infty}\left(\left(|x|^2+|\xi-\eta|^2\right)^2+w'^2\right)^{-\frac{n+1-s}{2}p'}dw'd\eta\right)^{1/p'}dx.$$

We split the x-integral into two parts and estimate them separately.

$$\int_{|x|\leq\mu}\Big(\int_{\{|\xi-\eta|>\mu\}}\int_{-\infty}^{\infty}\Big((|x|^2+|\xi-\eta|^2)^2+w'^2\Big)^{-\frac{n+1-s}{2}p'}dw'd\eta\Big)^{1/p'}dx$$
$$\leq C\int_{|x|\leq\mu}\Big(\int_{\{|\eta|>\mu\}}(|x|^2+|\eta|^2)^{-(n+1-s)p'+1}d\eta\Big)^{1/p'}dx.$$

The above integral can be estimated by

$$\int_{|x|\leq\mu}\Big(\int_{\{|\eta|>\mu\}}|\eta|^{-2(n+1-s)p'+2}d\eta\Big)^{1/p'}dx\leq C\mu^{-\frac{n+2}{p}+2s}=C\mu^{-\frac{n+2}{q}}.$$

Proceeding as above we also have

$$\int_{|x|>\mu}\Big(\int_{|\xi-\eta|>\mu}\int_{-\infty}^{\infty}\Big((|x|^2+|\xi-\eta|^2)^2+w'^2\Big)^{-\frac{n+1-s}{2}p'}dw'd\eta\Big)^{1/p'}dx$$
$$\leq C\int_{|x|>\mu}\Big(\int_{|\eta|>\mu}(|x|^2+|\eta|^2)^{-(n+1-s)p'+1}d\eta\Big)^{1/p'}dx$$

which in turn is bounded by

$$\int_{|x|>\mu}\Big(\int_{|\eta|>\mu}|x|^{-2(n+1-s)p'+2+n}(1+|\eta|^2)^{-(n+1-s)p'+1}d\eta\Big)^{1/p'}dx$$
$$\leq C\int_{|x|>\mu}|x|^{-2(n+1-s)+\frac{n+2}{p'}}dx\leq C\mu^{-\frac{n+2}{q}}.$$

Combining the two estimates we have proved our claim with a suitable constant C and consequently, we have

$$\left|\int_{\mathbb{R}^{n+1}}K_s^{\infty}(\xi,\eta,w-w')f(\eta,w')d\eta dw\right|\leq C\mu^{-\frac{n+2}{q}}\|f\|_p.$$

We can now easily prove that the operator G_{-s} is weak type (p,q). Indeed, given $\lambda>0$ and f with $\|f\|_p=1$ we choose μ in such a way that $C\mu^{-\frac{n+2}{q}}=\frac{\lambda}{2}$. With this choice we have

$$\left|\int_{\mathbb{R}^{n+1}}K_s^{\infty}(\xi,\eta,w-w')f(\eta,w')d\eta dw\right|\leq\frac{\lambda}{2}$$

and therefore,

$$\left|\{|G_{-s}f(\xi,w)|>\lambda\}\right| \leq \left|\left\{\left|\int_{\mathbb{R}^{n+1}} K_s^0(\xi,\eta,w-w')f(\eta,w')d\eta dw'\right| > \frac{\lambda}{2}\right\}\right|.$$

The estimate on the kernel K_s^0 gives us the inequality

$$\left|\{|G_{-s}f(\xi,w)|>\lambda\}\right| \leq \left(\frac{2C\mu^{2s}}{\lambda}\right)^p \leq C\left(\frac{\|f\|_p}{\lambda}\right)^q.$$

By Marcinkiewicz interpolation theorem we get the boundedness of G_{-s} from L^p into L^q. □

From the above theorem we obtain the following Hardy–Littlewood–Sobolev inequality for the operator $G_{s/2}$.

Corollary 17 *For* $0 < s < 2$, *we have the inequality*

$$\left(\int_{\mathbb{R}^{n+1}} |f(\xi,w)|^{\frac{2(n+2)}{n+2-2s}} d\xi dw\right)^{\frac{n+2-2s}{(n+2)}} \leq C_{n,s}\langle G_s f, f\rangle.$$

Proof From Theorem 7.1 with $s/2$ in place of s and $G_{s/2}f$ in place of f we get the inequality

$$\left(\int_{\mathbb{R}^{n+1}} |f(\xi,w)|^{\frac{2(n+2)}{n+2-2s}} d\xi dw\right)^{\frac{n+2-2s}{(n+2)}} \leq C_{n,s}\langle G_{s/2} f, G_{s/2}f\rangle.$$

The corollary follows from the observation that G_s differs from $G_{s/2}^2$ by an operator that is bounded on $L^2(\mathbb{R}^{n+1})$. □

Remark 18 The Hardy–Littlewood–Sobolev inequality for the sublaplacian on the Heisenberg group reads as

$$\left(\int_{\mathbb{H}^n} |f(z,t)|^{\frac{(2n+2)}{n+1-s}} dzdt\right)^{\frac{n+1-s}{(n+1)}} \leq C_{n,s}\langle \mathcal{L}_s f, f\rangle.$$

In [9] the authors have calculated that sharp constant in the above inequality. More precisely, they have shown that

$$C_{n,s}^{-1} = \frac{\Gamma(\frac{n+1+s}{2})^2}{\Gamma(\frac{n+1-s}{2})^2}\omega_{2n+1}^{\frac{s}{n+1}}$$

where ω_{2n+1} is the surface area of the unit sphere in $\mathbb{R}^{2n+1}$. It would be interesting to see what the sharp constant is in our case. We conjecture that in our case

$$C_{n,s}^{-1} = \frac{\Gamma(\frac{n+2+2s}{4})^2}{\Gamma(\frac{n+2-2s}{4})^2}\omega_{n+1}^{\frac{2s}{n+2}}.$$

This conjecture is supported by the observation that when f is a radial function on $\mathbb{H}^n$, $\mathcal{L}_s f$ is the same as $G_s f$ where G is the Grushin operator on $\mathbb{R}^{2n+1}$ acting on $f(\xi, w)$ considered as a function on $\mathbb{R}^{2n+1}$ which is radial in the ξ variable.

Acknowledgments Part of this work was done while the second author (PB) visited Department of Mathematics, Indian Institute of Science, Bangalore. He and ST wish to thank Luz Roncal for several useful discussions. The authors are very grateful to the referee for the careful reading of the manuscript and useful comments which have been made use of in revising the original version.

Authors Pradeep Boggarapu and Sundaram Thangavelu were supported by the J. C. Bose Fellowship of Sundaram Thangavelu from the Department of Science and Technology, Government of India.

References

1. Balhara, R.: Hardy's inequality for fractional powers of Grushin operator. Proc. Indian Acad. Sci. **129**(3), 25–33 (2019)
2. Beckner, W.: Pitt's inequality and the fractional Laplacian: sharp error estimates. Forum Math. **24**, 177–209 (2012)
3. Ben Saïd, S., Oshima, T., Shimeno, N.: Fatou's theorems and Hardy-type spaces for eigenfunctions of invariant differential operators on symmetric spaces. Int. Math. Res. Notes **16**, 915–931 (2003)
4. Caffarelli, L., Silvestre, L.: An extension problem related to the fractional Laplacian. Comm. Partial Diff. Equ. **32**, 1245–1260 (2007)
5. Ciaurri, Ó., Roncal, L., Thangavelu, S.: Hardy-type inequalities for fractional powers of the Dunkl-Hermite operator. Proc. Edinburg Math. Soc. **61**(2), 513–544 (2018)
6. Cowling, M., Haagerup, U.: Completely bounded multipliers of the Fourier algebra of a simple Lie group of real rank one. Invent. Math. **96**, 507–549 (1989)
7. Damek, E., Kumar, P.: Eigenfunctions of the Laplace-Beltrami operator on harmonic NA groups. J. Geom. Anal. **26**, 1913–1924 (2016)
8. Frank, R.L., González, M.d.M., Monticelli, D.D., Tan, J.: An extension problem for the CR fractional Lapalcian. Adv. Math. **270**, 97–137 (2015)
9. Frank, R.L., Lieb, E.H.: Sharp constants in several inequalities on the Heisenberg group. Ann. Math. **176**(2), 349–381 (2012)
10. Herbst, I.W.: Spectral theory of the operator $(p^2+m^2)^{1/2} - Ze^2/r$. Commun. Math. Phys. **53**, 285–294 (1977)
11. Möllers, J., Ørsted, B., Zhang, G.: On boundary value problems for some conformally invariant differential operators. Comm. Partial Differ. Equ. **41**, 609–643 (2016)
12. Nikiforov, A., Uvarov, V.: *Special Functions of Mathematical Physics*. Birkhauser Verlag, Basel (1988)
13. Roncal, L., Thangavelu, S.: Hardy's Inequality for Fractional Powers of the Sublaplacian on the Heisenberg Group. Adv. Math. **302**, 106–158 (2016)

14. Roncal, L., Thangavelu, S.: An extension problem and trace Hardy inequality for sublaplacians on H-type groups. Int. Math. Res. Not. **14**, 4238–4294 (2020)
15. Stinga, P., Torrea, J.L.: Extension problem and Harnack's inequality for some fractional operators. Comm. Partial Diff. Equ. **35**, 2092–2122 (2010)
16. Szego, G.: *Orthogonal Polynomials*. American Mathematical Society Colloq. Pub., Rhode Island (1967)
17. Thangavelu, S.: *Lectures on Hermite and Laguerre Expansions*. Mathematical Notes, vol. 42. Princeton University Press, Princeton (1993)
18. Thangavelu, S.: *An Introduction to the Uncertainty Principle. Hardy's Theorem on Lie Groups. With a Foreword by Gerald B. Folland*. Progress in Mathematics, vol. 217. Birkhäuser, Boston (2004)
19. Weis, L.: Operator valued Fourier multiplier theorems and maximal L^p regularity. Math. Ann. **319**, 735–758 (2001)
20. Yafaev, D.: Sharp constants in the Hardy-Rellich inequalities. J. Funct. Anal. **168**, 121–144 (1999)

Sharp Local Smoothing Estimates for Fourier Integral Operators

David Beltran, Jonathan Hickman, and Christopher D. Sogge

Abstract The theory of Fourier integral operators is surveyed, with an emphasis on local smoothing estimates and their applications. After reviewing the classical background, we describe some recent work of the authors which established sharp local smoothing estimates for a natural class of Fourier integral operators. We also show how local smoothing estimates imply oscillatory integral estimates and obtain a maximal variant of an oscillatory integral estimate of Stein. Together with an oscillatory integral counterexample of Bourgain, this shows that our local smoothing estimates are sharp in odd spatial dimensions. Motivated by related counterexamples, we formulate local smoothing conjectures which take into account natural geometric assumptions arising from the structure of the Fourier integrals.

Keywords Local smoothing · Variable coefficient · Fourier integral operators · Decoupling inequalities

D. Beltran
Department of Mathematics, University of Wisconsin, Madison, WI, USA
e-mail: dbeltran@math.wisc.edu

J. Hickman (✉)
School of Mathematics, James Clerk Maxwell Building, The King's Buildings, Edinburgh, UK
e-mail: jeh25@st-andrews.ac.uk; jonathan.hickman@ed.ac.uk

C. D. Sogge
Department of Mathematics, Johns Hopkins University, Baltimore, MD, USA
e-mail: sogge@jhu.edu

P. Ciatti, A. Martini (eds.), *Geometric Aspects of Harmonic Analysis*,
Springer INdAM Series 45, https://doi.org/10.1007/978-3-030-72058-2_2

1 Basic Definitions and Examples of Fourier Integral Operators

1.1 Motivating Examples

This article explores aspects of the theory of *Fourier integral operators* (FIOs), a rich class of objects which substantially generalises the class of pseudo-differential operators. The genesis of the theory can be found in various early works on hyperbolic equations [17, 18, 28, 35, 41] but for the purposes of this article the study of FIOs began in earnest in the groundbreaking treaties of Hörmander [29] and Duistermaat–Hörmander [16].

For the majority of this discussion it will suffice to work with the following definition of a FIO, although below a more general and robust framework is recalled.

Preliminary definition A *Fourier integral operator* (or FIO) $\mathcal{F}$ *of order* $\mu \in \mathbb{R}$ is an operator, defined initially on the space of Schwartz functions $\mathcal{S}(\mathbb{R}^n)$, of the form

$$\mathcal{F}f(x) := \frac{1}{(2\pi)^n} \int_{\hat{\mathbb{R}}^n} e^{i\phi(x;\xi)} a(x;\xi) \hat{f}(\xi)\, d\xi \tag{1}$$

where

- The *phase* $\phi\colon \mathbb{R}^n \times \mathbb{R}^n \to \mathbb{R}$ is homogeneous of degree 1 in ξ and smooth away from $\xi = 0$ on the support of a.
- The *amplitude* $a\colon \mathbb{R}^n \times \mathbb{R}^n \to \mathbb{R}$ belongs to the symbol class S^μ; that is, a is smooth away from $\xi = 0$ and satisfies

$$|\partial_x^\beta \partial_\xi^\alpha a(x;\xi)| \lesssim_{\alpha,\beta} (1 + |\xi|)^{\mu - |\alpha|} \qquad \text{for all} (\alpha, \beta) \in \mathbb{N}_0^n \times \mathbb{N}_0^n.$$

Taking $\phi(x;\xi) := \langle x, \xi \rangle$, one immediately recovers the class of pseudo-differential operators associated to standard symbols (that is, symbols belonging to some class S^μ). For the purposes of this article this is a somewhat trivial case, however, and it is constructive to consider some more representative examples of FIOs.

Example 1 Prototypical FIOs arise from the (Euclidean) half-wave propagator, defined by

$$e^{it\sqrt{-\Delta}} f(x) := \frac{1}{(2\pi)^n} \int_{\hat{\mathbb{R}}^n} e^{i(\langle x,\xi\rangle + t|\xi|)} \hat{f}(\xi)\, d\xi. \tag{2}$$

Under suitable regularity hypotheses on f_0 and f_1, if $f_+ := \frac{1}{2}(f_0 - i(\sqrt{-\Delta})^{-1} f_1)$ and $f_- := \frac{1}{2}(f_0 + i(\sqrt{-\Delta})^{-1} f_1)$,[1] then the function

$$u(x,t) := e^{it\sqrt{-\Delta}} f_+(x) + e^{-it\sqrt{-\Delta}} f_-(x)$$

solves the Cauchy problem

$$\begin{cases} (\partial_t^2 - \Delta)u(x,t) = 0 \\ u(x,0) := f_0(x), \quad \partial_t u(x,0) := f_1(x) \end{cases}. \tag{3}$$

Up to a constant multiple, each term in the expression for $u(x,t)$ is of the form

$$\mathcal{F}_t f(x) := \frac{1}{(2\pi)^n} \int_{\hat{\mathbb{R}}^n} e^{i(\langle x,\xi\rangle \pm t|\xi|)} |\xi|^{-j} \hat{f}(\xi)\, d\xi$$

for either $j = 0$ or $j = 1$. These operators provide important examples of FIOs of order $-j$. Indeed, much of the motivation for the development of the theory of FIOs was to provide an effective counterpart to the theory of pseudo-differential operators to study hyperbolic, rather than elliptic, PDE, a fundamental example being the wave equation (3). The reader is referred to the original papers [16, 29] and the classical texts [32] and [15] for further discussion in this direction.

Example 2 One may also consider wave propagators on other Riemannian manifolds (M, g), defined with respect to the Laplace–Beltrami operator Δ_g. In particular, suppose (M, g) is a compact n-dimensional Riemannian manifold, in which case $-\Delta_g$ has a discrete, positive spectrum which may be ordered $0 = \lambda_0^2 < \lambda_1^2 \leq \lambda_2^2 \leq \dots$ (here the eigenvalues are enumerated with multiplicity). Thus, one may write $-\Delta_g = \sum_{j=0}^{\infty} \lambda_j^2 E_j$ where each E_j is the orthogonal projection in $L^2(M)$ onto a one-dimensional eigenspace associated to the eigenvalue λ_j^2. For proofs of these facts see, for instance, [52, 69].

Now consider the half-wave propagator

$$e^{it\sqrt{-\Delta_g}} f(x) := \sum_{j=0}^{\infty} e^{it\lambda_j} E_j f(x). \tag{4}$$

---

[1] In general, $m(i^{-1}\partial_x)$ denotes the Fourier multiplier operator (defined for f belonging to a suitable *a priori* class)

$$m(i^{-1}\partial_x) f(x) := \frac{1}{(2\pi)^n} \int_{\hat{\mathbb{R}}^n} e^{i\langle x,\xi\rangle} m(\xi) \hat{f}(\xi)\, d\xi$$

for any $m \in L^\infty(\hat{\mathbb{R}}^n)$. The operator $m(\sqrt{-\Delta_x})$ is then defined in the natural manner via the identity $-\Delta_x = i^{-1}\partial_x \cdot i^{-1}\partial_x$.

If u is defined as in the previous example (but now the initial data f_0, f_1 is defined on M and the multipliers are interpreted in terms of the spectral decomposition), then this function solves the Cauchy problem

$$\begin{cases} (\partial_t^2 - \Delta_g)u(x,t) = 0 \\ u(x,0) := f_0(x), \quad \partial_t u(x,0) := f_1(x) \end{cases}. \tag{5}$$

In local coordinates, one may construct a parametrix for the propagator (4) which is of the form of a Fourier integral operator. In particular, for some $t_0 > 0$ one may write

$$e^{it\sqrt{-\Delta_g}} f(x) = \int_{\hat{\mathbb{R}}^n} e^{i\phi(x;t;\xi)} a(x;t;\xi) \hat{f}(\xi)\,d\xi + R_t f(x) \qquad \text{for all } 0 < t < t_0$$

for some suitable choice of phase ϕ and 0-order symbol a, where R_t is a smoothing operator (that is, a pseudo-differential operator with rapidly decaying symbol). Here the Fourier transform of f is taken in the Euclidean sense, in the chosen coordinate domain. This construction is a special case of a general result concerning *strictly hyperbolic* equations (of arbitrary order) which dates back to Lax [35]; further discussion can be found in [15, Chapter 5] or [53, Chapter 4].

Example 3 Closely related to the wave propagator (2) are the convolution operators

$$A_t f(x) := f * \sigma_t(x), \qquad t > 0$$

where $\sigma = \sigma_1$ is the surface measure on the unit sphere $\mathbb{S}^{n-1}$ and σ_t is defined by

$$\int_{\mathbb{R}^n} f(x)\,d\sigma_t(x) := \int_{\mathbb{R}^n} f(tx)\,d\sigma(x).$$

When $n = 3$ the solution to (3) at time t is related to A_t via the classical Kirchhoff formula (see, for instance, [51, Chapter 1]). These averaging operators are also of significant interest in harmonic analysis and, in particular, the spherical maximal function of Stein [56] and Bourgain [4] is defined by $Mf(x) := \sup_{t>0} A_t|f|(x)$.

To see how such averages fall into the Fourier integral framework, recall that the method of stationary phase (see, for instance, [58, Chapter VIII] or [53, Chapter 1]) yields the formula

$$\hat{\sigma}(\xi) := \sum_{\pm} e^{\pm i|\xi|} a_{\pm}(\xi)$$

for the Fourier transform of the measure σ, where $a_\pm \in S^{-(n-1)/2}$ are smooth symbols of order $-(n-1)/2$. Thus, one may write

$$A_t f(x) = \sum_{\pm} \frac{1}{(2\pi)^n} \int_{\hat{\mathbb{R}}^n} e^{i(\langle x,\xi\rangle \pm t|\xi|)} a_\pm(t\xi) \hat{f}(\xi)\,d\xi;$$

note that the operators appearing in this formula agree with those arising in Example 1 except for the choice of symbol.

1.2 Distributions Defined by Oscillatory Integrals

The remainder of this section will be dedicated to describing a more general framework for the study of FIOs. For much of this article the preliminary definition given in the preceding subsection is sufficient; the refined definitions are included here in order to relate this survey to the perspective espoused in many of the references, and in particular in the classical works [16, 29].

In contrast with the discussion in the previous subsection, here the operators will be defined in terms of a kernel. Formally, the kernel of the FIO in (1) is given by

$$K(x;y) := \frac{1}{(2\pi)^n}\int_{\hat{\mathbb{R}}^n} e^{-i(\langle y,\xi\rangle - \phi(x;\xi))} a(x;\xi)\,\mathrm{d}\xi,$$

although without strong conditions on the symbol a this integral is not defined in any classical sense. To give precise meaning to such expressions one appeals to the theory of distributions; the relevant concepts from this theory are reviewed presently.

1.2.1 Distributions

Given $W \subseteq \mathbb{R}^d$ open, let $\mathcal{D}(W)$ denote the space of test functions on W; that is, $\mathcal{D}(W)$ is the space $C_c^\infty(W)$ of C^∞ functions with compact support in W under the topology defined by $f_j \to f$ as $j \to \infty$ for $f_j, f \in \mathcal{D}(W)$ if

(i) There exists a compact set $K \subset W$ containing $\operatorname{supp} f$ and $\operatorname{supp} f_j$ for all $j \in \mathbb{N}$;
(ii) $\partial_x^\alpha f_j \to \partial_x^\alpha f$ uniformly as $j \to \infty$ for all $\alpha \in \mathbb{N}_0^d$.

One then defines the space of distributions $\mathcal{D}'(W)$ on W to be the dual topological vector space to $\mathcal{D}(W)$ endowed with the weak* topology. With this definition $\mathcal{D}'(W)$ is complete.[2]

1.2.2 Homogeneous Oscillatory Integrals

Now let $\varphi\colon W \times (\mathbb{R}^N \setminus \{0\}) \to \mathbb{R}$ be a smooth function, $a \in S^\mu(W \times \mathbb{R}^N)$ and consider the oscillatory integral formally defined by

$$I[\varphi, a](w) := \int_{\hat{\mathbb{R}}^N} e^{i\varphi(w;\theta)} a(w;\theta)\,\mathrm{d}\theta \qquad \text{for} w \in W. \tag{6}$$

If $\mu < -N$, then this integral converges absolutely and, moreover, the resulting function of w defines a distribution on W (by integrating $I[\varphi, a]$ against a given test

[2]Here a sequence $(u_j)_{j=1}^n \subseteq \mathcal{D}'(W)$ is Cauchy if $(\langle u_j, f\rangle)_{j=1}^\infty$ is a Cauchy sequence of complex numbers for all $f \in \mathcal{D}(W)$.

function). If $\mu > -N$, then it is not clear that the expression (6) makes sense and additional hypotheses are required on φ to give the integral meaning. In particular, suppose that

$$\varphi \text{ is homogeneous of degree 1 in } \theta \tag{7}$$

and

$$\nabla_{w,\theta}\varphi(w;\theta) \neq 0 \text{ for all } (w;\theta) \in W \times (\mathbb{R}^N \setminus \{0\}) \tag{8}$$

where the gradient $\nabla_{w,\theta}$ is taken with respect to all the variables $(w;\theta)$. Now let $\beta \in C_c^\infty(\mathbb{R}^N)$ satisfy $\beta(0) = 1$ and consider the truncated integral

$$I^j[\varphi, a](w) := \int_{\hat{\mathbb{R}}^N} e^{i\varphi(w;\theta)}\beta(2^{-j}\theta)a(w;\theta)\,d\theta.$$

Each $I^j[\varphi, a]$ is a well-defined function which induces a distribution. Moreover, under the conditions (7) and (8), a simple integration-by-parts argument allows one to deduce that for any $K \subseteq W$ compact

$$|\langle I^j[\varphi, a], f\rangle| \leq C_K 2^{-j} \sum_{|\alpha|\leq k} \|\partial_x^\alpha f\|_{L^\infty(K)} \qquad \text{for all } f \in \mathcal{D}(W) \text{ with supp } f \subseteq K$$

where k satisfies $\mu < -N+k$ (see, for instance, [53, Theorem 0.5.1]). Since $\mathcal{D}'(W)$ is complete, one may therefore define $I[\varphi, a]$ to be the distribution given by the limit of the sequence of distributions $I^j[\varphi, a]$.

Definition 4 The distribution $I[\varphi, a]$, defined for φ satisfying (7) and (8), will be referred to as *(local)*[3] *homogeneous oscillatory integral*. By a slight abuse of notation, the distribution $I[\varphi, a]$ will also be denoted by the formal expression (6).

In what follows, it will be useful to assume a further condition on the phase φ.

Non-degeneracy hypothesis A smooth function $\varphi\colon W\times(\mathbb{R}^N\setminus\{0\}) \to \mathbb{R}$ satisfying (7) and (8) is a *non-degenerate phase function* if, in addition, it satisfies

$$\text{if } \partial_\theta\varphi(w;\theta) = 0, \text{ then } \bigwedge_{j=1}^{N} \nabla_{w,\theta}\partial_{\theta_j}\varphi(w;\theta) \neq 0. \tag{9}$$

The rationale behind this additional hypothesis will become apparent in Sect. 1.4.

[3]The terminology *local*, as opposed to *global*, will become clearer in Sect. 1.5.

1.3 *Local Fourier Integral Operators*

For $X \subseteq \mathbb{R}^n$ and $Y \subseteq \mathbb{R}^m$ open, any distribution $K \in \mathcal{D}'(X \times Y)$ defines a natural continuous linear mapping $T : \mathcal{D}(Y) \to \mathcal{D}'(X)$ given by

$$\langle T(f), g\rangle := \langle K, f \otimes g\rangle \qquad \text{for all } (f, g) \in \mathcal{D}(Y) \times \mathcal{D}(X). \tag{10}$$

In fact, a converse to this observation also holds, which is the content of the celebrated Schwartz kernel theorem (see, for instance, [31, §5.2]). In particular, given any continuous linear mapping $T : \mathcal{D}(Y) \to \mathcal{D}'(X)$ there exists a unique distribution $K \in \mathcal{D}'(X \times Y)$, referred to as the *(Schwartz) kernel* of T, such that (10) holds.

Definition 5 A continuous linear operator $\mathcal{F}: C_c^\infty(Y) \to \mathcal{D}'(X)$ is a *(local) Fourier integral operator* if the Schwartz kernel is given by a homogeneous oscillatory integral $I[\varphi, a]$ for some non-degenerate phase function $\varphi: X \times Y \times \mathbb{R}^N \setminus \{0\} \to \mathbb{R}$ and amplitude $a \in S^\mu(X \times Y \times \mathbb{R}^N)$.

Given a test function $f \in C_c^\infty(Y)$, by an abuse of notation the distribution $\mathcal{F}f$ will also be denoted by

$$\mathcal{F}f(x) = \int_{\mathbb{R}^n}\int_{\mathbb{R}^N} e^{i\varphi(x;y;\theta)} a(x; y; \theta)\, \mathrm{d}\theta\, f(y)\, \mathrm{d}y. \tag{11}$$

Example 6 The averaging operator from Example 3 can be expressed as

$$A_t f(x) = \sum_{\pm} \frac{1}{(2\pi)^n} \int_{\mathbb{R}^n}\int_{\hat{\mathbb{R}}^n} e^{i(\langle x-y,\xi\rangle \pm t|\xi|)} a_\pm(t\xi)\, \mathrm{d}\xi\, f(y)\, \mathrm{d}y, \tag{12}$$

where this formal expression is interpreted in the above distributional sense. Note that the phase $\varphi(x; y; \xi) := \langle x - y, \xi\rangle \pm t|\xi|$ satisfies the desired conditions (7), (8) and (9).

There are significant short-comings in defining FIOs in this way. In particular, there are fundamental problems regarding uniqueness: a given operator will admit many distinct representations of the form (11).

Example 7 Once again recall the operator $A_t f$ from Example 3, which can be interpreted as taking an average of f over the sphere $x + t\mathbb{S}^{n-1}$. For fixed x, t, this surface corresponds to the zero locus of the defining function

$$\Phi(x; t; y) := \frac{|x - y|^2}{t^2} - 1.$$

This allows one to rewrite the averaging operator as

$$A_t f(x) := \int_{\mathbb{R}^n} f(y) a(x; t; y)\, \delta(\Phi(x; t; y))\mathrm{d}y$$

where $\delta(\Phi(x;t;y))\mathrm{d}y$ is the normalised induced Lebesgue measure on $x + t\mathbb{S}^{n-1}$ (see, for instance, [58, Chapter XI, §3.1.2] or [59, Chapter VIII, §3]) and $a(x;t;y)$ is an appropriate choice of bump function. Using the heuristic identity

$$\delta(x) = \frac{1}{2\pi}\int_{\mathbb{R}} e^{i\lambda x}\,\mathrm{d}\lambda$$

for the Dirac δ function, this leads to the expression

$$A_t f(x) = \frac{1}{2\pi}\int_{\mathbb{R}^n}\int_{\hat{\mathbb{R}}} e^{i\lambda\Phi(x;t;y)} a(x;t;y)\,\mathrm{d}\lambda\, f(y)\,\mathrm{d}y. \tag{13}$$

Thus, one arrives at an alternative Fourier integral representation of the average $A_t f$ to that in (12). Although this argument has been presented as a heuristic, it is not difficult to make the details precise, provided (13) is interpreted correctly (that is, as converging in the sense of oscillatory integrals); the full details can be found, for instance, in [58, Chapter XI].

1.4 Wave Front Sets and Equivalence of Phase

Examples 6 and 7 show that very different phase/amplitude pairs $[\varphi, a]$ can define the same Fourier integral operator. It is natural to ask whether one can formulate a "coordinate-free" or "invariant" definition of FIOs which does not rely on fixing a choice of phase and amplitude. Such a *global* definition does indeed exist and is discussed in detail in [16, 29] as well as the texts [15, 53, 63]. The full details of the global theory of Fourier integral operators is, however, beyond the scope of this article, but nevertheless here some of the basic ideas are presented.

To arrive at a global definition of Fourier integral operators, it is necessary to analyse the geometry of the singularities of the underlying Schwartz kernel. This leads to the construction of a geometric object known as the *canonical relation* for a given FIO, which is in some sense independent of the choice of phase function used to define the kernel (this is the content of Hörmander's equivalence of phase theorem, discussed below). The idea is then to think of the FIO purely in terms of the canonical relation (and some order), without reference to a particular choice of pair $[\varphi, a]$.

To carry out the above programme, some basic definitions from microlocal analysis (which may be described as the geometric study of distributions) are required.

1.4.1 The Singular Support

Once again, let $W \subseteq \mathbb{R}^d$ be some fixed open set.

Definition 8 The *singular support* $\operatorname{sing\,supp} u$ of $u \in \mathcal{D}'(W)$ is defined to be the set of points in $w \in W$ for which there exists no open neighbourhood upon which u agrees with a C^∞ function.

The singular support identifies the location of the singularities of a distribution, but for a complete geometric description one must also understand the associated "directions" of the singularities.

Example 9 Let $\gamma\colon [0,1] \to \mathbb{R}^2$ be (a parametrisation of) a smooth curve and let μ be a smooth density on γ in $\mathbb{R}^2$, viewed as a measure (and therefore a distribution) on the plane. In particular, there exists some $a \in C_c^\infty(\mathbb{R})$ with support in $(0, 1)$ such that μ is defined by

$$\int_{\mathbb{R}^2} f \,\mathrm{d}\mu := \int_0^1 (f \circ \gamma)(t) a(t)\,\mathrm{d}t \qquad \text{for all } f \in C(\mathbb{R}^2).$$

It is immediate that the singular support of μ consists of the support of the measure (and is therefore a subset of the curve). Given $x_0 \in \operatorname{supp}\mu$, one expects that the singular direction should the normal to the curve at x_0. A rigorous formulation of this intuitive statement is discussed below.

To identify the singular directions, one appeals to the correspondence between regularity and the decay of the Fourier transform. For instance, recall that any $f \in C_c^\infty(\mathbb{R}^n)$ satisfies

$$|\hat{f}(\xi)| \lesssim_N (1 + |\xi|)^{-N} \qquad \text{for all } N \in \mathbb{N} \tag{14}$$

for $\xi \in \hat{\mathbb{R}}^n$ (and, moreover, the property $f \in C_c^\infty(\mathbb{R}^n)$ is completely characterised in terms of the decay of the Fourier–Laplace transform by the Paley–Weiner theorem [31, §7.3]). If $u \in \mathcal{D}'(W)$ and $w \notin \operatorname{sing\,supp} u$, then it follows that there exists some $\psi \in C_c^\infty(W)$ satisfying $\psi(w) \neq 0$ for which $f := \psi u$ is a C_c^∞ function satisfying (14). Given $w \in \operatorname{sing\,supp} u$ and $\psi \in C_c^\infty(W)$ satisfying $\psi(w) \neq 0$ as above, the idea is now to analyse the directions in which (14) fails for $f := \psi u$. Since in this case ψu is no longer guaranteed to be a function, the precise definition requires some facts about Fourier transforms of distributions, which are recalled presently.

1.4.2 Tempered Distributions and the Fourier Transform

If $\mathcal{S}(\mathbb{R}^n)$ denotes the Schwartz space, then recall that the space of *tempered distributions* is the dual $\mathcal{S}'(\mathbb{R}^n)$, which can be identified with a subspace of $\mathcal{D}'(\mathbb{R}^n)$. The Fourier transform $\hat{u} \in \mathcal{S}'(\mathbb{R}^n)$ of a tempered distribution $u \in \mathcal{S}'(\mathbb{R}^n)$ is defined by $\langle \hat{u}, f \rangle := \langle u, \hat{f} \rangle$ for all $f \in \mathcal{S}(\mathbb{R}^n)$. If $u \in \mathcal{D}'(W)$ is compactly-supported (in the sense that there exists a compact set $K \subset W$ such that $\langle u, f \rangle = 0$ whenever $f \in \mathcal{D}(W)$ is supported outside K), then u automatically extends to a tempered distribution and, moreover, the Fourier transform $\hat{u}$ is a C^∞ function which grows at most polynomially. For proofs of these facts see, for instance, [31].

1.4.3 The Wave Front Set

Let $u \in \mathcal{D}'(W)$ be compactly supported. Define $\Gamma(u)$ to be the set of points $\eta \in \hat{\mathbb{R}}^n \setminus \{0\}$ for which there does not exist an open conic neighbourhood C upon which

$$|\hat{u}(\xi)| \lesssim_N (1 + |\xi|)^{-N} \qquad \text{for all } N \in \mathbb{N} \text{ and all } \xi \in C.$$

Given $u \in \mathcal{D}'(W)$ and $w \in W$ one then defines

$$\Gamma_w(u) := \bigcap_{\substack{\psi \in C_c^\infty(W) \\ \psi(w) \neq 0}} \Gamma(\psi u).$$

By the discussion following (14), it is clear that if $\Gamma_w(u) \neq \emptyset$, then $w \in \operatorname{sing\,supp} u$.

Definition 10 (Wave Front Set) If $u \in \mathcal{D}'(W)$, then the *wave front set* $\mathrm{WF}(u)$ of u is defined

$$\mathrm{WF}(u) := \big\{(w; \xi) \in W \times (\mathbb{R}^d \setminus \{0\}) : \xi \in \Gamma_w(u)\big\}.$$

Example 11 Returning to the measure μ discussed in Example 9, fix $x_0 \in \operatorname{sing\,supp} \mu$ so that $x_0 = \gamma(t_0)$ for some $t_0 \in \operatorname{supp} a$. Suppose $\eta \in \hat{\mathbb{R}}^2 \setminus \{0\}$ does not lie in the linear subspace $N_{x_0}\gamma$ spanned by the normal to γ at x_0. It is not difficult to show that there exists a conic neighbourhood C of η and some $\varepsilon_0 > 0$ such that

$$|\langle \xi, \gamma'(t) \rangle| \geq \varepsilon_0 |\xi| \qquad \text{for all } \xi \in C \text{ and } |t - t_0| < \varepsilon_0.$$

If $\psi \in C_c^\infty(\mathbb{R}^2)$ is chosen to have support in a sufficiently small neighbourhood of x_0, it therefore follows by non-stationary phase (that is, repeated integration-by-parts) that $|(\psi\mu)\hat{}\,(\xi)| \lesssim_N (1 + |\xi|)^{-N}$ for all $N \in \mathbb{N}$ and $\xi \in C$ and, consequently, $\Gamma_{x_0}(\mu) \subseteq N_{x_0}\gamma$. On the other hand, if $\eta \in N_{x_0}\gamma \cap \mathbb{S}^1$ and $\psi \in C_c^\infty(\mathbb{R}^2)$ satisfies $\psi(x_0) \neq 0$, then the asymptotic expansion for oscillatory integrals (see, for instance, [58, Chapter VIII]) shows that $|(\psi\mu)\hat{}\,(\lambda\eta)|$ fails to decay rapidly in $\lambda \geq 1$.

For a homogeneous oscillatory integral $I[\varphi, a]$, as defined in Sect. 1.2, it is not difficult to show that the wave front set of this distribution is contained in

$$\Lambda_\varphi := \{(w, \partial_w\varphi(w; \theta)) \in W \times (\mathbb{R}^d \setminus \{0\}) : (w; \theta) \in \operatorname{supp} a,\ \partial_\theta\varphi(w; \theta) = 0\}.$$

Indeed, as a rough sketch of why this should hold, taking the Fourier transform of $I[\varphi, a] \cdot \psi$ for some $\psi \in C_c^\infty(\mathbb{R}^d)$ yields an oscillatory integral in the $(w; \theta)$ variables with phase $\varphi(w; \theta) - \langle w, \xi \rangle$. By non-stationary phase (that is, integration-by-parts), one expects rapid decay away from the set of $(w; \xi)$ for which the phase admits a $(w; \theta)$-stationary point. Since the w-gradient of the phase is given by $\partial_w \varphi(w; \theta) - \xi$ and the θ-gradient is $\partial_\theta \varphi(w; \theta)$, this naturally leads one to consider the set Λ_φ. The full details can be found, for instance, in [31, Theorem 8.1.9].

1.4.4 Non-degeneracy and Lagrangian Manifolds

If the phase function φ is non-degenerate in the sense that (9) holds, then it follows from the implicit function theorem that

$$\Sigma_\varphi := \big\{(w, \theta) \in W \times (\mathbb{R}^N \setminus \{0\}) : \partial_\theta \varphi(w, \theta) = 0\big\}$$

is a smooth d-dimensional submanifold. Moreover, one may readily verify that the map $\kappa \colon \Sigma_\varphi \to W \times (\mathbb{R}^d \setminus \{0\})$ given by $\kappa(w, \theta) := (w, \partial_w \varphi(w, \theta))$ is an immersion with image Λ_φ. Typically, in this situation one identifies $W \times (\mathbb{R}^d \setminus \{0\})$ with $T^*W \setminus 0$, the cotangent bundle of W with the zero section removed. The rationale behind this is that, under the above identification, Λ_φ has a special property defined with respect to the natural symplectic structure on T^*W. Concepts from symplectic geometry form an important part of the analysis of FIOs, but will only be mentioned in passing here (see, for instance, [15] for a thorough introduction to symplectic differential geometry and its connection to Fourier integral theory).

Definition 12 A smooth (immersed) submanifold $\Lambda \subseteq T^*W \setminus 0$ is *conic* if it is conic in the fibres: that is, $(w, t\xi) \in \Lambda$ for all $t > 0$ whenever $(w, \xi) \in \Lambda$. Such a Λ is a *Lagrangian submanifold* if it is d-dimensional and the restriction of the *canonical 1-form* $\omega := \sum_{j=1}^d \xi_j \mathrm{d}w_j$ on T^*W to Λ is identically zero.

It is not difficult to show that Λ_φ is a (conic) Lagrangian submanifold.[4] Conversely, given any conic Lagrangian submanifold $\Lambda \subset T^*W \setminus 0$, one can show that locally Λ agrees with Λ_φ for some non-degenerate phase function φ (see, for instance, [29, §3.1]).

[4]The pull-back $\kappa^*\omega$ of ω is given by

$$\sum_{j=1}^d \partial_{w_j} \varphi(w, \theta) \mathrm{d}w_j = \mathrm{d}\phi - \sum_{i=1}^N \partial_{\theta_i} \varphi(w, \theta) \mathrm{d}w_j.$$

This vanishes identically on Λ_φ since $\partial_\theta \varphi(w, \theta) = 0$ and the homogeneity of φ with respect to θ implies $\varphi(w, \theta) = \theta \partial_\theta \varphi(w, \theta) = 0$.

1.4.5 Equivalence of Phase

The correspondence between conic Lagrangian submanifolds and non-degenerate phase functions described above is clearly not unique: for instance, one may compose the phase function φ with any fibre-preserving diffeomorphism $(w, \theta) \mapsto (w, \tilde{\theta}(w, \theta))$ to obtain a new phase function $\tilde{\varphi}$ which satisfies $\Lambda_\varphi = \Lambda_{\tilde{\varphi}}$. However, in this case it follows by the change of variables formula that $I[\varphi, a] = I[\tilde{\varphi}, \tilde{a}]$ where $\tilde{a}(w, \tilde{\theta}(w, \theta)) = a(w, \theta)|\partial_\theta \tilde{\theta}(w, \theta)|^{-1}$. Thus, provided the symbols are appropriately defined, the phases φ and $\tilde{\varphi}$ define the same homogeneous oscillatory integral.

Now suppose φ, $\tilde{\varphi}$ are two phase functions which satisfy $\Lambda_\varphi = \Lambda_{\tilde{\varphi}}$, but are not necessarily related by a fibre-preserving diffeomorphism. What can be said about the homogeneous oscillatory integrals in this case? A typical example of this situation has already appeared above.

Example 13 Fixing $t \in \mathbb{R} \setminus \{0\}$, consider the phase function $\varphi_t : \mathbb{R}^n \times \mathbb{R}^n \times (\mathbb{R}^n \setminus \{0\}) \to \mathbb{R}$ featured in Example 1 and Example 3, given by $\varphi_t(x, y; \xi) := \langle x - y, \xi \rangle + t|\xi|$. Then a simple computation shows that

$$\Lambda_{\varphi_t} = \Big\{(x, x + t\frac{\xi}{|\xi|}, \xi, -\xi) : (x, \xi) \in \mathbb{R}^n \times \mathbb{R}^n \setminus \{0\}\Big\}.$$

Now consider the phase function $\tilde{\varphi}_t : \mathbb{R}^n \times \mathbb{R}^n \times (\mathbb{R} \setminus \{0\}) \to \mathbb{R}$ given by

$$\tilde{\varphi}_t(x, y; \lambda) := \lambda\Big(\frac{|x - y|^2}{t^2} - 1\Big),$$

which is featured in the alternative representation for the averaging operator from Example 7. A simple computation shows that

$$\Lambda_{\tilde{\varphi}_t} = \Big\{\Big(x, y, 2\lambda\frac{x - y}{t^2}, -2\lambda\frac{x - y}{t^2}\Big) : |x - y| = t\Big\}.$$

However, making the substitution $\xi = 2\lambda\frac{x-y}{t^2}$, this set agrees precisely with that in Example 13.

Note that the fibres of φ_t and $\tilde{\varphi}_t$ have different dimensions so clearly the two phases cannot be related via a fibre-preserving change of variables.

It is still true that, for suitable choices of amplitude function, the phases discussed in Example 13 define the same homogeneous oscillatory integral (indeed, both phases are used to represent the same averaging operator A_t). These observations suggest the possibility of a *unique* correspondence between conic Lagrangian manifolds Λ and homogeneous oscillatory integrals. This correspondence is the subject of the following fundamental result of Hörmander [29].

Theorem 14 (Hörmander's Equivalence of Phase Theorem [29]) *Suppose φ and $\tilde{\varphi}$ are non-degenerate phase functions defined on a neighbourhood of $(w_0, \theta_0) \in$*

$W \times \mathbb{R}^N$ and $(w_0, \tilde{\theta}_0) \in W \times \mathbb{R}^{\tilde{N}}$, respectively, which define the same Lagrangian submanifold there.

Then every homogeneous oscillatory integral $I[\varphi, a]$ with $a \in S^{\mu+(d-2N)/4}(W \times \mathbb{R}^N)$ and $\operatorname{supp} a$ in a sufficiently small θ-conic neighbourhood[5] *of (w_0, θ_0) can also be written as $I[\tilde{\varphi}, \tilde{a}]$ for some $\tilde{a} \in S^{\mu+(d-2\tilde{N})/4}(W \times \mathbb{R}^{\tilde{N}})$ with $\operatorname{supp} \tilde{a}$ in a small $\tilde{\theta}$-conic neighbourhood of $(w_0, \tilde{\theta}_0)$.*

The equivalence of phase theorem suggests that rather than thinking of the distribution $I[\varphi, a]$ as defined by some choice of phase/amplitude pair $[\varphi, a]$, one should think of the distribution as determined by the conic Lagrangian submanifold Λ. This perspective is described in the following subsection.

1.5 Global Theory

The equivalence of phase theorem allows much of the analysis of the previous sections to be lifted to the more general setting of smooth manifolds W. Given a conic Lagrangian submanifold $\Lambda \subset T^*W$, one roughly defines $I^\mu(W, \Lambda)$ to be the class of homogeneous oscillatory integrals which can be locally represented as some $I[\varphi, a]$ for some non-degenerate phase function φ for which Λ_φ locally agrees with Λ and symbol a belonging to the class $S^{\mu+(d-2N)/4}$. Here N is the dimension of the fibres (that is, the number of Fourier variables θ) in this local representation. To give more precise details of this definition requires a brief review of basic concepts pertaining to analysis on manifolds.

1.5.1 Distributions on Manifolds

Let W be a d-dimensional smooth manifold so that W is equipped with a system of coordinate charts $\kappa_\alpha \colon W_\alpha \to \tilde{W}_\alpha$, each of which is a diffeomorphism from some open subset $W_\alpha \subseteq W$ to an open subset $\tilde{W}_\alpha \subseteq \mathbb{R}^d$.

Definition 15 A *distribution u on W* is an assignment of a distribution $u_\alpha \in \mathcal{D}'(\tilde{W}_\alpha)$ to every coordinate chart which satisfies the following consistency property: given two charts $\kappa_{\alpha_j} \colon W_{\alpha_j} \to \tilde{W}_{\alpha_j}$, $j = 1, 2$, the identity

$$\langle u_{\alpha_2}, f \rangle = \langle u_{\alpha_1}, f \circ \alpha_1 \circ \alpha_2^{-1} \rangle$$

holds whenever $f \in \mathcal{D}(\tilde{W}_{\alpha_2})$ is supported inside $\kappa_{\alpha_2}(W_{\alpha_1} \cap W_{\alpha_2})$. The space of all distributions on W is denoted by $\mathcal{D}'(W)$.

[5]Here a θ-conic neighbourhood of $(w_0, \theta_0) \in W \times \mathbb{R}^N \setminus \{0\}$ is an open neighbourhood $U \subseteq W \times \mathbb{R}^N \setminus \{0\}$ of (w_0, θ_0) such that $(w, t\theta) \in U$ for all $t > 0$ whenever $(w, \theta) \in U$.

Remark 16 It makes perfect sense, in analogy with the definition in the Euclidean case, to consider the dual of the space of test functions on W. Elements of this dual are slightly different objects to the distributions defined above, however, and are known as *distribution densities* (see, for instance, [31, §6.3]).

1.5.2 Global Homogeneous Oscillatory Integrals

Let $\Lambda \subseteq T^*W$ be a closed, conic Lagrangian submanifold and $\mu \in \mathbb{R}$. A *global homogeneous oscillatory integral of order* μ is a distribution $u \in \mathcal{D}'(W)$ such that for every coordinate chart $\kappa_\alpha : W_\alpha \to \tilde{W}_\alpha$ the associated distribution $u_\alpha \in \mathcal{D}'(\tilde{W}_\alpha)$ is of the form $u_\alpha = I[\varphi_\alpha, a_\alpha]$ where:

(i) Each φ_α is a non-degenerate phase function defined on a θ-conic open subset $U_\alpha \subseteq \tilde{W}_\alpha \times \mathbb{R}^{N_\alpha}$ for some integer $N_\alpha \in \mathbb{N}$. Furthermore, an open neighbourhood of Λ is diffeomorphically mapped to

$$\Lambda_{\varphi_\alpha} := \{(w, \partial_w \varphi(w, \theta)) : (w, \theta) \in U_\alpha, \quad \partial_\theta \varphi(w, \theta) = 0\}$$

under the coordinates induced by κ_α.[6]

(ii) Each $a_\alpha \in S^{\mu+(d-2N_\alpha)/4}(\mathbb{R}^d \times \mathbb{R}^{N_\alpha})$ is supported in U_α and has compact w-support.

1.5.3 Global FIOs and Canonical Relations

Fix a pair of manifolds X and Y with $\dim X = n$ and $\dim Y = m$ and let $\Lambda \subseteq T^*X \setminus 0 \times T^*Y \setminus 0$ be a conic Lagrangian submanifold. The (global) homogeneous oscillatory integrals in $I^\mu(X \times Y, \Lambda)$ define (global) Fourier integral operators via the Schwartz kernel on each coordinate chart. The resulting collection of operators is denoted by $I^\mu(X, Y, C)$ where C is what is known as the *canonical relation*: it is the rotated and reflected copy of Λ given by

$$C := \big\{(x, \xi, y, -\eta) \in T^*X \setminus 0 \times T^*Y \setminus 0 : (x, y, \xi, \eta) \in \Lambda\big\}.$$

One typically works with the canonical relation C rather than Λ since it is often easier (notationally speaking) to formulate various hypotheses over C and C arises naturally in the composition calculus for FIOs (see, for instance, [53, Chapter 6]). Of course, C inherits the symplectic structure of Λ and, in particular, if $\omega_X := \sum_{j=1}^n \xi_j \mathrm{d}x_j$ and $\omega_Y := \sum_{j=1}^n \eta_j \mathrm{d}y_j$ denote the canonical 1-forms on X and Y, respectively, then $\omega_X - \omega_Y$ vanishes identically on C.

[6] In particular, if $\pi : T^*W \to W$ denotes the projection onto the base point, then for each chart $\kappa : W_\alpha \to \tilde{W}_\alpha$ one may define the induced local coordinates on the tangent bundle $\tilde{\kappa} : \pi^{-1}(W_\alpha) \to \tilde{W}_\alpha \times \mathbb{R}^d$ by $\tilde{\kappa}(w, \xi) := (\kappa(w), (\mathrm{d}\kappa_w)^{-\top}\xi)$.

1.5.4 Global Versus Local Theory

In the *global* approach to Fourier integral theory one typically frames the hypotheses on the operator in terms of geometric properties of the canonical relation (some examples of this will be given in Sect. 2, where the *rotational* and *cinematic* curvature conditions are discussed). For the majority of this article, however, it will suffice to work with a concrete representation of the operator given by a choice of phase and amplitude as in (1). In this *local* approach, the hypotheses on the operator are framed in terms of properties of the choice of phase function ϕ and its derivatives.

The local approach will in fact afford no loss of generality, since the problems under consideration are all of a local nature and it is always possible to locally express any FIO as an operator of the form of (1). Indeed, if $\dim X = \dim Y$, given any FIO as above, basic results in symplectic geometry (see, for instance, [15, Proposition 3.7.3] or [53, Proposition 6.2.4]) guarantee that the canonical relation C can be expressed locally as a graph (modulo a reflection) of the form

$$(x, \eta) \mapsto (x, S(x, \eta), T(x, \eta), -\eta).$$

Moreover, it is not difficult to show that $S = \partial_x \phi$ and $T = \partial_\eta \phi$ for some *generating function* ϕ. Indeed, the canonical 1-form $\omega_X - \omega_Y$ is given in the above coordinates by

$$\sum_{j=1}^{n} \big[S_j(x, \eta) - \sum_{i=1}^{n} \eta_i \partial_{x_j} T_i(x, \eta)\big]\mathrm{d}x_j - \sum_{j=1}^{n} \big[\sum_{i=1}^{n} \eta_i \partial_{\eta_j} T_i(x, \eta)\big]\mathrm{d}\eta_j.$$

By the Lagrangian property of C, the coefficient functions must all vanish identically on the domain, which implies that $\phi(x, \eta) := \langle \eta, T(x, \eta)\rangle$ is a suitable generating function. Thus, the canonical relation induced by the phase function $\varphi(x, y, \xi) := \phi(x, \xi) - \langle y, \xi\rangle$ agrees locally with C and consequently, by the equivalence of phase theorem, the FIO admits a local expression of the form (1).

2 Local Smoothing Estimates

This survey is primarily concerned with the continuity of FIOs as maps between certain function spaces. Such problems are inherently local in nature and, consequently, for the majority of the discussion it will suffice to work FIOs of the form (1). In particular, let $\mathcal{F}$ be an operator given by

$$\mathcal{F}f(x) := \frac{1}{(2\pi)^n} \int_{\hat{\mathbb{R}}^n} e^{i\phi(x;\xi)} a(x;\xi) \hat{f}(\xi)\,\mathrm{d}\xi \tag{15}$$

for a choice of phase ϕ and symbol $a \in S^{\mu}(\mathbb{R}^n \times \mathbb{R}^n)$ satisfying the conditions described in Sect. 1.1. We are interested in two kinds of estimates:

1. L^p-Sobolev bounds

$$\|\mathcal{F}f\|_{L^p_s(\mathbb{R}^n)} \lesssim \|f\|_{L^p(\mathbb{R}^n)}. \tag{16}$$

   Here $L^p_s(\mathbb{R}^n)$ denotes the standard Sobolev (Bessel potential) space defined with respect to the Fourier multipliers $(1+|\xi|^2)^{s/2}$ (see, for instance, [55, Chapter V]).
2. Given a 1-parameter family of FIOs $(\mathcal{F}_t)_{t\in I}$ for $I \subseteq \mathbb{R}$ a compact interval, we will consider inequalities of the form

$$\Big(\int_I \|\mathcal{F}_t f\|^p_{L^p_s(\mathbb{R}^n)} \,\mathrm{d}t\Big)^{1/p} \lesssim \|f\|_{L^p(\mathbb{R}^n)}. \tag{17}$$

A prototypical case which motivates (17) is given by the family of half-wave propagators $\mathcal{F}_t := e^{it\sqrt{-\Delta}}$. In this case, taking $\mathcal{F} = \mathcal{F}_t$ for a given t (or the composition of this operator with a pseudo-differential operator) in (16) leads to fixed-time estimates for solutions to the wave equation; owing to this, such L^p_s bounds will often be referred to as *fixed-time estimates* (regardless of whether the operator involves a time parameter). On the other hand, (17) is a *"space-time" estimate*.

Clearly, if one has a uniform bound of the kind (16) for every operator belonging to a 1-parameter family $(\mathcal{F}_t)_{t\in I}$, then (17) follows directly by integrating these estimates over the time interval I. However, in many situations averaging over time has an additional smoothing effect; this allows for stronger space-time estimates than those obtained trivially by averaging fixed-time inequalities. This phenomenon is referred to as *local smoothing*.

In this section known and conjectured local smoothing properties of FIOs are described. In contrast with the fixed time case, the necessary conditions on p for which an inequality of the form (17) can hold can be quite subtle, depending on various geometric properties of the phase. An indication of the key considerations is provided below.

It transpires that local smoothing estimates are substantially stronger than their fixed-time counterparts and have many applications and implications in harmonic analysis and PDE. For instance, as is well-known, the sharp range of local smoothing inequalities for the wave propagator is known to imply numerous major open problems in harmonic analysis, including the Bochner–Riesz problem on convergence of Fourier series and the Kakeya conjecture. Non-sharp local smoothing estimates can also be very useful, and provide powerful tools for studying a wide range of maximal and variational problems in harmonic analysis (see, for instance, [2, 23, 24] for recent examples of this). An introduction to the vast array of applications of local smoothing estimates is provided in Sects. 3–4.

It appears that sharp local smoothing inequalities are extremely difficult to prove and, indeed, in the prototypical case of the half-wave propagator $e^{it\sqrt{-\Delta}}$ in the plane obtaining the sharp range of exponents remains a challenging open problem (although there are numerous partial results: see Sect. 2.3 below and the appendix). However, there is a fairly complete understanding in cases where the operator has a particularly badly behaved underlying geometry. For these, somewhat pathological, examples, geometric considerations place rather stringent constraints on the range of admissible p values; consequently, it has been possible to obtain the full range of L^p local smoothing estimates. This was observed recently in [1] and relies heavily on fundamental work of Wolff [67] on the local smoothing problem and an important extension of Wolff's work due to Bourgain–Demeter [8]. These topics are discussed in detail in Sects. 5–6.

2.1 *Fixed-Time Estimates for FIOs*

Before discussing local smoothing in detail, it is instructive to first consider fixed-time estimates (16) for FIOs, which are somewhat easier to understand and help motivate the local smoothing theory. Consider a Fourier integral operator $\mathcal{F}$ of order μ as in (15) and suppose the symbol a is compactly supported in the x variable. In order to obtain a non-trivial L^p theory, it is necessary to impose some further conditions on the phase.

Mixed Hessian condition The phase ϕ satisfies

$$\det \partial^2_{x\xi}\phi(x;\xi) \neq 0 \qquad \text{for all } (x;\xi) \in \operatorname{supp} a. \tag{18}$$

An obvious example of a phase function satisfying (18) is $\phi(x,\xi) := \langle x,\xi\rangle$, corresponding to the case of pseudo-differential operators. The phase function appearing in the Euclidean wave propagator in Example 1 also satisfies this hypothesis, as do those arising in the manifold setting in Example 2. It transpires that (18) has a natural geometric interpretation in terms of the canonical relation C introduced in Sect. 1.5; this is described below in Sect. 2.6.

It was shown by Eskin [19] and Hörmander [29] that FIOs of order 0 satisfying the above hypotheses are bounded on L^2. In general, for $p \neq 2$, they are not bounded on L^p but L^p-Sobolev estimates do hold with some (necessary) loss in regularity. The sharp range of estimates of this form were established by Seeger, Stein and the third author [48].

Theorem 17 ([48]) *If $\mathcal{F}$ is a FIO of order μ satisfying the mixed Hessian condition, then for all $1 < p < \infty$ the fixed-time estimate*

$$\|\mathcal{F}f\|_{L^p_{-\mu-\bar{s}_p}(\mathbb{R}^n)} \lesssim \|f\|_{L^p(\mathbb{R}^n)} \tag{19}$$

holds for $\bar{s}_p := (n-1)\left|\frac{1}{p}-\frac{1}{2}\right|$.

The proof of (19) in [48] follows from a $H^1(\mathbb{R}^n)$ to $L^1(\mathbb{R}^n)$ bound for FIOs of order $-\frac{n-1}{2}$ and interpolation with the aforementioned $L^2(\mathbb{R}^n)$ estimate for FIOs of order 0; this yields the results for $1 < p \leq 2$ and the results for $2 < p < \infty$ follow by duality.

As an example of an application of this theorem, one may apply the estimate to the FIOs arising in the parametrix for the half-wave propagator $e^{it\sqrt{-\Delta_g}}$ on a compact Riemannian manifold (M, g). If u is the solution to (5), then one obtains the bound

$$\|u(\cdot\,,t)\|_{L^p_{s-\bar{s}_p}(M)} \lesssim_{M,g} \|f_0\|_{L^p_s(M)} + \|f_1\|_{L^p_{s-1}(M)} \tag{20}$$

for all $s \in \mathbb{R}$. Here $L^p_s(M)$ denotes the standard Sobolev (or Bessel potential) space, defined with respect to the spectral multiplier $(1+\lambda^2)^{s/2}$ (see, for instance, [53, Chapter 4]). Moreover, provided t avoids a discrete set of times, the estimate (20) is sharp for all $1 < p < \infty$ in the sense that one cannot replace $\bar{s}_p$ with $\bar{s}_p - \sigma$ for any $\sigma > 0$. This provides an analogue of earlier bounds for solutions to the Euclidean wave equation from [43, 46]. Theorem 17 can also be applied to solutions of more general strictly-hyperbolic equations, of any order: see [48] for further details.

Remark 18 (Sharpness of Fixed-Time Estimates) An integration-by-parts argument shows that for any $\alpha > 0$ the (distributional) inverse Fourier transform of $e^{-i|\xi|}(1+|\xi|^2)^{-\alpha/2}$ agrees with a function f_α. Moreover, f_α is rapidly decaying for $|x| \geq 2$ and for $|x| \leq 2$ satisfies

$$|f_\alpha(x)| \sim |1-|x||^{-(n+1)/2+\alpha}.$$

Thus, if $\alpha > \frac{n+1}{2} - \frac{1}{p}$, then $f_\alpha \in L^p(\mathbb{R}^n)$. On the other hand,

$$|(1-\Delta)^{-s/2} e^{i\sqrt{-\Delta}} f_\alpha(x)| \gtrsim |x|^{-(n-\alpha-s)} \qquad \text{for } |x| \lesssim 1.$$

Thus, if $\alpha \leq n - \frac{n}{p} - s$, then $e^{i\sqrt{-\Delta}} f_\alpha \notin L^p_{-s}(\mathbb{R}^n)$. Comparing these two conditions shows that Theorem 17 is optimal for $2 \leq p < \infty$, in the sense that $\bar{s}_p$ cannot be replaced with any smaller exponent. The range $1 < p \leq 2$ then follows by duality. See [58, Chapter IX, §6.13] for further details.

Remark 19 (Sharpness in the Range $1 < p \leq 2$) In fact, one may also deduce the sharpness of Theorem 17 in the regime $1 < p \leq 2$ from a direct example rather than from a duality argument. Reasoning as in Remark 18, given $\alpha > 0$, let g_α be a function whose distributional Fourier transform is $(1+|\xi|^2)^{-\alpha/2}$. The function g_α is rapidly decreasing at infinity and

$$|g_\alpha(x)| \sim |x|^{-(n-\alpha)} \qquad \text{for } |x| \lesssim 1;$$

thus $g_\alpha \in L^p(\mathbb{R}^n)$ if $\alpha > n - \frac{n}{p}$. On the other hand,

$$|(1-\Delta)^{-s/2}e^{i\sqrt{-\Delta}}g_\alpha(x)| \gtrsim |1-|x||^{-(n+1)/2+\alpha+s},$$

so $e^{i\sqrt{-\Delta}}g_\alpha \notin L^p_{-s}(\mathbb{R}^n)$ if $\alpha < \frac{n+1}{2} - s - \frac{1}{p}$. Combining both conditions on α yields $\bar{s}_p$ in Theorem 17 cannot be replaced with any smaller exponent if $1 < p \leq 2$.

2.2 *Local Smoothing Estimates for FIOs*

We now turn to the subject of local smoothing estimates. Recall that the problem is to analyse a 1-parameter family $(\mathcal{F}_t)_{t\in I}$ of FIOs, the prototypical example being the wave semigroup $e^{it\sqrt{-\Delta}}$. It is convenient to formulate the problem in terms of a single Fourier integral operator mapping functions on $\mathbb{R}^n$ to functions on $\mathbb{R}^{n+1}$. In particular, consider an operator

$$\mathcal{F}f(x,t) := \frac{1}{(2\pi)^n}\int_{\hat{\mathbb{R}}^n} e^{i\phi(x,t;\xi)}a(x,t;\xi)\hat{f}(\xi)\,\mathrm{d}\xi \qquad (x,t)\in\mathbb{R}^{n+1} \tag{21}$$

where the symbol $a \in S^\mu(\mathbb{R}^{n+1}\times\mathbb{R}^n)$ is compactly supported in x and t and the phase function ϕ is homogeneous of degree 1 in ξ and smooth away from $\xi = 0$. The conditions on the phase are now formulated with respect to the space-time variables (x,t) and the analogous condition to (18) reads as follows.

Mixed Hessian condition The phase ϕ satisfies:

(H1) rank $\partial^2_{\xi z}\phi(x,t;\xi) = n$ for all $(x,t;\xi) \in \operatorname{supp} a \setminus 0$.

Here and below z is used to denote vector in $\mathbb{R}^{n+1}$ comprised of the space-time variables (x,t).

Trivially, under these hypotheses Theorem 17 implies the space-time estimate

$$\Big(\int_{\mathbb{R}} \|\mathcal{F}f(\,\cdot\,,t)\|^p_{L^p_{-\mu-\bar{s}_p}(\mathbb{R}^n)}\,\mathrm{d}t\Big)^{1/p} \lesssim \|f\|_{L^p(\mathbb{R}^n)}. \tag{22}$$

This range of exponents, which follows from fixed-time estimates alone, does not encapsulate any additional smoothing arising from taking the average in time. Moreover, without further conditions on the phase, no such additional smoothing is possible, in general, and the above range is in fact sharp (a standard example which demonstrates this is given by the Radon transform in the plane: see Example 37 below or [53, Chapter 6]). In order to establish non-trivial local smoothing estimates, one restricts to the class of phases satisfying the following additional hypothesis.

Curvature condition The phase ϕ satisfies:

(H2) The generalised Gauss map, defined by $G(z;\xi) := \frac{G_0(z;\xi)}{|G_0(z;\xi)|}$ for all $(z;\xi) \in \operatorname{supp} a \setminus 0$ where

$$G_0(z;\xi) := \bigwedge_{j=1}^{n} \partial_{\xi_j} \partial_z \phi(z;\xi),$$

satisfies

$$\operatorname{rank} \partial^2_{\eta\eta} \langle \partial_z \phi(z;\eta), G(z;\xi) \rangle|_{\eta=\xi} = n-1$$

for all $(z;\xi) \in \operatorname{supp} a \setminus 0$.

Geometrically, the curvature condition means that for fixed z_0 the cone

$$\Gamma_{z_0} := \{\partial_z \phi(z_0;\eta) : \eta \in \mathbb{R}^n \setminus 0 \text{ in a conic neighbourhood of } \eta_0\} \tag{23}$$

is a smooth (conic) manifold of dimension n with $n-1$ non-vanishing principal curvatures at every point. One may readily verify that the phase featured in the prototypical example of the half-wave propagator $e^{it\sqrt{-\Delta}}$ satisfies both conditions (H1) and (H2). The same is also true for the phases arising from the parametrix construction for $e^{it\sqrt{-\Delta_g}}$ in Example 2.

Under the conditions (H1) and (H2), it is possible to show that for $2 < p < \infty$ there exists some $\sigma(p) > 0$ such that inequality (22) holds with $\bar{s}_p$ replaced with $\bar{s}_p - \sigma(p)$. This corresponds to a regularity gain over the estimate (19) and is an example of the local smoothing phenomenon. The existence of local smoothing estimates of the type (22) was first observed by the third author [50] in the context of the Euclidean half-wave propagator $e^{it\sqrt{-\Delta}}$. Shortly after, Mockenhaupt, Seeger and the third author [44, 45] established stronger local smoothing estimates in the general context of Fourier integral operators satisfying (H1) and (H2).

2.3 *The Local Smoothing Conjecture*

A natural question is to quantify the precise range of exponents for which (22) holds for a given FIO $\mathcal{F}$ satisfying the hypotheses (H1) and (H2). It transpires that this is a difficult and largely unresolved problem, involving a subtle dependence on certain geometric properties of $\mathcal{F}$.

2.3.1 The Euclidean Wave Semigroup

To begin the discussion, we first consider the prototypical case of the wave semigroup $e^{it\sqrt{-\Delta}}$. In [50], the following conjecture was formulated.

Conjecture 20 (Local Smoothing Conjecture) For $n \geq 2$ the inequality

$$\Big(\int_1^2 \|e^{it\sqrt{-\Delta}} f\|^p_{L^p_{-\bar{s}_p+\sigma}(\mathbb{R}^n)} \mathrm{d}t\Big)^{1/p} \lesssim \|f\|_{L^p(\mathbb{R}^n)} \tag{24}$$

holds for all $\sigma < 1/p$ if $\frac{2n}{n-1} \leq p < \infty$ and $\sigma < \bar{s}_p$ if $2 < p \leq \frac{2n}{n-1}$.

Note that the order of the half-wave propagator $e^{it\sqrt{-\Delta}}$ is $\mu = 0$, so the conjecture claims a σ-regularity gain with respect to the fixed time estimates (19) in Theorem 17. This conjecture is open in all dimensions, although there have been numerous partial results which establish (24) either for:

- A restricted range of regularity [6, 37, 44, 50, 62] or
- A sharp gain in regularity for a restricted range of Lebesgue exponent [8, 21, 22, 26, 34, 36, 67].

It is remarked that in [26] a strengthened version of the conjecture was in fact established, involving estimates with the endpoint regularity index $\sigma = 1/p$ (for a restricted range of p). The history of the problem is discussed in more detail in the appendix.

For $p = 2$, Plancherel's theorem implies the energy conservation identity

$$\|e^{it\sqrt{-\Delta}} f\|_{L^2(\mathbb{R}^n \times [1,2])} = \|f\|_{L^2(\mathbb{R}^n)} \tag{25}$$

whilst for $p = \infty$ one may show

$$\|e^{it\sqrt{-\Delta}} f\|_{L^\infty_{-\frac{(n-1)}{2}-\varepsilon}(\mathbb{R}^n \times [1,2])} \lesssim \|f\|_{L^\infty(\mathbb{R}^n)}. \tag{26}$$

The estimate (26) follows by bounding certain localised pieces of the kernel in L^1; this kind of argument is described in detail in Sect. 5.5. On a heuristic level, (26) can be understood by comparison with the averaging operators A_t from Example 3 which are automatically bounded on L^∞ and roughly correspond to the composition of $e^{it\sqrt{-\Delta}}$ with a multiplier in $S^{-(n-1)/2}$.

As a consequence of these simple estimates, (24) is strongest at $p = \frac{2n}{n-1}$: the estimates for all other p follow from the $p = \frac{2n}{n-1}$ case via interpolation with (25) and (26). Thus, Conjecture 20 amounts to the assertion that $e^{it\sqrt{-\Delta}}$ is essentially (that is, modulo a necessary loss of $\varepsilon > 0$ derivatives) bounded on $L^{2n/(n-1)}(\mathbb{R}^{n+1})$ locally in time.

Remark 21 Historically, the local smoothing phenomenon was first observed in the context of L^2-type bounds for dispersive equations [14, 33, 49, 64]. Here the setup is somewhat different. For instance, in the case of the Schrödinger equation a gain of $1/2$ a derivative is obtained when one integrates the solution locally with respect to time *over a compact spatial region*:

$$\Big(\int_1^2 \|e^{-it\Delta} f\|_{L^2(B(0,1))}^2 \,\mathrm{d}t\Big)^{1/2} \lesssim \|f\|_{L^2_{1/2}(\mathbb{R}^n)}. \tag{27}$$

Of course, the spatial localisation is essential in (27): the estimate cannot hold with a global $L^2(\mathbb{R}^n)$-norm owing to conservation of energy. In the case of the wave equation, the operator $e^{it\sqrt{-\Delta}}$ is local at scale t (as is clear either from the Kirchhoff formula for the solution (see, for instance, [51, Chapter 1]) or by analysing the kernel of $e^{it\sqrt{-\Delta}}$ via (non-) stationary phase). Consequently, local and global L^2 estimates for the half-wave propagator are essentially equivalent and, thus, conservation of energy prohibits any analogous inequality of the form (27) for $e^{it\sqrt{-\Delta}}$.

It is worthwhile examining the examples which dictate the numerology appearing in Conjecture 20. First of all, it is clear that no local smoothing is possible for $p = 2$ for reasons described in Remark 21 above. Furthermore, the example in Remark 19 showing the sharpness of the fixed-time estimates if $1 < p < 2$ immediately yields that no local smoothing estimates hold in this regime. The situation is different if $2 < p < \infty$.

Remark 22 (Sharpness of Local Smoothing Conjecture) The example used in Remark 18 can be used to show that a gain of 1/p derivatives in the local smoothing conjecture would be best possible. In particular, let f_α be as defined in Remark 18, so that if $\alpha > \frac{n+1}{2} - \frac{1}{p}$, then $f_\alpha \in L^p(\mathbb{R}^n)$. Moreover, one may show that

$$|(1-\Delta)^{-s/2} e^{it\sqrt{-\Delta}} f_\alpha(x)| \gtrsim |x|^{-(n-1)/2} |t - 1 - |x||^{-(n+1)/2+\alpha+s} \quad \text{if } t \geq 2|x|+1$$

whenever $|x| \lesssim 1$. Thus, if $\alpha \leq n - \frac{n+1}{p} - s$, then

$$\Big(\int_1^2 \|e^{it\sqrt{-\Delta}} f_\alpha\|_{L^p_{-s}(\mathbb{R}^n)}^p \,\mathrm{d}t\Big)^{1/p} = \infty.$$

Comparing the two conditions on α shows that Conjecture 20 is optimal in the sense that $1/p$ cannot be replaced with any larger number.

2.3.2 Wave Equations on Manifolds

Given a compact n-dimensional Riemannian manifold (M, g) one may also consider the local smoothing problem for the propagator $e^{it\sqrt{-\Delta_g}}$, as defined in Example 2. It

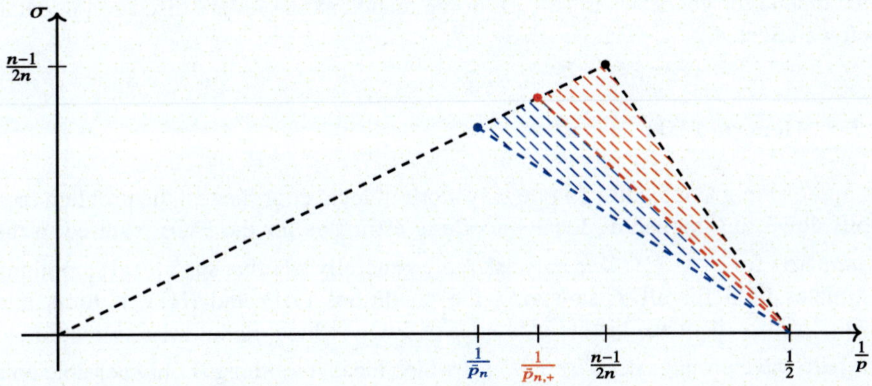

Fig. 1 Exponents for various formulations of the local smoothing conjecture. In contrast to the Euclidean case, for wave propagators on certain compact manifolds the red region is inadmissible. There exist FIOs for which the blue region is also inadmissible

is perhaps tempting to conjecture that (24) should also hold for $e^{it\sqrt{-\Delta_g}}$ for the same range of exponents as described in Conjecture 20. This turns out to be somewhat naïve, however. In particular, Minicozzi and the third author [42] identified compact manifolds (M, g) for which local smoothing fails to hold for all orders $\sigma < 1/p$ whenever $p < \bar{p}_{n,+}$ where

$$\bar{p}_{n,+} := \begin{cases} \frac{2(3n+1)}{3n-3} & \text{if } n \text{ is odd} \\ \frac{2(3n+2)}{3n-2} & \text{if } n \text{ is even} \end{cases} ;$$

see Fig. 1. Furthermore, (M, g) may be taken to be an arbitrarily small smooth perturbation of the Euclidean metric on $\mathbb{R}^n$. Thus, one is led to the following conjecture.

Conjecture 23 (Local Smoothing Conjecture: Compact Manifolds) For $n \geq 2$ and (M, g) a compact Riemannian manifold of dimension n, the inequality

$$\Big(\int_1^2 \|e^{it\sqrt{-\Delta_g}} f\|^p_{L^p_{-\bar{s}_p+\sigma}(M)} \mathrm{d}t\Big)^{1/p} \lesssim \|f\|_{L^p(M)} \qquad (28)$$

holds for all $\sigma < 1/p$ if $\bar{p}_{n,+} \leq p < \infty$.

Note that the conjecture would automatically imply bounds of the form (28) in the $2 \leq p \leq \bar{p}_{n,+}$ range via interpolation with the L^2 energy estimate. For simplicity, the values for σ in this range of p are omitted.

The counterexamples in [42] were inspired by earlier work of Bourgain [5, 7] in the context of oscillatory integral operators and are geometric in nature. In particular, obstacles to (28) arise owing to so-called Kakeya/Nikodym compression phenomena for geodesics in (M, g). Some related examples are discussed in detail below in Sect. 4.2.

2.3.3 General FIOs

In approaching Conjecture 23 one may work in local coordinates; the problem then boils down to establishing local smoothing estimates for the FIOs featured in the parametrix for $e^{it\sqrt{-\Delta_g}}$. One may ask more generally whether such local smoothing estimates hold for all $\mathcal{F}$ satisfying the conditions (H1) and (H2). It turns out, however, that there are further examples of FIOs (which do not arise in relation to the half-wave propagators $e^{it\sqrt{-\Delta_g}}$) for which local smoothing is only possible on a strictly smaller range of exponents than $p \geq \bar{p}_{n,+}$. These examples are of a slightly indirect nature. In particular, in [1] it was shown that local smoothing estimates (22) imply certain oscillatory integral bounds: see Theorem 47. Counterexamples of Bourgain [5, 7] in the latter context can then be applied to the problem; the details of this argument are reviewed below in Sect. 4. In particular, there exist choices of $\mathcal{F}$ satisfying (H1) and (H2) for which local smoothing fails to hold for all orders $\sigma < 1/p$ whenever $p < \bar{p}_n$ where

$$\bar{p}_n := \begin{cases} \frac{2(n+1)}{n-1} & \text{if } n \text{ is odd} \\ \frac{2(n+2)}{n} & \text{if } n \text{ is even} \end{cases} ; \tag{29}$$

see Fig. 1. Thus, one is led to the following general conjecture.

Conjecture 24 (Local Smoothing Conjecture: FIOs) Suppose $\mathcal{F}$ is a FIO with symbol of order μ satisfying conditions (H1) and (H2) above. The inequality

$$\Big(\int_1^2 \|\mathcal{F}f(\,\cdot\,,t)\|^p_{L^p_{-\mu-\bar{s}p+\sigma}(\mathbb{R}^n)}\,\mathrm{d}t\Big)^{1/p} \lesssim \|f\|_{L^p(\mathbb{R}^n)} \tag{30}$$

holds for all $\sigma < 1/p$ if $\bar{p}_n \leq p < \infty$.

As in the case of the wave semigroup, the conjecture automatically implies bounds of the form (28) in the $2 \leq p \leq \bar{p}_n$ range, this time via interpolation with the L^2 bounds of Eskin [19] and Hörmander [29].

It will be useful to introduce the following terminology.

Definition 25 Given $2 < p < \infty$, we say there is $1/p-$ *local smoothing* (or *local smoothing of order* $1/p-$) for a FIO $\mathcal{F}$ as above if (30) holds for all $\sigma < 1/p$.

In this language, the above conjectures may be stated succinctly as follows:

Conjecture 20 There is $1/p-$ local smoothing for $e^{it\sqrt{-\Delta}}$ for all $\frac{2n}{n-1} \leq p < \infty$.

Conjecture 23 There is $1/p-$ local smoothing for $e^{it\sqrt{-\Delta_g}}$ for all $\bar{p}_{n,+} \leq p < \infty$.

Conjecture 24 If $\mathcal{F}$ is a FIO of order μ satisfying H1) and H2), then there is $1/p-$ local smoothing for $\mathcal{F}$ for all $\bar{p}_n \leq p < \infty$.

The examples in [1] show that Conjecture 24 would be sharp *across the entire class* of FIOs satisfying (H1) and (H2), but there are many situations where one expects a better range of estimates to hold (not least of all the case of the wave propagators described above). One may, in fact, formulate a more refined conjecture, which combines both Conjecture 23 and Conjecture 24 into a single statement, by considering finer geometric properties of the phase function and working with a more precise version of the hypothesis (H2). This is discussed below in Sect. 2.5.

2.4 Positive Results

Recently in [1], the odd dimensional case of Conjecture 24 was established.

Theorem 26 *Let $\mathcal{F}$ be a FIO as in (21) satisfying H1) and H2) and with symbol of order μ. There is $1/p-$ local smoothing for $\mathcal{F}$ for all $\frac{2(n+1)}{n-1} \leq p < \infty$.*

This result extends earlier work of Wolff [67] and Bourgain–Demeter [8] which establishes the theorem in the special case of the Euclidean wave semigroup.

Theorem 26 is, up to endpoints, sharp across the entire class of FIOs in odd dimensions, in terms of both the regularity and the Lebesgue exponents. The question of what happens at the endpoint regularity index remains open; see [38] for partial results in this direction. Thus, in order to prove estimates for a wider range of exponents than those provided by Theorem 26 one must assume additional hypotheses on $\mathcal{F}$. In view of this, some natural refinements of the condition H2) are discussed in the following subsection.

The method used to establish Theorem 26 follows Wolff's approach to local smoothing [67]. This relies on establishing variable coefficient counterparts to the sharp $\ell^p(L^p)$ *Wolff-type* (or *decoupling*) inequalities of Bourgain–Demeter [8]. It is a remarkable fact that the aforementioned decoupling inequalities are stable under smooth perturbations of the phase in the underlying operator, leading to the results in [1]. A detailed review of this argument is provided in Sect. 5. An interesting aspect of this analysis is that the variable coefficient decoupling estimates can be derived rather directly as a consequence of the constant coefficient estimates, via an induction-on-scale argument. This is discussed in Sect. 6.

2.5 Formulating a Local Smoothing Conjecture for General FIOs

Comparing Conjectures 20, 23 and 24, it is natural to ask what the special properties of the half-wave propagators $e^{it\sqrt{-\Delta_g}}$ and $e^{it\sqrt{-\Delta}}$ are which distinguish them from general FIOs and allow one to conjecture a larger range of local smoothing estimates in these cases.

It is first remarked that the stronger numerology in the Euclidean conjecture (Conjecture 20) is related to deep questions in geometric measure theory. In particular, it is well-known that Conjecture 20 implies the Kakeya conjecture[7] concerning the Hausdorff dimension of Kakeya sets in $\mathbb{R}^n$; for a discussion of the relationship between these and other important problems in harmonic analysis and geometric measure theory see, for instance, [42, 60, 66] and the following section. A similar relationship holds when one considers wave propagators on manifolds and, moreover, general FIOs. In particular, both Conjectures 23 and 24 imply bounds on the dimension of certain *Kakeya* (or, more precisely, *Nikodym*) *sets of curves*. For instance, when dealing with the propagator $e^{it\sqrt{-\Delta_g}}$ the curves in question are geodesics in (M, g). The precise definition of a Kakeya/Nikodym set of curves will not be recalled here, but the interested reader is directed to [11, 25, 42, 65] or [53, Chapter 9] for further details. The key observation is that, for certain examples, the families of curves which arise in this manner **fail** the Kakeya/Nikodym conjecture. More precisely, the curves can be arranged so that they lie in a set of small Hausdorff dimension; see Fig. 2.[8] Such geometric configurations can be used to preclude local smoothing estimates near $\frac{2n}{n-1}$ for certain propagators $e^{it\sqrt{-\Delta_g}}$ and lead to the numerology in Conjecture 23.

It remains to explain the difference in numerology between Conjecture 23 for wave propagators on manifolds and Conjecture 24 for general FIOs. Recall that the necessary condition in Conjecture 24 arises from counterexamples of Bourgain [5, 7] for bounds for oscillatory integral operators; this is discussed in detail below in Sect. 4.2. It is remarked that one key feature of Bourgain's examples is that they give rise to *hyperbolic* cones Γ_{z_0}: that is, the non-vanishing principal curvatures of Γ_{z_0} have *different signs*. Moreover, the analysis can be refined to give necessary conditions which depend on the difference between the number of positive and number of negative principal curvatures [27].

Definition 27 Suppose $\mathcal{F}$ is an FIO which satisfies the conditions H1) and H2). We say $\mathcal{F}$ has *signature* κ for some integer $0 \leq \kappa \leq n-1$ if each of the cones Γ_{z_0} satisfies

$$\kappa = |\#\text{ positive principal curvatures} - \#\text{ negative principal curvatures}|$$

at every point.

[7] This conjecture states that if $K \subseteq \mathbb{R}^n$ is compact and contains a unit line segment in every direction, then K should have Hausdorff dimension n.

[8] In the worst case scenario, the curves can be arranged to lie in a set of dimension $\lceil \frac{n+1}{2} \rceil$ where n is the ambient dimension [5, 7, 11].

Fig. 2 An example of a *Kakeya/Nikodym set of curves*, arising from Bourgain's example [5, 7] (see also [25]). Here a large family of distinct parabolas lies inside a two-dimensional set (a hyperbolic paraboloid)

With this definition, and in light of the modified versions of Bourgain's examples, one may formulate a refined version of Conjecture 24. In particular, letting

$$\bar{p}_{n,\kappa} := \begin{cases} 2 \cdot \dfrac{\kappa + 2(n+1)}{\kappa + 2(n-1)} & \text{if } n \text{ is odd} \\ 2 \cdot \dfrac{\kappa + 2n + 3}{\kappa + 2n - 1} & \text{if } n \text{ is even} \end{cases},$$

the new conjecture reads thus.

Conjecture 28 (Local Smoothing Conjecture: FIOs) Suppose $\mathcal{F}$ is a FIO with symbol of order μ satisfying conditions (H1) and (H2) and that $\mathcal{F}$ has signature κ. There is $1/p-$ local smoothing for $\mathcal{F}$ for all $\bar{p}_{n,\kappa} \leq p < \infty$.

Since there are $n - 1$ non-vanishing principal curvatures, in the worst case scenario the signature is given by

$$\kappa = \begin{cases} 0 & \text{if } n \text{ is odd} \\ 1 & \text{if } n \text{ is even} \end{cases}.$$

Substituting these values into the formula for $\bar{p}_{n,\kappa}$, one recovers the exponent $\bar{p}_n$ from (29) and therefore Conjecture 28 subsumes Conjecture 24. On the other hand,

	n odd	n even
$n-1$ non-vanishing curvatures	$\frac{2(n+1)}{n-1}$	$\frac{2(n+2)}{n}$
$n-1$ positive curvatures	$\frac{2(3n+1)}{3n-3}$	$\frac{2(3n+2)}{3n-2}$

Fig. 3 Conjectured endpoint values for the exponent p for the sharp local smoothing estimates (22) under various signature hypotheses on the phase. Theorem 26 establishes the odd dimensional case under the hypothesis of $n-1$ non-vanishing principal curvatures

in the best case scenario the principal curvatures all have the same sign and $\kappa = n-1$. In this case, we see that $\bar{p}_{n,n-1}$ agrees with the exponent $\bar{p}_{n,+}$ from Conjecture 23. Furthermore, it is indeed the case that the Fourier integral operators associated to the wave semigroups $e^{it\sqrt{-\Delta_g}}$ have signature $n-1$ (see, for instance, [42] or [53, Chapter 4]). Thus, Conjecture 28 also subsumes Conjecture 23. See Fig. 3.

From the above discussion it is not at all clear *why* the signature should be important in the analysis of these operators, other than it is a consideration in the construction of counterexamples. In the case of oscillatory integral operators, the precise rôle of the signature is fairly well understood and is discussed in detail in [25] (see also [11, 27, 36]). It is highly likely that the signature will play a similar rôle in the analysis of FIOs.

2.6 The Geometric Conditions in Terms of the Canonical Relation

To round off this section, we describe how the local results of the previous subsections can be transcribed into the broader setting of global FIOs. In particular, we provide a natural geometric interpretation of the mixed Hessian and curvature conditions in terms of the canonical relation C.

2.6.1 L^p Estimates and Canonical Graphs

Fix X, Y smooth manifolds of dimension n and a canonical relation $C \subseteq T^*X \setminus 0 \times T^*Y \setminus 0$. Theorem 17 can easily be extended to the setting of global FIOs $\mathcal{F} \in I^\mu(X, Y; C)$ once the mixed Hessian condition is correctly interpreted in terms of the geometry of C.

We first observe that in the specific context of a local operator $\mathcal{F}$ given by (15), with $\dim X = \dim Y = n$ and a symbol $a \in S^{\mu}(\mathbb{R}^n \times \mathbb{R}^n)$, the order of $\mathcal{F}$ corresponds to the order μ of the symbol, and therefore $I^{\mu}(X, Y; C)$ is the correct class to work in if one wishes to extend the local fixed-time results described above. Indeed, by the convention established in Sect. 1.5 (which is motivated by the equivalence of phase theorem), the order m of the operator satisfies

$$m = \mu - \frac{d - 2N}{4} \tag{31}$$

where d is the number of $(x; y)$ variables and N is the number of Fourier variables. In the case of (15), we have $d = 2n$ and $N = n$, and so m and μ coincide.

We now turn to describing the appropriate generalisation of the mixed Hessian condition.

Projection condition The natural projection mappings $\Pi_{T^*X} : C \to T^*X \setminus 0$ and $\Pi_{T^*Y} : C \to T^*Y \setminus 0$ are local diffeomorphisms.

$$\begin{array}{ccccc} & & C & & \\ & \swarrow^{\Pi_{T^*X}} & & \searrow^{\Pi_{T^*Y}} & \\ T^*X \setminus 0 & & & & T^*Y \setminus 0 \end{array}.$$

The projection condition clearly forces $\dim X = \dim Y$. It is also not difficult to show that if $\dim X = \dim Y$ and either one of the projections Π_{T^*X} or Π_{T^*Y} is a local diffeomorphism, then so too is the other.[9] Thus, for instance, an equivalent formulation of the projection condition is that $\dim X = \dim Y$ and

$$\operatorname{rank} d\Pi_{T^*Y} = 2n, \tag{32}$$

where n is the common dimension of X and Y. Yet another way to interpret this property, which will be useful later in the discussion, is that C is *locally a canonical graph*. In particular, for every $\gamma_0 = (x_0, \xi_0, y_0, \eta_0) \in C$ there exists a symplectomorphism χ defined on a neighbourhood of $(x_0, \xi_0) \in T^*X \setminus 0$ and mapping into $T^*Y \setminus 0$ such that on this neighbourhood C is given by the graph

$$\{(x, \xi, y, \eta) : (y, \eta) = \chi(x, \xi)\}.$$

With this definition, the global variant of Theorem 17 reads thus.

[9]This can be seen by expressing the operator in local coordinates: see the proof of Lemma 33 below for a very similar argument.

Theorem 29 ([48]) *If $\mathcal{F} \in I^{\mu}(X, Y; C)$ is a global FIO where C satisfies the projection condition, then for all $1 < p < \infty$ the fixed-time estimate*[10]

$$\|\mathcal{F} f\|_{L^p_{-\mu-\bar{s}_p,\mathrm{loc}}(\mathbb{R}^n)} \lesssim \|f\|_{L^p_{\mathrm{comp}}(\mathbb{R}^n)}$$

holds.

Using the theory described in Sect. 1, it is not difficult to deduce Theorem 29 as a fairly direct consequence of its local counterpart Theorem 17. In particular, since the result is inherently local, one may assume that $\mathcal{F} \in I^{\mu}(X, Y; C)$ is given in local coordinates by some kernel

$$K(x; y) = \int_{\hat{\mathbb{R}}^N} e^{i\varphi(x,y;\theta)} a(x, y; \theta)\, \mathrm{d}\theta,$$

where $a \in S^{\mu}(\mathbb{R}^n \times \mathbb{R}^n \times \hat{\mathbb{R}}^N)$ and $\varphi : \mathbb{R}^n \times \mathbb{R}^n \times \hat{\mathbb{R}}^N \setminus \{0\} \to \mathbb{R}$ is a non-degenerate phase function. Thus, one may write

$$C = \{(x, \partial_x\varphi(x, y; \theta), y, -\partial_y\varphi(x, y; \theta)) : \partial_\theta\varphi(x, y; \theta) = 0\} \tag{33}$$

and it follows that if C is a local canonical graph, then there exist smooth solutions to the equations

$$\xi = \partial_x\varphi(x, y; \theta), \qquad \partial_\theta\varphi(x, y; \theta) = 0$$

in (y, θ). By the inverse function theorem, this amounts to the condition that the Jacobian of the map $(y, \theta) \mapsto (\partial_x\varphi(x, y; \theta), \partial_\theta\varphi(x, y; \theta))$ is non-vanishing:

$$\det \begin{pmatrix} \partial^2_{xy}\varphi & \partial^2_{x\theta}\varphi \\ \partial^2_{\theta y}\varphi & \partial^2_{\theta\theta}\varphi \end{pmatrix} (x, y; \theta) \neq 0 \qquad \text{whenever } \partial_\theta\varphi(x, y; \theta) = 0. \tag{34}$$

As described in Sect. 1.5, one may further assume that $N = n$ and φ has the special form $\varphi(x, y; \eta) = \langle y, \eta\rangle - \phi(x; \eta)$, where ϕ is smooth and homogeneous of degree 1 in η. In this case, the condition (34) then becomes

$$\det \partial^2_{x\eta}\phi \neq 0,$$

which corresponds precisely with the mixed Hessian condition from (18).

Example 30 (Variable Coefficient Averaging Operators) The class of FIOs of order $-\frac{n-1}{2}$ which satisfy the projection condition includes averaging operators over

[10]Here, an $L^p_{\mathrm{comp}}(\mathbb{R}^n) \to L^p_{s,\mathrm{loc}}(\mathbb{R}^n)$ bound is interpreted as follows: for any pair of compact sets $\Omega_1, \Omega_2 \subseteq \mathbb{R}^n$ the *a priori* estimate $\|\mathcal{F} f\|_{L^p_s(\Omega_2)} \lesssim_{\Omega_1,\Omega_2} \|f\|_{L^p(\Omega_1)}$ holds whenever $f \in C^\infty_c(\Omega_1)$.

variable families of hypersurfaces which satisfy the *rotational curvature* condition of Phong and Stein [47] (see also [58, Chapter XI §3.1]). Indeed, consider the family of hypersurfaces

$$S_{x,t} = \{y \in \mathbb{R}^n : \Phi_t(x; y) = 0\}$$

where Φ_t is a smooth defining function of $(t, x, y) \in [1, 2] \times \mathbb{R}^n \times \mathbb{R}^n$. We say that Φ_t satisfies the rotational curvature condition if the Monge–Ampère matrix associated to Φ_t is non-singular on $\Phi_t = 0$: that is,

$$\mathrm{RotCurv}(\Phi_t)(x; y) := \det \begin{pmatrix} \Phi_t & \partial_y \Phi_t \\ \partial_x \Phi_t & \partial^2_{xy} \Phi_t \end{pmatrix} (x; y) \neq 0 \qquad \text{whenever } \Phi_t(x; y) = 0. \tag{35}$$

As in Example 3, the averaging operator

$$A_t f(x) := \int_{\mathbb{R}^n} f(y) a(t, x, y) \delta(\Phi_t(x; y))\, \mathrm{d}y$$

may be written as

$$A_t f(x) := \frac{1}{2\pi} \int_{\mathbb{R}^n} \int_{\mathbb{R}} e^{i\theta \Phi_t(x;y)} a(t, x, y) f(y)\, \mathrm{d}\theta\, \mathrm{d}y;$$

here $a \in S^0(\mathbb{R} \times \mathbb{R}^n \times \mathbb{R}^n)$. By Theorem 14, A_t is a FIO of order $-\frac{n-1}{2}$ and one may readily verify that if Φ_t satisfies (35), then the phase function $\varphi_t(x, y; \theta) = \theta \Phi_t(x; y)$ satisfies the condition (34).

Example 31 (Spherical Averages) As a special case of the previous example, let A_t denote the averaging operator associated to the family of spheres $x + t\mathbb{S}^{n-1}$, with defining function

$$\Phi_t(x; y) = \frac{|x - y|^2}{t^2} - 1.$$

In this case, $\mathrm{RotCurv}(\Phi_t)(x; y) = (-2)^{n+1} t^{-2n}$, which is non-vanishing. In general, whenever the operator is *translation-invariant*, in the sense that the family of hypersurfaces is given by $x \mapsto x + tS_0$ for some fixed S_0, the rotational curvature is nonvanishing if and only if the Gaussian curvature of S_0 is non-vanishing.

Example 32 (Radon Transform) As another special case of Example 30, consider the *Radon transform* A_t which is the averaging operator with defining function $\Phi_t(x; y) = \langle x, y\rangle - t$ for some $t \neq 0$. Observe that $\mathrm{Rot\,Curv}(\Phi_t)(x; y) = -\langle x, y\rangle$, so that the rotational curvature condition is satisfied. However, in contrast with Example 31, each hyperplane $S_{x,t} = \{y \in \mathbb{R}^n : \Phi_t(x; y) = 0\}$ has zero Gaussian curvature. In this case, the rotational curvature is capturing the rotation of the planes $S_{x,t}$ as x varies, rather than curvature of the planes themselves.

2.6.2 Local Smoothing Estimates and Cinematic Curvature Condition

Fix Y and Z smooth manifolds of dimension n and $n+1$ respectively, with $n \geq 2$, and let C be a canonical relation in $T^*Z\backslash 0 \times T^*Y\backslash 0$. Thus, C is a conic submanifold of dimension $2n+1$ which is Lagrangian with respect to the 1-form $\omega_Z - \omega_Y = \sum_{j=1}^{n+1} \zeta_j \mathrm{d}z_j - \sum_{i=1}^{n} \eta_i \mathrm{d}y_i$. The local smoothing estimates in Theorem 26 hold for global FIOs $\mathcal{F} \in I^{\mu-1/4}(Z, Y; C)$ which satisfy certain conditions on C.

Note, in contrast with the fixed-time estimates described above, here one works with operators of order $\mu - 1/4$ so that the FIO in that class admit the local expression (21) with a symbol $a \in S^{\mu}(\mathbb{R}^{n+1} \times \mathbb{R}^n)$ of order μ. This is a quirk of the order convention from Sect. 1.5. Indeed, if we consider the local operator (21), which is interpreted as mapping functions of n variables to functions of $n+1$ variables, the number d of (x, t, y) variables is equal to $2n+1$ whilst the number N of Fourier variables is n. Thus, recalling (31), we see the order m of the operator (21) is indeed $\mu - 1/4$.[11]

We now turn to describing the hypotheses on the canonical relation C which generalise properties (H1) and (H2) from the local theory. The first hypothesis corresponds to the mixed hessian condition (H1) and is the natural analogue of the projection condition featured in Theorem 29 (see (32)).

Projection condition If $\Pi_{T^*Y} \colon C \to T^*Y \setminus 0$ denotes the natural projection mapping, then

$$\operatorname{rank} \mathrm{d}\Pi_{T^*Y} = 2n. \tag{36}$$

$$\begin{array}{ccccc} & & C & & \\ & \swarrow^{\Pi_{T^*Y}} & \downarrow \Pi_Z & \searrow^{\Pi_{T^*_{z_0}Z}} & \\ T^*Y \setminus 0 & & Z & & T^*_{z_0}Z \setminus 0 \end{array}$$

Geometrically, this condition has the following interpretation. Fix $z_0 \in \Pi_Z(C)$ and let $\Pi_{T^*_{z_0}Z}$ denote the projection $C \to T^*_{z_0}Z\backslash 0$. Define

$$\Gamma_{z_0} := \Pi_{T^*_{z_0}Z}(C),$$

[11] If $\mathcal{F}$ is viewed as a 1-parameter family of operators $(\mathcal{F}_t)_{t\in I}$, then each $\mathcal{F}_t$ is a FIO of order μ.

which is a conic subset of $T^*_{z_0}Z\backslash 0$. The projection condition implies that Γ_{z_0} is in fact a *smooth n-dimensional surface*. Indeed, this is a consequence of the following lemma.

Lemma 33 *The condition* (36) *implies that* $\mathrm{d}\Pi_{T^*_{z_0}Z}$ *has constant rank* n.

Proof Recall from Sect. 1.5 that the operator can be expressed locally in the form (15), in which case the phase function φ appearing in the expression (33) is given by $\varphi(x, y; \xi) := \phi(x; \xi) - \langle y, \xi\rangle$. In particular, local coordinates may be chosen so that C is locally parametrised as a graph (modulo a reflection)

$$(z, \eta) \mapsto (z, \partial_z\phi(z; \eta), \partial_\eta\phi(z; \eta), -\eta), \tag{37}$$

where ϕ is homogeneous in η. Thus, computing the differential of Π_{T^*Y} in these coordinates, the condition (36) implies that the map $(z, \eta) \mapsto (\partial_\eta\phi(z; \eta), \eta)$ is a submersion; this of course reduces to

$$\operatorname{rank} \partial^2_{z\eta}\phi(z, \eta) = n. \tag{38}$$

By combining (37) and (38), it immediately follows that the differentials of $\Pi_{T^*_{z_0}Z}$ must have rank n, as required. □

The second condition concerns the curvature of the cones Γ_{z_0}.

Cone condition For every $z_0 \in \Pi_Z(C)$ the cone Γ_{z_0} has $n - 1$ non-vanishing principal curvatures at every point.

If C satisfies both the projection and the cone condition, then, following [50], it is said to satisfy the *cinematic curvature condition*.

Theorem 34 ([1]) *Suppose* $\mathcal{F} \in I^{\mu-1/4}(Y, Z; C)$ *is a global FIO where* $C \subset T^*Y \setminus 0 \times T^*Z \setminus 0$ *satisfies the cinematic curvature condition. If* $\frac{2(n+1)}{n-1} \leq p < \infty$, *then*

$$\Big(\int_1^2 \|\mathcal{F}f(\,\cdot\,, t)\|^p_{L^p_{-\mu-\bar{s}_p+\sigma,\,\mathrm{loc}}(\mathbb{R}^n)}\,\mathrm{d}t\Big)^{1/p} \lesssim \|f\|_{L^p_{\mathrm{comp}}(\mathbb{R}^n)}$$

holds for all $\sigma < 1/p$.

Once again, it is not difficult to deduce Theorem 34 as a direct consequence of its local counterpart, Theorem 26. Most of this argument has already been described in the proof of the first claim above. In particular, in local coordinates one may express C as a graph as in (37). The projection condition then implies (38), which is precisely the condition H1) in the local theorem. On the other hand, the cones $\Gamma_{z_0} = \Pi_{T^*_{z_0}Z(C)}$ take the form (23), and so the cone condition clearly amounts to H2).

Example 35 (Variable Coefficient Averaging Operators) We return to the variable hypersurfaces $S_{x,t}$ and associated averaging operators A_t discussed in Example 30.

Suppose that the defining function Φ_t satisfies the rotational curvature condition (35) for all t in the t-support of a. Thus, $A_t \in I^{-(n-1)/2}(X, Y; C_t)$ for a canonical relation C_t which is locally a canonical graph. Note that the rotational curvature condition applies to each A_t individually and, in particular, does not take into account how the family of surfaces $S_{x,t}$ vary in t.

The cinematic curvature condition, on the other hand, provides additional information about the behaviour of the $S_{x,t}$ under changes of t. Indeed, let C denote the canonical relation associated to the *family* of averages A_t (viewed as an operator taking functions on $\mathbb{R}^n$ to functions on $\mathbb{R}^{n+1}$). It follows from the rotational curvature hypothesis that

$$C = \{(x, t, \xi, \tau, y, \eta) : (y, \eta) = \chi_t(x, \xi), \tau = q(x, t, \xi)\}$$

where:

- χ_t is a symplectomorphism
- the function q is homogeneous of degree 1 in ξ and smooth if $\xi \neq 0$.

Indeed, the function χ_t arises from the canonical graph property, satisfied by each C_t. Note that the variable τ may be written in terms of x, t and ξ because χ_t is a diffeomorphism and C is a $2n + 1$ dimensional manifold. Moreover, q is necessarily homogeneous of degree 1 in ξ due to the conic nature of C in the η variable. Having written the canonical relation in the above form, the cone condition requires that

$$\operatorname{rank} \partial^2_{\xi\xi} q = n - 1,$$

which is the maximum possible rank in view of the homogeneity of q. This additional hypothesis takes into account the change in t.

Finally, if one represents the averaging operator using a single Fourier variable, as in Example 7, then it is possible to obtain a formula for computing the function q. Indeed, the phase function is given by $\varphi(x, t, y; \theta) = \theta\Phi_t(x; y)$ and so in C we have

$$\tau = \partial_t\varphi(x, t, y; \theta) = \theta\partial_t\Phi_t(x; y) \quad \text{and} \quad \xi = \partial_x\varphi(x, t, y; \theta) = \theta\partial_x\Phi_t(x; y).$$

The condition $\tau = q(x, t, \xi)$ therefore becomes

$$q(x, t, \partial_x\Phi_t(x; y)) = \partial_t\Phi_t(x; y) \qquad \text{whenever} \quad \Phi_t(x; y) = 0,$$

due the homogeneity of q in the ξ variable.

Example 36 (Spherical Averages) For $t > 0$ let A_t denote the averaging operator associated to the defining function $\Phi_t(x; y) = \frac{|x-y|^2}{t^2} - 1$. It was observed in Example 31 that each A_t satisfies the rotational curvature condition. Moreover, the family of operators satisfies the cinematic curvature condition, since $q(x, t; \xi) = -|\xi|$.

Example 37 (Radon Transform) For $t \neq 0$ let A_t denote the averaging operator associated to the defining function $\Phi_t(x; y) = \langle x, y\rangle - t$. It was observed in Example 32 that each A_t satisfies the rotational curvature condition. However, the cinematic curvature condition is violated, as there is no change in the curvatures of the $S_{x,t}$ as t varies. In particular, $q(x, t; \xi) = -\frac{\langle x,\xi\rangle}{t}$, so that $\partial^2_{\xi\xi} q = 0$.

Incidentally, for $n = 2$ this example can also be used to show the necessity of the cinematic curvature hypothesis for local smoothing (see, for instance, [53, Chapter 6]).

3 Local Smoothing and Maximal Estimates

In the next two sections we investigate some of the many applications of local smoothing estimates to problems in harmonic analysis. Here we review connections with (maximal) Bochner–Riesz multipliers and circular maximal function theorems.

3.1 *Bochner–Riesz Estimates*

Recall that the Bochner–Riesz multipliers of order $\delta > 0$ are defined by

$$S_t^\delta f(x) := \frac{1}{(2\pi)^n} \int_{\hat{\mathbb{R}}^n} e^{i\langle x,\xi\rangle}(1 - |t\xi|)_+^\delta \hat{f}(\xi)\, d\xi \qquad \text{for } t > 0.$$

A classical problem in harmonic analysis is to determine whether these multipliers constitute a Fourier summation method: in particular, one is interested in whether

$$S_t^\delta f \to f \quad \text{as} \quad t \to 0_+$$

for a given mode of convergence (typically convergence in L^p or almost everywhere convergence). By a simple rescaling argument, together with some standard functional analysis, the L^p convergence question is equivalent to determining the range of L^p boundedness for the operators $S^\delta := S_1^\delta$ (see, for instance, [58, Chapter IX] for further details).

Conjecture 38 (Bochner–Riesz Conjecture) Let $1 \leq p \leq \infty$. If $\delta > \delta(p) := \max\{n|\frac{1}{2} - \frac{1}{p}| - \frac{1}{2}, 0\}$, then

$$\|S^\delta f\|_{L^p(\mathbb{R}^n)} \lesssim \|f\|_{L^p(\mathbb{R}^n)}. \tag{39}$$

It is known that $\delta > \delta(p)$ is a necessary condition for (39) to hold whenever $p \neq 2$. The results for $p = \infty$ are trivial and it is also well known that one would obtain this conjecture for all $1 \le p \le \infty$ by interpolation and duality if the bounds held for $p \ge \frac{2n}{n-1}$.

It was observed by the third author [50] that the local smoothing conjecture for $e^{it\sqrt{-\Delta}}$ formally implies Conjecture 38.

Proposition 39 *Let $\frac{2n}{n-1} \le p < \infty$ be given. If there is $1/p-$ local smoothing for $e^{it\sqrt{-\Delta}}$, then the Bochner–Riesz estimate (39) holds for all $\delta > \delta(p)$.*

It is remarked that the Bochner–Riesz conjecture is itself known to imply the Fourier restriction conjecture for spheres and paraboloids, which in turn implies the Kakeya conjecture: see [60, 66] for a discussion of these problems and the relationships between them. Thus, we see that the local smoothing conjecture sits at the top of a chain of implications relating important central questions in harmonic analysis and geometric measure theory.

Local smoothing ⇒ Bochner–Riesz ⇒ Restriction ⇒ Kakeya.

Proof *(of Proposition 39)* Note that

$$\bar{s}_p - 1/p = \delta(p) \quad \text{if} \quad p \ge \frac{2n}{n-1} \tag{40}$$

and that one may write

$$(1 - |\xi|)_+^\delta = r(|\xi|) + \sum_{k=1}^{\infty} 2^{-k\delta} \psi\big(2^k(1 - |\xi|)\big), \tag{41}$$

where $r = r_\delta \in C_0^\infty([0, \infty))$ and $\psi = \psi_\delta \in C_0^\infty([1/2, 2])$.

Since r is smooth and compactly supported the Fourier multiplier operator associated with $r(|\xi|)$ is bounded on $L^p(\mathbb{R}^n)$ for all $1 \le p \le \infty$, and one concludes that (39) would follow for a given $p \ge \frac{2n}{n-1}$ if the inequality

$$\Big\| \int_{\hat{\mathbb{R}}^n} e^{i\langle x,\xi\rangle} \psi\big(2^k(1 - |\xi|)\big)\, \hat{f}(\xi)\, \mathrm{d}\xi \Big\|_{L^p(\mathbb{R}^n)} \lesssim_\varepsilon 2^{k(\delta(p)+\varepsilon)} \|f\|_{L^p(\mathbb{R}^n)}$$

holds for all $k \in \mathbb{N}$ and all $\varepsilon > 0$. By a simple change of variables argument, the inequality in the above display holds if and only if

$$\|A^\lambda f\|_{L^p(\mathbb{R}^n)} \lesssim_\varepsilon \lambda^{\delta(p)+\varepsilon} \|f\|_{L^p(\mathbb{R}^n)} \qquad \text{for all } \lambda \gg 1 \tag{42}$$

where

$$A^\lambda f(x) := \int_{\hat{\mathbb{R}}^n} e^{i\langle x,\xi\rangle}\, \psi(\lambda - |\xi|)\, \hat{f}(\xi)\, \mathrm{d}\xi.$$

In proving (42) for a given $\lambda \gg 1$, since supp $\psi \in [1/2, 2]$, one may assume that

$$\operatorname{supp} \hat{f} \subset \{\xi : |\xi| \in [\lambda/2, 2\lambda]\}. \tag{43}$$

Also, if one writes

$$\psi(\lambda - |\xi|) = (2\pi)^{-1} \int_{\mathbb{R}} \check{\psi}(t) e^{-i\lambda t} e^{it|\xi|} \, dt,$$

then Hölder's inequality in the t-variable after multiplying and dividing by $(1 + |t|)$ implies that

$$\begin{aligned} \|A^\lambda f\|_{L^p(\mathbb{R}^n)} &\lesssim \Big\| (1 + |t|)\check{\psi}(t) \int_{\hat{\mathbb{R}}^n} e^{i\langle x, \xi\rangle} e^{it|\xi|} \hat{f}(\xi) \, d\xi \Big\|_{L^p(\mathbb{R}^n \times \mathbb{R})} \\ &\lesssim_N \|(1 + |t|)^{-N} e^{it\sqrt{-\Delta}} f\|_{L^p(\mathbb{R}^n \times \mathbb{R})}. \end{aligned} \tag{44}$$

for all $N \in \mathbb{N}$ and one may write

$$\|A^\lambda f\|_{L^p(\mathbb{R}^n)} \lesssim_N \|e^{it\sqrt{-\Delta}} f\|_{L^p(\mathbb{R}^n \times [-1,1])} + \sum_{k \in \mathbb{N}} \|(1 + |t|)^{-N} e^{it\sqrt{-\Delta}} f\|_{L^p(\mathbb{R}^n \times I_k)}, \tag{45}$$

where $I_k := [-2^{k-1}, -2^k] \cup [2^{k-1}, 2^k]$. In view of (43) and (40), $1/p-$ local smoothing for $e^{it\sqrt{-\Delta}}$ implies that

$$\|e^{it\sqrt{-\Delta}} f\|_{L^p(\mathbb{R}^n \times [-1,1])} \lesssim_\varepsilon \lambda^{\delta(p)+\varepsilon} \|f\|_{L^p(\mathbb{R}^n)}. \tag{46}$$

Thus, the first term in the right-hand-side of (45) is controlled by the right-hand side of (42). For the remaining terms, the rapid decay in (44) together with (46) and a simple change of variables argument yields

$$\|(1 + |t|)^{-N} e^{it\sqrt{-\Delta}} f\|_{L^p(\mathbb{R}^n \times I_k)} \lesssim 2^{-k} \lambda^{\delta(p)+\varepsilon} \|f\|_p$$

uniformly in $k \in \mathbb{N}$, and then the desired result just follows from summing a geometric series in $k \in \mathbb{N}$. □

3.2 *Maximal Bochner–Riesz Estimates*

When studying almost everywhere convergence of the Bochner–Riesz summation method, one naturally considers the maximal estimates

$$\Big(\int_{\mathbb{R}^n} \sup_{t>0} |S_t^\delta f(x)|^p \, dx \Big)^{1/p} \lesssim \|f\|_{L^p(\mathbb{R}^n)} \tag{47}$$

for the operators S_t^δ. It transpires that $1/p-$ local smoothing for $e^{it\sqrt{-\Delta}}$ also implies inequalities of this form.

Proposition 40 *Let $\frac{2n}{n-1} \le p < \infty$ be given. If there is $1/p-$ local smoothing for $e^{it\sqrt{-\Delta}}$, then the maximal Bochner–Riesz estimate (47) holds for all $\delta > \delta(p)$.*

To prove this, note that if $r(|\xi|)$ is as in (41), then

$$\sup_{t>0}\Big|\int_{\hat{\mathbb{R}}^n} e^{i\langle x,\xi\rangle} r(t|\xi|)\,\hat{f}(\xi)\,\mathrm{d}\xi\Big| \lesssim Mf(x),$$

where M denotes the Hardy–Littlewood maximal function. Since $\|Mf\|_p \lesssim \|f\|_p$ for all $p > 1$, the previous arguments reveal that (47) would follow if one can show that the maximal version of (42) holds. Explicitly, it suffices to show that

$$\Big(\int_{\mathbb{R}^n} \sup_{t>0} |A_t^\lambda f(x)|^p\,\mathrm{d}x\Big)^{1/p} \lesssim_\varepsilon \lambda^{\delta(p)+\varepsilon}\|f\|_{L^p(\mathbb{R}^n)} \tag{48}$$

holds for all $\lambda \gg 1$ and all $\varepsilon > 0$ where

$$A_t^\lambda f(x) := \frac{1}{(2\pi)^n}\int_{\hat{\mathbb{R}}^n} e^{i\langle x,\xi\rangle}\psi\big(\lambda - |t\xi|\big)\,\hat{f}(\xi)\,\mathrm{d}\xi.$$

To prove this, we appeal to the following simple lemma.

Lemma 41 *Suppose that*

$$\operatorname{supp} m \subset \{\xi \in \hat{\mathbb{R}}^n : |\xi| \in (\lambda_0/2, 2\lambda_0)\}$$

for some fixed $\lambda_0 > 0$ and set

$$A_t f(x) := \frac{1}{(2\pi)^n}\int_{\hat{\mathbb{R}}^n} e^{i\langle x,\xi\rangle} m(t\xi)\,\hat{f}(\xi)\,\mathrm{d}\xi, \quad t > 0.$$

If $2 \le p < \infty$ and the inequality

$$\Big(\int_{\mathbb{R}^n} \sup_{t\in[1,2]} |A_t f(x)|^p\,\mathrm{d}x\Big)^{1/p} \le \bar{C}_p\|f\|_{L^p(\mathbb{R}^n)}, \tag{49}$$

holds for some fixed constant $\bar{C}_p > 0$, then it follows that

$$\Big(\int_{\mathbb{R}^n} \sup_{t>0} |A_t f(x)|^p\,\mathrm{d}x\Big)^{1/p} \lesssim C_p\,\|f\|_{L^p(\mathbb{R}^n)}.$$

Proof Let $k_0 \in \mathbb{Z}$ be the unique integer such that $\lambda_0 \in [2^{k_0}, 2^{k_0+1})$. Next, choose a Littlewood–Paley bump function $\beta \in C_0^\infty((1/2, 2))$ satisfying $\sum_{-\infty}^{\infty}\beta(2^{-j}r) = 1$,

$r > 0$, and define $P_\ell f$ by $(P_\ell f)\widehat{\ }(\xi) := \beta(2^{-\ell}|\xi|)\hat{f}(\xi)$. Then, by Littlewood–Paley theory (see, for instance, [55, Chapter IV]),

$$\Big\| \Big(\sum_{\ell=-\infty}^{\infty} |P_\ell f|^2 \Big)^{1/2} \Big\|_{L^p(\mathbb{R}^n)} \lesssim \|f\|_{L^p(\mathbb{R}^n)} \quad \text{for all } 1 < p < \infty. \tag{50}$$

To use this, note that

$$A_t f = A_t\Big(\sum_{\{\ell \in \mathbb{Z}:\, |\ell-(k+k_0)| \le 10\}} P_\ell f \Big) \quad \text{if } t \in [2^{-k}, 2^{-k+1}],$$

since it is assumed that $m(\xi) = 0$ if $|\xi| \notin (2^{k_0-2}, 2^{k_0+2})$. By this observation together with a scaling argument, the assumption (49) yields

$$\begin{aligned}
\int_{\mathbb{R}^n} \sup_{t \in [2^{-k}, 2^{-k+1}]} |A_t f(x)|^p \,\mathrm{d}x &= \int_{\mathbb{R}^n} \sup_{t \in [2^{-k}, 2^{-k+1}]} \Big| A_t\Big(\sum_{|\ell-(k+k_0)| \le 10} P_\ell f(x) \Big) \Big|^p \,\mathrm{d}x \\
&\le \bar{C}_p^p \int_{\mathbb{R}^n} \Big| \sum_{|\ell-(k+k_0)| \le 10} P_\ell f(x) \Big|^p \,\mathrm{d}x \\
&\lesssim \bar{C}_p^p \sum_{|\ell-(k+k_0)| \le 10} \int_{\mathbb{R}^n} |P_\ell f(x)|^p \,\mathrm{d}x
\end{aligned}$$

for all $k \in \mathbb{Z}$. Consequently,

$$\begin{aligned}
\int_{\mathbb{R}^n} \sup_{t>0} |A_t f(x)|^p \,\mathrm{d}x &\le \sum_{k=-\infty}^{\infty} \int_{\mathbb{R}^n} \sup_{t \in [2^{-k}, 2^{-k+1}]} |A_t f(x)|^p \,\mathrm{d}x \\
&\lesssim \bar{C}_p^p \sum_{k=-\infty}^{\infty} \sum_{|\ell-(k+k_0)| \le 10} \int_{\mathbb{R}^n} |P_\ell f(x)|^p \,\mathrm{d}x \\
&\lesssim \bar{C}_p^p \int \Big(\sum_{\ell} |P_\ell f(x)|^2 \Big)^{p/2} \,\mathrm{d}x \\
&\lesssim \bar{C}_p^p \|f\|_{L^p(\mathbb{R}^n)}^p.
\end{aligned}$$

Here it was used that $p \ge 2$ yields $\ell^2 \subseteq \ell^p$ in the second to last inequality and (50) in the last inequality. □

Proof (of Proposition 40) The preceding observations together with the above lemma reduce the proof of (48) to showing that for $\lambda \gg 1$ one has

$$\Big(\int_{\mathbb{R}^n} \sup_{1 \le t \le 2} \big| A_t^\lambda f(x) \big|^p \,\mathrm{d}x \Big)^{1/p} \lesssim_\varepsilon \lambda^{\delta(p)+\varepsilon} \|f\|_{L^p(\mathbb{R}^n)}.$$

This in turn would follow by showing that

$$\left(\int_{\mathbb{R}^n} |(A^\lambda_{t(x)} f)(x)|^p \, dx \right)^{1/p} \lesssim_\varepsilon \lambda^{\delta(p)+\varepsilon} \|f\|_{L^p(\mathbb{R}^n)} \tag{51}$$

holds for any measurable function $t(x) : \mathbb{R}^n \to [1, 2]$.

To adapt the earlier argument, write

$$\begin{aligned} \psi(\lambda - |t(x)\xi|) &= (2\pi)^{-1} \int_{\mathbb{R}} \check{\psi}(s)\, e^{-i\lambda s} e^{ist(x)|\xi|} \, ds \\ &= (2\pi t(x))^{-1} \int_{\mathbb{R}} \check{\psi}(s/t(x))\, e^{-i\lambda s/t(x)} e^{is|\xi|} \, ds. \end{aligned}$$

Since it is assumed that $1 \le t(x) \le 2$ and since $\check{\psi}$ is rapidly decreasing, one can use Hölder's inequality and argue as in (44) to see that the left-hand-side of (51) is dominated by

$$\|(1 + |s|)^{-N} e^{is\sqrt{-\Delta}} f\|_{L^p(\mathbb{R}^n \times \mathbb{R})}$$

for all $N \in \mathbb{N}$ (with a constant depending on N).

Repeating the earlier arguments, one sees that the $1/p-$ local smoothing estimates (46) imply that this last expression is dominated by $\lambda^{\delta(p)+\varepsilon}\|f\|_{L^p(\mathbb{R}^n)}$. This establishes the desired estimate (51) and thereby finishes the proof of Proposition 40. □

3.3 *Circular Maximal Function Estimates*

One may use local smoothing estimates for the half-wave propagator in two spatial dimensions to give an alternative proof of Bourgain's celebrated circular maximal function theorem [4]. Letting σ_t denote the normalised Lebesgue measure on the dilated circle $t\mathbb{S}^1$, recall that this theorem states the following:

Theorem 42 (Bourgain [4]) *For all $p > 2$,*

$$\left(\int_{\mathbb{R}^2} \sup_{t>0} |f * \sigma_t(x)|^p \, dx \right)^{1/p} \lesssim \|f\|_{L^p(\mathbb{R}^2)}. \tag{52}$$

Theorem 42 extends Stein's [56] earlier spherical maximal theorem which states that for $n \ge 3$ the maximal operator associated with spherical averages in $\mathbb{R}^n$ is bounded on $L^p(\mathbb{R}^n)$ for $p > \frac{n}{n-1}$. Stein [56] also showed that these bounds fail for any $p \le \frac{n}{n-1}$. In particular, the circular maximal function featured in (52) is not bounded on $L^2(\mathbb{R}^2)$; this partially accounts for the added difficulties in two dimensions (which were later overcome by Bourgain [4]).

It is remarked that, in contrast to the applications featured in the previous sections, to prove the sharp maximal function result from local smoothing estimates one does *not* require a sharp gain in regularity in the hypothesised local smoothing estimates; in fact, as will be discussed in Remark 44 below, *any* non-trivial gain in regularity over the fixed-time estimate for *any* $2 < p < \infty$ yields the sharp maximal inequality (however, for concreteness, we shall work with the L^6 local smoothing estimate). That local smoothing estimates can be used to give an alternative proof of (52) was first observed by Mockenhoupt, Seeger and the third author in [44].

Proof (of Theorem 42) It suffices to prove the maximal estimates for nonnegative f. Also, since the result is trivial when $p = \infty$, it suffices to prove the bounds under this assumption when $2 < p < \infty$.

Let $\beta \in C_0^\infty((1/2, 2))$ be the Littlewood–Paley bump function occurring in the proof of Lemma 41. As discussed in Example 3, the Fourier transform of the arc length measure σ on $\mathbb{S}^1$ may be written as

$$\hat{\sigma}(\xi) = \sum_{\pm} a_\pm(|\xi|)e^{\pm i|\xi|}$$

where $a_\pm \in S^{-1/2}$. Consequently, if $\beta_0(|\xi|) := 1 - \sum_{j=1}^{\infty} \beta(2^{-j}|\xi|) \in C_0^\infty$, one may write

$$f*\sigma_t(x) = (2\pi)^{-2}\int_{\hat{\mathbb{R}}^2} e^{i\langle x,\xi\rangle}\beta_0(t|\xi|)\hat{\sigma}(t\xi)\,\hat{f}(\xi)\,\mathrm{d}\xi + (2\pi)^{-2}\sum_{\pm}\sum_{j=1}^{\infty} F_\pm^j(x,t) \tag{53}$$

where

$$F_\pm^j(x,t) := \int_{\hat{\mathbb{R}}^2} e^{i\langle x,\xi\rangle \pm it|\xi|}\beta(2^{-j}t|\xi|)\,a_\pm(t|\xi|)\,\hat{f}(\xi)\,\mathrm{d}\xi.$$

Note that the $F_\pm^j$ correspond to the half-wave propagator $e^{it\sqrt{-\Delta}}$ except for the choice of symbol and the frequency localisation to the dyadic scale 2^j.

Since $\beta_0\hat{\sigma} \in C_0^\infty$, it follows that the maximal operator associated with the first term in the right-hand side of (53) is dominated by the Hardy–Littlewood maximal function of f and thus bounded on all $L^p(\mathbb{R}^2)$ for $p > 1$. It therefore suffices to show that for all $2 < p < \infty$ the remaining terms satisfy

$$\Big(\int_{\mathbb{R}^2} \sup_{t>0} |F_\pm^j(x,t)|^p\,\mathrm{d}x\Big)^{1/p} \lesssim 2^{-j\varepsilon_p}\|f\|_{L^p(\mathbb{R}^2)}$$

for some $\varepsilon_p > 0$. On account of the support properties of β and Lemma 41, it suffices to show that

$$\Big(\int_{\mathbb{R}^2} \sup_{1\le t\le 2} |F_\pm^j(x,t)|^p \mathrm{d}x\Big)^{1/p} \lesssim 2^{-j\varepsilon_p}\|f\|_{L^p(\mathbb{R}^2)}. \tag{54}$$

To prove this, we appeal to the following elementary lemma.

Lemma 43 *Suppose that $F \in C^1(\mathbb{R})$ and that $p > 1$. Then*

$$\sup_{1\le t\le 2} |F(t)|^p \le |F(1)|^p + p\Big(\int_1^2 |F(t)|^p\,\mathrm{d}t\Big)^{(p-1)/p}\Big(\int_1^2 |F'(t)|^p\,\mathrm{d}t\Big)^{1/p}. \tag{55}$$

Proof The proof of (55) is very simple. If one first writes

$$|F(t)|^p = |F(1)|^p + p\int_1^t |F(s)|^{p-1}\cdot F'(s)\,\mathrm{d}s,$$

then (55) follows via Hölder's inequality. □

To use this lemma to prove (54) we shall exploit the fact that the operators $F_\pm^j$ have symbols of order $-1/2$ which are localised to frequencies $|\xi| \sim 2^j$. Consequently, the fixed-time estimates (19) for $e^{it\sqrt{-\Delta}}$ give

$$\Big(\int_{\mathbb{R}^2} |F_\pm^j(x,1)|^p\,\mathrm{d}x\Big)^{1/p} \lesssim 2^{-j(1/2-\bar{s}_p)}\|f\|_{L^p(\mathbb{R}^2)} \quad \text{for all } 1 < p < \infty.$$

Note that in two dimensions $1/2 - \bar{s}_p > 0$. As a result, by Hölder's inequality after integrating (55) in the x-variables, it suffices to prove that for all $2 < p < \infty$ the inequality

$$\Big(\int_1^2\int_{\mathbb{R}^2} |F_\pm^j(x,t)|^p \mathrm{d}x\,\mathrm{d}t\Big)^{(p-1)/p}\Big(\int_1^2\int_{\mathbb{R}^2} \Big|\frac{d}{dt}F_\pm^j(x,t)\Big|^p \mathrm{d}x\,\mathrm{d}t\Big)^{1/p} \lesssim 2^{-jp\varepsilon_p}\|f\|_{L^p(\mathbb{R}^2)} \tag{56}$$

holds for some $\varepsilon_p > 0$.

Using, for instance, the $1/6-$ L^6 local smoothing estimates for the half-wave operators $e^{it\sqrt{-\Delta}}$ in $\mathbb{R}^2$ from Theorem 26 one has

$$\Big(\int_1^2\int_{\mathbb{R}^2} |F_\pm^j(x,t)|^6\,\mathrm{d}x\,\mathrm{d}t\Big)^{1/6} \lesssim_\varepsilon 2^{(-\frac{1}{3}+\varepsilon)j}\|f\|_{L^6(\mathbb{R}^2)} \quad \text{for } \varepsilon > 0. \tag{57}$$

Here we use the fact that $F^j_\pm$ incorporates a symbol of order $-1/2$, is frequency localised at scale 2^j and $-1/2 + (\bar{s}_p - 1/p) = -1/3$ if $p = 6$. Since $\frac{d}{dt}F^j_\pm$ is as $e^{it\sqrt{-\Delta}}$ with symbol of order $1/2$, one similarly obtains

$$\Big(\int_1^2 \int_{\mathbb{R}^2} \Big|\frac{d}{dt}F^j_\pm(x,t)\Big|^6\,dx\,dt\Big)^{1/6} \lesssim_\varepsilon 2^{(\frac{2}{3}+\varepsilon)j}\|f\|_{L^6(\mathbb{R}^2)} \quad \text{for } \varepsilon > 0, \tag{58}$$

using the fact that $2/3 = (\bar{s}_p - 1/p) + 1/2$ if $p = 6$. Clearly (57) and (58) together imply that (56) holds for $p = 6$ and any $0 < \varepsilon_6 < 1/6$.

Note that, by Plancherel's theorem and Lemma 43, if $p = 2$ we have (54) for $\varepsilon_2 = 0$. Also, the kernels of the operators $f \to F^j_\pm(t,\cdot)$ are easily seen to be in $L^1(\mathbb{R}^n)$ uniformly in $t > 0$ and $j \in \mathbb{N}$. This yields the analogue of (54) with $p = \infty$ and $\varepsilon_\infty = 0$. Interpolating between these two easier cases and the non-trivial bounds for $p = 6$, one obtains (54) for any $2 < p < \infty$, which completes the proof of Bourgain's circular maximal function theorem.

Remark 44 Note that any $\varepsilon_6 > 0$ suffices to obtain $\varepsilon_p > 0$ for $2 < p < \infty$ in (56) after interpolating with $\varepsilon_2 = \varepsilon_\infty = 0$. Thus, as is remarked at the beginning of this subsection, the full strength of $1/6 - L^6$ local smoothing for $e^{it\sqrt{-\Delta}}$ is not needed here (nor is the particular choice of exponent $p = 6$): *any* non-trivial local smoothing suffices. This is in contrast with Sects. 3.1–3.2. Similar considerations will apply for the variable coefficient variants in the next subsection.

3.4 Variable Coefficient Circular Maximal Function Estimates

Using local smoothing estimates for general Fourier integral operators (as opposed to simply the Euclidean half-wave propagators $e^{it\sqrt{-\Delta}}$), one may modify the argument in Sect. 3.3 to obtain a generalization of Bourgain's circular maximal function theorem for geodesic circles on Riemannian surfaces. This was originally shown by the third author in [50].

Before describing the results, it is perhaps useful to review the relevant concepts from Riemannian geometry. If (M, g) is a Riemannian manifold, then for any point $x \in M$ and tangent vector $v \in T_xM$ there exists a unique geodesic γ_v such that $\gamma_v(0) = x$ and $\gamma_v'(0) = v$. Moreover, there exists some open neighbourhood $U \subseteq T_xM$ of the origin such that the *exponential map* $\exp_x : U \to M$ taking $v \in U$ to $\exp_x(v) := \gamma_v(1)$ is well-defined. The *injectivity radius* $\operatorname{Inj}_x M > 0$ *of* M *at* x is the supremum over all $r > 0$ for which $\exp_x$ may be defined on $B(0,r) \subset T_xM$. The *injectivity radius* $\operatorname{Inj} M \geq 0$ *of* M is then defined to be the infimum of $\operatorname{Inj}_x M$ over all $x \in M$. If M is compact, then $\operatorname{Inj} M > 0$ and given any $x \in M$ and $0 < t < \operatorname{Inj} M$ one may define the *geodesic circle*

$$S_{x,t} := \big\{\exp_x(v) : v \in T_xM \text{ such that } |v| = t\big\}.$$

Note that in the case $M = \mathbb{S}^2$, a geodesic circle amounts to a great circle. See, for instance, [13, Chapter III] for more details.

Now suppose (M, g) is a two-dimensional compact Riemannian manifold. Define the average over the geodesic circle $S_{x,t}$ about $x \in M$ of radius $0 < t <$ Inj M by

$$A_t f(x) := \int_{S_{x,t}} f(y)\, \mathrm{d}\sigma_{x,t}(y),$$

where $\sigma_{x,t}$ denotes the normalised (to have unit mass) arc length measure on $S_{x,t}$. Fixing $0 < r_0 <$ Inj M, a natural variable coefficient version of Theorem 42 is as follows:

Theorem 45 ([50]) *With the above definitions, for all $p > 2$,*

$$\Big(\int_M \sup_{0<t<r_0} |A_t f(x)|^p\, \mathrm{d}x \Big)^{1/p} \lesssim_p \|f\|_{L^p(M)}. \tag{59}$$

Here $\mathrm{d}x$ *is the volume element on (M, g) and the L^p-norm on the right is associated with this measure.*

Proof To prove these general maximal inequalities, it suffices to establish the analogue of (59) where the norm is taken over $\Omega \subset M$, a relative compact subset of a coordinate patch and f is assumed to be supported in Ω; of course the estimate should be established uniformly over all such Ω. Working in local coordinates, and if β is the Littlewood–Paley bump function used before, for $0 < t < r_0$ and $x \in \Omega$ and supp $f \subset \Omega$ one may write

$$A_t f(x) = A_t^0 f(x) + \sum_{j=1}^{\infty} A_t^j f(x),$$

where $A_t^0 f$ is dominated by the Hardy–Littlewood maximal function of f and

$$A_t^j f(x) := \int K_j(x, t; y)\, f(y)\, \mathrm{d}y$$

for all $j \in \mathbb{N}$, where

$$K_j(x, t; y) := \int_{\hat{\mathbb{R}}^2} \hat{\sigma}_{x,t}(\xi)\, \beta(2^{-j} t|\xi|)\, e^{i\langle x-y,\xi\rangle}\, \mathrm{d}\xi.$$

Here, and in what follows, the Fourier transforms are taken with respect to the local coordinates in which we are working.

The maximal operator associated with A_t^0 is trivial to handle. Thus, (59) would follow if one can show that for all $p > 2$ and all $j \in \mathbb{N}$ the inequality

$$\left(\int_M \sup_{0<t<r_0} |A_t^j f(x)|^p \, \mathrm{d}x \right)^{1/p} \lesssim 2^{-j\varepsilon_p} \|f\|_{L^p(M)} \tag{60}$$

holds for some $\varepsilon_p > 0$. If, as before, $\hat{f}_\ell(\xi) = \beta(2^{-\ell}|\xi|)\hat{f}(\xi)$, then

$$A_t^j f = \sum_{|\ell-(k+j)|\leq 10} A_t^j f \qquad \text{for } t \in [2^{-k}, 2^{-k+1}] \cap (0, \text{Inj } M).$$

Based on this, one may adapt the earlier arguments of Lemma 41 to see that (60) would follow from favourable bounds for the maximal operators associated with dyadic intervals: in particular, it suffices to show that for all $p > 2$ there exists some $\varepsilon_p > 0$ such that

$$\left(\int \sup_{t\in[2^{-k},2^{-k+1}]\cap(0,r_0]} |A_t^j f(x)|^p \, \mathrm{d}x \right)^{1/p} \lesssim 2^{-j\varepsilon_p} \|f\|_p. \tag{61}$$

If 2^{-k} is bounded away from zero, then the operators $f \to A_t f(x)$ for $t \in [2^{-k}, 2^{-k}] \cap (0, \text{Inj } M)$ are a family of Fourier integral operators of order $-1/2$ satisfying the cinematic curvature condition (see [53]). Thus, (61) easily follows from the $1/p-$ local smoothing estimates for FIOs when $p \geq 6$ (that is, Theorem 34) and the above arguments. One may also handle the case where $2^{-k} \ll r_0$ by using a dilation argument and the local smoothing estimates in Theorem 34. This is due to the fact that for $x, y \in \Omega$ the Riemannian distance function in our local coordinates satisfies

$$d_g(x; y) = \sqrt{\sum_{1\leq j,k\leq 2} g_{jk}(x)(x_j - y_j)(x_k - y_k) + O(|x - y|^2)},$$

where $g_{jk}(x)\mathrm{d}x^j\mathrm{d}x^k$ is the Riemannian metric written in our local coordinates. See [45] or [53] for more details. □

3.5 *Maximal Bounds for Half-Wave Propagators*

Consider half-wave propagators $e^{it\sqrt{-\Delta_g}}$ either on Euclidean space or on a compact Riemannian manifold (M, g) of dimension $n \geq 2$. By the previous arguments one has

$$\left(\int_M \sup_{0<t<1} \left|e^{it\sqrt{-\Delta_g}} f(x)\right|^p \mathrm{d}x \right)^{1/p} \lesssim_\varepsilon \|f\|_{L^p_{\bar{s}_p+\varepsilon}(M)} \tag{62}$$

if there is $1/p-$ local smoothing for $e^{it\sqrt{-\Delta_g}}$; see Sect. 2.1 for the relevant definitions. Thus, Theorem 34 yields the following:

Theorem 46 *Under the above assumptions* (62) *holds for all* $p \geq \frac{2(n+1)}{n-1}$. *Consequently, for this range of exponents,*

$$e^{it\sqrt{-\Delta_g}} f(x) \to f(x) \text{ a.e. if } f \in L^p_s \quad \text{with } s > \bar{s}_p.$$

It is noted that for a given p, (62) is sharp. For instance, in the Euclidean case if $s < \bar{s}_p$ and $t \neq 0$ there are $f \in L^p_s$ for which $e^{it\sqrt{-\Delta}} f \notin L^p(\mathbb{R}^n)$ by a counterexample of Littman [40], and, in the manifold case, the same is true when t avoids a discrete set of times (see [48]).

4 Local Smoothing and Oscillatory Integral Estimates

The aim of this section is to explore connections between local smoothing for Fourier integral operators and L^p bounds for oscillatory integrals. As a consequence of this investigation, we will establish the necessary conditions for Conjecture 24.

4.1 L^p Estimates for Oscillatory Integrals Satisfying the Carleson–Sjölin Condition

Consider oscillatory integral operators of the form

$$T_\lambda f(x) := \int_{\mathbb{R}^{n-1}} e^{i\lambda\varphi(x;y')} a(x;y') f(y')\,\mathrm{d}y', \tag{63}$$

sending functions of $(n-1)$ variables to functions of n variables. Here $\varphi \in C^\infty(\mathbb{R}^n \times \mathbb{R}^{n-1})$ is assumed to be real-valued and $a \in C_0^\infty(\mathbb{R}^n \times \mathbb{R}^{n-1})$. It is also assumed that the phase functions satisfy the *Carleson–Sjölin condition*, which has two parts:

Mixed Hessian condition

$$\operatorname{rank} \partial^2_{xy'}\varphi(x;y') \equiv n-1 \quad \text{for all } (x;y') \in \operatorname{supp} a. \tag{64}$$

Provided the support of a is sufficiently small, this non-degeneracy condition ensures that for every x_0 in the x-support of a the gradient graph

$$\Sigma_{x_0} := \{\nabla_x\varphi(x_0;y') : a(x_0;y') \neq 0\} \subset T^*_{x_0}\mathbb{R}^n \tag{65}$$

is a smooth hypersurface.

The other part of the Carleson–Sjölin condition is the following curvature assumption.

Curvature condition For each x_0 in the x-support of a, the hypersurface Σ_{x_0} has non-vanishing Gaussian curvature at every point.

Under these assumptions, a problem of Hörmander [29] is to determine for which $p \geq \frac{2n}{n-1}$ the estimate

$$\|T_\lambda\|_{L^p(\mathbb{R}^{n-1}) \to L^p(\mathbb{R}^n)} = O_\varepsilon(\lambda^{-n/p+\varepsilon}) \tag{66}$$

holds for all $\varepsilon > 0$ (simple examples show that the constraint $p \geq \frac{2n}{n-1}$ is necessary). There are somewhat stronger formulations for $p > \frac{2n}{n-1}$ where $\varepsilon = 0$ and $L^p(\mathbb{R}^{n-1})$ is replaced by $L^r(\mathbb{R}^{n-1})$ for exponents $r < p$ satisfying $\frac{n+1}{n-1}r' = p$; however, we shall focus on the formulation in (66) and its relation with local smoothing estimates.

Theorem 47 ([1]) *Suppose that for a given $\frac{2n}{n-1} \leq p < \infty$ there is local smoothing of order $1/p-$ for all Fourier integral operators satisfying the cinematic curvature condition. Then (66) holds for the same exponent p for all phase functions φ satisfying the Carleson–Sjölin condition.*

As Theorem 34 ensures that there is local smoothing of order $1/p-$ for all $\frac{2(n+1)}{n-1} \leq p < \infty$ whenever the cinematic curvature condition holds, it follows that (66) is valid for this range of exponents. This recovers a slightly weaker version of Stein's [57] oscillatory integral theorem which says that the stronger $L^p - L^r$ estimates hold for $p \geq \frac{2(n+1)}{n-1}$ with $\varepsilon = 0$.

Proof (of Theorem 47) One may assume, of course, that a is supported in a small neighbourhood of the origin in $\mathbb{R}^{n-1} \times \mathbb{R}^n$. Also, since replacing φ by $\varphi(x; y') + Bx + Cy'$ where $B : \mathbb{R}^n \to \mathbb{R}^n$ and $C : \mathbb{R}^{n-1} \to \mathbb{R}^{n-1}$ are linear does not change the operator norm of T_λ, one may also assume that

$$\nabla_{x;y'}\varphi(0;0) = 0 \qquad \text{and} \qquad \det \partial^2_{x'y'}\varphi(0,0) \neq 0. \tag{67}$$

If we set

$$\Phi(x;y) := \varphi(x;y') + x_n + y_n, \qquad y = (y', y_n), \tag{68}$$

and if the support of a is small enough, then the Monge–Ampère determinant of Φ satisfies

$$\det \begin{pmatrix} 0 & \partial_y \Phi \\ \partial_x \Phi & \partial^2_{xy}\Phi \end{pmatrix} \neq 0 \qquad \text{for } (x; y') \in \text{supp a}. \tag{69}$$

If $\rho \in C_0^\infty(\mathbb{R})$ satisfies $\rho \geq 0$ and $\rho(0) = 1$, then this implies that

$$K(x, t; y) := a(x, y')\rho(y_n)\delta_0\big(t - \Phi(x; y)\big) \tag{70}$$

is the kernel of a non-trivial Fourier integral of order $-(n-1)/2$ for each fixed t near 0. Moreover, for each $t \in \operatorname{supp} \rho$, the associated Fourier integral operator satisfies the projection condition since (69) is equivalent to the fact that it has a canonical relation which is a canonical graph; see Example 30. For later use, note also that K vanishes if $|t|$ is large.

Based on this, the canonical relation

$$C \subset T^*\mathbb{R}^{n+1} \setminus 0 \times T^*\mathbb{R}^n \setminus 0$$

arising from the Fourier integral operator with kernel as in (70), regarded as an operator sending smooth functions of y to smooth functions of (x, t), satisfies the projection condition in the cinematic curvature hypothesis; see Example 35. The cone condition must also be valid since the image of the projection onto the fibers $T^*_{x_0,t_0}\mathbb{R}^{n+1} \setminus 0$ for (x_0, t_0) in the (x, t) support of the kernel are just the cones

$$\begin{aligned} \Gamma_{x_0,t_0} &= \big\{\tau(\nabla_x\Phi(x_0; y), -1) : \tau \in \mathbb{R} \setminus 0,\ \Phi(x_0; y) = t_0,\ y \in \operatorname{supp} a\,\rho\big\} \\ &= \big\{\tau(z, -1) : z \in \Sigma_{x_0}\big\} \subset T^*_{x_0,t_0}\mathbb{R}^{n+1} \setminus 0 \end{aligned} \tag{71}$$

and these have $(n-1)$ non-vanishing principal curvatures in view of the curvature condition.

Thus, the Fourier integral operators

$$f \in C_0^\infty(\mathbb{R}^n) \to \mathcal{F}_s f(x, t) := \big(\sqrt{I - \Delta_x}\big)^{(n-1)/2-s}\Big(\int_{\mathbb{R}^n} K(x, t; y) f(y)\,\mathrm{d}y\Big) \tag{72}$$

are Fourier integral operators of order $-s$ for each fixed t and the resulting family of Fourier integral operators satisfies the cinematic curvature hypothesis. As by hypothesis it is assumed that there is $1/p-$ local smoothing for such FIOs, one has

$$\|\mathcal{F}_s f\|_{L^p(\mathbb{R}^n\times\mathbb{R})} \lesssim_s \|f\|_{L^p(\mathbb{R}^n)} \quad \text{if } s > \bar{s}_p - 1/p.$$

To see how this leads to (66), observe first that since ρ is non-trivial and Φ differs from φ by terms which are linear in x and y (namely, $x_n + y_n$) one must have that

$$\|T_\lambda\|_{L^p(\mathbb{R}^{n-1})\to L^p(\mathbb{R}^n)} \approx \|S_\lambda\|_{L^p(\mathbb{R}^n)\to L^p(\mathbb{R}^n)} \tag{73}$$

for

$$S_\lambda f(x) := \int_{\mathbb{R}^n} e^{i\lambda\Phi(x;y)} a(x;y')\rho(y_n)\, f(y)\,\mathrm{d}y.$$

Next, let $m \in C^\infty(\mathbb{R})$ satisfy $m(r) = 1$ if $r < 1$ and $m(r) = 0$ if $r > 2$. Then, since the Monge–Ampère condition (69) implies that $\nabla_x\Phi \neq 0$ on the support of the oscillatory integral, a simple integration-by-parts argument shows that

$$\| m(\sqrt{-\Delta_x}/c_o\lambda) \circ S_\lambda \|_{L^p(\mathbb{R}^n)\to L^p(\mathbb{R}^n)} = O_N(\lambda^{-N}) \tag{74}$$

for all $N \in \mathbb{N}$ if $c_0 > 0$ is chosen to be sufficiently small. Furthermore,

$$\| (I - m(\sqrt{-\Delta_x}/c_0\lambda)) \circ (\sqrt{I-\Delta_x})^{-\gamma} \|_{L^p(\mathbb{R}^n)\to L^p(\mathbb{R}^n)} = O(\lambda^{-\gamma}), \quad \text{if } \gamma \geq 0.$$

Therefore, by (73) and (74), for such γ one has

$$\|T_\lambda\|_{L^p(\mathbb{R}^{n-1})\to L^p(\mathbb{R}^n)} \lesssim \lambda^{-\gamma} \| (\sqrt{I-\Delta_x})^{\gamma} \circ S_\lambda \|_{L^p(\mathbb{R}^n)\to L^p(\mathbb{R}^n)} + O(\lambda^{-N}). \tag{75}$$

On the other hand, if K is as in (70) then

$$\int_{\mathbb{R}} e^{i\lambda t} K(x,t;y)\,\mathrm{d}t = e^{i\lambda\Phi(x;y)} a(x;y')\rho(y_n).$$

Since the right-hand-side is the kernel of the oscillatory integral S_λ, one can use Hölder's inequality in t to see that if $\mathcal{F}_s$ is as in (72), then

$$\|(\sqrt{I-\Delta_x})^{(n-1)/2-s} \circ S_\lambda\|_{L^p(\mathbb{R}^n)\to L^p(\mathbb{R}^n)} \lesssim \|\mathcal{F}_s\|_{L^p(\mathbb{R}^n)\to L^p(\mathbb{R}^n\times\mathbb{R})}.$$

Therefore, by (75), if $(n-1)/2 - s \geq 0$, then

$$\|T_\lambda\|_{L^p(\mathbb{R}^{n-1})\to L^p(\mathbb{R}^n)} \lesssim \lambda^{-(n-1)/2+s}\, \|\mathcal{F}_s\|_{L^p(\mathbb{R}^n)\to L^p(\mathbb{R}^n\times\mathbb{R})}. \tag{76}$$

Taking $s = \bar{s}_p - 1/p + \varepsilon$ with $\varepsilon > 0$ small, (76) yields

$$\|T_\lambda\|_{L^p(\mathbb{R}^{n-1})\to L^p(\mathbb{R}^n)} = O(\lambda^{-(n-1)/2+\bar{s}_p-1/p+\varepsilon}) = O(\lambda^{-n/p+\varepsilon}),$$

as $\bar{s}_p = (n-1)(\frac{1}{2} - \frac{1}{p})$. Since this is (66), the proof is complete. □

4.2 Necessary Conditions in Conjecture 24: Sharpness of Theorem 34 in Odd Dimensions

Bourgain identified in [7] counterexamples to the estimates (66) for $p < \bar{p}_n$, where the exponent $\bar{p}_n$ is as defined in (29). Using this and Theorem 47, it follows that there are Fourier integral operators satisfying the cinematic curvature hypothesis for which there cannot be local smoothing of order $1/p-$ for any $p < \bar{p}_n$, leading to the range of exponents featured in Conjecture 24. In particular, this shows that the local smoothing estimates in Theorem 34 are sharp in odd dimensions. Furthermore, Theorem 47 may be used to formulate the more refined Conjecture 28 through appropriate refined examples; details of the last fact will be omitted here and the reader is referred, for instance, to [25, §2.1] for the heuristics behind such examples.

Bourgain's counterexample to (66) if $p < \frac{2(n+1)}{n-1}$ and $n \geq 3$ is odd is recalled presently. The construction makes use of the symmetric matrices

$$A(s) = \begin{pmatrix} 1 & s \\ s & s^2 \end{pmatrix},$$

which depends on the real parameter s. Observe that the matrices consisting of the derivatives of each component satisfy

$$\det A'(s) \equiv -1,$$

while, on the other hand,

$$\operatorname{Rank} A(s) \equiv 1.$$

Using these matrices, if $x' = (x_1, \ldots, x_{n-1})$, define the phase function $\varphi(x; y')$ on $\mathbb{R}^n \times \mathbb{R}^{n-1}$ by

$$\varphi(x; y') = \langle x', y' \rangle + \frac{1}{2} \sum_{j=0}^{(n-3)/2} \big\langle A(x_n)(y_{2j+1}, y_{2j+2}), (y_{2j+1}, y_{2j+2}) \big\rangle$$

and let Φ be as in (68). If T_λ is as in (63), then stationary phase arguments yield (see, for example, [53])

$$\lambda^{-\frac{n-1}{4} - \frac{n-1}{2p}} \lesssim \|T_\lambda\|_{L^\infty(\mathbb{R}^{n-1}) \to L^p(\mathbb{R}^n)} \lesssim \|T_\lambda\|_{L^p(\mathbb{R}^{n-1}) \to L^p(\mathbb{R}^n)}$$
$$\text{if } \lambda \gg 1 \text{ and } p \geq 2, \quad \text{provided that } a(0; 0) \neq 0. \tag{77}$$

Clearly φ satisfies the Carleson–Sjölin conditions (64) and (65). Indeed, the surfaces Σ_{x_0} in (65) are, up to linear transformations, the hyperbolic paraboloids in $\mathbb{R}^n$ parametrised by the graph

$$\Big(y', \frac{1}{2}\sum_{j=0}^{(n-1)/2} y_{2j+1}y_{2j+2}\Big).$$

The counterexample now turns into a FIO counterexample for local smoothing following the proof of Proposition 47. Since (67) is also valid, if $\mathcal{F}_s$ is defined to be the Fourier integral operators in (72), then, by (76) and (77), one must have

$$\lambda^{-\frac{n-1}{4}-\frac{n-1}{2p}} \lesssim \lambda^{-\frac{n-1}{2}+s}\, \|\mathcal{F}_s\|_{L^p(\mathbb{R}^n)\to L^p(\mathbb{R}^n\times\mathbb{R})}.$$

Since

$$\tfrac{n-1}{4} + \tfrac{n-1}{2p} < \tfrac{n-1}{2} - \big(\bar{s}_p - \tfrac{1}{p}\big) \quad \text{if } p < \tfrac{2(n+1)}{n-1},$$

one concludes that it does not hold that for sufficiently small $\sigma_p > 0$

$$\|\mathcal{F}_s\|_{L^p(\mathbb{R}^n)\to L^p(\mathbb{R}^n\times\mathbb{R})} < \infty \;\text{ if }\; p < \tfrac{2(n+1)}{n-1} \;\text{ and }\; s < (\bar{s}_p - 1/p) + \sigma_p,$$

which means that the $1/p-$ local smoothing bounds break down for these Fourier integral operators for this range of exponents.

The construction can be modified to produce certain negative results when $n \geq 4$ is even. In this case one takes

$$\varphi(x, y') = \langle x', y'\rangle + \tfrac{1}{2}\sum_{j=0}^{(n-4)/2} \big\langle A(x_n)(y_{2j+1}, y_{2j+2}), (y_{2j+1}, y_{2j+2})\big\rangle + \tfrac{1}{2}(1+x_n)y_{n-1}^2$$

and defines Φ as in (68). The lower bound for the resulting oscillatory integrals S_λ in (77) changes to be

$$\lambda^{-\frac{n}{4}-\frac{n-2}{2p}} \lesssim \|T_\lambda\|_{L^p(\mathbb{R}^{n-1})\to L^p(\mathbb{R}^n)},$$

which, in turn, leads to the lower bound

$$\lambda^{-\frac{n}{4}-\frac{n-2}{2p}} \lesssim \lambda^{-\frac{n-1}{2}+s}\|\mathcal{F}_s\|_{L^p(\mathbb{R}^n)\to L^p(\mathbb{R}^n\times\mathbb{R})}$$

for the Fourier integrals as in (72). A simple calculation, as in the case of odd dimensions, now shows that in the case of *even* $n \geq 4$ these Fourier integral operators, which satisfy the cinematic curvature hypothesis, cannot have $1/p-$ local smoothing when $p < \frac{2(n+2)}{n}$.

Remark 48 (Odd Versus Even Dimensional Case) As is mentioned in Sect. 2.5, the difference between the counterexamples for even and odd dimensions can be explained by the geometry of the cones (71) associated to the Fourier integral operators. When n is odd the cones involve dilates of hyperbolic paraboloids and the number of positive principal curvatures exactly matches the number of negative ones: both equal $\frac{n-1}{2}$. This is not possible when n is even and in this case there are only $\frac{n-2}{2}$ pairs of opposite signs; consequently, the resulting counterexamples involve larger exponents than those for odd n.

It should be noted that the positive results of Stein [57] showing that (66) holds for $p \geq \frac{2(n+1)}{n-1}$ are sharp in odd dimensions in view of Bourgain's counterexample. More recently, Bourgain and Guth [11] obtained the positive results for $p \geq \frac{2(n+2)}{n}$ in the even dimensional case $n \geq 4$. Results for $n = 2$ were obtained much earlier by Carleson and Sjölin [12].

Remark 49 (Signature Hypothesis) In view of Bourgain's counterexample, one should expect the estimate (66) to hold for $p < \bar{p}_n$ for oscillatory integrals T_λ satisfying additional hypothesis on the signature of the associated hypersurfaces Σ_{x_0}. When all principal curvatures of Σ_{x_0} are assumed to be of the same sign at each point, recent results of Guth, Ilioupoulou and the second author [25] show the favourable bounds for T_λ in (66) for $p \geq \bar{p}_{n,+}$, which are sharp in view of previous counterexamples of Minicozzi and the third author [42] (see also [11, 65]).

In view of the connections between local smoothing estimates and oscillatory integral theorems explored in this section, the results in [25] suggest that if the cones arising in the cinematic curvature hypothesis have $(n-1)$ principal curvatures of the same sign, one should have $1/p-$ local smoothing for all $\bar{p}_{n,+} \leq p < \infty$; this corresponds to Conjecture 28 with $\kappa = n - 1$ and would imply Conjecture 23. Observe that Conjecture 28 for $\kappa = n - 1$ formally implies the results in [25] via Theorem 47.

4.3 Maximal Oscillatory Integral Estimates

The above arguments also lead to maximal estimates for a natural class of oscillatory integral operators, including ones arising in spectral theory. As in the case of Bochner–Riesz operators, minor modifications of the proof that local smoothing implies oscillatory integral bounds yield corresponding maximal versions.

Consider oscillatory integrals of the form

$$S_\lambda f(x) := \int_{\mathbb{R}^n} e^{i\lambda\Phi(x;y)} a(x;y) f(y)\, dy$$

where $a \in C_0^\infty(\mathbb{R}^n \times \mathbb{R}^n)$ and the real smooth phase function Φ satisfies the $n \times n$ *Carleson–Sjölin condition*, which has two parts:

Rank condition

$$\operatorname{rank} \partial^2_{xy}\Phi(x; y) \equiv n - 1 \quad \text{for all } (x; y) \in \text{supp a}. \tag{78}$$

This implies that for every fixed x_0 in the x-support of a the gradient graph

$$\Sigma_{x_0} := \{ \nabla_x \Phi(x_0; y) : \ a(x_0; y) \neq 0 \} \subset T^*_{x_0}\mathbb{R}^n$$

is a smooth immersed hypersurface.

The other part of the $n \times n$ Carleson–Sjölin condition is identical to the curvature assumption which appeared earlier in this section.

Curvature condition For each x_0 in the x-support of a, the hypersurface Σ_{x_0} has non-vanishing Gaussian curvature at every point.

To be able to use local smoothing estimates we shall also assume that the **Monge–Ampère condition** (69) holds on the support of a.

Example 50 The class of oscillatory integral operators satisfying these conditions includes ones arising in harmonic analysis on Riemannian manifolds. In particular, if d_g is the Riemannian distance function, then away from the diagonal $\Phi(x; y) := d_g(x; y)$ satisfies (78) and the curvature condition and has non-vanishing Monge–Ampère determinant.

The bounds (66) imply the corresponding bounds

$$\|S_\lambda\|_{L^p(\mathbb{R}^n) \to L^p(\mathbb{R}^n)} = O_\varepsilon(\lambda^{-n/p+\varepsilon}) \tag{79}$$

for the same exponents. Specifically, if (66) is valid for a given p and all oscillatory integral satisfying the Carleson–Sjölin condition and possibly an additional assumption on the geometry of the hypersurfaces Σ_{x_0} associated with φ, then (79) must be valid for the same exponent p for operators satisfying the $n \times n$ Carleson–Sjölin condition along with the same additional geometric condition on the hypersurfaces $\Sigma_{x_0} \subset T^*_{x_0}\mathbb{R}^n$ associated with Φ.

As a consequence of the preceding observation, the bounds (66) of Stein [57] and Bourgain–Guth [11] for the oscillatory integral T_λ imply that (79) holds for $p \geq \bar{p}_n$ if the $n \times n$ Carleson–Sjölin condition is satisfied. Note that the counterexample of Bourgain [7] described in the previous subsection also applies to operators S_λ, so such bounds are optimal in the sense that there are S_λ for which (79) cannot hold if $p < \frac{2(n+1)}{n-1}$ and n is odd or $p < \frac{2(n+2)}{n}$ and $n \geq 4$ is even.

The local smoothing estimates in Theorem 34 can be used to prove a maximal version of Stein's result for S_λ which, by the previous discussion, is sharp in odd dimensions. It should be noted that the argument presented below would also yield maximal estimates for $p < \frac{2(n+1)}{n-1}$ in the even dimensional case or under stronger curvature hypotheses if one had the corresponding local smoothing estimates.

Theorem 51 *Suppose that S_λ satisfies the $n \times n$ Carleson–Sjölin condition for all $\lambda \geq 1$ and that the phase function Φ satisfies the Monge–Ampère condition (69) on the support of a. For $p \geq \frac{2(n+1)}{n-1}$ the maximal estimate*

$$\Big(\int_{\mathbb{R}^n} \sup_{\mu \in [\lambda, 2\lambda]} |S_\mu f(x)|^p \, \mathrm{d}x\Big)^{1/p} \leq C_\varepsilon \lambda^{-n/p+\varepsilon} \|f\|_{L^p(\mathbb{R}^n)} \tag{80}$$

holds for all $\varepsilon > 0$.

Proof It suffices to show that, given $\varepsilon > 0$ and $\frac{2(n+1)}{n-1} \leq p < \infty$, there is a constant C_ε such that

$$\Big(\int_{\mathbb{R}^n} |S_{\mu(x)} f(x)|^p \, \mathrm{d}x\Big)^{1/p} \leq C_\varepsilon \lambda^{-n/p+\varepsilon} \|f\|_{L^p(\mathbb{R}^n)}$$

holds whenever $\mu(x) : \mathbb{R}^n \to [\lambda, 2\lambda]$ is measurable. To prove this, note that if

$$K(x, t; y) := a(x; y)\, \delta_0\big(t - \Phi(x; y)\big)$$

then

$$\mathcal{F} f(x, t) := \int_{\mathbb{R}^n} K(x, t; y)\, f(y)\, \mathrm{d}y$$

forms a one-parameter family of Fourier integrals of order $-\frac{n-1}{2}$ satisfying the cinematic curvature condition. As in Sect. 4.2, the projection condition follows from the assumption that the Monge–Ampère determinant associated with Φ never vanishes, whilst the cone condition, as before, follows from the curvature condition. Thus, Theorem 34 implies that

$$\|\mathcal{F} f\|_{L^p(\mathbb{R}^n \times \mathbb{R})} \lesssim_\varepsilon \|f\|_{L^p_{-n/p+\varepsilon}(\mathbb{R}^n)}, \tag{81}$$

since

$$-\tfrac{n-1}{2} + \big(\bar{s}_p - \tfrac{1}{p}\big) = -\tfrac{n}{p}.$$

Next, let $h \in C^\infty(\mathbb{R})$ satisfy $h(r) = 0$ for $r < 1/2$ and $h(r) = 1$ for $r \geq 1$. Then, since the Monge–Ampère condition (69) implies that $\nabla_y \Phi \neq 0$ on the support of a, a simple integration-by-parts argument shows that, if $c_0 > 0$ is chosen to be sufficiently small,

$$S_\mu f(x) = S_\mu(f_\lambda)(x) + O(\lambda^{-N}\|f\|_p)$$
$$\text{for} \quad f_\lambda := h(\sqrt{-\Delta_x}/c_o\lambda) f \quad \text{and} \quad \mu \in [\lambda, 2\lambda], \tag{82}$$

for each $N \in \mathbb{N}$. This is because for $\mu \approx \lambda$ the kernel of $S_\mu \circ (I - h(\sqrt{-\Delta_x}/c_0\lambda))$ is $O(\lambda^{-N}(1+|y|)^{-N})$ for any N.

Next, use the fact that

$$S_{\mu(x)}g(x) = \int_{\mathbb{R}}\int_{\mathbb{R}^n} e^{i\mu(x)t}\, K(x,t;y)\, g(y)\,\mathrm{d}y\,\mathrm{d}t = \int_{\mathbb{R}} e^{i\mu(x)t}\, \mathcal{F}g(x,t)\,\mathrm{d}t.$$

Since $\mathcal{F}g(x,t)$ is compactly supported in t, one may use Hölder's inequality and (82) to deduce that

$$|S_{\mu(x)}f(x)| \lesssim \Big(\int_{\mathbb{R}} |\mathcal{F}f_\lambda(x,t)|^p\,\mathrm{d}t\Big)^{1/p} + O(\lambda^{-N}\|f\|_p).$$

Since $S_{\mu(x)}f(x)$ vanishes for large $|x|$, this along with (81) yields

$$\begin{aligned}\Big(\int_{\mathbb{R}^n} |S_{\mu(x)}f(x)|^p\,\mathrm{d}x\Big)^{1/p} &\lesssim \|\mathcal{F}f_\lambda\|_{L^p(\mathbb{R}^n\times\mathbb{R})} + \lambda^{-N}\|f\|_{L^p(\mathbb{R}^n)} \\ &\lesssim \|f_\lambda\|_{L^p_{-n/p+\varepsilon}(\mathbb{R}^n)} + \lambda^{-N}\|f\|_{L^p(\mathbb{R}^n)} \\ &\lesssim \lambda^{-n/p+\varepsilon}\|f\|_{L^p(\mathbb{R}^n)},\end{aligned}$$

provided $N > n/p$. Here, in the last inequality, we used the fact that $\hat{f}_\lambda(\xi) = 0$ when $|\xi|$ is smaller than a fixed multiple of λ. This completes the proof of Theorem 51.

Remark 52 Under the assumption that the principal curvatures of Σ_{x_0} are of the same sign, the counterexamples in [42] show that (80) need not hold for $p < \bar{p}_{n,+}$. Furthermore, these counterexamples involve the model case where $\Phi(x;y) := d_g(x;y)$ for (certain choices of) Riemannian metrics g. In this setting, the recent results of Guth, Ilioupoulou and the second author [25] concerning T_λ suggest that (80) may hold for $p \geq \bar{p}_{n,+}$. It is not clear whether the additional hypothesis concerning the Monge–Ampère determinant of Φ is necessary, since the results of [25] obtain (79) without this assumption. In any case, the proof of Theorem 51 required this assumption in order to be able to invoke the local smoothing estimates.

5 Wolff's Approach to Local Smoothing Estimates

The remaining sections of this survey discuss the proof of Theorem 26. Here we describe Wolff's approach which reduces local smoothing estimates to so-called *decoupling inequalities* (see Theorem 53 below). His method has its roots in several ideas extensively used in harmonic analysis which go back to the work of Fefferman on the ball multiplier [20].

5.1 Preliminary Observations

Of course, Theorem 26 follows from establishing

$$\|\mathcal{F}f\|_{L^p(\mathbb{R}^{n+1})} \lesssim \|f\|_{L^p(\mathbb{R}^n)} \tag{83}$$

for $\bar{p}_n \leq p < \infty$ and $\mu < -\alpha(p) := -\bar{s}_p + 1/p$, where $\mathcal{F}$ is the operator (21). We work with the representation of $\mathcal{F}$ in terms of an integral kernel: explicitly,

$$\mathcal{F}f(x,t) = \int_{\mathbb{R}^n} K(x,t;y) f(y)\,\mathrm{d}y$$

where

$$K(x,t;y) := \int_{\hat{\mathbb{R}}^n} e^{i(\phi(x,t;\xi)-\langle y,\xi\rangle)} b(x,t;\xi)(1+|\xi|^2)^{\mu/2}\,\mathrm{d}\xi \tag{84}$$

and $b \in S^0(\mathbb{R}^{n+1} \times \mathbb{R}^n)$ is compactly supported in x and t.

By the principle of stationary phase, one expects K to be singular for those $(x,t;y)$ satisfying

$$\nabla_\xi[\phi(x,t;\xi) - \langle y,\xi\rangle] = 0$$

for some $\xi \in \mathrm{supp}(b)$. In the prototypical case of the half-wave propagator $e^{it\sqrt{-\Delta}}$, for fixed (x,t) this observation identifies the singular set of $K(x,t;\,\cdot)$ as lying in

$$\Big\{y \in \mathbb{R}^n : y - x = t\frac{\xi}{|\xi|} \text{ for some } \xi \in \mathrm{supp}\,\mathrm{b}\Big\} \tag{85}$$

and therefore inside the sphere

$$\Sigma_{(x,t)} := \{y \in \mathbb{R}^n : |x-y| = t\}.$$

For general $\mathcal{F}$, as the map $\xi \mapsto \nabla_\xi\phi(x,t;\xi)$ is homogeneous of degree 0, the associated singular set for each fixed (x,t) is typically an $(n-1)$-dimensional manifold. The relative complexity of the geometry of the singular sets places the study of such operators $\mathcal{F}$ well outside the classical Calderón–Zygmund theory; this is in contrast, for instance, with pseudo-differential operators, where the singularity occurs at an isolated point.

The fundamental approach to understanding the kernel K is to perform multiple decompositions of the ξ-support of b and thereby break K into pieces with a much simpler underlying geometry.

5.2 Basic Dyadic Decomposition

The first step is to break up $\mathcal{F}$ into pieces which are Fourier supported on dyadic annuli. Fix $\beta \in C_c^\infty(\mathbb{R})$ with supp $\beta \in [1/2, 2]$ and such that $\sum_{\lambda>0:\,\mathrm{dyadic}} \beta(r/\lambda) = 1$ for $r \neq 0$. Let $\mathcal{F}^\lambda := \mathcal{F} \circ \beta(\sqrt{-\Delta}/\lambda)$, so that $\mathcal{F}^\lambda f$ has kernel K^λ given by introducing a $\beta(|\xi|/\lambda)$ factor into the symbol in (84), and decompose $\mathcal{F}f$ as

$$\mathcal{F}f =: \mathcal{F}^{\lesssim 1} f + \sum_{\lambda \in \mathbb{N}:\,\mathrm{dyadic}} \mathcal{F}^\lambda f.$$

It is not difficult to verify that $\mathcal{F}^{\lesssim 1}$ is a pseudo-differential operator of order 0 and therefore bounded on L^p for all $1 < p < \infty$ by standard theory (see, for instance, [58, Chapter VI, §5]). Thus, the problem is further reduced to showing that for any arbitrarily small $\varepsilon > 0$ the estimate

$$\|\mathcal{F}^\lambda f\|_{L^p(\mathbb{R}^{n+1})} \lesssim \lambda^{\alpha(p)+\mu+\varepsilon} \|f\|_{L^p(\mathbb{R}^n)}$$

holds for all $\lambda \geq 1$; letting $\varepsilon := -\frac{\mu+\alpha(p)}{2} > 0$ the estimate (83) would then follow from summing a geometric series.

The remaining pieces $\mathcal{F}^\lambda$ (for λ large) are more complicated objects. The uncertainty principle tells us that the singularity present in K should have been "resolved to scale λ^{-1}" in K^λ. For instance, in the case of the wave propagator $e^{it\sqrt{-\Delta}}$ the kernel K^λ should no longer be singular along $\Sigma_{(x,t)}$ but should satisfy:

(i) $K^\lambda(x, t;\, \cdot)$ is concentrated in a λ^{-1}-neighbourhood of $\Sigma_{(x,t)}$, given by

$$\big\{y \in \mathbb{R}^n : \big||x - y| - t\big| \lesssim \lambda^{-1}\big\}; \tag{86}$$

(ii) $\|K^\lambda(x, t;\, \cdot)\|_\infty \lesssim \lambda^\mu \lambda^n$.

Here property (i) is an uncertainty heuristic, whilst the second property trivially follows from the formula (84) for the kernel. These two features combine to give the crude estimate

(iii) $\int_{\mathbb{R}^n} |K^\lambda(x, t; y)|\, \mathrm{d}y \lesssim \lambda^\mu \lambda^{n-1}$,

which in turn yields an $L^\infty \to L^\infty$ bound for $\mathcal{F}^\lambda$. However, one may obtain a significant gain in the λ exponent by subjecting K^λ to a more refined stationary phase analysis. The method of stationary phase requires a uniform lower bound for $|\nabla_\xi \phi(x, t; \xi)|$ on $|\xi| \sim \lambda$; as $\xi \mapsto \nabla_\xi \phi(x, t; \xi)$ is homogeneous of degree 0, one should therefore decompose the angular variables into small regions in which $|\nabla_\xi \phi(x, t; \xi)|$ does not vary too much.

5.3 Angular Decomposition

For λ fixed, let $\{\xi_\nu^\lambda\}_{\nu\in\Theta_{\lambda^{-1/2}}}$ be a maximal $\lambda^{-1/2}$-separated subset of $\mathbb{S}^{n-1}$, so that the indexing set satisfies $\#\Theta_{\lambda^{-1/2}} \sim \lambda^{(n-1)/2}$. Let

$$\Gamma_\nu^\lambda := \{\xi \in \hat{\mathbb{R}}^n : |\pi_{\xi_\nu^\lambda}^\perp \xi| \lesssim \lambda^{-1/2}|\xi|\}$$

denote the sector of aperture $\sim \lambda^{-1/2}$ whose central direction is ξ_ν^λ; here $\pi_{\xi_\nu^\lambda}^\perp$ is the orthogonal projection onto the hyperplane perpendicular to ξ_ν^λ. Let $\{\chi_\nu^\lambda\}_{\nu\in\Theta_{\lambda^{-1/2}}}$ be a smooth partition of unity, homogeneous of degree 0, adapted to the Γ_ν^λ, with $|D^\alpha \chi_\nu^\lambda(\xi)| \lesssim \lambda^{|\alpha|/2}$ for $\xi \in \mathbb{S}^{n-1}$ and all $\alpha \in \mathbb{N}_0^n$. Setting $b_\nu^\lambda(x,t;\xi) := b(x,t;\xi)\beta(|\xi|/\lambda)\chi_\nu^\lambda(\xi)$, the resulting operators $\mathcal{F}_\nu^\lambda$ have corresponding kernels

$$K_\nu^\lambda(x,t;y) := \int_{\mathbb{R}^n} e^{i(\phi(x,t;\xi)-\langle y,\xi\rangle)} b_\nu^\lambda(x,t;\xi)(1+|\xi|^2)^{\mu/2}\,\mathrm{d}\xi.$$

To understand the effect of this frequency localisation on the spatial side, we once again consider the prototypical case of $e^{it\sqrt{-\Delta}}$. Recalling (85), it follows from the choice of localisation that K_ν^λ should now be concentrated on the angular sector

$$\big\{y \in \mathbb{R}^n : |\pi_{\xi_\nu^\perp}(x-y)| \lesssim \lambda^{-1/2}|x-y|\big\}.$$

Combining this with property (i) from the basic dyadic decomposition, it follows that:

(i′) $K_\nu^\lambda(x,t;\,\cdot\,)$ is concentrated in a $t\lambda^{-1/2}$ cap on the fattened sphere (86), centred at $t\xi_\nu^\lambda$ (see Fig. 4);
(ii′) $\|K_\nu^\lambda(x,t;\,\cdot\,)\|_\infty \lesssim \lambda^\mu \lambda^{(n+1)/2}$.

It is not difficult to make these heuristics precise and, moreover, extend these observations to general variable-coefficient operators $\mathcal{F}$. In particular, the dyadic and annular decompositions allow one to linearise the phase $\phi(x,t;\xi)$ in the ξ-variable; this permits a standard stationary phase argument (see [48] or [58, Chapter IX §§4.5-4.6]) which reveals that the associated kernel K_ν^λ of $\mathcal{F}_\nu^\lambda$ satisfies the pointwise bound

$$|K_\nu^\lambda(x,t;y)| \lesssim \frac{\lambda^\mu \lambda^{(n+1)/2}}{(1+\lambda|\pi_{\xi_\nu^\lambda}[y-\nabla_\xi\phi(x,t,\xi_\nu^\lambda)]| + \lambda^{1/2}|\pi_{\xi_\nu^\lambda}^\perp[y-\nabla_\xi\phi(x,t;\xi_\nu^\lambda)]|)^N} \tag{87}$$

for all $N \geq 0$, where $\pi_{\xi_\nu^\lambda}$ denotes the projection onto the direction ξ_ν^λ and $\pi_{\xi_\nu^\lambda}^\perp$ its perpendicular projection. Note that (87) immediately yields $\|K_\nu^\lambda(x,t;\,\cdot\,)\|_1 \lesssim \lambda^\mu$, which together with the triangle inequality implies that

(iii′) $\int_{\mathbb{R}^n} |K^\lambda(x,t;y)|\,\mathrm{d}y \lesssim \lambda^\mu \lambda^{(n-1)/2}$;

note the square root gain over (iii) obtained via the angular decomposition.

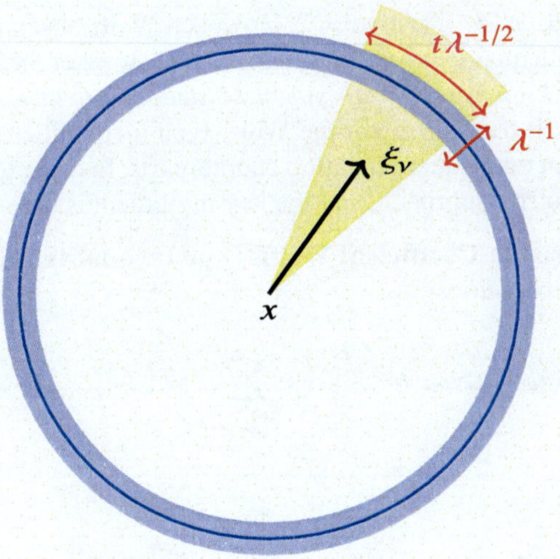

Fig. 4 For fixed (x, t), the kernel $K^\lambda(x, t;\ \cdot)$ associated to the half-wave propagator $e^{it\sqrt{-\Delta}}$ is concentrated on an annulus around the circle $x + t\mathbb{S}^{n-1}$ of thickness $\sim \lambda^{-1}$ (denoted here in blue). The piece $K_\nu^\lambda(x, y;\ \cdot)$ is further localised to an angular sector with angle $\lambda^{-1/2}$ (denoted here in yellow)

5.4 Decoupling into Localised Pieces

Having found a natural decomposition of the operator

$$\mathcal{F}^\lambda = \sum_{\nu \in \Theta_{\lambda^{-1/2}}} \mathcal{F}_\nu^\lambda,$$

the problem is to effectively separate the contributions to $\|\mathcal{F}^\lambda f\|_{L^p(\mathbb{R}^{n+1})}$ coming from the individual the pieces. Since each $\mathcal{F}_\nu^\lambda f$ carries some oscillation, one may attempt to prove a square function estimate of the form

$$\|\mathcal{F}^\lambda f\|_{L^p(\mathbb{R}^{n+1})} \lesssim_\varepsilon \lambda^\varepsilon \Big\| \Big(\sum_{\nu \in \Theta_{\lambda^{-1/2}}} |\mathcal{F}_\nu^\lambda f|^2 \Big)^{1/2} \Big\|_{L^p(\mathbb{R}^{n+1})}; \qquad (88)$$

here the appearance of the ℓ^2 expression (rather than the ℓ^1 norm which arises trivially from the triangle inequality) encapsulates the cancellation between the $\mathcal{F}_\nu^\lambda f$. Inequalities of the form (88) were established in [44, 45], albeit with a unfavourable dependence on λ, and these results have subsequently been refined by various authors [6, 37, 39, 62].

Unfortunately, establishing sharp versions (88) appears to be an extremely difficult problem: indeed, the question is open even in the simplest possible case of the wave propagator $e^{it\sqrt{-\Delta}}$ with $n = 2$. However, Wolff observed in [67] that sharp local smoothing inequalities can be obtained via a weaker variant of the estimate (88) which is now known as a *Wolff-type* or *ℓ^p-decoupling* inequality. Although still highly non-trivial, it transpires that the Wolff-type inequalities are nevertheless far easier to prove than their square function counterparts. In order to prove Theorem 26 we will pursue Wolff's approach, and the key ingredient is the following estimate.

Theorem 53 (Variable-Coefficient Wolff-Type Inequality [1]) *Let* $2 \leq p \leq \infty$. *For all* $\varepsilon > 0$ *the inequality*

$$\|\mathcal{F}^\lambda f\|_{L^p(\mathbb{R}^{n+1})} \lesssim_{p,\varepsilon} \lambda^{\alpha(p)+\varepsilon}\Big(\sum_{\nu\in\Theta_{\lambda^{-1/2}}} \|\mathcal{F}^\lambda_\nu f\|^p_{L^p(\mathbb{R}^{n+1})}\Big)^{1/p} \tag{89}$$

holds[12], *where*

$$\alpha(p) := \begin{cases} \frac{\bar{s}_p}{2} & \text{if } 2 \leq p \leq \frac{2(n+1)}{n-1}, \\ \bar{s}_p - \frac{1}{p} & \text{if } \frac{2(n+1)}{n-1} \leq p < \infty. \end{cases}$$

Remark 54

(1) The value of $\alpha(p)$ coincides with that in Sect. 5.1, which was only defined in the $\frac{2(n+1)}{n-1} \leq p < \infty$.
(2) A necessary condition on p for the square function estimate (88) to hold is that $2 \leq p \leq \frac{2n}{n-1}$. For this range (88) is stronger than (89), as can be seen by a simple application of Minkowski's and Hölder's inequalities.
(3) It is instructive to compare (89) with estimates obtained via trivial means. In particular, the triangle and Hölder's inequality imply that (89) holds with the exponent $\alpha(p)$ replaced with $\frac{n-1}{2}(1-\frac{1}{p}) = \bar{s}_p + \frac{n-1}{2p}$. Thus, the gain in the λ-power present in (89) provides a measurement of the cancellation between the $\mathcal{F}^\lambda_\nu f$ arising from their oscillatory nature.

Theorem 53 can be combined with simple estimates for the localised pieces (see (92) below) to deduce the desired estimate

$$\|\mathcal{F}^\lambda f\|_{L^p(\mathbb{R}^{n+1})} \lesssim_{s,p,\varepsilon} \lambda^{\alpha(p)+\mu+\varepsilon}\|f\|_{L^p(\mathbb{R}^n)}; \tag{90}$$

the details of this argument are discussed in the following subsection.

[12] Some slight technicalities have been suppressed in the statement of this theorem. In particular, the precise formulation includes some innocuous error terms: see [1] for further details.

Theorem 53 is an extension of the result for the constant-coefficient operators

$$e^{ith(D)}f(x) := \frac{1}{(2\pi)^n}\int_{\hat{\mathbb{R}}^n} e^{i(\langle x,\xi\rangle + th(\xi))}\hat{f}(\xi)\,\mathrm{d}\xi, \tag{91}$$

which is a celebrated theorem of Bourgain–Demeter [8, 9]; in line with our previous hypotheses on the phase, h is assumed to be homogeneous of degree 1, smooth away from $\xi = 0$ and such that the cone parametrised by $\xi \mapsto (\xi, h(\xi))$ has everywhere $(n-1)$ non-vanishing principal curvatures.

Theorem 55 (Bourgain–Demeter [8, 9]) *For all* $2 \leq p \leq \infty$ *and all* $\varepsilon > 0$ *the estimate*

$$\|e^{ith(D)}f\|_{L^p(\mathbb{R}^{n+1})} \lesssim_{\varepsilon,h} \lambda^{\alpha(p)+\varepsilon}\Big(\sum_{\nu\in\Theta_{\lambda^{-1/2}}} \|e^{ith(D)}f_\nu\|^p_{L^p(\mathbb{R}^{n+1})}\Big)^{1/p}$$

holds for all $\lambda \geq 1$ *and functions* f *such that* $\mathrm{supp}(\widehat{f}) \subseteq \{\xi \in \mathbb{R}^n : \lambda \leq |\xi| \leq 2\lambda\}$.

The proof of Theorem 55 is difficult and deep and relies on tools from multilinear harmonic analysis (in particular, the Bennett–Carbery–Tao multilinear Kakeya theorem [3] and the Bourgain–Guth method [11]). These important ideas will not be addressed in this survey, and the interested reader is referred to the original papers [8, 9] or the study guide [10] for further information.

It transpires that the variable coefficient Theorem 53 can be deduced as a consequence of the constant coefficient Theorem 55 via a relatively simple induction-on-scales and approximation argument. A sketch of the proof of Theorem 53 (avoiding many of the technical details) will be given in the next section.

5.5 *Bounding the Localised Pieces*

Given the variable-coefficient Wolff-type inequality from Theorem 53, to conclude the proof of Theorem 26 it suffices to show the localised pieces satisfy

$$\Big(\sum_{\nu\in\Theta_{\lambda^{-1/2}}} \|\mathcal{F}^\lambda_\nu f\|^p_{L^p(\mathbb{R}^{n+1})}\Big)^{1/p} \lesssim \lambda^{\mu}\|f\|_{L^p(\mathbb{R}^n)} \tag{92}$$

for $2 \leq p \leq \infty$. Indeed, combining this inequality with Theorem 53 one immediately obtains (90), as required.

The inequality (92) is a simple consequence of the basic properties of the localised operators and, in particular, the kernel estimate (87). By real interpolation, it suffices to prove the bounds only at the endpoints $p = \infty$ and $p = 2$.

5.5.1 L^∞-Bounds

Observe that (87) immediately implies

$$\max_{\nu\in\Theta_{\lambda^{-1/2}}} \sup_{(x,t)\in\mathbb{R}^{n+1}} \|K_\nu^\lambda(x,t;\,\cdot\,)\|_{L^1(\mathbb{R}^n)} \lesssim \lambda^\mu.$$

From this, one deduces that

$$\max_{\nu\in\Theta_{\lambda^{-1/2}}} \|\mathcal{F}_\nu^\lambda f\|_{L^\infty(\mathbb{R}^n)} \lesssim \lambda^\mu \|f\|_{L^\infty(\mathbb{R}^n)}.$$

5.5.2 L^2-Bounds

Useful estimates are also available at the L^2-level. For instance, the wave propagator $e^{it\sqrt{-\Delta}}$ satisfies the conservation of energy identity

$$\|e^{it\sqrt{-\Delta}} f\|_{L^2(\mathbb{R}^n)} = (2\pi)^{-n/2}\|f\|_{L^2(\mathbb{R}^n)} \qquad \text{for each fixed time } t\in\mathbb{R}, \tag{93}$$

which, indeed, is a trivial consequence of Plancherel's theorem. This observation can be extended to the general variable coefficient setting at the expense of relaxing the equality to an inequality. In particular, a theorem of Hörmander [30] (see also [58, Chapter IX §1.1]) implies the bound

$$\|\mathcal{F}_\nu^\lambda f(\,\cdot\,,t)\|_{L^2(\mathbb{R}^n)} \lesssim \|f_\nu^\lambda\|_{L^2(\mathbb{R}^n)} \qquad \text{for each fixed time } t\in\mathbb{R}, \tag{94}$$

where $\widehat{f_\nu^\lambda} := \hat{f}\chi_\nu^\lambda$ is a piece of f given by localising the frequencies to Γ_ν^λ. Of course, in the general variable coefficient case Plancherel's theorem cannot be directly applied as in the proof of (93); nevertheless, (94) can be established via a simple T^*T argument and standard oscillatory integral techniques.

One may now obtain space-time estimates for the $\mathcal{F}_\nu^\lambda f$ simply by integrating both sides of (94) over a (compact) time interval containing the t-support of b. The almost orthogonality of the f_ν^λ, given by Plancherel's theorem and the almost disjointness of Γ_ν^λ, then readily implies that

$$\Big(\sum_{\nu\in\Theta_{\lambda^{-1/2}}} \|\mathcal{F}_\nu^\lambda f\|_{L^2(\mathbb{R}^{n+1})}^2\Big)^{1/2} \lesssim \lambda^\mu \|f\|_{L^2(\mathbb{R}^n)}.$$

This concludes the proof of Theorem 26.

6 Variable-Coefficient Wolff-Type Inequalities

In the previous section the proof of the local smoothing estimate in Theorem 26 was reduced to establishing the variable-coefficient Wolff-type inequalities in Theorem 53. In this section we sketch the proof of Theorem 53, which is in fact a consequence of the constant-coefficient case (that is, Theorem 55).

6.1 *Preliminaries*

It suffices to consider the case $\mu = 0$ (the general case then follows by writing any given operator as a composition of a pseudo-differential operator and an operator of order 0). By the homogeneity of $\phi(x, t; \cdot)$ and rescaling, Theorem 53 follows from its analogous statement when $|\xi| \sim 1$ and $(x, t) \in B(0, \lambda)$. Namely, it suffices to prove (89) for the rescaled operators

$$\mathcal{F}^\lambda f(x,t) := \int_{\hat{\mathbb{R}}^n} e^{i\phi^\lambda(x,t;\xi)} b^\lambda(x,t;\xi)\hat{f}(\xi)\,\mathrm{d}\xi$$

where

$$\phi^\lambda(x,t;\omega) := \lambda\phi(x/\lambda, t/\lambda;\omega) \qquad \text{and} \qquad b^\lambda(x,t;\xi) := b(x/\lambda, t/\lambda, \xi)$$

and b is supported in $B^{n+1} \times \Gamma$. Here Γ is a conic domain of the type

$$\Gamma := \{\xi \in \hat{\mathbb{R}}^n : 1 \le |\xi| \le 2 \text{ and } |\xi/|\xi| - e| \lesssim 1\}$$

for a unit vector $e \in \mathbb{S}^{n-1}$. Note that the notation $\mathcal{F}^\lambda$ is *not* consistent with that used in the previous section.

6.2 *Inductive Setup*

The proof will involve an induction-on-scale procedure. To this end, an additional spatial scale parameter R is introduced: it will be shown that for $1 \le R \le \lambda$ the inequality

$$\|\mathcal{F}^\lambda f\|_{L^p(B_R)} \le \bar{C}(\varepsilon, p) R^{\alpha(p)+\varepsilon}\Big(\sum_{\nu\in\Theta_{R^{-1/2}}} \|\mathcal{F}^\lambda_\nu f\|^p_{L^p(B_R)}\Big)^{1/p} \tag{95}$$

holds for a suitable choice of constant $\bar{C}(\varepsilon, p)$. Here $B_R \subseteq B(0, \lambda)$ is a ball of radius R so that Theorem 53 follows by setting $R = \lambda$.

By the trivial argument described in Remark 54, the inequality

$$\|\mathcal{T}^\lambda f\|_{L^p(B_R)} \lesssim R^{(n-1)/2p'}\Big(\sum_{\nu\in\Theta_{R^{-1/2}}} \|\mathcal{T}^\lambda_\nu f\|^p_{L^p(B_R)}\Big)^{1/p}$$

holds. This settles the desired decoupling inequality (95) for $R \sim 1$, and thereby establishes the base case for the induction.

Fix $1 \ll R \leq \lambda$ and assume the following induction hypothesis:

Radial induction hypothesis Assume (95) holds whenever (R, λ) is replaced with (R', λ') for any $1 \leq R' \leq R/2$ and $\lambda' \geq R'$.

In fact, one must work with a slightly more sophisticated induction hypothesis which involves not just a single operator $\mathcal{T}^\lambda$ but a whole class of related operators $\widetilde{\mathcal{T}}^\lambda$ which is closed under certain rescaling operations. The precise details are omitted here: see [1] for further information.

6.3 *Key Ingredients of The Proof*

The proof of the inductive step comes in three stages:

1. At sufficiently small scales $1 \ll K \ll \lambda^{1/2}$, the operator $\mathcal{T}^\lambda$ may be effectively approximated by constant coefficient operators (91).
2. For each of the approximating constant-coefficient operators, one may use the Bourgain–Demeter theorem at the small scale K.
3. The inherent symmetries of the inequality (95) allow one to propagate the gain arising from the constant-coefficient Bourgain–Demeter theorem at the small scale K to larger scales. This is achieved via a parabolic rescaling argument, together with an application of the radial induction hypothesis.

Further details of this simple programme are provided in the forthcoming subsections.

6.4 *Approximation by Constant Coefficient Operators*

Let $\mathcal{B}_K$ be a cover of B_R by balls of radius K for some value of $1 \ll K \ll \lambda^{1/2}$ to be determined later. Consider the spatially localised norm $\|\mathcal{T}^\lambda f\|_{L^p(B_K)}$ for $B_K = B(\bar{z}, K) \in \mathcal{B}_K$. By the uncertainty principle, localising to a spatial ball of radius K should induce frequency uncertainty at the reciprocal scale K^{-1}. To understand what this means for our operator, we return once again to the prototypical case of the wave propagator. Observe that for any test function $\varphi \in C_c(\hat{\mathbb{R}}^{n+1})$ one has

$$\int_{\mathbb{R}^{n+1}} e^{it\sqrt{-\Delta}} f(x)\check{\varphi}(x,t)\,\mathrm{d}x\mathrm{d}t = \int_{\hat{\mathbb{R}}^n} \varphi(\xi,|\xi|)\hat{f}(\xi)\,\mathrm{d}\xi \tag{96}$$

and therefore the space-time Fourier transform of $e^{it\sqrt{-\Delta}}f$ is distributionally supported on the light cone. For the general variable-coefficient case, the Fourier support properties of $\mathcal{F}^\lambda f$ involve a whole varying family of conic hypersurfaces $\Sigma_z\colon \xi \mapsto \partial_z\phi^\lambda(z;\xi)$, parametrised by $z \in \mathbb{R}^{n+1}$, and there is no clean distributional identity analogous to (96). However, note that for $z \in B(\bar{z}, K)$ one has

$$|\partial_z\phi^\lambda(z;\xi) - \partial_z\phi^\lambda(\bar{z};\xi)| \lesssim \frac{|z-\bar{z}|}{\lambda} \leq K^{-1}$$

provided $K \ll \lambda^{1/2}$, and so the uncertainty principle tells us that the surfaces Σ_z, and $\Sigma_{\bar{z}}$ should be essentially indistinguishable once the operator is spatially localised to B_K. It in fact follows that on B_K the operator $\mathcal{F}^\lambda$ can be effectively approximated by a constant coefficient operator

$$T_{\bar{z}}g(z) := \int_{\hat{\mathbb{R}}^n} e^{i\langle\partial_z\phi^\lambda(\bar{z};\xi),z\rangle} a(\xi)\hat{g}(\xi)\,\mathrm{d}\xi \tag{97}$$

associated to surface $\Sigma_{\bar{z}}$, where a is a suitable choice of cut-off function.

An alternative and slightly more accurate way to understand this approximation is to consider the first order Taylor expansion of the phase function

$$\phi^\lambda(z;\xi) - \phi^\lambda(\bar{z};\xi) = \langle\partial_z\phi^\lambda(\bar{z};\xi), z-\bar{z}\rangle + O(\lambda^{-1}|z-\bar{z}|^2).$$

Since $\lambda^{-1}|z-\bar{z}|^2 \ll 1$ for $z \in B_K$, the error term in the right-hand side does not contribute significantly to the oscillation induced by the phase. Consequently, over the ball B_K one may safely remove this error and thereby replace the phase ϕ^λ by its linearisation $\phi^\lambda(\bar{z};\xi) + \langle\partial_z\phi^\lambda(\bar{z};\xi), z-\bar{z}\rangle$. Observations of this kind lead to a statement of the form

$$\|\mathcal{F}^\lambda f\|_{L^p(B_K)} \sim \|T_{\bar{z}} f_{\bar{z}}\|_{L^p(B(0,K))}$$

where $f_{\bar{z}}$ is defined by $\widehat{f_{\bar{z}}} := e^{i\phi^\lambda(\bar{z};\,\cdot\,)}\widehat{f}$ and $T_{\bar{z}}$ is as in (97).

In practice, there are significant technical complications which arise in making these heuristics precise: the full details may be found in [1].

6.5 Application of Constant-Coefficient Decoupling

The above approximation allows one to take advantage of the sharp ℓ^p-decoupling theorem of Bourgain–Demeter for the constant coefficient operators $T_{\bar{z}}$ at scale K. In particular, on each $B_K = B(\bar{z}, K)$ one has

$$\begin{aligned}\|\mathcal{F}^\lambda f\|_{L^p(B_K)} &\sim \|T_{\bar{z}} f_{\bar{z}}\|_{L^p(B(0,K))} \\ &\lesssim_\varepsilon K^{\alpha(p)+\varepsilon/2}\Big(\sum_{\sigma\in\Theta_{K^{-1/2}}} \|T_{\bar{z}} f_{\bar{z},\sigma}\|^p_{L^p(B(0,K))}\Big)^{1/p} \\ &\sim K^{\alpha(p)+\varepsilon/2}\Big(\sum_{\sigma\in\Theta_{K^{-1/2}}} \|\mathcal{F}^\lambda_\sigma f\|^p_{L^p(B_K)}\Big)^{1/p},\end{aligned}$$

where the first inequality is due to Theorem 55. Summing over $B_K \subset B_R$, it follows that

$$\|\mathcal{F}^\lambda f\|_{L^p(B_R)} \lesssim K^{\alpha(p)+\varepsilon/2}\Big(\sum_{\sigma\in\Theta_{K^{-1/2}}} \|\mathcal{F}^\lambda_\sigma f\|^p_{L^p(B_R)}\Big)^{1/p}. \qquad (98)$$

Thus, we have succeeded in decoupling $\mathcal{F}^\lambda f$ into scale $K^{-1/2}$ pieces, but we are still far from achieving the required decoupling at scale $R^{-1/2}$.

At this point it is perhaps instructive to explain some of the ideas behind the proof, before fleshing out the details in the remaining subsections. The next step is to treat each of the summands on the right-hand side of (98) individually. This is (essentially) done by repeating the above argument to successively pass down from decoupling at scale $K^{-1/2}$ to decoupling at scales K^{-1}, $K^{-3/2}$, ... until we reach the small scale $R^{-1/2}$. The key difficulty is to keep control of the constants in the inequalities which would otherwise build up over repeated application of the preceding arguments.[13]

To control the constant build up, we assume a slightly different perspective. In particular, as in [8], we apply a parabolic rescaling in each stage of the iteration; this converts the improvement in the size of the decoupling regions to an improvement in the spatial localisation. In particular, (98) can be thought of as passing from decoupling at scale 1 (the left-hand side) to decoupling at scale $K^{-1/2}$ (the right-hand side); after rescaling it roughly corresponds to passing from spatial localisation at scale R to spatial localisation at scale R/K. The idea is then to iterate until we are spatially localised to ~ 1 scales, at which point the desired inequality becomes trivial. An advantage of the rescaling is that the repeatedly rescaled operators get closer and closer to constant coefficient operators over the course of the iterations.

[13]In the proof we will take $K \sim 1$: it is therefore necessary to iterate roughly $\log R$ times to pass all the way down to scale $R^{-1/2}$. If at each iteration we iterate we pick up an admissible constant C, then over all the iterations we pick up an inadmissible constant $C^{\log R} = R^{\log C}$.

Thus, we find ourselves are in a more and more favourable situation as the argument progresses and this prevents a constant build up.

We shall see that the iteration argument sketched above can be succinctly expressed using our radial induction hypothesis.

6.6 *Parabolic Rescaling*

By a parabolic rescaling argument, one can scale Γ_σ^K to Γ, so that the support of $\widehat{f_\sigma}$ is at unit scale. This essentially reduces the spatial scale from R to R/K and anticipates an appeal to the radial induction hypothesis in Sect. 6.7.

6.6.1 A Prototypical Example

To illustrate the rescaling procedure, we consider the model operator $e^{ith_{\mathrm{par}}(D)}$ where

$$h_{\mathrm{par}}(\xi) := \frac{|\xi'|^2}{\xi_n} \qquad \text{for} \xi = (\xi', \xi_n) \in \hat{\mathbb{R}}^n;$$

this is a close cousin of the classical half-wave propagator $e^{it\sqrt{-\Delta}}$, but $e^{ith_{\mathrm{par}}(D)}$ enjoys some additional symmetries which make it slightly easier to analyse.

Without loss of generality, one may interpret ξ' as the angular variable; in particular, it is assumed that Γ_σ^K is a sector of the form

$$\big\{(\xi', \xi_n) \in \hat{\mathbb{R}}^n : 1/2 \le \xi_n \le 2 \text{ and } |\xi'/\xi_n - \omega_\sigma| \le K^{-1/2}\big\}$$

for some $\omega_\sigma \in B^{n-1}(0,1)$. The sector Γ_σ^K is therefore mapped to Γ under the transformation $(\Psi_\sigma^K)^{-1}$ where $\Psi_\sigma^K : (\xi', \xi_n) \mapsto (K^{-1/2}\xi' + \omega_\sigma \xi_n, \xi_n)$: see Fig. 5.

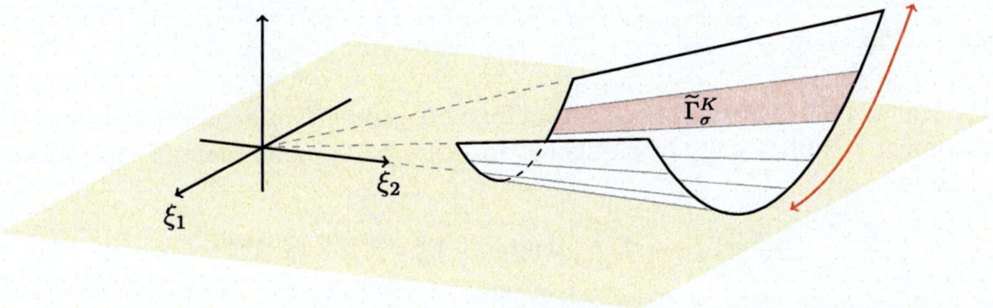

Fig. 5 The parabolic rescaling phenomenon for the phase $\phi(x, t; \xi) = x_1\xi_1 + x_2\xi_2 + t\xi_1^2/\xi_2$. Here $\widetilde{\Gamma}_\sigma^K$ denotes the image of Γ_σ^K under the map $\xi \mapsto (\xi, h_{\mathrm{par}}(\xi))$

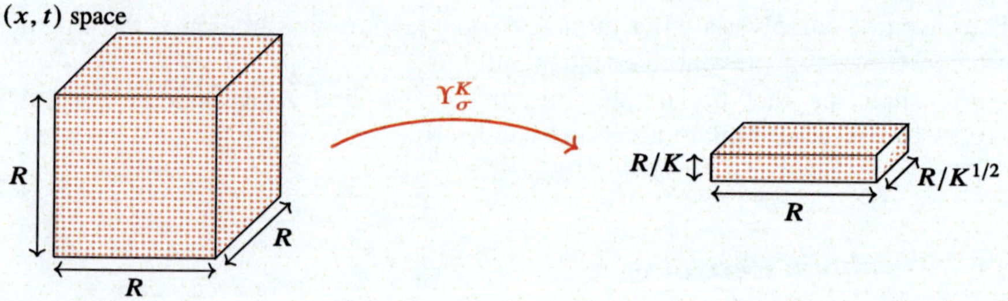

Fig. 6 The parabolic rescaling effect on the (x, t)-variables for the phase $\phi(x, t; \xi) = x_1\xi_1 + x_2\xi_2 + t\xi_1^2/\xi_2$

Let ϕ_{par} be the phase associated to the operator $e^{ith_{\mathrm{par}}(D)}$. The scaling in the frequency variables can be transferred onto the spatial variables via the identity

$$\phi_{\mathrm{par}}(x, t; \Psi_\sigma^K(\xi)) = \phi_{\mathrm{par}}(\Upsilon_\sigma^K(x, t); \xi) \tag{99}$$

where $\Upsilon_\sigma^K : (x, t) \mapsto (K^{-1/2}(x' + 2t\omega_\sigma), \langle x', \omega_\sigma\rangle + x_n + t|\omega_\sigma|^2, K^{-1}t)$. Consequently,

$$\|e^{ith_{\mathrm{par}}(D)} f\|_{L^p(B_R)} = K^{(n+1)/2p}\|e^{ith_{\mathrm{par}}(D)} \tilde{f}_\sigma\|_{L^p(\Upsilon_\sigma^K(B_R))}$$

where $\tilde{f}_\sigma$ is defined by

$$[\tilde{f}_\sigma]\widehat{\ } := K^{-(n-1)/2}\widehat{f_\sigma} \circ \Psi_\sigma^K. \tag{100}$$

Observe that the set $\Upsilon_\sigma^K(B_R)$ is contained in an $R \times R/K^{1/2} \times \cdots \times R/K^{1/2} \times R/K$ box: see Fig. 6. The longest side, which is of length R, points in the $(w_\sigma, 1)$ direction whilst the shortest side, which of length R/K, points in the time direction. The remaining sides, which are of length $R/K^{1/2}$, point in spatial directions orthogonal to the long and short sides.

6.6.2 The General Case

The scaling procedure can be carried out for more general phases, albeit with notable additional complications. In particular, for each σ one may identify changes of variable

$$\Psi_\sigma^K : \Gamma_\sigma^K \to \Gamma \qquad \text{and} \qquad \Upsilon_\sigma^K : \mathbb{R}^{n+1} \to \mathbb{R}^{n+1}$$

and a suitable Fourier integral operator $\widetilde{\mathcal{F}}_\sigma^{\lambda/K}$ at scale λ/K such that

$$\|\mathcal{F}_\sigma^\lambda f\|_{L^p(B_R)} \sim K^{(n+1)/2p}\|\widetilde{\mathcal{F}}_\sigma^{\lambda/K} \tilde{f}_\sigma\|_{L^p(\Upsilon_\sigma^K(B_R))},$$

where $\tilde{f}_\sigma$ is defined as in (100) and $\Upsilon_\sigma^K(B_R)$ is contained in a rectangular box of dimensions $R \times R/K^{1/2} \times \cdots \times R/K^{1/2} \times R/K$. This situation is somewhat more involved than the prototypical case described above, due to the lack of any simple scaling identity (99). In particular, the mapping Υ_σ^K will often be non-linear and the operators $\tilde{\mathcal{F}}^\lambda$ may not agree with the original $\mathcal{F}^\lambda$ (although they will of course be related). In order to deal with the latter point, it is necessary to formulate an induction hypothesis which applies to *an entire class of FIOs* which is closed under the relevant rescalings. The (somewhat technically involved) details of the rigorous realisation of this strategy can be found in [1].

6.7 Applying the Induction Hypothesis

Noting that $R' := R/K \leq R/2$, the (general) radial induction hypothesis implies that

$$\|\tilde{\mathcal{F}}_\sigma^{\lambda/K} \tilde{f}_\sigma\|_{L^p(B_{R/K})} \leq \bar{C}(p,\varepsilon)(R/K)^{\alpha(p)+\varepsilon}\Big(\sum_{\nu \in \Theta_{(R/K)^{-1/2}}} \|\tilde{\mathcal{F}}_\sigma^{\lambda/K}(\tilde{f}_\sigma)_\nu\|_{L^p(B_{R/K})}^p\Big)^{1/p}$$

for any ball $B_{R/K}$ of radius R/K. Take a finitely-overlapping cover of $\Upsilon_\sigma^K(B_R)$ by such balls and apply the above inequality to each member of this cover. Taking p-powers, summing over each member of the cover and taking p-roots, one deduces that

$$\|\tilde{\mathcal{F}}_\sigma^{\lambda/K} \tilde{f}_\sigma\|_{L^p(\Upsilon_\sigma^K(B_R))} \lesssim \bar{C}(p,\varepsilon)(R/K)^{\alpha(p)+\varepsilon}\Big(\sum_{\nu \in \Theta_{(R/K)^{-1/2}}} \|\tilde{\mathcal{F}}_\sigma^{\lambda/K}(\tilde{f}_\sigma)_\nu\|_{L^p(\Upsilon_\sigma^K(B_R))}^p\Big)^{1/p}.$$

Applying the rescaling argument to both sides of this inequality yields

$$\|\mathcal{F}_\sigma^\lambda f\|_{L^p(B_R)} \lesssim \bar{C}(p,\varepsilon)(R/K)^{\alpha(p)+\varepsilon}\Big(\sum_{\substack{\nu \in \Theta_{R^{-1/2}} \\ \Gamma_\nu^R \subseteq \Gamma_\sigma^K}} \|\mathcal{F}_\sigma^\lambda f_\nu\|_{L^p(B_R)}^p\Big)^{1/p}$$

and one may sum this estimate over all $K^{-1/2}$-sectors Γ_σ^K to obtain

$$\Big(\sum_{\sigma \in \Theta_{K^{-1/2}}} \|\mathcal{F}_\sigma^\lambda f\|_{L^p(B_R)}^p\Big)^{1/p} \lesssim \bar{C}(p,\varepsilon)(R/K)^{\alpha(p)+\varepsilon}\Big(\sum_{\nu \in \Theta_{R^{-1/2}}} \|\mathcal{F}_\nu^\lambda f\|_{L^p(B_R)}^p\Big)^{1/p}. \tag{101}$$

Finally, by combining (98) with (101), it follows that

$$\|\mathcal{F}^{\lambda} f\|_{L^p(B_R)} \lesssim_{\varepsilon} \bar{C}(p,\varepsilon) K^{-\varepsilon/2} R^{\alpha(p)+\varepsilon} \Big(\sum_{\nu \in \Theta_{R^{-1/2}}} \|\mathcal{F}^{\lambda}_{\nu} f\|^p_{L^p(B_R)} \Big)^{1/p}.$$

If C_ε denotes the implicit constant appearing in the above inequality, then the induction can be closed simply by choosing K large enough so that $C_\varepsilon K^{-\varepsilon/2} \leq 1$.

Acknowledgments D.B. was supported by: the ERCEA Advanced Grant 2014 669689 – HADE, the MINECO project MTM2014-53850-P, the Basque Government project IT-641-13, the Basque Government through the BERC 2018-2021 program and by the Spanish Ministry of Science, Innovation and Universities: BCAM Severo Ochoa accreditation SEV-2017-0718. J.H. was supported by NSF Grant DMS-1440140 and EPSRC standard grant EP/R015104/1. C.D.S. was supported by NSF Grant DMS-1665373 and a Simons Fellowship.

Appendix: Historical Background on the Local Smoothing Conjecture

The Euclidean Wave Equation

The local smoothing conjecture for the Euclidean half-wave propagator $e^{it\sqrt{-\Delta}}$, that is Conjecture 20, was formulated by the third author [50] in 1991. Moreover, he showed qualitative existence of the local smoothing phenomenon for $n = 2$, showing that there is some $\varepsilon_0 > 0$ such that (24) holds for $0 < \sigma < \varepsilon_0$ if $p = 4$. Note that by interpolation with L^2 and L^∞, this also shows that there is $\varepsilon(p) > 0$ such that (24) holds for all $0 < \sigma < \varepsilon(p)$ if $2 < p < \infty$. Shortly after, Mockenhoupt, Seeger and the third author [44] obtained a quantitative estimate at the critical Lebesgue exponent $p = 4$ through a square function estimate approach. Further refinements at $p = 4$ were later obtained by Bourgain [6] and Tao and Vargas [62]. In particular, the work of Tao and Vargas established a way to transfer bilinear Fourier restriction estimates into estimates for the square function; thus, the best results via their method are obtained through the sharp bilinear restriction estimates for the cone by Wolff [68] (see also the endpoint result of Tao [61]). In higher dimensions, Mockenhoupt, Seeger and the third author [45] also established existence of local smoothing estimates, although in this case their results are concerned with estimates at $p = \frac{2(n+1)}{n-1}$ rather than at the critical Lebesgue exponent $p = \frac{2n}{n-1}$.

All the initial results discussed in the previous paragraph did not imply sharp estimates in terms of the regularity exponent σ for any $2 < p < \infty$. A striking advance was made by Wolff [67] in 2000 when he introduced the decoupling inequalities discussed in Sect. 5 and established that in the plane $1/p-$ local smoothing holds for all $p > 74$. His result was later extended to higher dimensions by Łaba and Wolff [34]. Subsequent works of Garrigós and Seeger [21] and

Garrigós, Seeger and Schlag [22] improved the Lebesgue exponent p in the sharp[14] decoupling inequalities, and therefore the Lebesgue exponent for which Conjecture 20 holds. The sharp decoupling inequalities were finally established by Bourgain and Demeter [8] in 2015, which imply $1/p-$ local smoothing estimates for all $\frac{2(n+1)}{n-1} \leq p < \infty$.

In parallel progress obtained via decoupling inequalities, Heo, Nazarov and Seeger [26] introduced in 2011 a different approach to the problem, which in particular yields local smoothing estimates at the endpoint regularity $\sigma = 1/p$ if $\frac{2(n-1)}{n-3} < p < \infty$ for $n \geq 4$.

Finally, some further progress has been obtained for $n = 2$. In 2012, S. Lee and Vargas [39] proved local smoothing estimates for all $\sigma < \bar{s}_p$ if $p = 3$ via a sharp square function estimate in $L^3(\mathbb{R}^2)$. This is the first and only time sharp local smoothing estimates (up to the regularity endpoint) have been obtained in the range $2 < p < \frac{2n}{n-1}$. More recently, J. Lee [37] has further improved the square function estimate at $p = 4$ using the $L^6(\mathbb{R}^2)$ decoupling inequalities of Bourgain–Demeter [8], showing that (24) holds for all $\sigma < 3/16$ when $p = 4$.

The precise numerology and historical progress on the Euclidean local smoothing conjecture have been outlined[15] in Fig. 7 and Table 1 for $n = 2$ and Fig. 8 and Table 2 for $n \geq 3$.

Fourier Integral Operators

Shortly after the formulation of the local smoothing conjecture for the Euclidean wave equation, Mockenhoupt, Seeger and the third author [45] considered the analogous problem for wave equations on compact manifolds and general classes of Fourier integral operators. They established positive partial results at the critical Lebesgue exponent $p = 4$ for $n = 2$, and at the subcritical exponent $p = \frac{2(n+1)}{n-1}$ for $n \geq 3$. In 1997, Minicozzi and the third author [42] provided examples of compact Riemannian manifolds (M, g) for which the operator $e^{it\sqrt{-\Delta_g}}$ does not demonstrate $1/p-$ local smoothing for $p < \bar{p}_{n,+}$ (see also [1]). This revealed a difference in the local smoothing phenomenon between the Euclidean and variable-coefficient cases.

The next positive results were obtained by Lee and Seeger in [38], where they extended the endpoint regularity results in [26] to general Fourier integral operators; as in the Euclidean case, these results only hold for $n \geq 4$. Except for the question of endpoint regularity, the best known results have recently been obtained by the authors in [1], extending to the variable coefficient case the results of Bourgain and Demeter [8]. Moreover, and as discussed in Sect. 2, the authors also showed that their results are best possible in odd dimensions in the general context of

[14]Here the word *sharp* refers to the sharp dependence of the constant in terms of the number of pieces featuring in the decoupling inequality; more precisely, the optimal dependence on λ in (89).

[15]For simplicity, the intermediate progress of Bourgain [6] and Tao and Vargas [60] at $p = 4$ has not been sketched in Fig. 7; see Table 1 for their contribution to the problem.

Conjecture 24, although one expects to go beyond these exponents in the even dimensional case and in the case of solutions arising from wave equations on compact manifolds.

The precise numerology and historical progress on Conjectures 23 and 24 have been outlined in Fig. 9 and Table 3.

Figure and Table for the Euclidean Wave Equation for $n = 2$

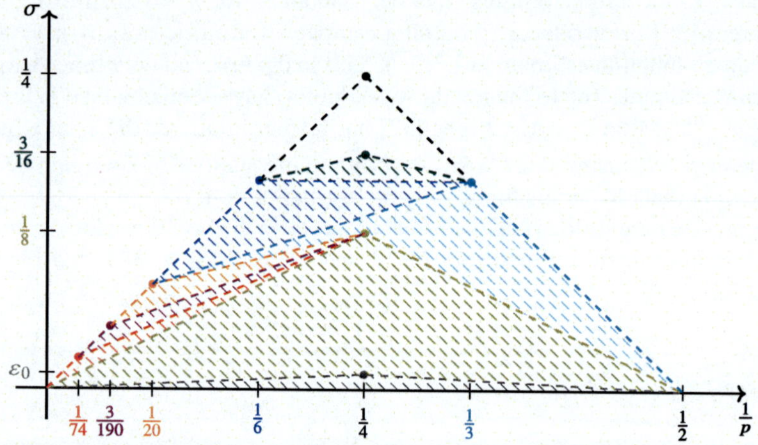

Fig. 7 Chronological progress on Conjecture 24 for $n = 2$. Each new result can be interpolated against the L^2 and L^∞ estimates and the previous results in order to yield a new region in the conjectured triangle. The current best results follow from interpolating [8, 39] and [37] and the L^2 and L^∞ estimates. The white region remains open

Table 1 Chronological progress on Conjecture 20 for $n = 2$. The notation p_0+ means that the estimate (24) holds for all $p > p_0$, whilst the notation σ_0- means that the estimate holds for all $\sigma < \sigma_0$. Otherwise, the equalities for the Lebesgue and regularity exponents are admissible. In the table $\varepsilon_0 > 0$ is a small, unspecified constant. The method of J. Lee can be applied away from the $p = 4$ exponent to give improved estimates in a slightly larger convex region than that given by interpolation; this was pointed out to us by Pavel Zorin–Kranich

	p	σ
S [54]	4	ε_0
Mockenhoupt–Seeger–S [44]	4	1/8
Bourgain [6]	4	$1/8 + \varepsilon_0$
Tao–Vargas [62] + Wolff [68]	4	$1/8 + 1/88-$
Wolff [67]	74+	$1/p-$
Garrigós–Seeger [21]	190/3+	$1/p-$
Garrigós–Seeger–Schlag [21]	20+	$1/p-$
S. Lee–Vargas [39]	3	$1/6-$
Bourgain–Demeter [8]	6	$1/6-$
J. Lee [37]	4	$3/16-$

Figure and Table for the Euclidean Wave Equation for $n \geq 3$

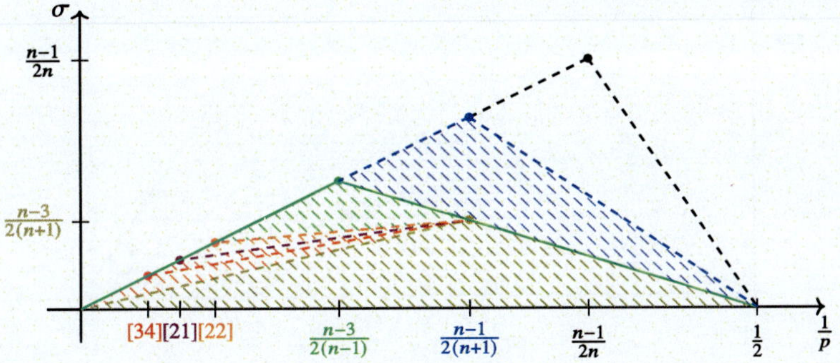

Fig. 8 Chronological progress on Conjecture 20 for high dimensions ($n \geq 5$). Each new result can be interpolated against the L^2 and L^∞ estimates and the previous results in order to yield a new region in the conjectured triangle. The best known results follow from interpolation between [8] and the L^2 and L^∞ estimates, together with the strengthened results [26] at the regularity endpoint $\sigma = 1/p$ if $p > \frac{n-3}{2(n-1)}$. The white region remains open

Table 2 Chronological progress on Conjecture 20 for $n \geq 3$. The notation + and − is used in a similar fashion to Table 1

	$n = 3$		$n = 4$		$n \geq 5$	
	p	σ	p	σ	p	σ
Mockenhoupt–Seeger–S [44]	4	$1/2p-$	10/3	$1/3p$	$\frac{2(n+1)}{n-1}$	$\frac{n-3}{n-1}\frac{1}{p}$
Łaba–Wolff [34]	18+	$1/p-$	8.4+	$1/p-$	$\frac{2(n+1)}{n-3}+$	$1/p-$
Garrigós–Seeger [21]	15+	$1/p-$	7.28+	$1/p-$	$\frac{2(n-1)(n+3)}{(n+1)(n-3)}+$	$1/p-$
Garrigós–Seeger–Schlag [22]	9+	$1/p-$	5.6+	$1/p-$	$\frac{2n(n+3)}{(n-1)(n-2)}+$	$1/p-$
Heo–Nazarov–Seeger [26]			6	$1/6$	$\frac{2(n-1)}{n-3}+$	$1/p$
Bourgain–Demeter [8]	4	$1/4-$	10/3	$3/10-$	$\frac{2(n+1)}{n-1}$	$1/p-$

Figure and Table for Fourier Integrals

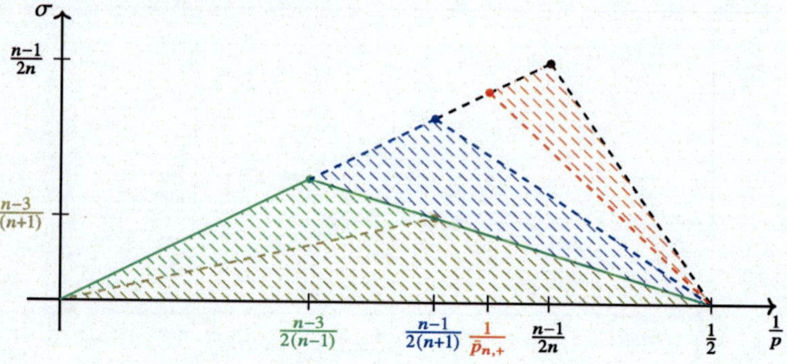

Fig. 9 Chronological progress on Conjecture 23 in high dimensions ($n \geq 4$). The red region is inadmissible. The best known results follow from interpolation between [1] and the L^2 and L^∞ estimates, together with the strengthened results [38] at the regularity endpoint $\sigma = 1/p$ if $p > \frac{n-3}{2(n-1)}$. The white region remains open. In the case of Conjecture 24 the red region extends to $p = \bar{p}_n$ and there is no white open region in odd dimensions

Table 3 Chronological progress on Conjectures 23 and 24 for $n \geq 2$. The notation + and − is used in a similar fashion to Table 1

	$n = 2$		$n = 3$		$n \geq 4$	
	p	σ	p	σ	p	σ
Mockenhoupt–Seeger–S [44]	4	$1/2p-$	4	$1/2p-$	$\frac{2(n+1)}{n-1}$	$\frac{n-3}{n-1}\frac{1}{p}$
Lee–Seeger [38]					$\frac{2(n-1)}{n-3}+$	$1/p$
B–H–S [1]	6	$1/6-$	4	$1/4-$	$\frac{2(n+1)}{n-1}$	$1/p-$

References

1. Beltran, D., Hickman, J., Sogge, C.D.: Variable coefficient Wolff-type inequalities and sharp local smoothing estimates for wave equations on manifolds. Anal. PDE **13**(2), 403–433 (2020)
2. Beltran, D., Ramos, J.P., Saari, O.: Regularity of fractional maximal functions through Fourier multipliers. J. Funct. Anal. **276**(6), 1875–1892 (2019). MR 3912794
3. Bennett, J., Carbery, A., Tao, T.: On the multilinear restriction and Kakeya conjectures. Acta Math. **196**(2), 261–302 (2006). MR 2275834
4. Bourgain, J.: Averages in the plane over convex curves and maximal operators. J. Anal. Math. **47**, 69–85 (1986). MR 874045
5. Bourgain, J.: L^p-estimates for oscillatory integrals in several variables. Geom. Funct. Anal. **1**(4), 321–374 (1991). MR 1132294
6. Bourgain, J.: *Estimates for Cone Multipliers*. Geometric Aspects of Functional Analysis (Israel, 1992–1994). Operator Theory: Advances and Applications, vol. 77. Birkhäuser, Basel (1995), pp. 41–60. MR 1353448

7. Bourgain, J.: *Some New Estimates on Oscillatory Integrals*. Essays on Fourier Analysis in Honor of Elias M. Stein (Princeton, NJ, 1991). Princeton Math. Ser., vol. 42, Princeton University Press, Princeton (1995), pp. 83–112. MR 1315543
8. Bourgain, J., Demeter, C.: The proof of the l^2 decoupling conjecture. Ann. Math. (2) **182**(1), 351–389 (2015). MR 3374964
9. Bourgain, J., Demeter, C.: Decouplings for curves and hypersurfaces with nonzero Gaussian curvature. J. Anal. Math. **133**, 279–311 (2017). MR 3736493
10. Bourgain, J., Demeter, C.: A study guide for the l^2 decoupling theorem. Chin. Ann. Math. Ser. B **38**(1), 173–200 (2017). MR 3592159
11. Bourgain, J., Guth, L.: Bounds on oscillatory integral operators based on multilinear estimates. Geom. Funct. Anal. **21**(6), 1239–1295 (2011). MR 2860188
12. Carleson, L., Sjölin, P.: Oscillatory integrals and a multiplier problem for the disc. Studia Math. **44**, 287–299 (1972) (errata insert). Collection of articles honoring the completion by Antoni Zygmund of 50 years of scientific activity, III. MR 0361607
13. Chavel, I.: *Riemannian Geometry*, 2nd edn., Cambridge Studies in Advanced Mathematics, vol. 98. Cambridge University Press, Cambridge (2006). A modern introduction. MR 2229062
14. Constantin, P., Saut, J.-C.: Local smoothing properties of dispersive equations. J. Amer. Math. Soc. **1**(2), 413–439 (1988). MR 928265
15. Duistermaat, J.J.: *Fourier Integral Operators*. Modern Birkhäuser Classics. Birkhäuser, Springer, New York (2011). Reprint of the 1996 edition [MR1362544], based on the original lecture notes published in 1973 [MR0451313]. MR 2741911
16. Duistermaat, J.J., Hörmander, L.: Fourier integral operators. II. Acta Math. **128**(3-4), 183–269 (1972). MR 0388464
17. Egorov, J.V.: The canonical transformations of pseudodifferential operators. Uspehi Mat. Nauk **24**(5)(149), 235–236 (1969). MR 0265748
18. Èskin, G.I.: The Cauchy problem for hyperbolic convolution equations. Mat. Sb. (N.S.) **74**(116), 262–297 (1967). MR 0509918
19. Èskin, G.I.: Degenerate elliptic pseudodifferential equations of principal type. Mat. Sb. (N.S.) **82**(124), 585–628 (1970). MR 0510219
20. Fefferman, C.: The multiplier problem for the ball. Ann. Math. (2) **94**, 330–336 (1971). MR 0296602 (45 #5661)
21. Garrigós, G., Seeger, A.: On plate decompositions of cone multipliers. Proc. Edinb. Math. Soc. (2) **52**(3), 631–651 (2009). MR 2546636
22. Garrigós, G., Seeger, A.: *A Mixed Norm Variant of Wolff's Inequality for Paraboloids*. Harmonic analysis and partial differential equations. Contemporary Mathematics, vol. 505. American Mathematical Society, Providence (2010), pp. 179–197. MR 2664568
23. Guo, S., Hickman, J., Lie, V., Roos, J.: Maximal operators and Hilbert transforms along variable non-flat homogeneous curves. Proc. Lond. Math. Soc. (3) **115**(1), 177–219 (2017). MR 3669936
24. Guo, S., Roos, J., Yung, P.-L.: Sharp variation-norm estimates for oscillatory integrals related to Carleson's theorem. Anal. PDE **13**(5), 1457–1500 (2020)
25. Guth, L., Hickman, J., Iliopoulou, M.: Sharp estimates for oscillatory integral operators via polynomial partitioning. To appear. Acta Math. arxiv.org/abs/1710.10349
26. Heo, Y., Nazarov, F., Seeger, A.: Radial Fourier multipliers in high dimensions. Acta Math. **206**(1), 55–92 (2011). MR 2784663
27. Hickman, J., Iliopoulou, M.: Sharp L^p estimates for oscillatory integral operators of arbitrary signature. To appear. arxiv.org/abs/2006.01316
28. Hörmander, L.: The spectral function of an elliptic operator. Acta Math. **121**, 193–218 (1968). MR 0609014
29. Hörmander, L.: Fourier integral operators. I. Acta Math. **127**, 79–183 (1971)
30. Hörmander, L.: Oscillatory integrals and multipliers on FL^p. Ark. Mat. **11**, 1–11 (1973). MR 0340924

31. Hörmander, L.: *The Analysis of Linear Partial Differential Operators. I.* Classics in Mathematics. Springer, Berlin (2003). Distribution theory and Fourier analysis, Reprint of the second (1990) edition [Springer, Berlin; MR1065993 (91m:35001a)]. MR 1996773
32. Hörmander, L.: *The Analysis of Linear Partial Differential Operators. IV.* Classics in Mathematics. Springer, Berlin (2009). Fourier integral operators, Reprint of the 1994 edition. MR 2512677
33. Kato, T.: *On the Cauchy Problem for the (Generalized) Korteweg-de Vries Equation.* Studies in Applied Mathematics. Adv. Math. Suppl. Stud., vol. 8. Academic Press, New York (1983), pp. 93–128. MR 759907
34. Łaba, I., Wolff, T.: A local smoothing estimate in higher dimensions. J. Anal. Math. **88**, 149–171 (2002). Dedicated to the memory of Tom Wolff. MR 1956533
35. Lax, P.D.: Asymptotic solutions of oscillatory initial value problems. Duke Math. J. **24**, 627–646 (1957). MR 0097628
36. Lee, S.: Linear and bilinear estimates for oscillatory integral operators related to restriction to hypersurfaces, J. Funct. Anal. **241**(1), 56–98 (2006). MR 2264247
37. Lee, J.: Trilinear approach to square function and local smoothing estimates for the wave operator (2016). arxiv.org/abs/1607.08426
38. Lee, S., Seeger, A.: Lebesgue space estimates for a class of Fourier integral operators associated with wave propagation, Math. Nachr. **286**(7), 743–755 (2013). MR 3060843
39. Lee, S., Vargas, A.: On the cone multiplier in $\mathbb{R}^3$. J. Funct. Anal. **263**(4), 925–940 (2012). MR 2927399
40. Littman, W.: *$L^p - L^q$-Estimates for Singular Integral Operators Arising from Hyperbolic Equations.* Partial Differential Equations (Proc. Sympos. Pure Math., Vol. XXIII, Univ. California, Berkeley, Calif., 1971), American Mathematical Society, Providence, (1973), pp. 479–481. MR 0358443
41. Maslov, V.P.: *Theory of Perturbations and Asymptotic Methods.* Moskov. Gos. Univ., Moscow (1965). Russian
42. Minicozzi, W.P., II, Sogge, C.D.: Negative results for Nikodym maximal functions and related oscillatory integrals in curved space. Math. Res. Lett. **4**(2–3), 221–237 (1997). MR 1453056
43. Miyachi, A.: On some estimates for the wave equation in L^p and H^p. J. Fac. Sci. Univ. Tokyo Sect. IA Math. **27**(2), 331–354 (1980). MR 586454
44. Mockenhaupt, G., Seeger, A., Sogge, C.D.: Wave front sets, local smoothing and Bourgain's circular maximal theorem. Ann. Math. (2) **136**(1), 207–218 (1992). MR 1173929
45. Mockenhaupt, G., Seeger, A., Sogge, C.D.: *Local Smoothing of Fourier Integral Operators and Carleson-Sjölin Estimates*. J. Amer. Math. Soc. **6**(1), 65–130 (1993). MR 1168960
46. Peral, J.C.: L^p estimates for the wave equation. J. Funct. Anal. **36**(1), 114–145 (1980). MR 568979
47. Phong, D.H., Stein, E.M.: Hilbert integrals, singular integrals, and Radon transforms. I. Acta Math. **157**, 99–157 (1986)
48. Seeger, A., Sogge, C.D., Stein, E.M.: Regularity properties of Fourier integral operators. Ann. Math. (2) **134**(2), 231–251 (1991). MR 1127475
49. Sjölin, P.: Regularity of solutions to the Schrödinger equation. Duke Math. J. **55**(3), 699–715 (1987). MR 904948
50. Sogge, C.D.: Propagation of singularities and maximal functions in the plane. Invent. Math. **104**(2), 349–376 (1991). MR 1098614
51. Sogge, C.D.: *Lectures on Non-linear Wave Equations*, 2nd ed. International Press, Boston (2008). MR 2455195
52. Sogge, C.D.: *Hangzhou Lectures on Eigenfunctions of the Laplacian.* Annals of Mathematics Studies, vol. 188, Princeton University Press, Princeton (2014). MR 3186367
53. Sogge, C.D.: *Fourier Integrals in Classical Analysis*, 2nd edn., Cambridge Tracts in Mathematics. Cambridge University Press, Cambridge (2017)
54. Sogge, C.D., Xi, Y., Xu, H.: On instability of the Nikodym maximal function bounds over Riemannian manifolds. J. Geom. Anal. **28**(3), 2886–2901 (2018)

55. Stein, E.M.: *Singular Integrals and Differentiability Properties of Functions*. Princeton Mathematical Series, No. 30, Princeton University Press, Princeton (1970). MR 0290095
56. Stein, E.M.: Maximal functions. I. Spherical means. Proc. Nat. Acad. Sci. U.S.A. **73**(7), 2174–2175 (1976). MR 0420116
57. Stein, E.M.: *Oscillatory Integrals in Fourier Analysis*. Beijing Lectures in Harmonic Analysis (Beijing, 1984). Annals of Mathematics Studies, vol. 112. Princeton University Press, Princeton (1986), pp. 307–355. MR 864375
58. Stein, E.M.: *Harmonic Analysis: Real-Variable Methods, Orthogonality, and Oscillatory Integrals*. Princeton Mathematical Series, vol. 43, Princeton University Press, Princeton (1993), With the assistance of Timothy S. Murphy, Monographs in Harmonic Analysis, III. MR 1232192
59. Stein, E.M., Shakarchi, R.: *Functional Analysis*. Princeton Lectures in Analysis, vol. 4. Princeton University Press, Princeton (2011), Introduction to further topics in analysis. MR 2827930
60. Tao, T.: The Bochner-Riesz conjecture implies the restriction conjecture. Duke Math. J. **96**(2), 363–375 (1999). MR 1666558
61. Tao, T.: Endpoint bilinear restriction theorems for the cone, and some sharp null form estimates. Math. Z. **238**(2), 215–268 (2001). MR 1865417
62. Tao, T., Vargas, A.: A bilinear approach to cone multipliers. II. Applications. Geom. Funct. Anal. **10**(1), 216–258 (2000). MR 1748921
63. Trèves, F.: *Introduction to Pseudodifferential and Fourier Integral Operators*, vol. 2. Plenum Press, New York (1980). Fourier integral operators, The University Series in Mathematics. MR 597145
64. Vega, L.: Schrödinger equations: pointwise convergence to the initial data. Proc. Amer. Math. Soc. **102**(4), 874–878 (1988). MR 934859
65. Wisewell, L.: Kakeya sets of curves. Geom. Funct. Anal. **15**(6), 1319–1362 (2005). MR 2221250
66. Wolff, T.: *Recent Work Connected with the Kakeya Problem*. Prospects in Mathematics (Princeton, NJ, 1996). American Mathematical Society, Providence (1999), pp. 129–162. MR 1660476
67. Wolff, T.: Local smoothing type estimates on L^p for large p. Geom. Funct. Anal. **10**(5), 1237–1288 (2000). MR 1800068
68. Wolff, T.: A sharp bilinear cone restriction estimate. Ann. Math. (2) **153**(3), 661–698 (2001). MR 1836285
69. Zelditch, S.: *Eigenfunctions of the Laplacian on a Riemannian Manifold*, CBMS Regional Conference Series in Mathematics, vol. 125. Published for the Conference Board of the Mathematical Sciences, Washington. American Mathematical Society, Providence (2017). MR 3729409

On the Hardy–Littlewood Maximal Functions in High Dimensions: Continuous and Discrete Perspective

Jean Bourgain, Mariusz Mirek, Elias M. Stein, and Błażej Wróbel

Dedicated to Fulvio Ricci on the occasion of his 70th birthday.

Abstract This is a survey article about recent developments in dimension-free estimates for maximal functions corresponding to the Hardy–Littlewood averaging operators associated with convex symmetric bodies in $\mathbb{R}^d$ and $\mathbb{Z}^d$.

Keywords Hardy–Littlewood maximal functions · Fourier multipliers · Oscillatory integrals · Convex bodies · Dimension-free estimates

Jean Bourgain was supported by NSF grant DMS-1800640. Mariusz Mirek was partially supported by the Schmidt Fellowship and the IAS School of Mathematics and by the National Science Center, Poland grant DEC-2015/19/B/ST1/01149. Elias M. Stein was partially supported by NSF grant DMS-1265524. Błażej Wróbel was partially supported by the National Science Centre, Poland grant Opus 2018/31/B/ST1/00204.

J. Bourgain
School of Mathematics, Institute for Advanced Study, Princeton, NJ, USA

M. Mirek (✉)
Department of Mathematics, Rutgers University, Piscataway, NJ, USA

Instytut Matematyczny, Uniwersytet Wrocławski, Wrocław, Poland
e-mail: mariusz.mirek@rutgers.edu

E. M. Stein
Department of Mathematics, Princeton University, Princeton, NJ, USA

B. Wróbel
Instytut Matematyczny, Uniwersytet Wrocławski, Wrocław, Poland
e-mail: blazej.wrobel@math.uni.wroc.pl

P. Ciatti, A. Martini (eds.), *Geometric Aspects of Harmonic Analysis*, Springer INdAM Series 45, https://doi.org/10.1007/978-3-030-72058-2_3

1 Introduction

1.1 Statement of the Results: Continuous Perspective

Let G be a convex symmetric body in $\mathbb{R}^d$, which is simply a bounded closed and symmetric convex subset of $\mathbb{R}^d$ with non-empty interior. In the literature it is usually assumed that a symmetric convex body $G \subset \mathbb{R}^d$ is open. In fact, in $\mathbb{R}^d$ there is no difference whether we assume G is closed or open, since the boundary of a convex set has Lebesgue measure zero. However, in the discrete case, if $G \cap \mathbb{Z}^d$ is considered, it matters. Therefore, later on in order to avoid some technicalities, we will assume that a symmetric convex body $G \subset \mathbb{R}^d$ is always closed.

For every $t > 0$ and for every $x \in \mathbb{R}^d$ we define the Hardy–Littlewood averaging operator

$$M_t^G f(x) = \frac{1}{|G_t|} \int_{G_t} f(x-y)\mathrm{d}y \quad \text{for} \quad f \in L^1_{\mathrm{loc}}(\mathbb{R}^d), \tag{1}$$

where $G_t = \{y \in \mathbb{R}^d : t^{-1}y \in G\}$ denotes a dilate of the body $G \subset \mathbb{R}^d$.

For $p \in (1, \infty]$, let $C_p(d, G) > 0$ be the best constant such that the following maximal inequality

$$\| \sup_{t>0} |M_t^G f| \|_{L^p(\mathbb{R}^d)} \le C_p(d, G) \|f\|_{L^p(\mathbb{R}^d)} \tag{2}$$

holds for every $f \in L^p(\mathbb{R}^d)$.

The question we shall be concerned with, in this survey, is to decide whether the constant $C_p(d, G)$ can be estimated independently of the dimension $d \in \mathbb{N}$ for every $p \in (1, \infty]$.

If $p = \infty$, then (2) holds with $C_p(d, G) = 1$, since M_t^G is an averaging operator. By appealing to a covering argument for $p = 1$, and a simple interpolation with $p = \infty$, we can conclude that $C_p(d, G) < \infty$ for every $p \in (1, \infty)$ and for every convex symmetric body $G \subset \mathbb{R}^d$. However, then the implied upper bound for $C_p(d, G)$ depends on the dimension, since the interpolation with a weak type $(1, 1)$ estimate does not give anything reasonable in these kind of questions, and generally it is better to work with $p \in (1, \infty)$ to obtain any non-trivial result concerning the behavior of $C_p(d, G)$ as $d \to \infty$.

The problem about estimates of $C_p(d, G)$, as $d \to \infty$, has been extensively studied by several authors for nearly four decades. The starting point was the work of the third author [33], where, in the case of the Euclidean balls $G = B^2$, it was shown that $C_p(d, B^2)$ is bounded independently of the dimension for every $p \in (1, \infty]$. Not long afterwards it was proved by the first author, in [4] for $p = 2$, that $C_p(d, G)$ is bounded by an absolute constant, which is independent of the underlying convex symmetric body $G \subset \mathbb{R}^d$. This result was extended in [5], and independently by Carbery [13], for all $p \in (3/2, \infty]$.

It is conjectured that the inequality in (2) holds for all $p \in (1, \infty]$ and for all convex symmetric bodies $G \subset \mathbb{R}^d$ with $C_p(d, G)$ independent of $d \in \mathbb{N}$. It is reasonable to believe that this is true, since it was verified for a large class of convex symmetric bodies.

For $q \in [1, \infty]$, let B^q be a q-ball in $\mathbb{R}^d$ defined by

$$B^q = \Big\{x \in \mathbb{R}^d : |x|_q = \Big(\sum_{1\le k\le d} |x_k|^q\Big)^{1/q} \le 1\Big\} \quad \text{for} \quad q \in [1, \infty),$$
$$B^\infty = \{x \in \mathbb{R}^d : |x|_\infty = \max_{1\le k\le d} |x_k| \le 1\}. \tag{3}$$

For the q-balls $G = B^q$ the full range $p \in (1, \infty]$ of dimension-free estimates for $C_p(d, B^q)$ was established by Müller in [26] (for $q \in [1, \infty)$) and in [8] (for cubes $q = \infty$) with constants depending only on q. More about the current state of the art and papers [4, 5, 8, 26, 33] will be given in Sect. 2.

The general case is beyond our reach at this moment. However, the approach undertaken in the present article permits us to provide a new simple proof of dimension-free estimates for the Hardy–Littlewood maximal functions associated with symmetric convex bodies $G \subset \mathbb{R}^d$, which independently were the subject of [5] and [13]. We prove the following theorem.

Theorem 1 *Let $p \in (3/2, \infty]$, then there exists a constant $C_p > 0$ independent of dimension $d \in \mathbb{N}$ and a symmetric convex body $G \subset \mathbb{R}^d$ such that the constant $C_p(d, G)$ defined in (2) satisfies*

$$C_p(d, G) \le C_p. \tag{4}$$

Moreover, a dyadic variant of (4) remains true for all $p \in (1, \infty]$. More precisely, for every $p \in (1, \infty]$ there exists a constant $C_p > 0$ independent of dimension $d \in \mathbb{N}$ and a symmetric convex body $G \subset \mathbb{R}^d$ such that for every $f \in L^p(\mathbb{R}^d)$ we have

$$\big\| \sup_{n\in\mathbb{Z}} |M^G_{2^n} f| \big\|_{L^p(\mathbb{R}^d)} \le C_p \|f\|_{L^p(\mathbb{R}^d)}. \tag{5}$$

The proof of Theorem 1 will be presented in Sect. 4 using a new flexible approach, which recently resulted in dimension-free bounds in r-variational and jump inequalities corresponding to the operators M^G_t from (1), see [9] and [25]. An important feature of this method is that it is also applicable to the discrete settings, see [10] and [25]. The method is described in Sect. 3, the proof of Theorem 1 is given in Sect. 4. Our aim is to continue the investigations in the discrete settings as well. Similar types of questions were recently studied by the authors [10] for the discrete analogues of the operators M^G_t in $\mathbb{Z}^d$.

1.2 Statement of the Results: Discrete Perspective

For every $t > 0$ and for every $x \in \mathbb{Z}^d$ we define the discrete Hardy–Littlewood averaging operator

$$\mathcal{M}_t^G f(x) = \frac{1}{|G_t \cap \mathbb{Z}^d|} \sum_{y \in G_t \cap \mathbb{Z}^d} f(x-y) \quad \text{for} \quad f \in \ell^1(\mathbb{Z}^d). \tag{6}$$

We note that the operator $\mathcal{M}_t^G$ is a discrete analogue of M_t^G from (1).

For $p \in (1, \infty]$, let $C_p(d, G) > 0$ be the best constant such that the following maximal inequality

$$\|\sup_{t>0} |\mathcal{M}_t^G f|\|_{\ell^p(\mathbb{Z}^d)} \le C_p(d, G)\|f\|_{\ell^p(\mathbb{Z}^d)} \tag{7}$$

holds for every $f \in \ell^p(\mathbb{Z}^d)$.

Arguing in a similar way as in (2) we conclude that $C_p(d, G) < \infty$ for every $p \in (1, \infty]$ and for every convex symmetric body $G \subset \mathbb{R}^d$. The question now is to decide whether $C_p(d, G)$ can be bounded independently of the dimension d for every $p \in (1, \infty)$.

In [10] the authors examined this question in the case of the discrete cubes $B^\infty \cap \mathbb{Z}^d$, and showed that for every $p \in (3/2, \infty]$ there is a constant $C_p > 0$ independent of the dimension such that $C_p(d, B^\infty) \le C_p$. It was also shown in [10] that if the supremum in (7) is restricted to the dyadic set $\mathbb{D} = \{2^n : n \in \mathbb{N} \cup \{0\}\}$, then (7) holds for all $p \in (1, \infty]$ and $C_p(d, G)$ is independent of the dimension.

The general case in much more complicated. However, it is not difficult to show [10] that for every symmetric convex body $G \subset \mathbb{R}^d$ there exists $t_G > 0$ with the property that the norm of the discrete maximal function $\sup_{t>t_G} |\mathcal{M}_t^G f|$ is controlled by a constant multiple of the norm of its continuous counterpart, and the implied constant is independent of the dimension. This is a simple comparison argument yielding dimension-free estimates for $\sup_{t>t_G} |\mathcal{M}_t^G f|$ as long as the corresponding dimension-free bounds are available for their continuous analogues. As a corollary, for q-balls $G = B^q$, if $p \in (1, \infty]$ and $q \in [1, \infty]$, we obtain that there is a constant $C_{p,q} > 0$ independent of the dimension $d \in \mathbb{N}$ such that for all $f \in \ell^p(\mathbb{Z}^d)$ we have

$$\|\sup_{t \ge d^{1+1/q}} |\mathcal{M}_t^{B^q} f|\|_{\ell^p(\mathbb{Z}^d)} \le C_{p,q}\|f\|_{\ell^p(\mathbb{Z}^d)}. \tag{8}$$

At this stage, the whole difficulty lies in estimating $\sup_{0<t\le t_G} |\mathcal{M}_t^G f|$, where the things are getting more complicated. Nevertheless, as we shall see below, in some cases improvements are possible.

We show that in the case of $\mathcal{M}_t^{B^2}$, which together with $\mathcal{M}_t^{B^\infty}$, is presumably the most natural setting for the discrete Hardy–Littlewood maximal functions, the range in (8) can be improved. Namely, the main discrete result of this paper is, an extension of (8) for $G = B^2$, stated below.

Theorem 2 *For each $p \in (1, \infty]$ there is a constant $C_p > 0$ independent of the dimension $d \in \mathbb{N}$ such that for every $f \in \ell^p(\mathbb{Z}^d)$ we have*

$$\big\| \sup_{t \geq Cd} |\mathcal{M}_t^{B^2} f| \big\|_{\ell^p(\mathbb{Z}^d)} \leq C_p \|f\|_{\ell^p(\mathbb{Z}^d)}, \tag{9}$$

for an appropriate absolute constant $C > 0$.

The proof of Theorem 2 is based on a delicate refinement of the arguments from [10], which in the end reduce the matters to the comparison of the norm of $\sup_{t \geq Cd} |\mathcal{M}_t^{B^2} f|$ with the norm of its continuous analogue, and consequently to the dimension-free estimates of $C_p(d, B^2)$ for all $p \in (1, \infty]$, that are guaranteed by Stein [33]. The proof of Theorem 2 is contained in Sect. 5.

Surprisingly, as it was shown in [10], the dimension-free estimates in the discrete case are not as broad as in the continuous setup and there is no obvious conjecture to prove. This is due to the fact that there exists a simple example of a convex symmetric body in $\mathbb{Z}^d$ for which maximal estimate (7) on $\ell^p(\mathbb{Z}^d)$, for every $p \in (1, \infty)$, involves the smallest constant $C_p(d, G) > 0$ unbounded in $d \in \mathbb{N}$. In order to carry out the construction it suffices to fix a sequence $1 \leq \lambda_1 < \ldots < \lambda_d < \ldots < \sqrt{2}$ and consider, as in [10], the ellipsoid

$$E(d) = \Big\{ x \in \mathbb{R}^d : \sum_{k=1}^{d} \lambda_k^2 x_k^2 \leq 1 \Big\}.$$

Then one can prove that for every $p \in (1, \infty)$ there is $C_p > 0$ such that for every $d \in \mathbb{N}$ one has

$$C_p(d, E(d)) \geq C_p (\log d)^{1/p}. \tag{10}$$

Inequality (10) shows that the dimension-free phenomenon for the Hardy–Littlewood maximal functions in the discrete setting is much more delicate, and the dimension-free estimates even in the Euclidean case for $C_p(d, B^2)$ may be very difficult. However, there is an evidence, gained recently by the authors in [11], in favor of the general problem, which makes the things not entirely hopeless. Namely, in [11] a dyadic variant of inequality (7) for $G = B^2$ was studied and we proved the following result.

Theorem 3 *For every $p \in [2, \infty]$ there exists a constant $C_p > 0$ independent of $d \in \mathbb{N}$ such that for every $f \in \ell^p(\mathbb{Z}^d)$ we have*

$$\big\| \sup_{t \in \mathbb{D}} |\mathcal{M}_t^{B^2} f| \big\|_{\ell^p(\mathbb{Z}^d)} \leq C_p \|f\|_{\ell^p(\mathbb{Z}^d)}. \tag{11}$$

All the aforementioned results give us strong motivation to understand the situation more generally. In particular, in the case of q-balls $G = B^q$ where $q \in [1, \infty)$, which is well understood in the continuous setup. More about the methods available in the discrete setting is in Sect. 3.

1.3 Notation

Here we fix some further notation and terminology.

1. Throughout the whole paper $d \in \mathbb{N}$ denotes the dimension and $C > 0$ denotes a universal constant, which does not depend on the dimension, but it may vary from occurrence to occurrence.
2. We write that $A \lesssim_\delta B$ ($A \gtrsim_\delta B$) to say that there is an absolute constant $C_\delta > 0$ (which possibly depends on $\delta > 0$) such that $A \le C_\delta B$ ($A \ge C_\delta B$), and we write $A \simeq_\delta B$ when $A \lesssim_\delta B$ and $A \gtrsim_\delta B$ hold simultaneously.
3. Let $\mathbb{N} = \{1, 2, \ldots\}$ be the set of positive integers let $\mathbb{N}_0 = \mathbb{N} \cup \{0\}$, and let $\mathbb{D} = \{2^n : n \in \mathbb{N}_0\}$ denote the set of all dyadic numbers. We set $\mathbb{N}_N = \{1, 2, \ldots, N\}$ for any $N \in \mathbb{N}$.
4. The Euclidean space $\mathbb{R}^d$ is endowed with the standard inner product
$$x \cdot \xi = \langle x, \xi \rangle = \sum_{k=1}^{d} x_k \xi_k$$
for every $x = (x_1, \ldots, x_d)$ and $\xi = (\xi_1, \ldots, \xi_d) \in \mathbb{R}^d$.
5. For a countable set $\mathcal{Z}$ (usually $\mathcal{Z} = \mathbb{Z}^d$) endowed with the counting measure we shall write that
$$\ell^p(\mathcal{Z}) = \{f : \mathcal{Z} \to \mathbb{C} : \|f\|_{\ell^p(\mathcal{Z})} < \infty\} \quad \text{for any} \quad p \in [1, \infty],$$
where for any $p \in [1, \infty)$ we have
$$\|f\|_{\ell^p(\mathcal{Z})} = \Big(\sum_{m \in \mathcal{Z}} |f(m)|^p\Big)^{1/p} \qquad \text{and} \qquad \|f\|_{\ell^\infty(\mathcal{Z})} = \sup_{m \in \mathcal{Z}} |f(m)|.$$
6. We shall abbreviate $\|\cdot\|_{L^p(\mathbb{R}^d)}$ to $\|\cdot\|_{L^p}$, and $\|\cdot\|_{\ell^p(\mathbb{Z}^d)}$ to $\|\cdot\|_{\ell^p}$.
7. Let $(X, \mathcal{B}, \mu)$ be a σ-finite measure space. Let $p \in [1, \infty]$ and suppose that $(T_t)_{t \in \mathbb{I}}$ is a family of linear operators such that T_t maps $L^p(X)$ to itself for every $t \in \mathbb{I} \subseteq (0, \infty)$. Then the corresponding maximal function will be denoted by
$$T_{*,\mathbb{I}} f = \sup_{t \in \mathbb{I}} |T_t f|, \quad \text{for every} \quad f \in L^p(X).$$
We shall abbreviate $T_{*,\mathbb{I}}$ to T_*, if $\mathbb{I} = (0, \infty)$.

8. Let $(B_1, \|\cdot\|_{B_1})$ and $(B_2, \|\cdot\|_{B_2})$ be Banach spaces. For a linear or sub-linear operator $T : B_1 \to B_2$ its norm is defined by

$$\|T\|_{B_1 \to B_2} = \sup_{\|f\|_{B_1} \le 1} \|T(f)\|_{B_2}.$$

9. Let $\mathcal{F}$ denote the Fourier transform on $\mathbb{R}^d$ defined for any function $f \in L^1(\mathbb{R}^d)$ as

$$\mathcal{F}f(\xi) = \int_{\mathbb{R}^d} f(x) e^{2\pi i \xi \cdot x}\, dx \quad \text{for any} \quad \xi \in \mathbb{R}^d.$$

If $f \in \ell^1(\mathbb{Z}^d)$ we define the discrete Fourier transform by setting

$$\hat{f}(\xi) = \sum_{x \in \mathbb{Z}^d} f(x) e^{2\pi i \xi \cdot x} \quad \text{for any} \quad \xi \in \mathbb{T}^d,$$

where $\mathbb{T}^d \equiv [-1/2, 1/2)^d$ is the d-dimensional torus. We shall denote by $\mathcal{F}^{-1}$ the inverse Fourier transform on $\mathbb{R}^d$ or the inverse Fourier transform (Fourier coefficient) on the torus $\mathbb{T}^d$. This will cause no confusions and the meaning will be always clear from the context.

2 A Review of the Current State of the Art

In the 1980s dimension-free estimates for the Hardy–Littlewood maximal functions over convex symmetric bodies had begun to be studied [33, 35] and went through a period of considerable changes and developments [4–6, 13, 26]. However, the dimension-free phenomenon in harmonic analysis had been apparent much earlier, see for instance [36, Chapter 14, §3 in Vol.II], as well as [31] and the references given there. We refer also to more recent results [1, 2, 8–11, 18, 25, 30] and the survey article [16] for a very careful and detailed exposition of the subject.

2.1 Dimension-Free Estimates for Semigroups

Consider the Poisson semigroup $(P_t)_{t \ge 0}$ defined on the Fourier transform side by

$$\mathcal{F}(P_t f)(\xi) = p_t(\xi)\mathcal{F}(f)(\xi),$$

for every $t \ge 0$ and $\xi \in \mathbb{R}^d$, with the symbol

$$p_t(\xi) = e^{-2\pi t L|\xi|},$$

involving an isotropic constant $L = L(G) > 0$ defined in (22). The dilation by the isotropic constant is a technical assumption, which will simplify our further discussion.

For every $x \in \mathbb{R}^d$ we introduce the maximal function

$$P_* f(x) = \sup_{t>0} |P_t f(x)|,$$

and the square function

$$g(f)(x) = \left(\int_0^\infty t \left| \frac{\mathrm{d}}{\mathrm{d}t} P_t f(x) \right|^2 \mathrm{d}t \right)^{1/2},$$

associated with the Poisson semigroup. From [31] we know that for every $p \in (1, \infty)$ there exists a constant $C_p > 0$, which does not depend on $d \in \mathbb{N}$, such that for every $f \in L^p(\mathbb{R}^d)$ we have

$$\|P_* f\|_{L^p} \le C_p \|f\|_{L^p}, \tag{12}$$

and

$$\|g(f)\|_{L^p} \le C_p \|f\|_{L^p}. \tag{13}$$

For the proof of (12) and (13) one has to check that $(P_t)_{t \ge 0}$ is a symmetric diffusion semigroup in the sense of [31, Chapter III]. For the convenience of the reader we recall the definition of a symmetric diffusion semigroup from [31, Chapter III, p.65]. Let $(X, \mathcal{B}(X), \mu)$ be a σ-finite measure space, and $(T_t)_{t \ge 0}$ be a strongly continuous semigroup on $L^2(X)$, which maps $L^1(X) + L^\infty(X)$ to itself for every $t \ge 0$. We say that $(T_t)_{t \ge 0}$ is a symmetric diffusion semigroup, if it satisfies for all $t \ge 0$ the following conditions:

1. *Contraction property:* for all $p \in [1, \infty]$ and $f \in L^p(X)$ we have $\|T_t f\|_{L^p(X)} \le \|f\|_{L^p(X)}$.
2. *Symmetry property:* each T_t is a self-adjoint operator on $L^2(X)$.
3. *Positivity property:* $T_t f \ge 0$, if $f \ge 0$.
4. *Conservation property:* $T_t 1 = 1$.

One major advantage of using the above-mentioned conditions is that the probabilistic techniques are applicable to understand properties of T_t. This is the reason why, in particular, inequalities (12) and (13) hold, see [31, Chapter III] for more details, and also [15] for an even more relaxed conditions. The semigroup P_t is closely linked to the averaging operator M_t^G. Namely, both operators are contractive on $L^p(\mathbb{R}^d)$ spaces for all $p \in [1, \infty]$, preserve the class of nonnegative functions, and satisfy $P_t 1 = M_t^G 1 = 1$.

Later on, we shall need a variant of the Littlewood–Paley inequality. For every $n \in \mathbb{Z}$ we define the Poisson projections S_n by setting

$$S_n = P_{2^{n-1}} - P_{2^n}.$$

Then, the sequence $(S_n)_{n\in\mathbb{Z}}$ is a resolution of the identity on $L^2(\mathbb{R}^d)$. Namely, we have

$$f = \sum_{n\in\mathbb{Z}} S_n f, \quad \text{for every} \quad f \in L^2(\mathbb{R}^d). \tag{14}$$

Observe that

$$S_n f(x) = -\int_{2^{n-1}}^{2^n} \frac{\mathrm{d}}{\mathrm{d}t} P_t f(x) \mathrm{d}t.$$

Then by the Cauchy–Schwarz inequality we obtain, for every $n \in \mathbb{Z}$ and $x \in \mathbb{R}^d$, the following bound

$$\begin{aligned}|S_n f(x)|^2 &\le \left(\int_{2^{n-1}}^{2^n} \left|\frac{\mathrm{d}}{\mathrm{d}t} P_t f(x)\right| \mathrm{d}t\right)^2\\ &\le 2^{n-1}\int_{2^{n-1}}^{2^n} \left|\frac{\mathrm{d}}{\mathrm{d}t} P_t f(x)\right|^2 \mathrm{d}t \le \int_{2^{n-1}}^{2^n} t\left|\frac{\mathrm{d}}{\mathrm{d}t} P_t f(x)\right|^2 \mathrm{d}t.\end{aligned}$$

Now summing over $n \in \mathbb{Z}$ and using (13) one shows that for every $p \in (1, \infty)$, there is a constant $C_p > 0$ independent of $d \in \mathbb{N}$ such that for every $f \in L^p(\mathbb{R}^d)$ the following Littlewood–Paley inequality holds

$$\Big\|\Big(\sum_{n\in\mathbb{Z}} |S_n f|^2\Big)^{1/2}\Big\|_{L^p} \le C_p \|f\|_{L^p}. \tag{15}$$

Inequality (15) will play an important role in the proof of Theorem 1.

We finish this subsection by showing a simple pointwise inequality between the Poisson semigroup and the Hardy–Littlewood maximal function associated with the Euclidean balls, which motivates, to some extent, the study of dimension-free estimates for the Hardy–Littlewood maximal functions. Namely, let K_t be the kernel corresponding to P_t, assume that $f \ge 0$ and observe that

$$\begin{aligned}P_t f(x) = K_t * f(x) &= \int_{\mathbb{R}^d}\int_0^{K_t(x-y)} \mathrm{d}s f(y)\mathrm{d}y\\ &= \int_0^\infty \int_{\{y\in\mathbb{R}^d : K_t(x-y)\ge s\}} f(y)\mathrm{d}y\mathrm{d}s.\end{aligned}$$

The set $\{y \in \mathbb{R}^d : K_t(x-y) \geq s\}$ is an Euclidean ball centered at $x \in \mathbb{R}^d$, since K_1 is radially decreasing. Thus

$$\begin{aligned} P_t f(x) &= K_t * f(x) \\ &\leq \left(\int_0^\infty |\{y \in \mathbb{R}^d : K_t(x-y) \geq s\}| \, \mathrm{d}s\right) M_*^{B^2} f(x) = \|K_t\|_{L^1} M_*^{B^2} f(x). \end{aligned}$$

Hence we conclude that

$$P_* f(x) \leq M_*^{B^2} f(x). \tag{16}$$

Inequality (12) gives us a bound independent of the dimension for $\|P_*\|_{L^p \to L^p}$, and in view of (16) we obtain $\|P_*\|_{L^p \to L^p} \leq C_p(d, B^2)$. Now a natural question arises whether $C_p(d, B^2)$ can be bounded independently of the dimension. This problem was investigated by the third author in [33].

2.2 *The Case of the Euclidean Balls [33, 35]*

The third author obtained in [33], see also the joint paper with Strömberg [35] for more details, that for every $p \in (1, \infty]$ there is a constant $C_p > 0$ independent of the dimension $d \in \mathbb{N}$ such that

$$C_p(d, B^2) \leq C_p. \tag{17}$$

Let us briefly describe the method used in [33] to prove (17). In $\mathbb{R}^d$, as $d \to \infty$, most of the mass of the unit ball B^2 concentrates at the unit sphere $\mathbb{S}^{d-1}$ in $\mathbb{R}^d$. In fact, if $\varepsilon \in (0, 1)$, we have

$$d \int_0^1 r^{d-1} \mathrm{d}r = 1, \qquad \text{while} \qquad \lim_{d \to \infty} d \int_0^{1-\varepsilon} r^{d-1} \mathrm{d}r = 0.$$

Therefore, the key idea is to use the spherical averaging operator, defined for any $r > 0$ and $x \in \mathbb{R}^d$ by

$$A_r^d f(x) = \int_{\mathbb{S}^{d-1}} f(x - r\theta) \mathrm{d}\sigma_{d-1}(\theta), \tag{18}$$

where σ_{d-1} denotes the normalized surface measure on $\mathbb{S}^{d-1}$. Using polar coordinates one easily sees that

$$M_t^{B^2} f(x) = d \int_0^1 r^{d-1} A_{tr}^d f(x) \mathrm{d}r,$$

which immediately implies

$$|M_*^{B^2} f(x)| \leq |A_*^d f(x)|. \tag{19}$$

By the earlier result of the third author [32], we know that for every $d \geq 3$ and for every $p > \frac{d}{d-1}$ there is a constant $C_{d,p} > 0$ such that for every $f \in L^p(\mathbb{R}^d)$ one has

$$\|A_*^d f\|_{L^p} \leq C_{d,p} \|f\|_{L^p}. \tag{20}$$

Inequality (20) is also true when $d = 2$, but this turned out to be a more difficult result, obtained by the first author in [3]. Now, the matters are reduced to show that the best constant in (20) can be taken to be independent of the dimension. For this purpose, the method of rotations enables one to view high-dimensional spheres as an average of rotated low-dimensional ones, and consequently one can conclude that for every $d \geq 3$ and $p > \frac{d}{d-1}$ we have

$$\|A_*^{d+1}\|_{L^p(\mathbb{R}^{d+1}) \to L^p(\mathbb{R}^{d+1})} \leq \|A_*^d\|_{L^p(\mathbb{R}^d) \to L^p(\mathbb{R}^d)}. \tag{21}$$

Hence the best constant in (20) is non-increasing, and in particular bounded, in $d \in \mathbb{N}$.

In order to prove (17) it suffices to take an integer $d_0 > \frac{p}{p-1}$. If $d \leq d_0$, then there is nothing to do. If $d > d_0$, taking into account (19) and (21), we see that

$$\|M_*^{B^2} f\|_{L^p} \leq \|A_*^{d_0}\|_{L^p(\mathbb{R}^{d_0}) \to L^p(\mathbb{R}^{d_0})} \|f\|_{L^p},$$

and we obtain (17) as claimed.

The method described above is limited to the Euclidean balls. The case of general convex symmetric bodies will require a different approach.

2.3 The L^2 Result for General Symmetric Bodies via Fourier Transform Methods [4]

In [4] the first author proposed a different approach, which is based on the estimates of the averaging operators M_t^G on the Fourier transform side. Before we present the main result from [4] we have to fix some notation and terminology. We begin with the most important definition of this paper.

Definition 4 We say that a convex symmetric body $G \subset \mathbb{R}^d$ is in the *isotropic position*, if it has Lebesgue measure $|G| = 1$, and there exists a constant $L = L(G) > 0$ depending only on G such that

$$\int_G \langle x, \xi\rangle^2 \mathrm{d}x = L(G)^2 |\xi|^2 \quad \text{for any} \quad \xi \in \mathbb{R}^d. \tag{22}$$

The constant $L(G)$ in (22) is called the *isotropic constant* of G.

From (22) one can deduce the following expression for the isotropic constant

$$L(G)^2 = \frac{1}{d}\int_G |x|^2 \mathrm{d}x. \tag{23}$$

Lemma 5 *For every convex symmetric body $G \subset \mathbb{R}^d$, there exists a linear transformation U of $\mathbb{R}^d$ such that $U(G)$ is in the isotropic position.*

Proof Observe that

$$M(\xi) = \int_G \langle x, \xi\rangle x \mathrm{d}x = \left(\int_G \langle x, \xi\rangle x_1 \mathrm{d}x, \ldots, \int_G \langle x, \xi\rangle x_d \mathrm{d}x\right)$$

is a positive operator on $\mathbb{R}^d$. Thus one can find a positive operator S such that $M = S^2$. Setting $U = c(G, S)S^{-1}$, where $c(G, S) = |\det S|^{1/d}|G|^{-1/d}$, we see that $|U(G)| = 1$ and

$$\begin{aligned}\int_{U(G)} \langle x, \xi\rangle^2 \mathrm{d}x &= c(G, S)^2|G|^{-1}\int_G \langle S^{-1}x, \xi\rangle^2 \mathrm{d}x\\ &= c(G, S)^2|G|^{-1}\langle M(S^{-1}\xi), S^{-1}\xi\rangle\\ &= c(G, S)^2|G|^{-1}|\xi|^2.\end{aligned}$$

Hence $U(G)$ is in the isotropic position, with the isotropic constant $L(U(G)) = c(G, S)|G|^{-1/2} > 0$. □

Observe that if the body G in (2) is replaced with any other set of the form $U(G)$, where U is an invertible linear transformation of $\mathbb{R}^d$, then the $L^p(\mathbb{R}^d)$ bounds from (2) remain unchanged and we have

$$C_p(d, G) = C_p(d, U(G)). \tag{24}$$

Indeed, considering an isometry U_p of $L^p(\mathbb{R}^d)$ given by

$$U_p f = |\det U|^{-1/p} f \circ U^{-1}, \quad \text{for any} \quad p \geq 1,$$

we obtain (24), since

$$U_p \circ M_t^G = M_t^{U(G)} \circ U_p.$$

In view of (24) the dimension-free estimates are unaffected by a change of the underlying body to an equivalent one. Therefore, from now on unless otherwise stated, we assume that $G \subset \mathbb{R}^d$ is in the isotropic position. For a symmetric convex body $G \subset \mathbb{R}^d$, let

$$m^G(t\xi) = \mathcal{F}(\mathbb{1}_G)(t\xi)$$

be the multiplier corresponding to the operator M_t^G from (1).

In [4] the first author provided the estimates for m^G and its derivatives in terms of the isotropic constant $L(G)$, see Theorem 6 below.

Theorem 6 ([4, eq. (10), (11), (12)]) *Let G be a symmetric convex body $G \subset \mathbb{R}^d$ which is in the isotropic position. Let $L = L(G)$ be the isotropic constant of G. Then for every $\xi \in \mathbb{R}^d \setminus \{0\}$ we have*

$$|m^G(\xi)| \le 150(L|\xi|)^{-1}, \qquad \textit{and} \qquad |m^G(\xi) - 1| \le 150(L|\xi|), \tag{25}$$

and

$$|\langle \xi, \nabla m^G(\xi)\rangle| \le 150. \tag{26}$$

In Sect. 4, for the sake of completeness, we provide a detailed proof of Theorem 6. In fact, as we shall see later on, the estimates in (25) and (26) will be the core of the proof of Theorem 1.

Using Theorem 6, as the main tool, it was proved in [4] that

$$C_2(d, G) \le C, \tag{27}$$

where $C > 0$ is a constant that does not depend neither on $d \in \mathbb{N}$ nor the underlying body $G \subset \mathbb{R}^d$. In view of the dimensional-free estimates for the Poisson semigroup (12) in order to prove (27) it suffices to obtain the following dimensional-free maximal estimate

$$\|\sup_{t>0} |(M_t^G - P_t) f|\|_{L^2} \le C\|f\|_{L^2}. \tag{28}$$

The estimate (28), in turn, was reduced, using some square function argument and the Plancherel theorem, to the uniform in $\xi \in \mathbb{R}^d$ estimate

$$\sum_{n\in\mathbb{Z}} \min\left\{(2^n L(G)|\xi|), (2^n L(G)|\xi|)^{-1}\right\} \le C, \tag{29}$$

where $C > 0$ is a universal constant independent of $d \in \mathbb{N}$ and the body $G \subset \mathbb{R}^d$. It is easy to see that (29) indeed holds. Moreover, it is true regardless of the exact value of the isotropic constant $L(G)$. Remarkably, we do not need to know whether $L(G)$ is comparable to a dimension-free constant.

2.4 Interlude: The Isotropic Conjecture

As we have already underlined, the approach from [4] does not require any information on the size of the isotropic constant $L(G)$. Recall at this point that $L(G)$ is known to be bounded from below by an absolute constant.

Proposition 7 *There is a universal constant $c > 0$ independent of the dimension such that for all convex symmetric bodies $G \subset \mathbb{R}^d$ in the isotropic position we have $L(G) \geq c$.*

Proof Let r_d be such that $|r_d B^2| = 1$. Then $r_d B^2$ is in the isotropic position and $r_d = |B^2|^{-1/d} \simeq d^{1/2}$. Using (23) and polar coordinates one has

$$L(r_d B^2)^2 = \frac{1}{d} \int_{r_d B^2} |x|^2 dx = \frac{|B^2| r_d^{d+2}}{d+2} = \frac{r_d^2}{d+2} \simeq 1.$$

Clearly, $|x| \geq r_d$ on $G \setminus r_d B^2$ and $|x| \leq r_d$ on $r_d B^2 \setminus G$. Thus, using (23) and the observation that $G \setminus r_d B^2$ and $r_d B^2 \setminus G$ have the same volume, we estimate

$$\begin{aligned} dL(G)^2 &= \int_G |x|^2 dx = \int_{G \cap r_d B^2} |x|^2 dx + \int_{G \setminus r_d B^2} |x|^2 dx \\ &\geq \int_{G \cap r_d B^2} |x|^2 dx + r_d^2 |G \setminus r_d B^2| \\ &\geq \int_{G \cap r_d B^2} |x|^2 dx + \int_{r_d B^2 \setminus G} |x|^2 dx = dL(r_d B^2)^2. \end{aligned}$$

Therefore, we see that $L(G) \geq L(r_d B^2) \geq c > 0$. This completes the proof. □

Conversely, it is not difficult to show the following upper bound.

Proposition 8 *There is a universal constant $C > 0$ independent of the dimension such that for all convex symmetric bodies $G \subset \mathbb{R}^d$ in the isotropic position we have $L(G) \leq Cd^{1/2}$.*

Proof If $r(G)$ is the largest radius $r > 0$ such that $rB^2 \subseteq G$ then there is an absolute constant $c > 0$ such that

$$cL(G) \leq r(G),$$

we refer to [12, Section 3.1, p.108] for more details. It follows that $cL(G)B^2 \subseteq G$ and

$$(cL(G))^d|B^2| \leq |G| = 1,$$

and consequently, using $|B^2|^{-1/d} \simeq d^{1/2}$ we obtain the desired claim. □

The estimate from Proposition 8 was improved by the first author in [7], where it was shown that $L(G) = O(d^{1/4} \log d)$. Klartag [19] proved that $L(G) = O(d^{1/4})$, and this is the best currently available general estimate for $L(G)$. However, the uniform bound from above for $L(G)$ is a well-known open problem with several equivalent formulations. More precisely, we are led to the following conjecture.

Conjecture 9 There is a constant $C > 0$ independent of $d \in \mathbb{N}$ such that for all convex symmetric bodies $G \subset \mathbb{R}^d$ in the isotropic position we have $L(G) \leq C$.

This conjecture was verified for various classes of convex symmetric bodies. To give an example we consider the class of 1-unconditional symmetric convex bodies. Let $\{e_1, \ldots, e_d\}$ denote the canonical basis in $\mathbb{R}^d$. We say that $G \subset \mathbb{R}^d$ is such a body, whenever, for every choice of signs $\varepsilon_1, \ldots, \varepsilon_d \in \{-1, 1\}$, we have

$$\Big\| \sum_{i=1}^{d} \varepsilon_i x_i e_i \Big\|_G = \|x\|_G, \quad \text{for all} \quad x = (x_1, \ldots, x_d) \in \mathbb{R}^d,$$

where $\|x\|_G = \inf\{t > 0 \colon x \in tG\}$ denotes the Minkowski norm associated with G.

Proposition 10 *There is a constant $C > 0$ independent of $d \in \mathbb{N}$ such that for all 1-unconditional convex bodies $G \subset \mathbb{R}^d$ in the isotropic position we have $L(G) \leq C$.*

For the proof of Proposition 10, and a more detailed exposition about the subject of geometry of isotropic convex bodies we refer to the monograph [12]. Interestingly, the issue of the isotropic constant did not impact the proofs of the dimension-free bounds for the Hardy–Littlewood maximal function (2) obtained in [4]. This gives us strong motivation to understand the role of the isotropic constant $L(G)$ in the estimates for $C_p(d, G)$ for all $p \in (1, \infty]$ for general convex symmetric bodies $G \subset \mathbb{R}^d$.

2.5 The L^p Results for $p \in (3/2, \infty]$ and Fractional Integration Method

The first author [5], and independently Carbery [13], extended the $L^2(\mathbb{R}^d)$ result from [4], and showed that for every $p \in (3/2, \infty]$ there exists a numerical constant $C_p > 0$, which does not depend on the dimension $d \in \mathbb{N}$ such that for every convex symmetric body $G \subset \mathbb{R}^d$ we have

$$C_p(d, G) \le C_p. \tag{30}$$

They also showed that if the supremum in (2) is restricted to the set of dyadic numbers $\mathbb{D}$, then inequality (30) remains valid for all $p \in (1, \infty]$. The methods used in these papers were completely different. We shall focus our attention merely on Carbery's paper [13], since it was an important starting point for the papers [26] and [8], which will be discussed in the next subsection.

The first main idea introduced in [13] reduced inequality (30) to proving that for every $p \in (3/2, 2]$ there exists $C_p > 0$ independent of $d \in \mathbb{N}$ such that for every convex symmetric body $G \subset \mathbb{R}^d$ we have

$$\sup_{n\in\mathbb{Z}} \Big\| \sup_{t\in[2^n,2^{n+1}]} |M_t^G f| \Big\|_{L^p} \le C_p \|f\|_{L^p}, \tag{31}$$

for every $f \in L^p(\mathbb{R}^d)$. This was achieved by appealing to an almost orthogonality principle, which combined with the Littlewood–Paley inequality (15) and inequality (12) for the Poisson semigroup, resulted in (30). The author of [13] adjusted an almost orthogonality principle to dimension-free setting from the unpublished notes of Christ, see also [14] for a more detailed discussion.

The second main idea of [13] relies on a fractional derivative/integration method, and it was used to prove (31). Let $\mathcal{F}_{\mathbb{R}}$ denote the one dimensional Fourier transform. For $\alpha \in (0, 1)$, let $\mathcal{D}^\alpha$ be the fractional derivative

$$\mathcal{D}^\alpha F(t) = \mathcal{D}_t^\alpha F(t) = \mathcal{D}_u^\alpha F(u)\big|_{u=t} = \mathcal{F}_{\mathbb{R}}\big((2i\pi\xi)^\alpha \mathcal{F}_{\mathbb{R}}^{-1}(F)(\xi)\big)(t), \quad \text{for} \quad t \in \mathbb{R}.$$

This formula gives a well defined tempered distribution on $\mathbb{R}$. Simple computations show that

$$\mathcal{D}_t^\alpha m^G(t\xi) = \int_G (2\pi i x \cdot \xi)^\alpha e^{-2\pi i t x\cdot\xi} \mathrm{d}x, \quad \text{for} \quad t > 0,$$

where $m^G(t\xi) = \mathcal{F}(\mathbb{1}_G)(t\xi)$. Moreover, [16, Lemma 6.6] guarantees that

$$\mathcal{D}_t^\alpha m^G(t\xi) = -\frac{1}{\Gamma(1-\alpha)} \int_t^\infty (u-t)^{-\alpha} \frac{\mathrm{d}}{\mathrm{d}u} m^G(u\xi)\mathrm{d}u, \quad \text{for} \quad \xi \in \mathbb{R}^d.$$

If $\mathcal{P}_u^\alpha$ is the operator associated with the multiplier

$$\mathfrak{p}_u^\alpha(\xi) = u^{\alpha+1} \mathcal{D}_v^\alpha \Big(\frac{m^G(v\xi)}{v}\Big)\Big|_{v=u}, \quad \text{for} \quad \xi \in \mathbb{R}^d,$$

then one can see that

$$M_t^G f(x) = \mathcal{F}^{-1}(m^G(t\xi)\mathcal{F}f)(x) = \frac{1}{\Gamma(\alpha)} \int_t^\infty \frac{t}{u}\left(1-\frac{t}{u}\right)^{\alpha-1} \mathcal{P}_u^\alpha f(x)\frac{du}{u}. \tag{32}$$

It was shown [13] that for general symmetric convex bodies one has

$$\|\mathcal{P}_1^\alpha f\|_{L^p} \lesssim \|f\|_{L^p} + \|T_{(\xi\cdot\nabla)^\alpha m^G} f\|_{L^p}, \tag{33}$$

where $T_{(\xi\cdot\nabla)^\alpha m^G} f$ is the multiplier operator associated with the symbol

$$(\xi\cdot\nabla)^\alpha m^G(\xi) = \mathcal{D}_t^\alpha m^G(t\xi)|_{t=1}.$$

The estimate from (33) immediately implies that

$$\sup_{u>0} \|\mathcal{P}_u^\alpha\|_{L^p\to L^p} \lesssim_p 1 + \|T_{(\xi\cdot\nabla)^\alpha m^G}\|_{L^p\to L^p}, \tag{34}$$

since the multipliers $\mathfrak{p}_u^\alpha$ are dilations of $\mathfrak{p}_1^\alpha$. Using (32) and (34) one controls

$$\Big\| \sup_{t\in[2^n,2^{n+1})} |M_t^G f|\Big\|_{L^p} \le C_p\big(1 + \|T_{(\xi\cdot\nabla)^\alpha m^G}\|_{L^p\to L^p}\big)\|f\|_{L^p}, \tag{35}$$

whenever $\alpha > 1/p$. Now, since $T_{(\xi\cdot\nabla)^1 m^G}$ is associated with the symbol $(\xi\cdot\nabla)m^G = \xi\cdot\nabla m^G(\xi)$, by Plancherel's theorem and (26) we have $\|T_{(\xi\cdot\nabla)^1 m^G}\|_{L^2\to L^2} \le C$. Clearly, $T_{(\xi\cdot\nabla)^0 m^G} = M_1$ is a contraction on $L^1(\mathbb{R}^d)$. Then by complex interpolation, as in [13], we get $\|T_{(\xi\cdot\nabla)^\alpha m^G}\|_{L^p\to L^p} \le C_\alpha$, whenever $\alpha < 2/p'$. In view of the restriction for $\alpha > 1/p$ in (35) we obtain (31) for $p \in (3/2, 2]$.

The above-mentioned method of fractional integration was exploited in [26] and [8].

2.6 *The L^p Result for $p \in (1, \infty]$, the Case of q-Balls*

Müller [26] proved, for all $p \in [1, \infty]$ and for every symmetric convex body $G \subset \mathbb{R}^d$, a remarkable upper bound for $C_p(d, G)$ in terms of certain geometric invariants. To be more precise, assuming that the body G is in the isotropic position, we define two constants, geometric invariants, by setting

$$\sigma(G)^{-1} = \max\big\{\varphi_\xi^G(0) : \xi \in \mathbb{S}^{d-1}\big\},$$

and

$$Q(G) = \max\big\{\mathrm{Vol}_{d-1}(\pi_\xi(G)) : \xi \in \mathbb{S}^{d-1}\big\},$$

where $\varphi_\xi^G(0) = \mathrm{Vol}_{d-1}(\{x \in G : x \cdot \xi = 0\})$, while $\pi_\xi : \mathbb{R}^d \to \xi^\perp$ denotes the orthogonal projection of $\mathbb{R}^d$ onto the hyperplane perpendicular to ξ. It follows from (57) that $\frac{1}{8}L(G) \le \sigma(G) \le 8L(G)$.

Using these two linear invariants $\sigma(G)$ and $Q(G)$ it was proved in [26] that for every $p \in (1, \infty]$ and for every symmetric convex body $G \subset \mathbb{R}^d$ there is a constant $C(p, \sigma(G), Q(G)) > 0$ independent of the dimension $d \in \mathbb{N}$ such that

$$C_p(d, G) \le C(p, \sigma(G), Q(G)). \tag{36}$$

In other words $C_p(d, G)$ may depend on $\sigma(G)$ and $Q(G)$, but not explicitly on the dimension $d \in \mathbb{N}$. For $p \in (3/2, \infty]$ inequality (36) is weaker than the estimates from [5] and [13], which show that $C_p(d, G)$ can be even chosen independently of d and G. However, using (36) it was proved that $C_p(d, B^q)$ is independent of the dimension for all $q \in [1, \infty)$, since $\sigma(B^q)$ and $Q(B^q)$ can be explicitly computed and they are independent of the dimension, but they depend on q. For the cubes $G = B^\infty$ it turned out that $\sigma(B^\infty)$ is independent of the dimension, but $Q(B^\infty) = d^{1/2}$, and at that time the cubes were thought of as candidates for a counterexample. However, the first author refined Müller's approach, and provided the dimensional-free bounds for $C_p(d, B^\infty)$ for all $p \in (1, \infty]$ as well. We shall now give a description of Müller's methods, which resulted in inequality (36).

As in [13], the proof of (36) in [26] is reduced to estimates of the $L^p(\mathbb{R}^d)$ norm of the operator $T_{(\xi.\nabla)^\alpha m^G}$. Recall, that the complex interpolation allowed Carbery to prove dimension-free $L^p(\mathbb{R}^d)$ bounds for $T_{(\xi.\nabla)^\alpha m^G}$ only in the restricted range of $\alpha < 2/p'$. Müller, by considering a suitable admissible family of Fourier multiplier operators, was able to prove that, for all $p \in (1, \infty)$ and for all $\alpha \in (1/2, 1)$, one has

$$\|T_{(\xi.\nabla)^\alpha m^G}\|_{L^p \to L^p} \le C_p(\alpha, \sigma(G), Q(G)).$$

More precisely, by using complex interpolation it was shown in [26] that

$$\|T_{(\xi.\nabla)^\alpha m^G}\|_{L^p \to L^p} \le C_\alpha\big(1 + \|T_{-2\pi|\xi|m^G(\xi)}\|_{L^p \to L^p}\big), \tag{37}$$

for $\alpha \in (1/2, 1)$, where $T_{-2\pi|\xi|m^G(\xi)}$ is the multiplier operator associated with the symbol $-2\pi|\xi|m^G(\xi)$.

Finally, (37) reduced the task to justifying

$$\|T_{-2\pi|\xi|m^G(\xi)}\|_{L^p \to L^p} \le C_p(\sigma(G), Q(G)), \tag{38}$$

for all $p \in (1, \infty)$. Since $T_{-2\pi|\xi|m^G(\xi)}$ is self-adjoint while proving (38) we can assume that $p \in [2, \infty)$. The key part of the proof of (38) in [26] is based on the following identity

$$-2\pi|\xi|m^G(\xi) = \sum_{j=1}^d \Big(-i\frac{\xi_j}{|\xi|}\Big)(-2\pi i\xi_j m^G(\xi)).$$

Thus, defining the measures $\mu_j = \frac{\mathrm{d}}{\mathrm{d}x_j}\mathbb{1}_G(x)$ we see that

$$T_{-2\pi|\xi|m^G(\xi)} = \sum_{j=1}^{d} R_j(\mu_j * f),$$

where R_j is the Riesz transform, corresponding to the multiplier $-i\xi_j/|\xi|$ for $j \in \{1,\dots,d\}$.

We now are at the stage, where the dimension-free estimates for the vector of Riesz transforms enter into the game. The third author [34] proved that for every $p \in (1,\infty)$ there is a constant $C_p > 0$ independent of the dimension $d \in \mathbb{N}$ such that the following estimate

$$\Big\|\Big(\sum_{j=1}^{d}|R_j f|^2\Big)^{1/2}\Big\|_{L^p} \le C_p\|f\|_{L^p}, \tag{39}$$

holds for every $f \in L^p(\mathbb{R}^d)$.

Then, dimension-free estimates for the vector of Riesz transforms (39) on $L^{p'}(\mathbb{R}^d)$, together with a duality argument reduce the problem to the following square function estimate

$$\Big\|\Big(\sum_{j=1}^{d}|\mu_j * f|^2\Big)^{1/2}\Big\|_{L^p} \le C_p(\sigma(G), Q(G))\|f\|_{L^p}, \tag{40}$$

for $p \in [2,\infty)$, which was achieved by interpolating between its $p = 2$ and $p = \infty$ endpoints.

As it has been mentioned above this approach resulted in dimension-free estimates for $C_p(d, B^q)$ for all $p \in (1,\infty]$ and $q \in [1,\infty)$, since in these cases the geometric invariants $\sigma(B^q)$ and $Q(B^q)$ turned out to be independent of $d \in \mathbb{N}$. For $q = \infty$ one obtains $Q(B^\infty) = d^{1/2}$, which resulted in no further progress for the Hardy–Littlewood maximal function for the cube.

However, for $q = \infty$ the first author observed [8] by a careful inspection of Müller's proof, that (40) for $p = 2$ can be estimated by a constant, which depends only on $\sigma(B^\infty)$, and the dependence on $Q(B^\infty)$ enters in (40) only for $p = \infty$. Therefore, instead of interpolating between $p = 2$ and $p = \infty$ in (40) it was natural to try, loosely speaking, to bound (40) for $p = q$ with large $q \ge 2$, and then interpolate with the improved estimate for $p = 2$, to obtain (40) with the implied constant depending only on p and $\sigma(B^\infty)$. In [8], in the proof of (40) for $p = q$ with large $q \ge 2$ the explicit formula for the multiplier

$$m^{B^\infty}(\xi) = \prod_{j=1}^{d}\frac{\sin(\pi\xi_j)}{\pi\xi_j}, \quad \text{for} \quad \xi \in \mathbb{R}^d$$

was essential. From Theorem 6 we have seen that $|m^{B^\infty}(\xi)| \le C|\xi|^{-1}$. However, $m^{B^\infty}(\xi)$, for most of ξ, decays much faster than $|\xi|^{-1}$ and the worst case happens only for ξ in narrow conical regions along the coordinate axes. This observation was implemented by making suitable localizations on the frequency space. An important ingredient, necessary to make these arguments rigorous in [8], was Pisier's holomorpic semigroup theorem [29]. The arguments presented in [8] are based on a very explicit analysis which does not immediately carry over to other convex symmetric bodies. Therefore, new methods will need to be invented to understand the growth of $C_p(d, G)$, as $d \to \infty$, in inequality (2) for general symmetric convex bodies $G \subset \mathbb{R}^d$ when $p \in (1, 3/2]$.

2.7 *Weak Type* (1, 1) *Considerations*

So far we have only discussed the question of dimension-free estimates on $L^p(\mathbb{R}^d)$ spaces for $p \in (1, \infty]$. However, one may ask about a dimension-free bound for the best constant $C_1(d, G)$ in the weak type (1, 1) estimate

$$\sup_{\lambda>0} \lambda \Big| \Big\{ x \in \mathbb{R}^d : \sup_{t>0} |M_t^G f(x)| > \lambda \Big\} \Big| \le C_1(d, G) \|f\|_{L^1}. \tag{41}$$

Appealing to the Vitali covering lemma one can easily show that $C_1(d, G) \le 3^d$. In [35] the third author and Strömberg proved that for general symmetric convex bodies $G \subset \mathbb{R}^d$ one has

$$C_1(d, G) \le C d \log d, \tag{42}$$

where $C > 0$ is a universal constant independent of $d \in \mathbb{N}$. This is the best known result to date, see also [28] for generalizations of (42). The proof of inequality (42) is based on a rather complicated variant of the Vitali covering idea. The authors in [35] were also able to sharpen this estimate in the case of the Euclidean balls by proving

$$C_1(d, B^2) \le C d, \tag{43}$$

with a universal constant $C > 0$ independent of the dimension. For justifying (43) the authors used a comparison with the heat semigroup together with the Hopf maximal ergodic theorem, see [31].

Now, in view of these results a natural question arises, whether we can take a dimension-free constant in (42) and (43). This was resolved in the case of the cube $G = B^\infty$ by Aldaz [1] who proved that

$$C_1(d, B^\infty) \ge C_d, \tag{44}$$

where C_d is a constant that tends to infinity as $d \to \infty$. The constant C_d was made more explicit by Aubrun [2], who proved (44) with $C_d \simeq_\varepsilon (\log d)^{1-\varepsilon}$ for every $\varepsilon > 0$, and by Iakolev and Strömberg [18], who considerably improved the latter lower bound by showing that $C_d \simeq d^{1/4}$. The arguments in the papers [1, 2, 18], were based on careful analysis of a discretized version of the initial problem. The function f realizing the supremum was then chosen as an appropriate sum of Dirac's deltas.

The case of the cube is the only one where we have a definitive answer on the size of $C_1(d, G)$ in (41). Remarkably even in the case of the Euclidean ball B^2 it is unknown whether the weak type $(1, 1)$ constant is dimension-free.

3 Overview of the Methods of the Paper

This section is intended to present a new flexible approach, which recently resulted in dimension-free bounds in r-variational and jump inequalities corresponding to the operators M_t^G from (4), see [9] and [25]. An important feature of this method is that it is also applicable to the discrete settings, see [10, 25].

For clarity of exposition we shall only be working with maximal functions on $L^p(\mathbb{R}^d)$ or $\ell^p(\mathbb{Z}^d)$. For a more abstract setting we refer to [25], and also [24].

3.1 Continuous Perspective

We shall briefly outline the method of the proof of Theorem 1. The proof of (4) is based on the following simple decomposition

$$\sup_{t>0} |M_t^G f| \le \sup_{n\in\mathbb{Z}} |M_{2^n}^G f| + \Big(\sum_{n\in\mathbb{Z}} \sup_{t\in[2^n,2^{n+1}]} |(M_t^G - M_{2^n}^G) f|^2\Big)^{1/2}. \tag{45}$$

In other words, the full maximal function corresponding to the operators M_t^G is controlled by the dyadic maximal function and the square function associated with maximal functions restricted to dyadic blocks.

The estimates, on $L^p(\mathbb{R}^d)$ for $p \in (1, \infty]$, of the dyadic maximal function (in fact inequality (5)) are based, upon comparing $\sup_{n\in\mathbb{Z}} |M_{2^n}^G f|$ with the Poisson semigroup P_t, see (61), on a variant of bootstrap argument. The idea of the bootstrap goes back to [27], where the context of differentiation in lacunary directions was studied. Later on, these ideas were used in many other papers [14, 17], including their applications in dimension-free estimates [13]. Recently, it turned out that certain variant of bootstrap arguments may be also used to obtain dimension-free estimates in r-variational inequalities [9, 10] and in jump inequalities [25]. In the latter paper applications to the operators of Radon type are discussed as well. The methods of [25], presented as a part of an abstract theory, immediately give the desired conclusion. However, in Sect. 4, for the sake of clarity, we give a simple direct proof and deduce (5) from inequality (62), which immediately leads to a bootstrap inequality in (63). In particular three tools, with dimension-free estimates, that we now highlight are used to obtain (62):

1. The maximal inequality (12) for the Poisson semigroup P_t.
2. The Littlewood–Paley inequality (15) associated with the Poisson projections S_n.
3. The estimates of the Fourier multiplies corresponding to M_t^G from Theorem 6.

The details are given in the second part of Sect. 4. In the third part of Sect. 4 we estimate, on $L^p(\mathbb{R}^d)$ for $p \in (3/2, 2]$, the square function from (45). In order to do so, we shall employ an elementary numerical inequality, as in [9, 10], see also [25], which asserts that for every $n \in \mathbb{Z}$ and for every function $\mathfrak{a} : [2^n, 2^{n+1}] \to \mathbb{C}$ we have

$$\sup_{t\in[2^n,2^{n+1}]} |\mathfrak{a}(t) - \mathfrak{a}(2^n)| \\ \leq \sqrt{2}\sum_{l\in\mathbb{N}_0}\Big(\sum_{m=0}^{2^l-1} |\mathfrak{a}(2^n + 2^{n-l}(m+1)) - \mathfrak{a}(2^n + 2^{n-l}m)|^2\Big)^{1/2}. \tag{46}$$

The inequality is the crucial new ingredient, which on the one hand, replaces the fractional integration argument from [13]. This is especially important in the discrete setting as it is not clear, due to the lack of the dilation structure on $\mathbb{Z}^d$, whether the fractional integration argument is available there. On the other hand, (46) reduces estimates for a supremum (or even for r-variations, see [25]) restricted to a dyadic block to the situation of certain square functions, where the division intervals over which differences are taken (in these square functions) are all of the same size, see inequality (68).

A variant of inequality (46) was proved by Lewko–Lewko [20, Lemma 13], and it was used to study variational Rademacher–Menshov type results for orthonormal systems. Inequality (46), essentially in this form, was independently obtained in [21, Lemma 1] by the second author and Trojan in the context of r-variational estimates for discrete Radon transforms, see also [22, 23].

Upon applying inequality (46) to control the square function from (45) the problem is reduced to control a new square function like in (69). The problem now is well suited to an application of the Fourier transform methods, and the estimates from Theorem 6 combined with the Littlewood–Paley inequality do the job and we obtain the desired claim.

The approach described above does not allow us to improve the range for $p \in (3/2, \infty]$ in the inequality from (4). To see this, it suffices to consider the maximal function corresponding to the spherical means in $\mathbb{R}^3$, see (18). Indeed, adopting the method from Sect. 4 we obtain that the spherical maximal function is bounded on $L^p(\mathbb{R}^3)$ for every $p \in (3/2, \infty]$, but unbounded on $L^{3/2}(\mathbb{R}^3)$, see [32].

Any extension of the range $p \in (3/2, \infty]$ in (4) will require more refined information besides the positivity of the operators M_t^G and estimates of the Fourier multipliers m_t^G from Theorem 6. To be more precise, assume that $p_0 \in (1, 2]$ and let $\alpha = 1/p_0 < 1$. Suppose that there is a constant $C_{p_0} > 0$ independent of the dimension $d \in \mathbb{N}$ such that for every $t > 0$ and $h \in (0, 1)$ and for every $f \in L^{p_0}(\mathbb{R}^d)$ the following Hölder continuity condition holds

$$\|(M_{t+h}^G - M_t^G)f\|_{L^{p_0}} \leq C_{p_0}\left(\frac{h}{t}\right)^{\alpha}\|f\|_{L^{p_0}}. \tag{47}$$

Then, as it was proved in [25] using a certain bootstrap argument, for every $p \in (p_0, 2]$ we have

$$C_p(d, G) \lesssim_p 1, \tag{48}$$

with the implicit constant independent of the dimension. Therefore, the general problem is reduced to understand (47). In the case of q-balls $G = B^q$ for $q \in [1, \infty]$, inequality (47), and consequently (48), can be verified as it was shown in [9, 25]. The general case is reduced, anyway, to understand the norm $\|T_{(\xi\cdot\nabla)^\alpha m^G}\|_{L^p\to L^p}$ as in Müller's proof [26]. But, as we said before, this will need new ideas.

3.2 Discrete Perspective

As we have seen in the introduction the dimension-free estimates in the discrete setting for $C_p(d, G)$ may be very hard, and in general there is no obvious conjecture to prove. However, for the q-balls $G = B^q$ as in (3), in view of the methods presented above, the problem may be reduced to estimates of the Fourier multipliers. For $q \in [1, \infty]$, let $\mathfrak{m}_N^{B^q}$ be the multiplier corresponding to the operator $\mathcal{M}_N^{B^q}$ as in (6). Let us define the proportionality factor

$$\kappa_q(d, N) = Nd^{-1/q},$$

which can be identified with the isotropic constant corresponding to B_N^q, if the normalization assumption $|B^q| = 1$ in definition (22) is dropped. If we could prove that there exists a constant $C_q > 0$ independent of the dimension $d \in \mathbb{N}$ such that for every $N \in \mathbb{N}$ and $\xi \in \mathbb{T}^d$ we have

$$\begin{aligned} |\mathfrak{m}_N^{B^q}(\xi) - 1| &\leq C_q\kappa_q(d, N)|\xi|, \\ |\mathfrak{m}_N^{B^q}(\xi)| &\leq C_q(\kappa_q(d, N)|\xi|)^{-1}, \\ |\mathfrak{m}_{N+1}^{B^q}(\xi) - \mathfrak{m}_N^{B^q}(\xi)| &\leq C_qN^{-1}, \end{aligned} \tag{49}$$

where $|\xi|$ denotes the Euclidean norm restricted to the torus $\mathbb{T}^d \equiv [-1/2, 1/2)^d$; then, using the methods from the proof of Theorem 1, we would be able to conclude that the best constant $C_p(d, B^q)$ in inequality (7) is bounded independently of the dimension for every $p \in (3/2, \infty]$.

Therefore, the problem of estimating $C_p(d, B^q)$ with bounds independent of the dimension is reduced to establishing (49). Even though, estimates (49) can be thought of as discrete analogues of the estimates for the continuous multipliers m_t^G, from Theorem 6 with $G = B^q$, the method of the proof of Theorem 6 is not applicable to derive (49). For $q \in [1, \infty)$ the question seems to be very hard due to the lack of reasonable estimates for the number of lattice points in the sets B_N^q.

However, if $q = \infty$ then $B_N^\infty = [-N, N]^d$ is a cube. Thus the number of lattice points is not a problem any more, and we easily have $|B_N^\infty \cap \mathbb{Z}^d| = (2N+1)^d$. This property distinguishes the cubes from the q-balls for $q \in [1, \infty)$. Using the product structure of the cubes we were able to analyze the behavior of the multiplier $\mathfrak{m}_N^{B^\infty}$ associated with the operator $\mathcal{M}_N^{B^\infty}$ and obtain (49), see [10] for more details. The multiplier $\mathfrak{m}_N^{B^\infty}$ is an exponential sum, which is the product of one dimensional Dirichlet's kernels. The explicit formula for $\mathfrak{m}_N^{B^\infty}$ in terms of the Dirichlet kernels was essential for our approach and permitted us to establish (49) for $q = \infty$ with $\kappa_\infty(d, N) = N$. Applying (49) we showed in [10], as it was mentioned in the introduction, that for every $p \in (3/2, \infty]$ there is a constant $C_p > 0$ independent of the dimension such that $C_p(d, B^\infty) \le C_p$. Moreover, if the supremum in (7) is restricted to the dyadic set $\mathbb{D}$, then (7) holds for all $p \in (1, \infty]$ and $C_p(d, B^\infty)$ is independent of the dimension as well. The inequalities in (49), for $q = \infty$, are based on elementary estimates, which are interesting in their own right. For this reason our method does not extend to discrete convex bodies other than B^∞. This is the second place which sets the operators $\mathcal{M}_N^{B^\infty}$ over the cubes apart from the operators $\mathcal{M}_N^{B^q}$ over the q-balls for $q \in [1, \infty)$.

Now it is desirable to understand whether inequalities (49) hold for $q \in [1, \infty)$. The absence of the product structure for $q \in [1, \infty)$ makes the estimates incomparably harder. However, using crude estimates for the number of lattice points in the q-balls B_N^q, if $p \in (1, \infty]$ and $q \in [1, \infty]$, we obtain, as in [10], that there is $C_{p,q} > 0$ independent of the dimension $d \in \mathbb{N}$ such that for all $f \in \ell^p(\mathbb{Z}^d)$ we have

$$\big\| \sup_{N \ge d^{1+1/q}} |\mathcal{M}_N^{B^q} f| \big\|_{\ell^p} \le C_{p,q} \|f\|_{\ell^p}. \tag{50}$$

Inequality (50) follows from a simple comparison argument, which permits us to dominate the $\ell^p(\mathbb{Z}^d)$ norm of the maximal function $\sup_{N \ge d^{1+1/q}} |\mathcal{M}_N^{B^q} f|$ by a constant multiple of $C_p(d, B^q)$, which we know is independent of the dimension for every $p \in (1, \infty]$ due to [26] for $q \in [1, \infty)$, and due to [8] for $q = \infty$.

In Sect. 5, for $q = 2$, we shall extend the range in the supremum in (50) and we show that $d^{1+1/q} = d^{3/2}$, (for $q = 2$), can be replaced by a constant multiple of d, see Theorem 2. Our argument is a subtle refinement of the arguments from

[10]. Even though, we will also use crude estimates for the number of lattice points in the balls B_N^2, the essential improvement comes from the fact that the Euclidean norm corresponds to the scalar product $|x|^2 = \langle x, x\rangle$. See Lemma 11 and Lemma 12, where this observation plays the key role. The rest of the argument reduces the problem to the comparison of the $\ell^p(\mathbb{Z}^d)$ norm of $\sup_{N\ge Cd} |\mathcal{M}_N^{B^2} f|$ with $C_p(d, B^2)$, which is independent of the dimension for all $p \in (1, \infty]$. Now the matters are reduced to understand $\sup_{1\le N\le Cd} |\mathcal{M}_N^{B^2} f|$.

In [11] the authors initiated investigations in this direction and the case of the discrete Euclidean balls with dyadic radii was studied. We obtained Theorem 3, which gives us some evidence that inequality (7) with dimension-free bounds in not entirely hopeless, at least for $q = 2$. The methods of the proof of Theorem 3 shed a new light on the general problem (7), but the best what we can do for the full maximal function at this moment is Theorem 2, and new methods will surely need to be invented to attack this case.

The proof of Theorem 3 is based on the estimates for $\mathfrak{m}_N^{B^2}$, which in turn are based on delicate combinatorial arguments that differ completely from the methods used to obtain estimates (49) for $\mathfrak{m}_N^{B^\infty}$. In particular, we proved analogues of the first two inequalities from (49) for $\mathfrak{m}_t^{B^2}$. However, the second inequality is perturbed by a negative power of $\kappa_2(d, N)$, which makes our method limited to the dyadic scales, and nothing reasonable beyond $\ell^2(\mathbb{Z}^d)$ theory can be said in (11). Our aim now is to understand whether the second estimate can be improved. If we succeeded in doing so, we could extend inequality (11) to $\ell^p(\mathbb{Z}^d)$ spaces for all $p \in (1, \infty]$. The second task, which seems to be quite challenging, is to obtain the third inequality in (49) for the multiplier $\mathfrak{m}_N^{B^2}$. This inequality, if proved, would allow us to think about dimension-free estimates of $C_p(d, B^2)$ for all $p \in (3/2, \infty]$. We refer to [11] for more details.

4 Continuous Perspective: Proof of Theorem 1

The purpose of this section is to provide dimension-free estimates on $L^p(\mathbb{R}^d)$, with $p \in (3/2, \infty]$, for the Hardy–Littlewood maximal function associated with convex symmetric bodies in $\mathbb{R}^d$. However, we begin with the proof of Theorem 6, which will allow us to build up the $L^2(\mathbb{R}^d)$ theory in Theorem 1.

4.1 Fourier Transform Estimates: Proof of Theorem 6

For $\zeta \in \mathbb{S}^{d-1}$ and $u \in \mathbb{R}$ we define the set

$$A_\zeta(u) = \{x \in G : x \cdot \zeta = u\},$$

and an even and compactly supported function by setting

$$\varphi_\zeta^G(u) = \mathrm{Vol}_{d-1}(A_\zeta(u)),$$

where Vol_{d-1} denotes $(d-1)$-dimensional Lebesgue measure. We observe that for all $\lambda \in [0, 1]$ and for all $u, v \in \mathbb{R}$ such that $\varphi_\zeta^G(u) \neq 0$ and $\varphi_\zeta^G(v) \neq 0$ we obtain

$$\mathrm{Vol}_{d-1}(A_\zeta(\lambda u + (1-\lambda)v))^{\frac{1}{d-1}} \geq \lambda \mathrm{Vol}_{d-1}(A_\zeta(u))^{\frac{1}{d-1}} + (1-\lambda)\mathrm{Vol}_{d-1}(A_\zeta(v))^{\frac{1}{d-1}}. \tag{51}$$

This can be verified using Brunn–Minkowski's inequality (in dimension $(d-1)$), since, by convexity of G, for every $u, v \in \mathbb{R}$ if $A_\zeta(u) \neq \emptyset$ and $A_\zeta(v) \neq \emptyset$ then

$$\lambda A_\zeta(u) + (1-\lambda)A_\zeta(v) \subseteq A_\zeta(\lambda u + (1-\lambda)v). \tag{52}$$

For $\zeta \in \mathbb{S}^{d-1}$ define $S_\zeta = \{x \in \mathbb{R} \colon \varphi_\zeta^G(x) \neq 0\}$. If $u_0 \in S_\zeta^c$ then, using (52), it is not difficult to see that for every $u \in \mathbb{R}$ such that $|u| > |u_0|$ we have $\varphi_\zeta^G(u) = 0$. This ensures that S_ζ is a symmetric interval contained in $[-u_\zeta, u_\zeta]$, where $u_\zeta = \sup\{x \geq 0 \colon \varphi_\zeta^G(x) \neq 0\}$. Taking $v = -u$ in (51) we obtain that $\varphi_\zeta^G((2\lambda - 1)u) \geq \varphi_\zeta^G(u)$ for all $\lambda \in [0, 1]$ and $u \in \mathbb{R}$. This implies that φ_ζ^G is decreasing on $S_\zeta \cap [0, \infty)$ as well as on $[0, \infty)$. Inequality (51) shows that the function $(\varphi_\zeta^G)^{\frac{1}{d-1}}$ is concave on S_ζ. In particular, φ_ζ^G is differentiable almost everywhere in $(-u_\zeta, u_\zeta)$, since it is absolutely continuous on each closed interval contained in $(-u_\zeta, u_\zeta)$. The inequality between the weighted arithmetic and geometric means together with (51) also implies the log-concavity of φ_ζ^G. Namely, for $\lambda \in [0, 1]$ and $u, v > 0$ we have

$$\varphi_\zeta^G(\lambda u + (1-\lambda)v) \geq \varphi_\zeta^G(u)^\lambda \varphi_\zeta^G(v)^{1-\lambda}.$$

Note that using Fubini's theorem we have, for $\xi \in \mathbb{R}^d \setminus \{0\}$, that

$$m^G(\xi) = \int_{\mathbb{R}} \varphi_{\xi/|\xi|}^G(u) e^{2\pi i |\xi| u} du. \tag{53}$$

More generally, for any $h \in L^\infty(\mathbb{R})$ and $\xi \in \mathbb{R}^d \setminus \{0\}$, one has

$$\int_G h(x \cdot \xi) dx = \int_{\mathbb{R}} \varphi_{\xi/|\xi|}^G(u) h(|\xi| u) du. \tag{54}$$

From the above properties of φ_ζ^G we shall deduce, as in [4, Lemma 1], that

$$\varphi_\zeta^G(u) \leq 2\varphi_\zeta^G(0) e^{-\varphi_\zeta^G(0)|u|}, \quad \text{for all} \quad u \in \mathbb{R}, \quad \text{and} \quad \zeta \in \mathbb{S}^{d-1}. \tag{55}$$

For this purpose, we fix $\zeta \in \mathbb{S}^{d-1}$ and let us consider the function $\psi_\zeta^G(u) = \varphi_\zeta^G(0)e^{-\varphi_\zeta^G(0)|u|}$, whose logarithm is a linear function. We have that $\varphi_\zeta^G(0) = \psi_\zeta^G(0)$, and suppose that there is a point $u_0 \in (0,\infty)$ such that $\varphi_\zeta^G(u_0) = \psi_\zeta^G(u_0)$. By the log-concavity we obtain that

$$\varphi_\zeta^G(u) \leq \psi_\zeta^G(u), \quad \text{for} \quad u > u_0,$$

and in this case there is nothing to do. Moreover, the log-concavity also gives

$$\varphi_\zeta^G(u) \geq \psi_\zeta^G(u), \quad \text{for} \quad 0 \leq u \leq u_0.$$

In this case, using (54) with $h(u) = \mathbb{1}_{[0,\infty)}(u)$ we obtain

$$\frac{1}{2} = \int_0^\infty \varphi_\zeta^G(u)\mathrm{d}u \geq \varphi_\zeta^G(0)\int_0^{u_0} e^{-u\varphi_\zeta^G(0)}\mathrm{d}u = \int_0^{u_0\varphi_\zeta^G(0)} e^{-u}\mathrm{d}u = 1 - e^{-u_0\varphi_\zeta^G(0)}, \tag{56}$$

and, consequently, $e^{-u_0\varphi_\zeta^G(0)} \geq 1/2$, so that $u_0\varphi_\zeta^G(0) \leq \log 2$. Hence, (55) follows, since

$$\varphi_\zeta^G(u) \leq \varphi_\zeta^G(0)e^{(u_0-u)\varphi_\zeta^G(0)} \leq 2\varphi_\zeta^G(0)e^{-u\varphi_\zeta^G(0)}, \quad \text{for} \quad 0 \leq u \leq u_0.$$

If $u = 0$ is the unique point such that $\varphi_\zeta^G(0) = \psi_\zeta^G(0)$, then $\varphi_\zeta^G(u) \leq \psi_\zeta^G(u)$ or $\varphi_\zeta^G(u) \geq \psi_\zeta^G(u)$ for all $u \in S_\zeta$. If the first inequality holds then we are done, so we may assume that the second inequality is true. Arguing in a similar way as in (56) with u_ζ in place of u_0 we obtain that $u_\zeta\varphi_\zeta^G(0) \leq \log 2$, and consequently

$$\varphi_\zeta^G(u) \leq \varphi_\zeta^G(0)e^{(u_\zeta-u)\varphi_\zeta^G(0)} \leq 2\varphi_\zeta^G(0)e^{-u\varphi_\zeta^G(0)}, \quad \text{for} \quad 0 \leq u \leq u_\zeta.$$

Hence (55) follows, since $\varphi_\zeta^G(u) = 0$ for $u \in S_\zeta^c$.

Since G is in the isotropic position we can also prove that $\varphi_\zeta^G(0)$ is of the same order, uniformly in $\zeta \in \mathbb{S}^{d-1}$. More precisely, as in [4, Lemma 2], we have

$$\frac{3}{16} \leq L\,\varphi_\zeta^G(0) \leq 3, \quad \text{for every} \quad \zeta \in \mathbb{S}^{d-1}, \tag{57}$$

where L is the isotropic constant. To prove the right-hand side inequality in (57) we show, with the aid of (54) (for $h(u) = u^2$) and (55), that

$$L^2 = \int_{\mathbb{R}} u^2\varphi_\zeta^G(u)\mathrm{d}u \leq 4\varphi_\zeta^G(0)\int_0^\infty u^2 e^{-\varphi_\zeta^G(0)u}\mathrm{d}u \leq 8\varphi_\zeta^G(0)^{-2}.$$

For the left-hand side inequality in (57) we calculate

$$1 = \int_{\mathbb{R}} \varphi_\zeta^G(u)\mathrm{d}u \le 4L\varphi_\zeta^G(0) + \frac{1}{4L^2}\int_{|u|\ge 2L} u^2\varphi_\zeta^G(u)\mathrm{d}u \le 4L\varphi_\zeta^G(0) + \frac{1}{4},$$

which implies $L\varphi_\zeta^G(0) \ge 3/16$, and (57) is justified. We now pass to the proof of Theorem 6.

Proof (of Theorem 6) We begin with the proof of inequalities in (25). For $\xi \in \mathbb{R}^d \setminus \{0\}$ we set $\zeta = \xi/|\xi|$, then integration by parts allows us to rewrite (53) as

$$\begin{aligned} m^G(\xi) &= \int_{\mathbb{R}} \varphi_\zeta^G(u)\cos(2\pi|\xi|u)\mathrm{d}u \\ &= \lim_{u\nearrow u_\zeta} \frac{\varphi_\zeta^G(u)\sin(2\pi|\xi|u)}{\pi|\xi|} - \frac{1}{2\pi|\xi|}\int_{-u_\zeta}^{u_\zeta} (\varphi_\zeta^G)'(u)\sin(2\pi|\xi|u)\mathrm{d}u. \end{aligned}$$

Then using (57) we obtain the first inequality in (25), since

$$\begin{aligned} |m^G(\xi)| &\le (\pi|\xi|)^{-1}\varphi_\zeta^G(0) + (2\pi|\xi|)^{-1}\int_{-u_\zeta}^{u_\zeta} |(\varphi_\zeta^G)'(u)|\mathrm{d}u \\ &= (\pi|\xi|)^{-1}\varphi_\zeta^G(0) - (\pi|\xi|)^{-1}\int_0^{u_\zeta} (\varphi_\zeta^G)'(u)\mathrm{d}u \\ &\le 6\pi^{-1}(L|\xi|)^{-1}. \end{aligned}$$

To prove the second inequality in (25), we use (55) and (57) to write

$$\begin{aligned} |m^G(\xi) - 1| &\le \int_{\mathbb{R}} \varphi_\zeta^G(u)|\cos(2\pi|\xi|u) - 1|\mathrm{d}u \\ &\le 4\pi|\xi|\int_0^\infty u\varphi_\zeta^G(u)\mathrm{d}u \\ &\le 8\pi|\xi|\varphi_\zeta^G(0)^{-1} \\ &\le 45\pi(L|\xi|). \end{aligned}$$

This completes the proof of (25). To justify (26), we use (54) and integrate by parts to get

$$\begin{aligned} \langle \xi, \nabla m^G(\xi)\rangle &= \int_G 2\pi i\langle x, \xi\rangle e^{2\pi i x\cdot\xi}\mathrm{d}x \\ &= \int_{-u_\zeta}^{u_\zeta} (2\pi i|\xi|e^{2\pi i u|\xi|})\,(u\varphi_\zeta^G(u))\mathrm{d}u \end{aligned}$$

$$= \lim_{u \nearrow u_\zeta} \big(e^{2\pi i u|\xi|} u\varphi_\zeta^G(u)\big) - \lim_{u \searrow -u_\zeta} \big(e^{2\pi i u|\xi|} u\varphi_\zeta^G(u)\big)$$

$$- \int_{-u_\zeta}^{u_\zeta} e^{2\pi i u|\xi|} \frac{\mathrm{d}}{\mathrm{d}u}(u\varphi_\zeta^G(u))\mathrm{d}u.$$

This leads, in view of (55), to the estimate

$$|\langle \xi, \nabla m^G(\xi)\rangle| \le 4u_\zeta \varphi_\zeta^G(0) e^{-\varphi_\zeta^G(0)u_\zeta} + \int_{-u_\zeta}^{u_\zeta} \varphi_\zeta^G(u)\mathrm{d}u + \int_{-u_\zeta}^{u_\zeta} |u||(\varphi_\zeta^G)'(u)|\mathrm{d}u$$

$$\le 5 - 2\int_0^{u_\zeta} u(\varphi_\zeta^G)'(u)\mathrm{d}u,$$

where we used the fact that $\varphi_\zeta^G(u)$ is decreasing in u. Hence, integrating by parts once again we reach $|\langle \xi, \nabla m^G(\xi)\rangle| \le 10$, which gives (26). The proof of Theorem 6 is completed. □

The approach we shall use to prove Theorem 1 was presented as a part of an abstract theory in [25]. The method has recently found many applications in r-variational and jump estimates (including dimension-free estimates) in the continuous and discrete settings, see [9, 10, 24, 25]. However here, for the sake of clarity, we shall only focus our attention on the maximal functions in the continuous setup.

Since we are working with a family of averaging operators only the range for $p \in (3/2, 2]$ will be interesting in Theorem 1. The range for $p \in (2, \infty]$ will follow then by a simple interpolation with the obvious $L^\infty(\mathbb{R}^d)$ bound. For instance, in order to prove dimension-free bounds for the dyadic maximal function, it will suffice to show that for every $p \in (1, 2]$ and for every $f \in L^p(\mathbb{R}^d)$ we have

$$\big\| \sup_{n\in\mathbb{Z}} |M_{2^n}^G f| \big\|_{L^p} \lesssim \|f\|_{L^p}. \tag{58}$$

In particular, (58) proves inequality (5) from Theorem 1. Then, in view of (45), the proof of inequality (4) will be completed, if we show that for $p \in (3/2, 2]$ and for every $f \in L^p(\mathbb{R}^d)$ we have

$$\Big\| \Big(\sum_{n\in\mathbb{Z}} \sup_{t\in[2^n, 2^{n+1}]} |(M_t^G - M_{2^n}^G) f|^2\Big)^{1/2} \Big\|_{L^p} \lesssim \|f\|_{L^p}. \tag{59}$$

In the next two subsections we prove inequalities (58) and (59) respectively.

4.2 *Proof of Inequality (58)*

We fix $N \in \mathbb{N}$ and define

$$B_p(N) = \sup_{\|f\|_{L^p} \le 1} \big\| \sup_{|n| \le N} |M^G_{2^n} f| \big\|_{L^p}.$$

We see that $B_p(N) \le 2N+1$ for every $N \in \mathbb{N}$, since M^G_t is an averaging operator. Our aim will be to show that for every $p \in (1,2]$ there is a constant $C_p > 0$ independent of the dimension and the underlying body $G \subset \mathbb{R}^d$ such that

$$\sup_{N \in \mathbb{N}} B_p(N) \le C_p. \tag{60}$$

Observe that, by (12), we have

$$\begin{aligned} \big\| \sup_{|n| \le N} |M^G_{2^n} f| \big\|_{L^p} &\le \big\| \sup_{t>0} |P_t f| \big\|_{L^p} + \big\| \sup_{|n| \le N} |(M^G_{2^n} - P_{2^n}) f| \big\|_{L^p} \\ &\lesssim \|f\|_{L^p} + \sum_{j \in \mathbb{Z}} \Big\| \Big(\sum_{|n| \le N} |(M^G_{2^n} - P_{2^n}) S_{j+n} f|^2 \Big)^{1/2} \Big\|_{L^p}, \end{aligned} \tag{61}$$

where in the last line we have used decomposition from (14). The proof of (58) will be completed, if we show that for every $p \in (1,2]$ there is $C'_p > 0$ independent of d, N, and the body $G \subset \mathbb{R}^d$ such that for every $j \in \mathbb{Z}$ and for every $f \in L^p(\mathbb{R}^d)$ we have

$$\Big\| \Big(\sum_{|n| \le N} |(M^G_{2^n} - P_{2^n}) S_{j+n} f|^2 \Big)^{1/2} \Big\|_{L^p} \le C'_p (1 + B_p(N))^{\frac{2-p}{2}} 2^{-\frac{(p-1)|j|}{2}} \|f\|_{L^p}. \tag{62}$$

Assume momentarily that (62) has been proven. Then combining (61) with (62) we obtain that

$$B_p(N) \lesssim_p 1 + (1 + B_p(N))^{\frac{2-p}{2}}, \tag{63}$$

with the implicit constant independent of d, N and the body $G \subset \mathbb{R}^d$. Thus we conclude, using (63), that (60) holds, and the proof of (58) and consequently (5) from Theorem 1 is completed.

4.2.1 Proof of Inequality (62) for $p = 2$

Using Theorem 6 we show that (62) holds for $p = 2$. Let $k(\xi) = m^G(\xi) - p_1(\xi) = m^G(\xi) - e^{-2\pi L|\xi|}$ be the multiplier associated with the operator $M^G_1 - P_1$. Observe that by Theorem 6 and the properties of $p_1(\xi)$ there exists a constant $C > 0$ independent of the dimension and the body $G \subset \mathbb{R}^d$ such that

$$|k(\xi)| \leq C \min\{L|\xi|, (L|\xi|)^{-1}\}, \tag{64}$$

where $L = L(G)$ is the isotropic constant as in (22). Now by (64) and Plancherel's theorem we get

$$\begin{aligned}
&\Big\|\Big(\sum_{n\in\mathbb{Z}} |(M_{2^n}^G - P_{2^n})S_{j+n} f|^2\Big)^{1/2}\Big\|_{L^2} \\
&\quad = \Big(\int_{\mathbb{R}^d} \sum_{n\in\mathbb{Z}} \big|k(2^n\xi)\big(e^{-2\pi 2^{n+j}L|\xi|} - e^{-2\pi 2^{n+j-1}L|\xi|}\big)\big|^2 |\mathcal{F}f(\xi)|^2 \mathrm{d}\xi\Big)^{1/2} \\
&\quad \lesssim 2^{-|j|/2}\Big(\int_{\mathbb{R}^d} \sum_{n\in\mathbb{Z}} \min\{2^n L|\xi|, (2^n L|\xi|)^{-1}\} |\mathcal{F}f(\xi)|^2 \mathrm{d}\xi\Big)^{1/2} \\
&\quad \lesssim 2^{-|j|/2} \|f\|_{L^2},
\end{aligned} \tag{65}$$

with the implicit constant independent of d, N and the body $G \subset \mathbb{R}^d$. This proves (62) for $p = 2$.

4.2.2 Proof of Inequality (62) for $p \in (1, 2)$

For $s \in (1, 2]$ and $r \in [1, \infty]$, let $A_N(s, r)$ be the smallest constant in the following inequality

$$\Big\|\Big(\sum_{|n|\leq N} |(M_{2^n}^G - P_{2^n})g_n|^r\Big)^{1/r}\Big\|_{L^s} \leq A_N(s, r)\Big\|\Big(\sum_{|n|\leq N} |g_n|^r\Big)^{1/r}\Big\|_{L^s}. \tag{66}$$

It is easy to see that $A_N(s, r) < \infty$. Let $u \in (1, p)$ be such that $\frac{1}{u} = \frac{1}{2} + \frac{1}{2p}$. Now it is not difficult to see that $A_N(1, 1) \lesssim 1$, since $\|(M_{2^n}^G - P_{2^n})f\|_{L^p} \leq 2\|f\|_{L^p}$. Moreover, by (12), if $g = \sup_{|n|\leq N} |g_n|$ then

$$\Big\| \sup_{|n|\leq N} |(M_{2^n}^G - P_{2^n})g_n|\Big\|_{L^p} \lesssim (B_p(N) + 1)\|g\|_{L^p}.$$

Hence by the complex interpolation we obtain

$$A_N(u, 2) \leq A_N(1, 1)^{1/2} A_N(p, \infty)^{1/2} \lesssim (B_p(N) + 1)^{1/2}.$$

Then by (66) and (15) we get

$$\Big\|\Big(\sum_{|n|\le N}|(M^G_{2^n}-P_{2^n})S_{j+n}f|^2\Big)^{1/2}\Big\|_{L^u}\le A_N(u,2)\Big\|\Big(\sum_{n\in\mathbb{Z}}|S_{j+n}f|^2\Big)^{1/2}\Big\|_{L^u}$$
$$\lesssim (B_p(N)+1)^{1/2}\|f\|_{L^u}. \tag{67}$$

We now take $\rho\in(0,1]$ satisfying $\frac{1}{p}=\frac{1-\rho}{u}+\frac{\rho}{2}$, then $\rho=p-1$ and $1-\rho=2-p$. Interpolation between (65) and (67) yields (62) for $p\in(1,2)$ as desired.

4.3 *Proof of Inequality (59)*

To estimate (59) we use (14) and (46) and obtain

$$\Big\|\Big(\sum_{n\in\mathbb{Z}}\sup_{t\in[2^n,2^{n+1}]}|(M^G_t-M^G_{2^n})f|^2\Big)^{1/2}\Big\|_{L^p}$$
$$\lesssim\sum_{l\ge0}\sum_{j\in\mathbb{Z}}\Big\|\Big(\sum_{n\in\mathbb{Z}}\sum_{m=0}^{2^l-1}|(M^G_{2^n+2^{n-l}(m+1)}-M^G_{2^n+2^{n-l}m})S_{j+n}f|^2\Big)^{1/2}\Big\|_{L^p}. \tag{68}$$

Our aim now is to show that for every $q\in(1,2)$ and $\theta\in[0,1]$ such that $\frac{1}{p}=\frac{\theta}{2}+\frac{1-\theta}{q}$ we have, for every $f\in L^p(\mathbb{R}^d)$, the following estimate

$$\Big\|\Big(\sum_{n\in\mathbb{Z}}\sum_{m=0}^{2^l-1}|(M^G_{2^n+2^{n-l}(m+1)}-M^G_{2^n+2^{n-l}m})S_{j+n}f|^2\Big)^{1/2}\Big\|_{L^p}$$
$$\lesssim 2^{-\theta l/2+(1-\theta)l}\min\big\{1,2^l2^{-|j|/2}\big\}^\theta\|f\|_{L^p}, \tag{69}$$

with the implicit constant independent of the dimension and the underlying body $G\subset\mathbb{R}^d$.

Assume momentarily that (69) has been proven. Then we combine (68) with (69) and obtain estimate (59), since the double series

$$\sum_{l\ge0}\sum_{j\in\mathbb{Z}}2^{-\theta l/2+(1-\theta)l}\min\big\{1,2^l2^{-|j|/2}\big\}^\theta\lesssim1$$

is summable, whenever $\theta/2-(1-\theta)>0$, which forces p to satisfy $\frac{3}{1+1/q}<p\le2$, due to $\theta=\frac{2}{p}\frac{p-q}{2-q}$. This completes the proof of (4) from Theorem 1.

4.3.1 Proof of Inequality (69) for $p = 2$

Using inequalities (25) and arguing in a similar way as in (65) we obtain

$$\Big\|\Big(\sum_{n\in\mathbb{Z}}\sum_{m=0}^{2^l-1}|(M^G_{2^n+2^{n-l}(m+1)} - M^G_{2^n+2^{n-l}m})S_{j+n}f|^2\Big)^{1/2}\Big\|_{L^2} \lesssim 2^{l/2}2^{-|j|/2}\|f\|_{L^2}. \tag{70}$$

Note that inequality (26) implies

$$\begin{aligned}|m^G((2^n+2^{n-l}(m+1))\xi) - m^G((2^n+2^{n-l}m)\xi)|\\ \leq \int_{2^n+2^{n-l}m}^{2^n+2^{n-l}(m+1)} |\langle t\xi, \nabla m^G(t\xi)\rangle|\frac{\mathrm{d}t}{t} \lesssim 2^{-l}.\end{aligned}$$

Therefore, by Plancherel's theorem

$$\begin{aligned}&\Big\|\Big(\sum_{n\in\mathbb{Z}}\sum_{m=0}^{2^l-1}|(M^G_{2^n+2^{n-l}(m+1)} - M^G_{2^n+2^{n-l}m})S_{j+n}f|^2\Big)^{1/2}\Big\|_{L^2}\\
&= \Big(\sum_{n\in\mathbb{Z}}\sum_{m=0}^{2^l-1}\|(M^G_{2^n+2^{n-l}(m+1)} - M^G_{2^n+2^{n-l}m})S_{j+n}f\|^2_{L^2}\Big)^{1/2}\\
&\lesssim \Big(\sum_{n\in\mathbb{Z}}2^{-l}\|S_{j+n}f\|^2_{L^2}\Big)^{1/2}\\
&\lesssim 2^{-l/2}\|f\|_{L^2}.\end{aligned} \tag{71}$$

Combining (70) and (71) we obtain

$$\begin{aligned}\Big\|\Big(\sum_{n\in\mathbb{Z}}\sum_{m=0}^{2^l-1}|(M^G_{2^n+2^{n-l}(m+1)} - M^G_{2^n+2^{n-l}m})S_{j+n}f|^2\Big)^{1/2}\Big\|_{L^2}\\
\lesssim 2^{-l/2}\min\big\{1, 2^l2^{-|j|/2}\big\}\|f\|_{L^2},\end{aligned} \tag{72}$$

which proves (69) for $p = 2$.

4.3.2 Proof of Inequality (69) for $p \in (3/2, 2)$

We begin with a general remark, a consequence of (5), which states that for every $q \in (1, \infty)$ there is a constant $C_q > 0$ independent of the dimension and the underlying body $G \subset \mathbb{R}^d$ such that for every sequence $(g_n)_{n\in\mathbb{Z}} \in L^q(\ell^2(\mathbb{R}^d))$ we have

$$\Big\|\Big(\sum_{n\in\mathbb{Z}} |M^G_{2^n} g_n|^2\Big)^{1/2}\Big\|_{L^q} \le C_q \Big\|\Big(\sum_{n\in\mathbb{Z}} |g_n|^2\Big)^{1/2}\Big\|_{L^q}. \tag{73}$$

Indeed, let $A(q, r)$ be the best constant in the following inequality

$$\Big\|\Big(\sum_{n\in\mathbb{Z}} |M^G_{2^n} g_n|^r\Big)^{1/r}\Big\|_{L^q} \le A(q,r) \Big\|\Big(\sum_{n\in\mathbb{Z}} |g_n|^r\Big)^{1/r}\Big\|_{L^q}.$$

By the complex interpolation and duality ($A(q, r) = A(q', r')$) and inequality (5) we obtain

$$A(q,2) \le A(q,1)^{1/2} A(q,\infty)^{1/2} = A(q',\infty)^{1/2} A(q,\infty)^{1/2} \le C_{q'}^{1/2} C_q^{1/2},$$

which implies (73). Observe that by (73) and (15), since $M^G_{2^n(1+t)} = M^{(1+t)G}_{2^n}$, we obtain

$$\begin{aligned}
&\Big\|\Big(\sum_{n\in\mathbb{Z}} \sum_{m=0}^{2^l-1} |(M^G_{2^n+2^{n-l}(m+1)} - M^G_{2^n+2^{n-l}m}) S_{j+n} f|^2\Big)^{1/2}\Big\|_{L^q} \\
&\qquad \lesssim 2^l \sup_{t\in[0,1]} \Big\|\Big(\sum_{n\in\mathbb{Z}} |M^G_{2^n(1+t)} S_{j+n} f|^2\Big)^{1/2}\Big\|_{L^q} \\
&\qquad \lesssim 2^l \Big\|\Big(\sum_{n\in\mathbb{Z}} |S_{j+n} f|^2\Big)^{1/2}\Big\|_{L^q} \\
&\qquad \lesssim 2^l \|f\|_{L^q}.
\end{aligned} \tag{74}$$

Interpolating (72) with (74) we obtain (69) as desired.

5 Discrete Perspective: Proof of Theorem 2

The main objective of this section is to provide dimensional-free estimates on $\ell^p(\mathbb{Z}^d)$, for $p \in (1, \infty]$, of the norm of the maximal function corresponding to the operators $\mathcal{M}^{B^2}_N$ from (6) with large scales $N \ge Cd$ for some $C > 0$, where $N > 0$ is a real number. The estimate in (9) will be deduced by comparison of $\sup_{N\ge Cd} |\mathcal{M}^{B^2}_N f|$ with its continuous analogue, for which we have dimension-free bounds provided by the third author in [33]. Namely, we know that for every

$p \in (1, \infty)$ there is $C_p > 0$ independent of the dimension such that for every $f \in L^p(\mathbb{R}^d)$ we have

$$\|M_*^{B^2} f\|_{L^p} \le C_p \|f\|_{L^p}. \tag{75}$$

Throughout this section, unless otherwise stated, $N > 0$ is always a real number and $Q = [-1/2, 1/2]^d$ denotes the unit cube. A fundamental role, in the proofs of this section, will be played by the fact that the Euclidean norm corresponds to the scalar product $|x|^2 = \langle x, x\rangle$. We begin with crude estimates for the number of lattice points the Euclidean balls B_N^2.

Lemma 11 *Let $N > 0$ and set $N_1 = (N^2 + d/4)^{1/2}$. Then*

$$|B_N^2 \cap \mathbb{Z}^d| \le 2|B_{N_1}^2|. \tag{76}$$

Moreover, if $N \ge Cd$ for some fixed $C > 0$, then we have

$$|B_N^2 \cap \mathbb{Z}^d| \le 2e^{1/(8C^2)}|B_N^2|. \tag{77}$$

Proof For $x \in B_N^2$ and $z \in Q$ we have

$$|x + z|^2 \le N^2 + \frac{d}{4} + 2\langle x, z\rangle.$$

Moreover, for all $x \in B_N^2$ we have

$$|\{z \in Q : \langle x, z\rangle \le 0\}| \ge \frac{1}{2}.$$

Hence

$$\begin{aligned}
|B_N^2 \cap \mathbb{Z}^d| &= \sum_{x \in B_N^2 \cap \mathbb{Z}^d} 1 \le 2 \sum_{x \in B_N^2 \cap \mathbb{Z}^d} \int_Q \mathbb{1}_{\{z \in Q:\, \langle x, z\rangle \le 0\}}(y)dy \\
&\le 2 \sum_{x \in B_N^2 \cap \mathbb{Z}^d} \int_Q \mathbb{1}_{\{z \in Q:\, |x+z| \le N_1\}}(y)dy \\
&\le 2 \sum_{x \in \mathbb{Z}^d} \int_Q \mathbb{1}_{B_{N_1}^2}(x + y)dy \\
&= 2 \sum_{x \in \mathbb{Z}^d} \int_{x+Q} \mathbb{1}_{B_{N_1}^2}(y)dy = 2|B_{N_1}^2|.
\end{aligned}$$

This proves (76). For (77) note that for $N \geq Cd$ we get

$$|B^2_{N_1}| = \frac{\pi^{d/2} N_1^d}{\Gamma(d/2+1)} = |B^2_N| \left(1 + \frac{d}{4N^2}\right)^{d/2} \leq |B^2_N| \left(1 + \frac{1}{4CN}\right)^{d/2} \leq e^{1/(8C^2)} |B^2_N|,$$

which proves (77). □

Lemma 12 *Assume that $N \geq Cd$ for some fixed $C > 0$ and let $t > 0$. Then for every $x \in \mathbb{R}^d$ such that $|x| \geq N(1+t/N)^{1/2}$ we have*

$$|Q \cap (B^2_N - x)| = |\{y \in Q \colon x + y \in B^2_N\}| \leq 2e^{-ct^2}, \tag{78}$$

where $c = \frac{7}{32} \frac{C^2}{(C+1)^2}$.

Proof Let $|x| \geq N(1+t/N)^{1/2}$. Then for $y \in Q$ and $x + y \in B^2_N$ we have

$$N^2 + Nt \leq |x|^2 = |x + y - y|^2 \leq N^2 - 2\langle x, y\rangle - |y|^2 \leq N^2 + 2|\langle x, y\rangle|.$$

Thus for $\bar{x} = x/|x|$ one has

$$|\langle \bar{x}, y\rangle| \geq \frac{1}{2} \frac{Nt}{|x|} \geq \frac{1}{2} \frac{Nt}{|x+y| + |y|} \geq \frac{1}{2} \frac{Nt}{N + d^{1/2}} \geq \frac{1}{2} \frac{Ct}{C+1},$$

and consequently we get

$$|\{y \in Q \colon x + y \in B^2_N\}| \leq |\{y \in Q : |\langle \bar{x}, y\rangle| \geq Ct/(2C+2)\}|. \tag{79}$$

We claim that for every unit vector $z \in \mathbb{R}^d$ and for every $s > 0$ we have

$$|\{y \in Q : \langle z, y\rangle \geq s\}| \leq e^{-\frac{7}{8}s^2}. \tag{80}$$

Taking $s = Ct/(2C+2)$ in (80) and coming back to (79) we complete the proof of (78) with $c = \frac{7}{32} \frac{C^2}{(C+1)^2}$.

In the proof of (80) we will appeal to the inequality $e^x + e^{-x} \leq 2e^{\frac{1}{2}x^2}$, which holds for all $x \geq 0$. Indeed, for every $\alpha > 0$ we get

$$\begin{aligned} e^{\alpha s} |\{y \in Q : \langle z, y\rangle \geq s\}| &\leq \int_Q e^{\alpha \sum_{j=1}^d z_j y_j} \mathrm{d}y \\ &= \prod_{j=1}^d \int_0^{1/2} e^{\alpha z_j y_j} + e^{-\alpha z_j y_j} \mathrm{d}y_j \\ &\leq \prod_{j=1}^d 2 \int_0^{1/2} e^{\frac{1}{2}\alpha^2 (z_j y_j)^2} \mathrm{d}y_j \end{aligned}$$

$$\le \prod_{j=1}^{d} e^{\frac{1}{8}\alpha^2 z_j^2}$$
$$= e^{\frac{1}{8}\alpha^2 \sum_{j=1}^{d} z_j^2}$$
$$= e^{\frac{1}{8}\alpha^2}.$$

Taking $\alpha = s$ in the inequality above and dividing by e^{s^2} we obtain (80) and the proof is completed. □

Lemma 13 *There are constants $C_1, C_2 > 0$ such that for every $N \ge C_1 d$ we have*

$$|B_N^2| \le C_2 |B_N^2 \cap \mathbb{Z}^d|. \tag{81}$$

Moreover, (81) combined with (77) from Lemma 11 yields

$$C_2^{-1} |B_N^2| \le |B_N^2 \cap \mathbb{Z}^d| \le 2e^{1/(8C_1^2)} |B_N^2|$$

for every $N \ge C_1 d$.

Proof We show that there is $J \in \mathbb{N}$ such that for every $M \ge d$ we have

$$|B_M^2| \le 2 |B_{M(1+J/M)^{1/2}}^2 \cap \mathbb{Z}^d|. \tag{82}$$

Assume momentarily that (82) is proven, then (81) follows. Indeed, for every $N \ge C_1 d$, where $C_1 = 2(1+J)$ we find $M \ge d$ such that $N = M(1+J/M)^{1/2}$, hence, (82) implies

$$|B_M^2| \le 2 |B_N^2 \cap \mathbb{Z}^d|. \tag{83}$$

On the other hand we have

$$|B_M^2| \le |B_N^2| \le (1+J/M)^{d/2} |B_M^2| \le e^J |B_M^2|,$$

since $M \ge d$. This estimate combined with (83) gives (81) with $C_2 = 2e^J$.

Our aim now is to prove (82). For this purpose let $J \in \mathbb{N}$ be a large number such that

$$\sum_{j \ge J} e^{-j^2/32} e^j \le \frac{1}{8e}.$$

Define $U_j = \{x \in \mathbb{R}^d : M\big(1+\frac{j}{M}\big)^{1/2} < |x| \le M\big(1+\frac{(j+1)}{M}\big)^{1/2}\}$ and observe that

$$\begin{aligned}
|B_M^2| &= \sum_{x\in\mathbb{Z}^d}\int_{x+Q}\mathbb{1}_{B_M^2}(y)\mathrm{d}y\\
&= \sum_{x\in\mathbb{Z}^d}\int_{Q}\mathbb{1}_{B_M^2}(x+y)\mathrm{d}y\\
&\le \sum_{x\in B_M^2\cap\mathbb{Z}^d}\int_{Q}\mathbb{1}_{B_M^2}(x+y)\mathrm{d}y + \sum_{j\ge 0}\sum_{x\in U_j\cap\mathbb{Z}^d}\int_{Q}\mathbb{1}_{B_M^2}(x+y)\mathrm{d}y\\
&\le |B_M^2\cap\mathbb{Z}^d| + \sum_{0\le j<J}\sum_{x\in U_j\cap\mathbb{Z}^d}|Q\cap(B_M^2-x)| + \sum_{j\ge J}\sum_{x\in U_j\cap\mathbb{Z}^d}|Q\cap(B_M^2-x)|\\
&\le |B^2_{M(1+J/M)^{1/2}}\cap\mathbb{Z}^d| + \sum_{j\ge J}\sum_{x\in U_j\cap\mathbb{Z}^d}|Q\cap(B_M^2-x)|.
\end{aligned}\tag{84}$$

By (77), since $M \ge d$, we get

$$\begin{aligned}
|B^2_{M(1+(j+1)/M)^{1/2}}\cap\mathbb{Z}^d| &\le 2e^{1/8}|B^2_{M(1+(j+1)/M)^{1/2}}|\\
&\le 2e^{1/8}\Big(1+\frac{j+1}{d}\Big)^{d/2}|B_M^2|\\
&\le 2e^{1/8}e^{(j+1)/2}|B_M^2|.
\end{aligned}$$

Using this estimate, the definition of the sets U_j and Lemma 12 we obtain for any $M \ge d$ that

$$\begin{aligned}
\sum_{j\ge J}\sum_{x\in U_j\cap\mathbb{Z}^d}|Q\cap(B_M^2-x)| &\le 2\sum_{j\ge J}e^{-j^2/32}|B^2_{M(1+(j+1)/M)^{1/2}}\cap\mathbb{Z}^d|\\
&\le 4e^{5/8}|B_M^2|\sum_{j\ge J}e^{-j^2/32}e^j\\
&\le \frac{1}{2}|B_M^2|.
\end{aligned}\tag{85}$$

Combining (85) with (84) we obtain (82) as desired. This completes the proof of Lemma 13. □

We now are ready to prove Theorem 2.

Proof (of Theorem 2) Let $f : \mathbb{Z}^d \to \mathbb{C}$ and define its extension $F : \mathbb{R}^d \to \mathbb{C}$ on $\mathbb{R}^d$ by setting

$$F(x) = \sum_{y \in \mathbb{Z}^d} f(y) \mathbb{1}_{y+Q}(x).$$

Then, clearly $\|F\|_{L^p(\mathbb{R}^d)} = \|f\|_{\ell^p(\mathbb{Z}^d)}$ for every $p \geq 1$.

From now on we assume that $f \geq 0$. For every $N \geq C_1 d$, with C_1 as in Lemma 13, we define $N_1 = (N^2 + d/4)^{1/2}$. Observe that for $z \in Q$ and $y \in B^2_N$ we have

$$|y + z|^2 = |y|^2 + |z|^2 + 2\langle z, y \rangle \leq N_1^2$$

on the set $\{z \in Q : \langle z, y \rangle \leq 0\}$, which has measure $1/2$. Then by Lemma 13 for all $x \in \mathbb{Z}^d$ we obtain

$$\begin{aligned}
\mathcal{M}_N^{B^2} f(x) &= \frac{1}{|B^2_N \cap \mathbb{Z}^d|} \sum_{y \in B^2_N \cap \mathbb{Z}^d} f(x + y) \mathbb{1}_{B^2_N}(y) \\
&\lesssim \frac{1}{|B^2_N|} \sum_{y \in \mathbb{Z}^d} f(x + y) \int_Q \mathbb{1}_{B^2_{N_1}}(y + z) \mathrm{d}z \\
&= \frac{1}{|B^2_N|} \sum_{y \in \mathbb{Z}^d} f(y) \int_{x + B^2_{N_1}} \mathbb{1}_{y+Q}(z) \mathrm{d}z \\
&= \frac{1}{|B^2_N|} \int_{x + B^2_{N_1}} F(z) \mathrm{d}z \\
&= \left(\frac{N_1}{N}\right)^d \frac{1}{|B^2_{N_1}|} \int_{B^2_{N_1}} F(x + z) \mathrm{d}z \\
&\lesssim \frac{1}{|B^2_{N_1}|} \int_{B^2_{N_1}} F(x + z) \mathrm{d}z \\
&= M_{N_1}^{B^2} F(x).
\end{aligned} \tag{86}$$

Finally, take $N_2 = (N_1^2 + d/4)^{1/2}$. Similarly as above, for $y \in Q$ and $z \in B^2_{N_1}$ we have

$$|y + z|^2 \leq |y|^2 + |z|^2 + 2\langle z, y \rangle \leq N_2^2$$

on the set $\{y \in Q \colon \langle z, y\rangle \le 0\}$, which has Lebesgue measure 1/2. Therefore, Fubini's theorem leads to

$$\begin{aligned} M_{N_1}^{B^2} F(x) &= \frac{1}{|B_{N_1}^2|}\int_{B_{N_1}^2} F(x+z)\mathrm{d}z \\ &\le \frac{2}{|B_{N_1}^2|}\int_{\mathbb{R}^d} F(x+z)\mathbb{1}_{B_{N_1}^2}(z)\int_Q \mathbb{1}_{B_{N_2}^2}(z+y)\mathrm{d}y\mathrm{d}z \\ &\lesssim \frac{1}{|B_{N_2}^2|}\int_Q\int_{\mathbb{R}^d} F(x+z-y)\mathbb{1}_{B_{N_2}^2}(z)\mathrm{d}z\mathrm{d}y \\ &= \int_{x+Q} M_{N_2}^{B^2} F(y)\mathrm{d}y. \end{aligned} \tag{87}$$

Combining (86) with (87), applying Hölder's inequality, and invoking (75) we arrive at

$$\begin{aligned} \| \sup_{N\ge C_1 d} |\mathcal{M}_N^{B^2} f|\|_{\ell^p(\mathbb{Z}^d)}^p &\lesssim \sum_{x\in\mathbb{Z}^d}\int_{x+Q} |\sup_{N\ge C_1 d} M_N^{B^2} F(y)|^p \mathrm{d}y \\ &= \| \sup_{N\ge C_1 d} M_N^{B^2} F\|_{L^p(\mathbb{R}^d)}^p \\ &\lesssim \|F\|_{L^p(\mathbb{R}^d)}^p \\ &= \|f\|_{\ell^p(\mathbb{Z}^d)}^p. \end{aligned}$$

This proves Theorem 2 with $C = C_1$. □

Acknowledgments The authors are grateful to the referees for careful reading of the manuscript and useful remarks that led to the improvement of the presentation.

References

1. Aldaz, J.M.: The weak type (1, 1) bounds for the maximal function associated to cubes grow to infinity with the dimension. Ann. Math. **173**(2), 1013–1023 (2011)
2. Aubrun, G.: Maximal inequality for high-dimensional cubes. Confluentes Math. **1**, 169–179 (2009)
3. Bourgain, J.: Averages in the plane over convex curves and maximal operators . J. Analyse Math. **46**(1), 69–85 (1986)
4. Bourgain, J.: On high dimensional maximal functions associated to convex bodies. Amer. J. Math. **108**, 1467–1476 (1986)
5. Bourgain, J.: On L^p bounds for maximal functions associated to convex bodies in $\mathbb{R}^n$. Israel J. Math. **54**, 257–265 (1986)

6. Bourgain, J.: On dimension free maximal inequalities for convex symmetric bodies in $\mathbf{R}^n$. In: Geometric Aspects of Functional Analysis, Israel Seminar (GAFA) (1985/86). Lecture Notes in Mathematics, vol. 1267, pp. 168–176. Springer, Berlin, Heidelberg (1987)
7. Bourgain, J.: On the distribution of polynomials on high dimensional convex sets. In: Geometric Aspects of Functional Analysis, Israel Seminar (GAFA) (1989/90). Lecture Notes in Mathematics, vol. 1469, pp. 127–137. Springer, Berlin, Heidelberg (1991)
8. Bourgain, J.: On the Hardy-Littlewood maximal function for the cube. Israel J. Math. **203**, 275–293 (2014)
9. Bourgain, J., Mirek, M., Stein, E.M., Wróbel, B.: Dimension-free variational estimates on $L^p(\mathbb{R}^d)$ for symmetric convex bodies. Geom. Funct. Anal. **28**(1), 58–99 (2018)
10. Bourgain, J., Mirek, M., Stein, E.M., Wróbel, B.: Dimension-free estimates for discrete Hardy–Littlewood averaging operators over the cubes in $\mathbb{Z}^d$. Amer. J. Math. **141**(4), 857–905 (2019)
11. Bourgain, J., Mirek, M., Stein, E.M., Wróbel, B.: On discrete Hardy–Littlewood maximal functions over the balls in $\mathbb{Z}^d$: dimension-free estimates. In: Klartag, B., Milman, E. (eds.) Geometric Aspects of Functional Analysis, Israel Seminar (GAFA), Lecture Notes in Mathematics 2256, Vol. I, pp, 127–169, Chapter 8, (2017–2019)
12. Brazitikos, S., Giannopoulos, A., Valettas, P., Vritsiou, B.-H.: Geometry of Isotropic Convex Bodies. Mathematical Surveys and Monographs, pp. 1–594. American Mathematical Society, Providence (2014)
13. Carbery, A.: An almost-orthogonality principle with applications to maximal functions associated to convex bodies. Bull. Amer. Math. Soc. **14**(2), 269–274 (1986)
14. Carbery, A.: Differentiation in lacunary directions and an extension of the Marcinkiewicz multiplier theorem. Ann. Inst. Fourier (Grenoble) **38**(1), 157–168 (1988)
15. Cowling, M.G.: Harmonic analysis on semigroups. Ann. Math. **117**(2), 267–283 (1983)
16. Delaval, L., Guédon, O., Maurey, B.: Dimension-free bounds for the Hardy-Littlewood maximal operator associated to convex sets. Ann. Fac. Sci. Toulouse Math. **27**(1), 1–198 (2018)
17. Duoandikoetxea, J., Rubio de Francia, J.L.: Maximal and singular integral operators via Fourier transform estimates. Invent. Math. **84**(3), 541–561 (1986)
18. Iakovlev, A.S., Strömberg, J.O.: Lower bounds for the weak type (1, 1) estimate for the maximal function associated to cubes in high dimensions. Math. Res. Lett. **20**, 907–918 (2013)
19. Klartag, B.: On convex perturbations with a bounded isotropic constant. Geom. Funct. Anal. (GAFA) **16**(6), 1274–1290 (2006)
20. Lewko, A., Lewko, M.: Estimates for the square variation of partial sums of Fourier series and their rearrangements. J. Funct. Anal. **262**(6), 2561–2607 (2012)
21. Mirek, M., Trojan, B.: Discrete maximal functions in higher dimensions and applications to ergodic theory. Amer. J. Math. **138**(6), 1495–1532 (2016)
22. Mirek, M., Stein, E.M., Trojan, B.: $\ell^p(\mathbb{Z}^d)$-estimates for discrete operators of Radon type: Variational estimates. Invent. Math. **209**(3), 665–748 (2017)
23. Mirek, M., Stein, E.M., Zorin–Kranich, P.: Jump inequalities for translation-invariant polynomial averages and singular integrals on $\mathbb{Z}^d$. Adv. Math. **365**(107065), 57 (2020)
24. Mirek, M., Stein, E.M., Zorin–Kranich, P.: Jump inequalities via real interpolation. Math. Annal. **376**(1–2), 797–819 (2020)
25. Mirek, M., Stein, E.M., Zorin–Kranich, P.: A bootstrapping approach to jump inequalities and their applications. Analysis & PDE **13**(2), 527–558 (2020).
26. Müller, D.: A geometric bound for maximal functions associated to convex bodies. Pacific J. Math. **142**(2), 297–312 (1990)
27. Nagel, A., Stein, E.M., Wainger, S.: Differentiation in lacunary directions. Proc. Nat. Acad. Sci. U. S. A. **75**(3), 1060–1062 (1978)
28. Naor, A., Tao, T.: Random martingales and localization of maximal inequalities. J. Funct. Anal. **259**(3), 731–779 (2010)
29. Pisier, G.: Holomorphic semi-groups and the geometry of Banach spaces. Ann. Math. **115**, 375–392 (1982)
30. Sommer, F.: Dimension free L^p-bounds of maximal functions associated to products of Euclidean balls (2018). Preprint. https://arxiv.org/abs/1703.07728

31. Stein, E.M.: Topics in Harmonic Analysis Related to the Littlewood-Paley Theory. Annals of Mathematics Studies, pp. 1–157. Princeton University Press, Princeton (1970)
32. Stein, E.M.: Maximal functions, spherical means. Proc. Nat. Acad. Sci. U. S. A. **73**(7), 2174–2175 (1976)
33. Stein, E.M.: The development of square functions in the work of A. Zygmund. Bull. Amer. Math. Soc. **7**, 359–376 (1982)
34. Stein, E.M.: Some results in harmonic analysis in $\mathbb{R}^n$, $n \to \infty$. Bull. Amer. Math. Soc. **9**, 71–73 (1983)
35. Stein, E.M., Strömberg, J.O.: Behavior of maximal functions in $\mathbb{R}^n$ for large n. Ark. Mat. **21**(1–2), 259–269 (1983)
36. Zygmund, A.: Trigonometric Series, Vols. I and II. 3rd edn. Cambridge Mathematical Library. Cambridge University Press, Cambridge (2002)

Potential Spaces on Lie Groups

Tommaso Bruno, Marco M. Peloso, and Maria Vallarino

Dedicated to Fulvio Ricci on the occasion of his 70th birthday.

Abstract In this paper we discuss function spaces on a general noncompact Lie group, namely the scales of Triebel–Lizorkin and Besov spaces, defined in terms of a sub-Laplacian with drift. The sub-Laplacian is written as the (negative) sum of squares of a collection of left-invariant vector fields satisfying Hörmander's condition. These spaces were recently introduced by the authors. In this paper we prove a norm characterization in terms of finite differences, the density of test functions, and related isomorphism properties.

Keywords Lie groups · Besov spaces · Triebel–Lizorkin spaces

All authors were partially supported by the grant PRIN 2015 *Real and Complex Manifolds: Geometry, Topology and Harmonic Analysis*, and are members of the Gruppo Nazionale per l'Analisi Matematica, la Probabilità e le loro Applicazioni (GNAMPA) of the Istituto Nazionale di Alta Matematica (INdAM). T. Bruno was also supported by the Research Foundation – Flanders (FWO) through the postdoctoral grant 12ZW120N.

T. Bruno
Dipartimento di Scienze Matematiche "Giuseppe Luigi Lagrange", Dipartimento di Eccellenza 2018–2022, Politecnico di Torino, Torino, Italy

Present Address: Department of Mathematics: Analysis, Logic and Discrete Mathematics, Ghent University, Ghent, Belgium
e-mail: tommaso.bruno@ugent.be

M. M. Peloso (✉)
Dipartimento di Matematica, Università degli Studi di Milano, Milano, Italy
e-mail: marco.peloso@unimi.it

M. Vallarino
Dipartimento di Scienze Matematiche "Giuseppe Luigi Lagrange", Dipartimento di Eccellenza 2018–2022, Politecnico di Torino, Torino, Italy
e-mail: maria.vallarino@polito.it

P. Ciatti, A. Martini (eds.), *Geometric Aspects of Harmonic Analysis*,
Springer INdAM Series 45, https://doi.org/10.1007/978-3-030-72058-2_4

1 Introduction

The theory of function spaces, regularity of integral operators, and of solutions of differential equations, began in the setting of Euclidean spaces, with smoothness measured in terms of Sobolev and Lipschitz norms, see e.g. [46]. A. Calderón and A. Zygmund developed the theory of singular integrals, proving their boundedness in the Lebesgue spaces, as well as regularity of solutions of classical differential equations, such as the Dirichlet and Neumann problems, in the case of a half-space and of smooth domains. Among the operators studied were the singular integrals, hence in particular the Hilbert and Riesz transforms, the Poisson integral, and the heat propagator. It is worth noticing that the singularities of the integral kernels of such operators, or better, of the level sets of their moduli, were naturally described in terms of the underlying Euclidean geometry. Such theory then included embedding and interpolation results for Lebesgue, Sobolev and Lipschitz spaces, see e.g. [3]. In this analysis, the Fourier series and transform played a crucial role, and a noticeable application of such techniques was the decomposition initially introduced by Littlewood and Paley, and later developed in depth by E. M. Stein [47]. The Littlewood–Paley decomposition was initially intended to provide a substitute for the Plancherel formula to the L^p-norms, with $p \neq 2$, but proved to be an invaluable tool in many other situations. The function spaces that naturally arose in studying the regularity properties of aforementioned operators were indeed, besides the Lebesgue spaces, the Sobolev and Lipschitz spaces, and also the Besov spaces. It became then natural to obtain other characterizations for such norms, and in this setting the Littlewood–Paley decomposition proved to be very useful, and was also used to define another, related, scale of spaces, the so-called Triebel–Lizorkin spaces, see e.g. [53], which include the Sobolev spaces as a special case.

While such theory was in its full development, L. Hörmander produced two breakthrough results, [25] and [26]. In [25] Hörmander extended a previous result by Mihlin, developing the theory of L^p-multipliers of the Laplacian. This approach also stimulated the study of a class of operators that naturally appear while solving partial differential equations involving the Laplacian — for instance the wave equation in the Euclidean space $\mathbb{R}^d$.

In [26] Hörmander showed that operators that are sum of squares of vector fields whose commutators up to a finite order span all directions of $\mathbb{R}^d$, although non-elliptic, enjoy many interesting and strong properties of elliptic operators, in particular hypoelliticity. Such phenomenon appeared for instance in the case of the Kohn-Laplacian on the boundary of the Siegel upper half-space in $\mathbb{C}^{d+1}$, in the works of A. Korányi and S. Vági [30], J. J. Kohn [28] and, with most relevance to this discussion and the present work, of G. B. Folland and Stein [13]. The operators that were considered in [13], that is the Kohn-Laplacian, the sub-Laplacian, the so-called Folland–Stein operators, their fundamental solutions, or the relative fundamental solutions in some cases, had the singularity that could be described in terms of a different underlying geometry. The boundary of the Siegel upper half-space can be identified with the Heisenberg group, and such geometry was more efficiently described using the nilpotent Lie group structure of the Heisenberg group. As a metric space, the Heisenberg group $\mathbb{H}_d$ is not equivalent to the Euclidean space $\mathbb{R}^{2d+1}$, and in fact the distance coincides with the Carnot–Carathéodory distance defined by the sub-Laplacian on $\mathbb{H}_d$. The Lie algebra of $\mathbb{H}_d$ can be written as the linear span of a family of vector fields $\mathbf{X} = \{X_1, \dots, X_{2d}\}$ and of their commutators, which reduce in fact to a single "transversal" vector field T. The sub-Laplacian on $\mathbb{H}_d$ is the (negative) sum of squares $-\sum_{j=1}^{2d} X_j^2$, and thus is of the type studied by Hörmander in [26]. The function spaces that better describe the smoothness of functions in this setting can be defined by their behaviour with respect to the action of only the vector fields $\mathbf{X}$. Such systems of vector fields were called *horizontal* and they were studied in [13] and [10] and again differed from their Euclidean analogues. In these papers, the authors proved analogue of embedding and interpolation results for the newly defined Sobolev and Lipschitz spaces, in the case of $\mathbb{H}_d$, and of Carnot–Carathéodory groups, respectively.[1]

These results gave tremendous impetus to the development of analysis on $\mathbb{H}_d$, and more in general on Carnot–Carathéodory groups. In a series of papers, F. Ricci and E. M. Stein [42–44] studied the boundedness of singular integrals on nilpotent Lie groups, exploring again the connection between the geometry of the metric balls, the size properties of the integral kernels, and the boundedness of the singular integral operators. In [48] R. Strichartz pointed out the importance of the role of the joint spectrum of the sub-Laplacian and T. In two fundamental papers, [35, 36] D. Müller, F. Ricci, and E. M. Stein then proved the boundedness of joint spectral multipliers of the sub-Laplacian and T on $\mathbb{H}_d$ and the closely related Heisenberg type groups – results that were effectively extended to more general groups, though in a slightly different way, by A. Martini [32, Theorem 5.7]. Other related results, on spaces of differential forms, in the spirit of this discussion are [37, 38] and [39, 40].

Thus, a common theme of this circle of ideas is that the underlying manifold, Riemannian or sub-Riemannian, and the collection of vector fields $\mathbf{X}$ satisfying Hörmander's condition and defining the corresponding sub-Laplacian determine a metric structure. The most efficient way to describe smoothness of functions and regularity of canonical operators is via a scale of spaces that are modeled by the sub-Laplacian, hence by $\mathbf{X}$, itself.

In the setting of Carnot–Carathéodory groups, and more in general of Lie groups of polynomial growth, endowed with the sub-Riemannian structure induced by a family $\mathbf{X}$ of vector fields satisfying Hörmander's condition, a Mihlin–Hörmander multiplier theorem holds. This fact allowed G. Furioli, C. Melzi and A. Veneruso [14] to introduce Besov spaces on such groups, which were later studied by I. Gallagher and Y. Sire [16]. The theory was recently extended to any unimodular Lie group by J. Feneuil [9].

This work aims to contribute to the analysis of function spaces on general noncompact Lie groups, hence including the nonunimodular groups, with Haar measures of exponential growth.

[1] In [10] the Carnot–Carathéodory groups were called *stratified nilpotent Lie groups*.

Concerning the function spaces, their algebra properties are of great importance, in particular in application to well-posedness results for nonlinear differential equations. In this direction, a remarkable paper is [7] by T. Coulhon, E. Russ and V. Tardivel-Nachev, where they proved algebra properties for the Sobolev spaces, in particular on any unimodular Lie group. The algebra properties were extended to the scale of Besov spaces on groups of polynomial growth in [16] and in [9] on unimodular Lie groups.

A number of the aforementioned results were also obtained in the context of doubling measure metric spaces with the reverse doubling property, see e.g. [22, 34], and in the setting of Riemannian manifolds of bounded geometry, see e.g. [49–51], [7], and references therein. On the other hand, not much is known in the setting of a sub-Riemannian manifold. This work is part of a program [41], [5] and [6], whose main long term goal is to address this type of questions on a sub-Riemannian manifold, and we started with the case of a general Lie group. The paper [41] studies Sobolev spaces with respect to the sum-of-squares sub-Laplacian, results then extended to Sobolev spaces with respect to sub-Laplacians with drift in [5], while in [6] we develop the theory of Besov and Triebel–Lizorkin spaces with respect to sub-Laplacians with drift, that we further analyse in this work.

We conclude this part of the introduction by pointing out that the literature in this area is extremely vast, and it is just impossible to give credit to all the authors that have contributed to its development. We apologise to everyone whom we did not explicitly mention.

Let G be a noncompact connected Lie group and let $\mathbf{X} = \{X_1, \dots, X_\ell\}$ be a family of linearly independent left-invariant vector fields on G satisfying Hörmander's condition. We denote by δ the modular function on G. Let ρ be a right Haar measure of G, let χ be a continuous positive character of G, and consider the measure μ_χ defined by the relation $d\mu_\chi = \chi d\rho$. Consider now the differential operator

$$\Delta_\chi = -\sum_{j=1}^{\ell}(X_j^2 + c_j X_j)\,, \tag{1}$$

with domain $C_c^\infty(G)$, where $c_j = (X_j\chi)(e)$, $j = 1, \dots, \ell$, and e is the identity of G.

This operator was introduced by W. Hebisch et al. in [23], where they showed that Δ_χ is essentially self-adjoint on $L^2(\mu_\chi)$. Moreover, they proved that if a sub-Laplacian with drift is symmetric on $L^2(\mu)$ for a positive measure μ on G, then necessarily $\mu = \mu_\chi$ for a positive character χ on G, and moreover the drift has the form $X := \sum_{j=1}^{\ell} c_j X_j$, where $c_j = (X_j\chi)(e)$, $j = 1, \dots, \ell$, as in (1). Notice that when the character χ is the modular function, $\mu_\delta = \lambda$ is a left Haar measure and the operator Δ_δ coincides with the intrinsic hypoelliptic Laplacian associated with the Carnot-Carathéodory metric induced on G by the vector fields $\mathbf{X}$—see [1]. The operator Δ_δ is the natural substitute of the Laplacian on a general Lie group G. This also reflects on the fact that the measure λ is privileged among the measures μ_χ. As shown in [1, 5], Δ_δ is not a sum-of-squares operator unless the group is unimodular. In this paper we continue the study of function spaces associated with Δ_χ for a generic positive continuous character χ. The more general treatment allows extra flexibility, see e.g. the embedding results, Theorems 1.1 and 4.4 in [5] and Theorems 5.2 and 5.3 in [6], and at the same time, highlights the naturality of Δ_δ.

In this paper we further develop the investigation of Besov and Triebel–Lizorkin spaces on G, defined in terms of Δ_χ, spaces that were introduced by the authors in the recent paper [6].

We prove characterizations of the norms in term of finite differences (Theorems 8 and 9), the density of the test functions in Besov and Triebel–Lizorkin spaces, and the boundedness of a simplified version of the local Riesz transforms (Theorem 13).

The plan of the paper is as follows. In the next section we recall the basic facts about our setting and in particular the heat semigroup generated by Δ_χ. In Sect. 3 we recall the definitions of Besov and Triebel–Lizorkin spaces, and the results of [6] needed in the present work. In Sect. 4 we prove finite difference characterizations for the Besov and Triebel–Lizorkin spaces. Such characterizations are then used in Sect. 5 to show that test functions are dense in such spaces, and in Sect. 6 we prove an isomorphism result and the boundedness of the aforementioned version of the local Riesz transforms for both scales of Besov and Triebel–Lizorkin spaces. We conclude by mentioning some directions for future work.

We shall denote by C a positive constant that may vary from place to place, and may depend on any factor quantified (implicitly or explicitly) before its occurrence, but not on factors quantified afterwards. For any quantities A and B, we write $A \lesssim B$ to indicate that there exists a constant $C > 0$ as above such that $A \leq C\,B$. If $A \lesssim B$ and $B \lesssim A$, we write $A \approx B$. In order to emphasize the dependence on a given parameter, say R, we write $\lesssim_R$, and analogously for the other cases.

Foreword by the Second Named Author Soon after getting my Ph. D., I obtained a position at the Politecnico in Torino, where Fulvio had been for a number of years. He was my main reason for seeking this position at the Politecnico. I immediately found myself immersed in a very pleasant environment, with Fulvio being the organiser of many activities, such as advanced courses, regular seminars, and the visits of many leading mathematicians. I was exposed to a flurry of recent and as well as ongoing research, on a variety of different topics. This gave me the possibility of meeting and interacting with many experts. Fulvio personally introduced me to this world, taking the time to explain to me a lot of mathematics, while advising and guiding me. I have always been very impressed by his poise, kindness, and, most of all, generosity in teaching all the younger mathematicians who had the fortune to interact with him. He has had a great impact on me, both professionally and personally.

I wish to express to Fulvio my most sincere gratitude for all he has taught me, and for his invaluable friendship.

2 Basic Facts and Definitions

Let G be any Lie group with identity element e. We denote by ρ a right Haar measure, and by δ the modular function. We let λ be the left Haar measure such that $d\lambda = \delta d\rho$. We recall that δ is a smooth positive character, that is, a smooth group homomorphism of G onto $\mathbb{R}^+$. If χ is any continuous positive character of G, then χ is automatically smooth. For any such χ, we define μ_χ to be the measure whose density with respect to ρ is χ, that is, $d\mu_\chi = \chi d\rho$. Notice that $\mu_1 = \rho$ and $\mu_\delta = \lambda$.

We fix once for all a family of left-invariant linearly independent vector fields $\mathbf{X} = \{X_1, \ldots, X_\ell\}$ satisfying Hörmander's condition. These vector fields induce the Carnot–Charathéodory distance, denoted by d_C, which turns out to be left-invariant. Then, for $x \in G$ we set $|x| = d_C(x, e)$ and we denote with $B(x, r)$ the ball centered at x and of radius $r > 0$. If $x = e$ and $r > 0$, we write $B_r = B(e, r)$, and define $V(r) = \rho(B_r)$. In general, we denote by $B(x_0, r)$ the ball with center x_0 and radius r, in the metric d_C.

It is known that there exist two constants $d, D > 0$ such that

$$\begin{cases} \rho(B_r) \approx r^d & \text{if } r \in (0, 1] \\ \rho(B_r) \lesssim e^{Dr} & \text{if } r \in (1, +\infty)\,, \end{cases} \tag{2}$$

see [21, 55]. It is worth pointing out that $d = d(\mathbf{X}, G)$, while $D = D(G)$.

We observe that, having fixed $R > 0$, every character χ satisfies the estimates

$$\chi(x) \approx_R \chi(y) \tag{3}$$

for all $x, y \in G$ such that $d_C(x, y) \leq R$. This equivalence easily implies that (G, d_C, μ_χ) is locally doubling, that is, for all $0 < r < R$ and $x_0 \in G$,

$$\mu_\chi(B(x_0, 2r) \lesssim_R \mu_\chi(B(x_0, r))\,. \tag{4}$$

Having fixed $\mathbf{X}$, we consider the operator Δ_χ defined in (1). With an abuse of notation, we still denote by Δ_χ its smallest closed extension on $L^p(\mu_\chi)$, where, for $p \in (1, +\infty)$, $L^p(\mu_\chi)$ denotes the standard Lebesgue space. The space L^∞ is the space of ρ-essentially bounded functions. We refer to [23] and [5] for further details about Δ_χ.

We set $\mathbb{J} = \{1, \ldots, \ell\}$ and we say that a multi-index $J = (j_1, \ldots, j_m) \in \mathbb{J}^m$ if $j_k \in \mathbb{J}$ for $k = 1, \ldots, m$. Moreover, we write

$$X_J = X_{j_1} \cdots X_{j_m}\,.$$

Next, we observe that, since Δ_χ is left-invariant, the associated heat semigroup admits a convolution kernel $p_t^\chi \in \mathcal{D}'(G)$, i.e.

$$e^{-t\Delta_\chi} f = f * p_t^\chi\,.$$

It is known (see [5]) that

$$p_t^\chi = e^{-c_\chi^2 t/4} \chi^{-1/2} p_t \,, \tag{5}$$

where $c_X = \left(\sum_{j=1}^{\ell}(X_j\chi(e))^2\right)^{1/2}$, and p_t is the convolution kernel of the heat semigroup in the case $\chi = 1$, for which the estimates in [56] are available.

We recall the expressions of the convolution on G. We have

$$\begin{aligned} f * g(x) &= \int_G f(xy^{-1})g(y)\,d\rho(y) = \int_G f(y^{-1})g(yx)\,d\rho(y) \\ &= \int_G f(y)g(y^{-1}x)\,d\lambda(y) = \int_G f(xy)g(y^{-1})\,d\lambda(y)\,. \end{aligned} \tag{6}$$

The following result is essentially Lemma 3.1 in [6].

Lemma 1 *The following properties hold:*

(i) $(e^{-t\Delta_X})_{t>0}$ *is a diffusion semigroup on* (G, μ_χ)*;*
(ii) *for every* $r > 0$, $\sup_{B_r} \chi = e^{c_X r}$*;*
(iii) *there exist two constants* $c_1, c_2 > 0$ *such that*

$$(\delta\chi^{-1})^{1/2}(x)\,V(\sqrt{t})^{-1}e^{-c_1|x|^2/t} \lesssim p_t^\chi(x) \lesssim (\delta\chi^{-1})^{1/2}(x)\,V(\sqrt{t})^{-1}e^{-c_2|x|^2/t}$$

for every $t \in (0, 1)$ *and* $x \in G$*;*
(iv) *given* $m \in \mathbb{N}$*, there exist a positive constant* $b = b_m$ *such that*

$$|X_J p_t^\chi(x)| \lesssim (\delta\chi^{-1})^{1/2}(x)t^{-m/2}\,V(\sqrt{t})^{-1}\,e^{-b|x|^2/t}$$

for every $x \in G$, $J \in \mathbb{J}^m$ *and* $t \in (0, 1)$*;*
(v) *there exists* $c_3 > 0$ *such that*

$$\left|\frac{\partial}{\partial t}\,p_t^\chi(x)\right| \lesssim (\delta\chi^{-1})^{1/2}(x)t^{-1}V(\sqrt{t})^{-1}e^{-c_3|x|^2/t}$$

for every $t \in (0, 1)$ *and* $x \in G$.

Definition 2 We define the space $\mathcal{S}(G)$ as the space of functions $\varphi \in C^\infty(G)$ such that for all $n, m \in \mathbb{N}$, $J \in \mathbb{J}^m$ the seminorms

$$\mathcal{N}_{J,n}(\varphi) = \sup_{x\in G} e^{n|x|}|X_J\varphi(x)|$$

are finite. The space $\mathcal{S}'(G)$ is defined as the dual space of $\mathcal{S}(G)$.

In particular, every function $g \in L^1_{loc}(G)$ and of most exponential growth, that is such that $e^{-c|x|}g \in L^\infty(G)$ for some $c > 0$, will be identified in the sequel with a distribution in $\mathcal{S}'(G)$ as

$$\langle g, \varphi \rangle = \int_G g(x)\varphi(x)\, d\rho(x) \qquad \forall \varphi \in \mathcal{S}(G)\,.$$

Observe moreover that, if G has polynomial volume growth, then $\mathcal{S}(G)$ is a subset of the usual Schwartz space on G. Indeed, if one defines $\mathcal{S}_\chi(G)$ as the space of functions $\varphi \in C^\infty(G)$ such that for all $n, m \in \mathbb{N}$, $J \in \mathbb{J}^m$ the seminorms

$$\mathcal{N}^\chi_{J,n}(\varphi) = \sup_{x \in G} \left(1 + \mu_\chi(B(e, |x|))\right)^n |X_J \varphi(x)|$$

are finite, then $\mathcal{S}(G) \subseteq \mathcal{S}_\chi(G)$ for any character χ by Lemma 1 (ii).

For simplicity of notation, we write $\mathcal{S}$ and $\mathcal{S}'$ in place of $\mathcal{S}(G)$ and $\mathcal{S}'(G)$ respectively. As a consequence of the Gaussian estimates (iv) in Lemma 1, we have the following simple lemma.

Lemma 3 *For all $t > 0$, $p^\chi_t \in \mathcal{S}$. Moreover, $e^{-t\Delta_\chi} : \mathcal{S} \to \mathcal{S}$ is bounded, with seminorms uniformly bounded for $t \in [\varepsilon, R]$, for any $0 < \varepsilon < R$. Therefore, $e^{-t\Delta_\chi}$ extends to a continuous map $e^{-t\Delta_\chi} : \mathcal{S}' \to \mathcal{S}'$, for all $t > 0$.*

Proof We indicate the argument for sake of completeness. Given $n \in \mathbb{N}$ and a multi-index J, we have

$$\begin{aligned} e^{n|x|}|X_J(\varphi * p^\chi_t)(x)| &= e^{n|x|}|(\varphi * X_J p^\chi_t)(x)| \\ &\le \int_G e^{n|xy^{-1}|}|\varphi(xy^{-1})| e^{n|y|}|X_J p^\chi_t(y)|\, d\rho(y) \\ &\le \mathcal{N}_{0,n}(\varphi) \int_G e^{n|y|}|X_J p^\chi_t(y)|\, d\rho(y) \\ &\lesssim t^{-|J|/2}\mathcal{N}_{0,n}(\varphi)\,, \end{aligned} \tag{7}$$

where the last inequality is obtained arguing as in the proof of Lemma 3.4 in [6]. The conclusions now follow easily. □

Definition 4 For $m \in \mathbb{N}$ and $t > 0$, we define the operator $W^{(m)}_t$ by setting

$$W^{(m)}_t = (t\Delta_\chi)^m e^{-t\Delta_\chi}\,.$$

For $m \in \mathbb{N}$ and $t > 0$, $W^{(m)}_t : \mathcal{S} \to \mathcal{S}$ is bounded, and therefore it extends to a continuous map $W^{(m)}_t : \mathcal{S}' \to \mathcal{S}'$. We also observe that, for $f \in \mathcal{S}'$, $W^{(m)}_t f$ is a C^∞ function, for all $t > 0$ and $m \in \mathbb{N}$.

We also recall the definition of the Littlewood–Paley–Stein g-function. Given a positive integer k, for $f \in \mathcal{S}$ we set

$$g_k(f) = \left(\int_0^{+\infty} |W_s^{(k)} f|^2 \, \frac{ds}{s} \right)^{1/2} . \tag{8}$$

Since Δ_χ generates a symmetric diffusion semigroup, if $p \in (1, +\infty)$ then g_k satisfies the estimate

$$\|g_k(f)\|_{L^p(\mu_\chi)} \approx \|f\|_{L^p(\mu_\chi)} \tag{9}$$

for any $f \in L^p(\mu_\chi)$, see [47], and also [33].

3 Triebel–Lizorkin and Besov Spaces on G

Here and in what follows, given a measure space (Ω, ν) and a Banach space $\mathcal{X}$, for $p \in [1, +\infty]$, we denote by $L^p\big(\Omega, \nu; \mathcal{X}\big)$ the space of measurable functions $f : \Omega \to \mathcal{X}$ such that

$$\|f\|_{L^p(\Omega,\nu;\mathcal{X})} := \left\{ \int_\Omega \|f(\omega)\|_{\mathcal{X}}^p \, d\nu(\omega) \right\}^{1/p} < \infty$$

when $p \in [1, +\infty)$, with the obvious modification if $p = +\infty$. We also denote by $[\tau]$ the integral part of $\tau \geq 0$.

We are now in the position to introduce the Triebel–Lizorkin and Besov spaces on G, defined in terms of the sub-Laplacian Δ_χ, see [6].

Definition 5 Let $p, q \in [1, +\infty]$, and $\alpha \geq 0$. Then we define:

(1) the Triebel–Lizorkin space $F_\alpha^{p,q}(\mu_\chi)$ as

$$F_\alpha^{p,q}(\mu_\chi) = \Big\{ f \in \mathcal{S}'(G) : t^{-\alpha/2} W_t^{([\alpha/2]+1)} f \in L^p\Big(G, \mu_\chi; L^q\big((0,1), dt/t\big)\Big) \text{ and } e^{-\frac{1}{2}\Delta_\chi} f \in L^p(\mu_\chi) \Big\}$$

endowed with the norm

$$\|f\|_{F_\alpha^{p,q}} := \mathbb{F}_\alpha^{p,q}(f) + \|e^{-\frac{1}{2}\Delta_\chi} f\|_{L^p(\mu_\chi)} , \tag{10}$$

where

$$\mathbb{F}_\alpha^{p,q}(f) := \left\| \left(\int_0^1 \left(t^{-\alpha/2} |W_t^{([\alpha/2]+1)} f| \right)^q \frac{dt}{t} \right)^{1/q} \right\|_{L^p(\mu_\chi)}$$

if $q < +\infty$, while

$$\mathbb{F}^{p,\infty}_{\alpha}(f) := \| \sup_{t\in(0,1)} t^{-\alpha/2}|W_t^{([\alpha/2]+1)} f| \|_{L^p(\mu_\chi)};$$

(2) the Besov space $B^{p,q}_{\alpha}(\mu_\chi)$ as

$$B^{p,q}_{\alpha}(\mu_\chi) = \Big\{ f \in \mathcal{S}'(G) : t^{-\alpha/2} W_t^{([\alpha/2]+1)} f \in L^q\Big((0,1),\, dt/t;\, L^p(G,\mu_\chi)\Big) \text{ and } e^{-\frac{1}{2}\Delta_\chi} f \in L^p(\mu_\chi) \Big\}$$

endowed with the norm

$$\|f\|_{B^{p,q}_{\alpha}} := \mathbb{B}^{p,q}_{\alpha}(f) + \|e^{-\frac{1}{2}\Delta_\chi} f\|_{L^p(\mu_\chi)}, \tag{11}$$

where

$$\mathbb{B}^{p,q}_{\alpha}(f) := \left(\int_0^1 \left(t^{-\alpha/2} \, \|W_t^{([\alpha/2]+1)} f\|_{L^p(\mu_\chi)} \right)^q \frac{dt}{t} \right)^{1/q}$$

if $q < +\infty$, while

$$\mathbb{B}^{p,\infty}_{\alpha}(f) := \sup_{t\in(0,1)} t^{-\alpha/2} \, \|W_t^{([\alpha/2]+1)} f\|_{L^p(\mu_\chi)}.$$

We emphasize that, when $p \in (1, +\infty)$ and $\alpha \geq 0$, the Triebel–Lizorkin space $F^{p,2}_{\alpha}(\mu_\chi)$ coincides with the Sobolev space $L^p_{\alpha}(\mu_\chi)$ defined in [5], with equivalence of norms, see [6, Theorem 5.2].

We now recall the main results in [6] about equivalence of norms in Besov and Triebel–Lizorkin spaces. The following is [6, Theorem 4.1].

Theorem 6 *Let $\alpha > 0$, $m > \alpha/2$ be an integer, $t_0 \in [0, 1)$ and $q \in [1, +\infty]$.*

(i) If $p \in (1, +\infty)$, then the norm $\|f\|_{F^{p,q}_{\alpha}}$ is equivalent to the norm

$$\left\| \left(\int_0^1 \left(t^{-\alpha/2} |W_t^{(m)} f| \right)^q \frac{dt}{t} \right)^{1/q} \right\|_{L^p(\mu_\chi)} + \|e^{-t_0\Delta_\chi} f\|_{L^p(\mu_\chi)}. \tag{12}$$

(ii) If $p \in [1, +\infty]$, then the norm $\|f\|_{B^{p,q}_{\alpha}}$ is equivalent to the norm

$$\left(\int_0^1 \left(t^{-\alpha/2} \|W_t^{(m)} f\|_{L^p(\mu_\chi)} \right)^q \frac{dt}{t} \right)^{1/q} + \|e^{-t_0\Delta_\chi} f\|_{L^p(\mu_\chi)}. \tag{13}$$

If $\alpha = 0$*, the norms* $\|f\|_{F^{p,q}_\alpha}$ *and* $\|f\|_{B^{p,q}_\alpha}$ *are equivalent to those in* (12) *and* (13) *respectively provided* $t_0 \in (0, 1)$.

The next result concerns a discretization of the norm that resembles the Littlewood–Paley characterization of Besov and Triebel–Lizorkin spaces in the classical cases. In our case, for $j \in \mathbb{N}$, the operators $W^{(m)}_{2^{-j}}$ play the role of the operators Δ_j in the classical Littlewood–Paley decomposition, while $e^{-t_0\Delta_X}$ plays the role of S_0; see e.g. [20] for such notation in the case of $\mathbb{R}^d$.

We point out that in the case of Δ_X the classical Littlewood–Paley characterization of Besov and Triebel–Lizorkin spaces cannot hold since any bounded spectral multiplier of Δ_X on $L^p(\mu_X)$, with $p \neq 2$, admits a holomorphic extension to a parabolic region in $\mathbb{C}$, see [23]. This is [6, Theorem 4.2].

Theorem 7 *Let* $\alpha > 0$*,* $m > \alpha/2$ *be an integer,* $t_0 \in [0, 1)$ *and* $q \in [1, +\infty]$*.*

(i) If $p \in (1, +\infty)$*, then the norm* $\|f\|_{F^{p,q}_\alpha}$ *is equivalent to the norm*

$$\Big\| \Big(\sum_{j=0}^{\infty} \big(2^{j\alpha/2} |W^{(m)}_{2^{-j}} f|\big)^q \Big)^{1/q} \Big\|_{L^p(\mu_X)} + \|e^{-t_0\Delta_X} f\|_{L^p(\mu_X)} . \tag{14}$$

(ii) If $p \in [1, +\infty]$*, then the norm* $\|f\|_{B^{p,q}_\alpha}$ *is equivalent to the norm*

$$\Big(\sum_{j=0}^{\infty} \Big(2^{j\alpha/2} \|W^{(m)}_{2^{-j}} f\|_{L^p(\mu_X)}\Big)^q \Big)^{1/q} + \|e^{-t_0\Delta_X} f\|_{L^p(\mu_X)} . \tag{15}$$

If $\alpha = 0$*, the norms* $\|f\|_{F^{p,q}_\alpha}$ *and* $\|f\|_{B^{p,q}_\alpha}$ *are equivalent respectively to those in* (14) *and* (15) *provided* $t_0 \in (0, 1)$.

4 Finite Differences Characterizations

In this section we prove characterizations for the spaces $F^{p,q}_\alpha$ and $B^{p,q}_\alpha$ in terms of finite differences. Such characterizations provide a key tool for the proof of the density lemma of Sect. 5. We begin by introducing the finite difference operator.

Given a measurable function f, for $x, y \in G$ we define

$$\mathrm{D}_y f(x) = f(xy^{-1}) - f(x) . \tag{16}$$

4.1 Characterization of Triebel–Lizorkin Norm by Differences

For $q \in [1, \infty]$ and $\alpha \in (0, 1)$, we define the functional

$$\mathbb{S}_\alpha^{\mathrm{loc},q} f(x) = \left(\int_0^1 \left[\frac{1}{u^\alpha V(u)} \int_{|y|<u} |\mathrm{D}_y f(x)| \, d\rho(y) \right]^q \frac{du}{u} \right)^{1/q} . \tag{17}$$

We point out that in the case $q = 2$, such functional coincides with the classical functional $\mathbb{S}_\alpha^{\mathrm{loc}}$ used to characterize the Sobolev norm, see [7] for the unimodular case, and [5] for the nonunimodular (and weighted) case.

The first result of this section is the characterization of the Triebel–Lizorkin norm of $F_\alpha^{p,q}$ in terms of the $L^p(\mu_\chi)$-integrability of the functional $\mathbb{S}_\alpha^{\mathrm{loc},q}$.

Theorem 8 *For every $p, q \in (1, \infty)$, $\alpha \in (0, 1)$, we have*

$$\|f\|_{F_\alpha^{p,q}} \approx \|\mathbb{S}_\alpha^{\mathrm{loc},q} f\|_{L^p(\mu_\chi)} + \|f\|_{L^p(\mu_\chi)}$$

for any $f \in F_\alpha^{p,q}(\mu_\chi)$.

Proof Set

$$H_{\alpha,q} f := \left(\int_0^1 \left(t^{-\alpha/2} |\mathrm{W}_t^{(1)} f| \right)^q \frac{dt}{t} \right)^{1/q} ,$$

and observe that, since $\alpha \in (0, 1)$,

$$\|f\|_{F_\alpha^{p,q}} \lesssim \|H_{\alpha,q} f\|_{L^p(\mu_\chi)} + \|f\|_{L^p(\mu_\chi)} .$$

Step 1 We shall prove that, for all $f \in F_\alpha^{p,q}$,

$$\|f\|_{F_\alpha^{p,q}} \lesssim \|\mathbb{S}_\alpha^{\mathrm{loc},q} f\|_{L^p(\mu_\chi)} + \|f\|_{L^p(\mu_\chi)} , \tag{18}$$

by showing that

$$\|H_{\alpha,q} f\|_{L^p(\mu_\chi)} \lesssim \|\mathbb{S}_\alpha^{\mathrm{loc},q} f\|_{L^p(\mu_\chi)} + \|f\|_{L^p(\mu_\chi)} . \tag{19}$$

We first notice that for every $t \in (0, 1)$ and $x \in G$, since $\frac{\partial}{\partial t} \int_G p_t^\chi \, d\rho = 0$, we have

$$\begin{aligned} |\Delta_\chi e^{-t\Delta_\chi} f(x)| &= \left| \frac{\partial}{\partial t} e^{-t\Delta_\chi} f(x) \right| \\ &= \left| \frac{\partial}{\partial t} \left(\int_G f(xy^{-1}) p_t^\chi(y) \, d\rho(y) - \int_G f(x) p_t^\chi(y) \, d\rho(y) \right) \right| \\ &\le \int_G |\mathrm{D}_y f(x)| \left| \frac{\partial p_t^\chi(y)}{\partial t} \right| d\rho(y). \end{aligned}$$

Using the estimates (v) of Lemma 1 we have

$$
\begin{aligned}
&(H_{\alpha,q} f(x))^q \\
&\lesssim \int_0^1 t^{-q\alpha/2} V(\sqrt{t})^{-q} \left(\int_{|y|<\sqrt{t}} |\,\mathrm{D}_y f(x)| (\delta\chi^{-1})^{1/2}(y) e^{-c_3|y|^2/t} d\rho(y) \right)^q \frac{dt}{t} \\
&\quad + \sum_{k=0}^{\infty} \int_0^1 t^{-q\alpha/2} V(\sqrt{t})^{-q} \\
&\quad \times \left(\int_{2^k\sqrt{t}<|y|<2^{k+1}\sqrt{t}} |\,\mathrm{D}_y f(x)| (\delta\chi^{-1})^{1/2}(y)\, e^{-c_3|y|^2/t} d\rho(y) \right)^q \frac{dt}{t} \\
&\lesssim \int_0^1 t^{-q\alpha/2} V(\sqrt{t})^{-q} \left(\int_{|y|<\sqrt{t}} |\,\mathrm{D}_y f(x)| d\rho(y) \right)^q \frac{dt}{t} \\
&\quad + \sum_{k=0}^{\infty} e^{-c_3 2^{2k}} \int_0^1 t^{-q\alpha/2} V(\sqrt{t})^{-q} \\
&\quad \times \left(\int_{|y|<2^{k+1}\sqrt{t}} |\,\mathrm{D}_y f(x)| (\delta\chi^{-1})^{1/2}(y) d\rho(y) \right)^q \frac{dt}{t}.
\end{aligned}
$$

By the change of variables $u = 2^{k+1}\sqrt{t}$ we obtain

$$
\begin{aligned}
&(H_{\alpha,q} f(x))^q \\
&\lesssim \int_0^1 \frac{1}{u^{q\alpha} V(u)^q} \left(\int_{|y|<u} |\,\mathrm{D}_y f(x)| d\rho(y) \right)^q \frac{du}{u} \\
&\quad + \sum_{k=0}^{\infty} e^{-c_3 2^{2k}} \int_0^{2^{k+1}} \frac{2^{(k+1)q\alpha}}{u^{q\alpha} V(2^{-k-1}u)^q} \\
&\quad \times \left(\int_{|y|<u} |\,\mathrm{D}_y f(x)| (\delta\chi^{-1})^{1/2}(y) d\rho(y) \right)^q \frac{du}{u} \\
&\lesssim (\mathbb{S}_\alpha^{\mathrm{loc},q} f(x))^q \\
&\quad + \sum_{k=0}^{\infty} e^{-c_3 2^{2k}} \int_0^1 \frac{2^{(k+1)q\alpha}}{u^{q\alpha} V(2^{-k-1}u)^q} \left(\int_{|y|<u} |\,\mathrm{D}_y f(x)|\, d\rho(y) \right)^q \frac{du}{u} \\
&\quad + \sum_{k=0}^{\infty} e^{-c_3 2^{2k}} \int_1^{2^{k+1}} \frac{2^{(k+1)q\alpha}}{u^{q\alpha} V(2^{-k-1}u)^q} \\
&\quad \times \left(\int_{|y|<u} |\,\mathrm{D}_y f(x)| (\delta\chi^{-1})^{1/2}(y) d\rho(y) \right)^q \frac{du}{u}.
\end{aligned}
$$

By the estimates (2), we obtain that

$$
\begin{aligned}
&(H_{\alpha,q} f(x))^q \\
&\lesssim (\mathbb{S}_\alpha^{\mathrm{loc},q} f(x))^q + (\mathbb{S}_\alpha^{\mathrm{loc},q} f(x))^q \sum_{k=0}^{\infty} e^{-c_3 2^{2k}} 2^{(k+1)(q\alpha+qd)} \\
&\quad + \sum_{k=0}^{\infty} e^{-c_3 2^{2k}} 2^{(k+1)(q\alpha+qd)} \int_1^{2^{k+1}} \frac{1}{u^{q\alpha+qd}} |f(x)|^q \\
&\quad \times \left(\int_{|y|<u} (\delta\chi^{-1})^{1/2}(y)\, d\rho(y) \right)^q \frac{du}{u} \qquad (20) \\
&\quad + \sum_{k=0}^{\infty} e^{-c_3 2^{2k}} 2^{(k+1)(q\alpha+qd)} \int_1^{2^{k+1}} \frac{1}{u^{q\alpha+qd}} \\
&\quad \times \left(\int_{|y|<u} |f(xy^{-1})| (\delta\chi^{-1})^{1/2}(y)\, d\rho(y) \right)^q \frac{du}{u} \\
&\lesssim (\mathbb{S}_\alpha^{\mathrm{loc},q} f(x))^q + \sum_{k=0}^{\infty} J_k(x) + \sum_{k=0}^{\infty} I_k(x)\,.
\end{aligned}
$$

By the growth estimates of characters in Lemma 1 (ii) and by (2) we deduce that there exists $C > 0$ such that

$$
\begin{aligned}
\sum_{k=0}^{\infty} J_k(x) &\lesssim \sum_{k=0}^{\infty} e^{-c_3 2^{2k}} 2^{(k+1)(q\alpha+qd)} \int_1^{2^{k+1}} \frac{1}{u^{q\alpha+qd}} |f(x)|^q e^{qCu} \frac{du}{u} \\
&\lesssim \sum_{k=0}^{\infty} |f(x)|^q e^{-c_3 2^{2k}} 2^{(k+1)(q\alpha+qd)} e^{qC2^{k+1}} \qquad (21) \\
&\lesssim |f(x)|^q\,.
\end{aligned}
$$

We now notice that there exists $c > 0$ such that

$$
\begin{aligned}
&\Big\| \Big(\sum_{k=0}^{\infty} I_k \Big)^{1/q} \Big\|_{L^p(\mu_\chi)} \\
&\lesssim \sum_{k=0}^{\infty} e^{-c 2^{2k}} \Big\| \Big(\int_1^{2^{k+1}} \Big(\int_{|y|<u} |f(\cdot\, y^{-1})| (\delta\chi^{-1})^{1/2}(y)\, d\rho(y) \Big)^q \Big)^{1/q} du \Big\|_{L^p(\mu_\chi)}.
\end{aligned}
$$

Here and in the rest of this work, we denote by $\mathbf{1}_E$ the characteristic function of the measurable set E. For every integer k, by Minkowski's integral inequality, we get

$$\begin{aligned}
&\left(\int_1^{2^{k+1}}\left(\int_G |f(xy^{-1})|\mathbf{1}_{B_u}(y)(\delta\chi^{-1})^{1/2}(y)\,d\rho(y)\right)^q du\right)^{1/q}\\
&\lesssim \int_G\left(\int_1^{2^{k+1}} |f(xy^{-1})|^q\mathbf{1}_{B_u}(y)(\delta\chi^{-1})^{q/2}(y)\,du\right)^{1/q} d\rho(y)\\
&\lesssim \int_{B_1}|f(xy^{-1})|(\delta\chi^{-1})^{1/2}(y)\left(\int_1^{2^{k+1}} du\right)^{1/q} d\rho(y)\\
&\quad+\int_{1<|y|<2^{k+1}}|f(xy^{-1})|(\delta\chi^{-1})^{1/2}(y)\left(\int_{|y|}^{2^{k+1}} du\right)^{1/q} d\rho(y)\\
&\lesssim 2^{k/q}\int_{B_1}|f(xz)|\,d\lambda(z)+2^{k/q}\int_{1<|z|<2^{k+1}}|f(xz)|(\delta^{-1}\chi)^{1/2}(z)\,d\lambda(z)\,.
\end{aligned}$$

By applying again Minkowski's inequality, we then obtain that

$$\begin{aligned}
&\left\|\left(\int_1^{2^{k+1}}\left(\int_G |f(\cdot\, y^{-1})|\mathbf{1}_{B_u}(y)(\delta\chi^{-1})^{1/2}(y)\,d\rho(y)\right)^q du\right)^{1/q}\right\|_{L^p(\mu_\chi)}\\
&\lesssim 2^{k/q}\int_{B_1}\left(\int_G |f(xz)|^p d\mu_\chi(x)\right)^{1/p} d\lambda(z)\\
&\quad+2^{k/q}\int_{1<|z|<2^{k+1}}\left(\int_G |f(xz)|^p\, d\mu_\chi(x)\right)^{1/p}(\delta^{-1}\chi)^{1/2}(z)\,d\lambda(z)\\
&\lesssim 2^{k/q}\|f\|_{L^p(\mu_\chi)}+2^{k/q}\|f\|_{L^p(\mu_\chi)}\int_{B_{2^{k+1}}}\chi^{1/p}(\delta\chi^{-1})^{1/2}\,d\rho\\
&\lesssim 2^{k/q}e^{C2^k}\|f\|_{L^p(\mu_\chi)}\,.
\end{aligned}$$

We then have

$$\left\|\left(\sum_{k=0}^{\infty} I_k\right)^{1/q}\right\|_{L^p(\mu_\chi)}\lesssim\sum_{k=0}^{\infty}e^{-c2^{2k}}2^{k/q+kd}e^{C2^k}\|f\|_{L^p(\mu_\chi)}\lesssim\|f\|_{L^p(\mu_\chi)}. \tag{22}$$

In conclusion, by (21) and (22) we get (19), as required.

It remains to show that for all $f\in F_\alpha^{p,q}$

$$\|\mathbb{S}_\alpha^{\mathrm{loc},q}f\|_{L^p(\mu_\chi)}\lesssim\|f\|_{F_\alpha^{p,q}}\,. \tag{23}$$

In order to prove this, we write $f = (f - e^{-\Delta_X} f) + e^{-\Delta_X} f$ and we estimate $\|\mathbb{S}_\alpha^{\mathrm{loc},q}(f - e^{-\Delta_X} f)\|_{L^p(\mu_X)}$ and $\|\mathbb{S}_\alpha^{\mathrm{loc},q} e^{-\Delta_X} f\|_{L^p(\mu_X)}$ separately.

Step 2 We prove that

$$\|\mathbb{S}_\alpha^{\mathrm{loc},q}(f - e^{-\Delta_X} f)\|_{L^p(\mu_X)} \lesssim \|f\|_{F_\alpha^{p,q}}\,. \tag{24}$$

Arguing similarly to [7, 2.1.2] we write

$$f - e^{-\Delta_X} f = -\sum_{m=1}^{+\infty} \int_{2^{-m}}^{2^{-m+1}} \frac{\partial}{\partial t} e^{-t\Delta_X} f\, dt =: \sum_{m=1}^{+\infty} f_m\,. \tag{25}$$

We then obtain

$$\begin{aligned}
&\big(\mathbb{S}_\alpha^{\mathrm{loc},q}(f - e^{-\Delta_X} f)(x)\big)^q \\
&= \int_0^1 \left(\frac{1}{u^\alpha V(u)} \int_{|y|<u} |\,\mathrm{D}_y(f - e^{-\Delta_X} f))(x)|\, d\rho(y)\right)^q \frac{du}{u} \\
&= \sum_{j=1}^{+\infty} \int_{2^{-j}}^{2^{-j+1}} \left(\frac{1}{u^\alpha V(u)} \int_{|y|<u} |\,\mathrm{D}_y(f - e^{-\Delta_X} f)(x)|\, d\rho(y)\right)^q \frac{du}{u} \\
&\lesssim \sum_{j=1}^{+\infty} 2^{jq\alpha} \left(2^{jd} \int_{|y|<2^{-j+1}} |\,\mathrm{D}_y(f - e^{-\Delta_X} f)(x)|\, d\rho(y)\right)^q \qquad (26) \\
&\lesssim \sum_{j=1}^{+\infty} 2^{jq\alpha} \left(\sum_{m=1}^{+\infty} 2^{jd} \int_{|y|<2^{-j+1}} |\,\mathrm{D}_y f_m(x)|\, d\rho(y)\right)^q \\
&= \sum_{j=1}^{+\infty} 2^{jq\alpha} \left(2^{jd} \left(\sum_{m=1}^{2j} + \sum_{m=2j+1}^{+\infty}\right) \int_{|y|<2^{-j+1}} |\,\mathrm{D}_y f_m(x)|\, d\rho(y)\right)^q,
\end{aligned}$$

where f_m is defined in (25). If $m > 2j$, then

$$2^{jd} \int_{|y|<2^{-j+1}} |\,\mathrm{D}_y f_m(x)|\, d\rho(y) \lesssim M g_{m+1}(x)\,, \tag{27}$$

where

$$g_{m+1} = \int_{2^{-m}}^{2^{-m+1}} \left|\frac{\partial}{\partial t} e^{-t\Delta_X} f\right| dt\,,$$

and M is the local maximal function with respect to the right Haar measure,

$$Mf(x) = \sup_{x\in B,\, r_B\leq 1} \frac{1}{\rho(B)} \int_B |f|\, d\rho\,, \tag{28}$$

which is bounded on $L^p(\mu_\chi)$ for every $p \in (1,\infty)$, see [5, Subsection 5.1].

In order to treat the case when $m \leq 2j$, we notice that for every $j \geq 1$, $y \in B_{2^{-j-1}}$ and $x \in G$

$$|\,\mathrm{D}_y\, f_m(x)| \leq 2^{-j+1} \sup\left\{|X_i\, f_m(w)| : i = 1,\dots,\ell,\ |w^{-1}x| \leq 2^{-j+1}\right\}. \tag{29}$$

Since

$$f_m = \int_{2^{-m-1}}^{2^{-m}} \frac{\partial}{\partial t}(e^{-2t\Delta_\chi} f)\, dt = 2\int_{2^{-m-1}}^{2^{-m}} e^{-t\Delta_\chi} \frac{\partial}{\partial t}(e^{-t\Delta_\chi} f)\, dt, \tag{30}$$

by applying the estimates of the heat kernel in Lemma 1 (v) and (iii), for every w such that $|w^{-1}x| \leq 2^{-j+1}$ we have

$$\begin{aligned}
&|X_i\, f_m(w)| \\
&\lesssim \int_{2^{-m-1}}^{2^{-m}} \int_G \left|\frac{\partial}{\partial t}(e^{-t\Delta_\chi} f)(z)\right| \left|X_i\, p_t^\chi(z^{-1}w)\right| d\lambda(z)\, dt \\
&\lesssim \int_G \int_{2^{-m-1}}^{2^{-m}} \left|\frac{\partial}{\partial t}(e^{-t\Delta_\chi} f)(z)\right| t^{-1/2} V(\sqrt{t})^{-1} (\chi^{-1}\delta)^{1/2}(z^{-1}w) e^{-b|z^{-1}w|^2/t}\, dt\, d\lambda(z) \\
&\lesssim 2^{m/2} 2^{md/2} \int_G g_m(z) (\delta\chi^{-1})^{1/2}(z^{-1}x) e^{-\frac{b}{2}2^m|z^{-1}x|^2}\, d\lambda(z) \\
&\lesssim 2^{m/2} e^{-c2^{-m}\Delta_\chi} g_m(x),
\end{aligned}$$

for a suitable constant c. From (29) it follows that

$$2^{jd} \int_{|y|<2^{-j+1}} |\,\mathrm{D}_y\, f_m(x)|\, d\rho(y) \lesssim 2^{-j+m/2} e^{-c2^{-m}\Delta_\chi} g_m(x)\,. \tag{31}$$

Thus, putting together (26), (27) and (31) we obtain

$$\begin{aligned}
&(\mathbb{S}_\alpha^{\mathrm{loc},q}(f - e^{-\Delta_\chi} f)(x))^q \\
&\lesssim \sum_{j=1}^{+\infty} 2^{jq\alpha} \left(\sum_{m=1}^{2j} 2^{-j+m/2} e^{-c2^{-m}\Delta_\chi} g_m(x) + \sum_{m=2j+1}^{+\infty} M g_{m+1}(x) \right)^q \\
&\lesssim \sum_{j=1}^{+\infty} 2^{jq\alpha} \left(\sum_{m=1}^{2j} 2^{-j+m/2} e^{-c2^{-m}\Delta_\chi} g_m(x) \right)^q + \sum_{j=1}^{+\infty} 2^{jq\alpha} \left(\sum_{m=2j+1}^{+\infty} M g_{m+1}(x) \right)^q \\
&=: \Sigma_1(x) + \Sigma_2(x)\,.
\end{aligned} \tag{32}$$

Now we apply Hölder's inequality to see that, for any $\varepsilon > 0$,

$$\Big(\sum_{m=1}^{2j} 2^{-j+m/2} e^{-c2^{-m}\Delta_X} g_m(x)\Big)^q$$

$$\leq \Big(\sum_{m=1}^{2j} 2^{\varepsilon m q'}\Big)^{q/q'} \sum_{m=1}^{2j} 2^{-jq+mq/2-\varepsilon mq}\big(e^{-c2^{-m}\Delta_X} g_m(x)\big)^q$$

$$\lesssim \sum_{m=1}^{2j} 2^{2j\varepsilon q-jq+mq/2-\varepsilon mq}\big(e^{-c2^{-m}\Delta_X} g_m(x)\big)^q .$$

Therefore, since $\alpha \in (0, 1)$, choosing $\varepsilon \in (0, (1-\alpha)/2)$ we obtain

$$\Sigma_1(x) \lesssim \sum_{m=1}^{+\infty} \sum_{j\geq m/2} 2^{-j(1-\alpha-2\varepsilon)q+mq/2-\varepsilon mq}\big(e^{-c2^{-m}\Delta_X} g_m(x)\big)^q$$

$$\lesssim \sum_{m=1}^{+\infty} \big(2^{m\alpha/2} e^{-c2^{-m}\Delta_X} g_m(x)\big)^q . \tag{33}$$

Analogously, using Hölder's inequality again, we see that, for $\varepsilon > 0$

$$\Big(\sum_{m=2j+1}^{+\infty} Mg_{m+1}(x)\Big)^q \lesssim \Big(\sum_{m=2j+1}^{+\infty} 2^{-\varepsilon m q'}\Big)^{q/q'} \sum_{m=2j+1}^{+\infty} \big(2^{\varepsilon m} Mg_{m+1}(x)\big)^q$$

$$\lesssim 2^{-2\varepsilon jq} \sum_{m=2j+1}^{+\infty} \big(2^{\varepsilon m} Mg_{m+1}(x)\big)^q ,$$

so that, if $\varepsilon < \alpha/2$

$$\Sigma_2(x) \lesssim \sum_{m=1}^{+\infty} \sum_{j\leq m/2} 2^{\varepsilon mq+jq(\alpha-2\varepsilon)}\big(Mg_{m+1}(x)\big)^q \lesssim \sum_{m=1}^{+\infty} \big(2^{m\alpha/2} Mg_{m+1}(x)\big)^q . \tag{34}$$

Now, from (32) we have

$$\big\|\mathbb{S}_\alpha^{\mathrm{loc},q}(f - e^{-\Delta_X} f)\big\|_{L^p(\mu_X)} \lesssim \big\|\Sigma_1^{1/q}\big\|_{L^p(\mu_X)} + \big\|\Sigma_2^{1/q}\big\|_{L^p(\mu_X)} .$$

Theorem 9 *Let $\alpha \in (0,1)$ and $p, q \in [1, +\infty]$. Then*

$$\|f\|_{B_\alpha^{p,q}} \approx \mathbb{A}_\alpha^{p,q}(f) + \|f\|_{L^p(\mu_\chi)} \tag{36}$$

for any $f \in B_\alpha^{p,q}(\mu_\chi)$.

Proof We separate the proof in three steps. The first step deals with some simple integral estimates relying on the classical Schur's test, while the second and third steps contain the inequalities $\lesssim$ and $\gtrsim$, respectively, in the statement.

Step 1 Let $a \in \mathbb{R}$, $s \geq 0$, $c > 0$ and define the integral kernel $K : (0,1) \times G \to [0, +\infty)$ by

$$K(t, y) = \chi^a(y) \left(\frac{|y|^2}{t}\right)^s \frac{V(|y|)}{V(\sqrt{t})} e^{-c|y|^2/t},$$

and the corresponding integral operator

$$T_K g(t) = \int_G K(t, y) g(y) \frac{d\rho(y)}{V(|y|)}.$$

Then, we show that for all $q \in [1, +\infty)$

$$T_K : L^q\big(G, d\rho/V(|\cdot|)\big) \to L^q\big((0,1), dt/t\big)$$

is bounded.

To this end, it suffices to apply Schur's test, see e.g. [11, Theorem 6.18], after showing that

$$\text{(a)} \int_G K(t, y) \frac{d\rho(y)}{V(|y|)} \lesssim 1, \qquad \text{(b)} \int_0^1 K(t, y) \frac{dt}{t} \lesssim 1. \tag{37}$$

Observe that

$$\int_G K(t, y) \frac{d\rho(y)}{V(|y|)} = \int_{|y|^2 \leq t} K(t, y) \frac{d\rho(y)}{V(|y|)} + \int_{|y|^2 > t} K(t, y) \frac{d\rho(y)}{V(|y|)} =: I + II.$$

It is easy to check that $I \lesssim 1$. Moreover, since by (2) for every $j \geq 0$ and $t \in (0,1)$

$$\frac{V(2^{j+1}\sqrt{t})}{V(\sqrt{t})} \lesssim 2^{d(j+1)} e^{D 2^{j+1}},$$

we have

$$II \lesssim \sum_{j=0}^{\infty} \int_{2^j\sqrt{t} \leq |y| < 2^{j+1}\sqrt{t}} \frac{1}{V(\sqrt{t})} 2^{2js} e^{ac\chi 2^{j+1}\sqrt{t}} e^{-c2^{2j}} \, d\rho(y) \lesssim 1.$$

Thus, condition (a) in (37) is satisfied. In order to prove (b), we separate two cases. If $|y| \geq 1$, then

$$\int_0^1 K(t,y)\,\frac{dt}{t} \lesssim e^{ac\chi|y| - \frac{c}{2}|y|^2} V(|y|) \int_0^1 \frac{1}{V(\sqrt{t})} e^{-\frac{c}{4}|y|^2/t}\,\frac{dt}{t} \lesssim 1$$

while, if $|y| \leq 1$, (hence $\chi(y) \lesssim 1$)

$$\int_0^1 K(t,y)\,\frac{dt}{t} \lesssim \int_0^1 \left(\frac{|y|^2}{t}\right)^{s+d/2} e^{-c|y|^2/t}\,\frac{dt}{t} = \int_{|y|^2}^{\infty} u^{s+d/2} e^{-cu}\,\frac{du}{u} \lesssim 1\,,$$

which proves (37). This completes Step 1.

Step 2 We show that, for $p, q \in [1, +\infty]$ and $\alpha > 0$,

$$\mathbb{B}_\alpha^{p,q}(f) \lesssim \mathbb{A}_\alpha^{p,q}(f) + \|f\|_{L^p(\mu_\chi)}\,. \tag{38}$$

We claim that there exists $c > 0$ such that, for all $f \in L^p(\mu_\chi)$,

$$\mathbb{B}_\alpha^{p,q}(f) \lesssim \left(\int_G \left(\frac{\|\,\mathrm{D}_y f\|_{L^p(\mu_\chi)} e^{-c|y|^2}}{|y|^\alpha}\right)^q \frac{d\rho(y)}{V(|y|)}\right)^{1/q}\,. \tag{39}$$

Assuming the claim, we prove the estimate (38). By the claim, it suffices to prove that

$$\left(\int_G \left(\frac{\|\,\mathrm{D}_y f\|_{L^p(\mu_\chi)} e^{-c|y|^2}}{|y|^\alpha}\right)^q \frac{d\rho(y)}{V(|y|)}\right)^{1/q} \lesssim \mathbb{A}_\alpha^{p,q}(f) + \|f\|_{L^p(\mu_\chi)}\,.$$

We split the integral on G as $\{|y| < 1\} \cup \{|y| \geq 1\}$. On the one hand, it is easy to see that

$$\left(\int_{|y|\leq 1} \left(\frac{\|\,\mathrm{D}_y f\|_{L^p(\mu_\chi)} e^{-c|y|^2}}{|y|^\alpha}\right)^q \frac{d\rho(y)}{V(|y|)}\right)^{1/q} \leq \mathbb{A}_\alpha^{p,q}(f)\,,$$

while since $\|\,\mathrm{D}_y f\|_{L^p(\mu_\chi)} \leq (1 + \chi^{1/p}(y))\|f\|_{L^p(\mu_\chi)}$,

$$\begin{aligned}
&\left(\int_{|y|\geq 1} \left(\frac{\|\,\mathrm{D}_y f\|_{L^p(\mu_\chi)} e^{-c|y|^2}}{|y|^\alpha}\right)^q \frac{d\rho(y)}{V(|y|)}\right)^{1/q} \\
&\lesssim \|f\|_{L^p(\mu_\chi)} \left(\int_{|y|\geq 1} (1 + e^{c\chi|y|}) e^{-cq|y|^2}\,d\rho(y)\right)^{1/q} \\
&\lesssim \|f\|_{L^p(\mu_\chi)}\,,
\end{aligned}$$

since the volume of balls grows at most exponentially (see (2)).

It remains to prove the claim (39). Since

$$\int_G \partial_t p_t^\chi(y)\, d\rho(y) = \partial_t \int_G p_t^\chi(y)\, d\rho(y) = 0,$$

we have

$$\begin{aligned}\frac{\partial}{\partial t} e^{-t\Delta_\chi} f(x) &= \int_G f(xy^{-1}) \frac{\partial p_t^\chi}{\partial t}(y)\, d\rho(y) \\ &= \int_G \frac{\partial p_t^\chi}{\partial t}(y)\, \mathrm{D}_y f(x)\, d\rho(y)\,.\end{aligned}$$

Thus,

$$\Big\| \frac{\partial}{\partial t} e^{-t\Delta_\chi} f \Big\|_{L^p(\mu_\chi)} \le \int_G \Big| \frac{\partial p_t^\chi}{\partial t}(y) \Big| \| \mathrm{D}_y f \|_{L^p(\mu_\chi)}\, d\rho(y)\,. \tag{40}$$

By Lemma 1 (v), (40) and [6, Lemma 3.3]

$$\begin{aligned}&\mathbb{B}_\alpha^{p,q}(f)^q \\ &\approx \int_0^1 \Big(t^{1-\alpha/2} \Big\| \frac{\partial}{\partial t} e^{-t\Delta_\chi} f \Big\|_{L^p(\mu_\chi)} \Big)^q \frac{dt}{t} \\ &\lesssim \int_0^1 \Big(t^{1-\alpha/2} \int_G (\delta\chi^{-1})^{1/2}(y) t^{-(d+2)/2} e^{-b|y|^2/t} \| \mathrm{D}_y f \|_{L^p(\mu_\chi)}\, d\rho(y) \Big)^q \frac{dt}{t} \\ &\lesssim \int_0^1 \Big(t^{-\alpha/2} \int_G (\delta\chi^{-1})^{1/2}(y) t^{-d/2} e^{-b'|y|^2/t} \| \mathrm{D}_y f \|_{L^p(\mu_\chi)} e^{-b'|y|^2}\, d\rho(y) \Big)^q \frac{dt}{t} \\ &= \int_0^1 \Big(\int_G K(t,y) g(y) \frac{d\rho(y)}{V(|y|)} \Big)^q \frac{dt}{t}\end{aligned}$$

with

$$b' = \frac{b}{2}, \quad g(y) = \frac{\| \mathrm{D}_y f \|_{L^p(\mu_\chi)}}{|y|^\alpha} e^{-b'|y|^2},$$

$$K(t,y) = \frac{V(|y|)}{t^{d/2}} \Big(\frac{|y|^2}{t} \Big)^{\alpha/2} (\delta\chi^{-1})^{1/2}(y) e^{-b'|y|^2/t}\,.$$

By Step 1 we obtain

$$\mathbb{B}_\alpha^{p,q}(f)^q \lesssim \int_G |g(y)|^q \frac{d\rho(y)}{V(|y|)} = \int_G \Big(\frac{\| \mathrm{D}_y f \|_{L^p(\mu_\chi)} e^{-b'|y|^2}}{|y|^\alpha} \Big)^q \frac{d\rho(y)}{V(|y|)}.$$

The claim is proved, and Step 2 is complete.

Step 3 We prove that, for $p, q \in [1, +\infty]$ and $\alpha \in (0, 1)$,

$$\mathbb{A}_{\alpha}^{p,q}(f) \lesssim \mathbb{B}_{\alpha}^{p,q}(f) + \|f\|_{L^p(\mu_\chi)} .$$

We write again f as $f = (f - e^{-\Delta_\chi} f) + e^{-\Delta_\chi} f$ and decompose $f - e^{-\Delta_\chi} f = \sum_{m=1}^{\infty} f_m$ as in (25). Then, using also (30), we have

$$\begin{aligned} f_m &= \int_{2^{-m}}^{2^{-m+1}} \frac{\partial}{\partial t} e^{-t\Delta_\chi} f\, dt = \int_{2^{-m}}^{2^{-m+1}} \Delta_\chi e^{-t\Delta_\chi} f\, dt = 2 \int_{2^{-m-1}}^{2^{-m}} \Delta_\chi e^{-2t\Delta_\chi} f\, dt \\ &= 2e^{-2^{-m-1}\Delta_\chi} \int_{2^{-m-1}}^{2^{-m}} e^{-(t-2^{-m-1})\Delta_\chi} \Delta_\chi e^{-t\Delta_\chi} f\, dt =: 2e^{-2^{-m-1}\Delta_\chi} h_m . \end{aligned}$$

We set

$$\sigma_m = \int_{2^{-m-1}}^{2^{-m}} \left\| \frac{\partial}{\partial t} e^{-t\Delta_\chi} f \right\|_{L^p(\mu_\chi)} dt$$

and observe that

$$\|f_m\|_{L^p(\mu_\chi)} \lesssim \sigma_{m+1} . \tag{41}$$

Hence,

$$\| \mathrm{D}_y f_m \|_{L^p(\mu_\chi)} \lesssim \left(1 + \chi^{1/p}(y)\right) \| f_m \|_{L^p(\mu_\chi)} \lesssim \sigma_{m+1} . \tag{42}$$

By [6, Lemma 3.3] it follows that for $i = 1, \dots, \ell$,

$$\begin{aligned} \|X_i f_m\|_{L^p(\mu_\chi)} &\lesssim 2^{m/2} \|h_m\|_{L^p(\mu_\chi)} \\ &\lesssim 2^{m/2} \int_{2^{-m-1}}^{2^{-m}} \left\| e^{-(t-2^{-m-1})\Delta_\chi} \frac{\partial}{\partial t} e^{-t\Delta_\chi} f \right\|_{L^p(\mu_\chi)} dt \lesssim 2^{m/2} \sigma_m . \end{aligned}$$

Therefore,

$$\| \mathrm{D}_y f_m \|_{L^p(\mu_\chi)} \lesssim |y| \sum_{i=1}^{\ell} \|X_i f_m\|_{L^p(\mu_\chi)} \lesssim |y| 2^{m/2} \sigma_m . \tag{43}$$

Since $\sigma_m \leq 2\sigma_{m+1}$, (43), (41) and (42) imply

$$\| \mathrm{D}_y f_m \|_{L^p(\mu_\chi)} \lesssim \begin{cases} |y| 2^{m/2} \sigma_m & \text{if } |y|^2 < 2^{-m} , \\ \sigma_{m+1} & \text{if } |y|^2 \geq 2^{-m} . \end{cases}$$

Therefore, by Lemma 4.3 (i) in [6], we have

$$
\begin{aligned}
\left(\mathbb{A}_\alpha^{p,q}(f-e^{-\Delta_X}f)\right)^q &\lesssim \sum_{j=1}^{\infty}\int_{2^{-j}\le|y|^2<2^{-j+1}} 2^{jq\alpha/2}\Big(\sum_{m=1}^{\infty}\|\,\mathrm{D}_y f_m\|_{L^p(\mu_\chi)}\Big)^q\,\frac{d\rho(y)}{V(|y|)}\\
&\lesssim \sum_{j=1}^{\infty}2^{jq\alpha/2}\Big(\sum_{m=1}^{j}2^{(m-j)/2}\sigma_m+\sum_{m=j+1}^{+\infty}\sigma_{m+1}\Big)^q\\
&\lesssim \sum_{j=1}^{\infty}2^{jq(\alpha-1)/2}\Big(\sum_{m=1}^{\infty}2^{\min\{j,m\}/2}(\sigma_{m+1}+\sigma_m)\Big)^q\\
&\lesssim \sum_{m=1}^{\infty}\Big(2^{m\alpha/2}(\sigma_{m+1}+\sigma_m)\Big)^q\\
&\lesssim \sum_{m=0}^{\infty}\Big(2^{m\alpha/2}\int_{2^{-m-1}}^{2^{-m}}\Big\|\frac{\partial}{\partial t}e^{-t\Delta_X}f\Big\|_{L^p(\mu_\chi)}dt\Big)^q\\
&\lesssim \sum_{m=0}^{\infty}2^{mq\alpha/2}2^{-m(q-1)}\int_{2^{-m-1}}^{2^{-m}}\Big\|\frac{\partial}{\partial t}e^{-t\Delta_X}f\Big\|_{L^p(\mu_\chi)}^q dt\\
&\lesssim \sum_{m=0}^{\infty}\int_{2^{-m-1}}^{2^{-m}}\Big(t^{1-\alpha/2}\Big\|\frac{\partial}{\partial t}e^{-t\Delta_X}f\Big\|_{L^p(\mu_\chi)}\Big)^q\frac{dt}{t}\\
&=\left(\mathbb{B}_\alpha^{p,q}(f)\right)^q.
\end{aligned}
$$

Therefore, by (9) we have

$$
\mathbb{A}_\alpha^{p,q}(f-e^{-\Delta_X}f)\lesssim \mathbb{B}_\alpha^{p,q}(f)\,.
$$

It remains to estimate $\mathbb{A}_\alpha^{p,q}(e^{-\Delta_X}f)$. As in (43), one can see that

$$
\|\,\mathrm{D}_y\,e^{-\Delta_X}f\|_{L^p(\mu_\chi)}\lesssim|y|\|f\|_{L^p(\mu_\chi)}
$$

so that, using the decomposition of the integral as sum of integrals over annuli,

$$
\mathbb{A}_\alpha^{p,q}(e^{-\Delta_X}f)\le\|f\|_{L^p(\mu_\chi)}\Big(\int_{|y|\le 1}|y|^{q(1-\alpha)}\,\frac{d\rho(y)}{V(|y|)}\Big)^{1/q}\lesssim\|f\|_{L^p(\mu_\chi)}\,.
$$

The proof of Step 3 is complete. This concludes the proof of Theorem 9.

□

5 A Density Result

The main goal of this section is to show that the smooth functions with compact support are dense in the Triebel–Lizorkin and Besov spaces on G. This is the analogue of the classical density result in the Euclidean setting [53]; we refer the reader to [22, 24, 52] for its counterpart in other settings.

To prove our density results we shall use the following version of Young's inequality which follows from [4, Proposition 12 of Chapter VIII, §4, No. 5].

Lemma 10 *If η has support in B_1, then*

$$\|\eta * f\|_{L^p(\mu_\chi)} \lesssim \|\eta\|_{L^1(\mu_\chi)}\|f\|_{L^p(\mu_\chi)} \tag{44}$$

for any $f \in L^p(\mu_\chi)$.

We now prove the density result.

Theorem 11 *Let $\alpha > 0$, $p, q \in (1, \infty)$ and let $X_\alpha^{p,q}$ denote either $F_\alpha^{p,q}(\mu_\chi)$ or $B_\alpha^{p,q}(\mu_\chi)$. Then, $C_c^\infty(G)$ is dense in $X_\alpha^{p,q}$.*

Proof We begin by observing that Lemma 3 implies that $\mathcal{S}$, hence C_c^∞, is contained in $X_\alpha^{p,q}$. Indeed, if $m, n \in \mathbb{N}$ with $m > [\alpha/2]$ and n to be chosen, for $\varphi \in \mathcal{S}$, using (7) we have

$$|W_t^{(m)}\varphi(x)| \le t^m e^{-n|x|} \sum_{|J|\le 2m} \mathcal{N}_{J,n}(\varphi)\,.$$

Then, in order to estimate the norms in (10) and (11), it suffices to chose n large enough so that $\int_G e^{-pn|x|} d\mu_\chi(x)$ is finite.

Step 1 We first prove that we can approximate functions in $F_\alpha^{p,q}$ with functions having compact support when $\alpha \in (0, 1)$.

Let $\eta \in C_c^\infty(B_1)$, $\eta \ge 0$, $\int_G \eta\, d\rho = 1$. Given any $R > 2$, define $\eta_R = \mathbf{1}_{B_R} * \eta$. Then, $\eta_R \in C_c^\infty(B_{R+1})$ and $\eta_R(x) = 1$ on B_{R-1}. We observe that $\|X_J\eta_R\|_\infty \lesssim 1$, for $|J| \le n$. Indeed, the definition of the convolution implies that for any $J \in \mathbb{J}^k$, with $k \le n$, we have that for all $x \in G$

$$|X_J\eta_R(x)| \le \int_G \mathbf{1}_{B_R}(y^{-1})|X_J\eta(yx)|d\rho(y) \le \|X_J\eta\|_{L^1(\rho)} \lesssim 1\,.$$

Hence

$$\|X_J\eta_R\|_\infty \lesssim 1\,. \tag{45}$$

Then, let $f \in F_\alpha^{p,q}$ be given. We shall estimate the norm $\|f - f\eta_R\|_{F_\alpha^{p,q}}$ using the $\mathbb{S}_\alpha^{\mathrm{loc},q}$-functional. Since $\alpha \in (0, 1)$, $f \in L^p(\mu_\chi)$ and $\|f - f\eta_R\|_{L^p(\mu_\chi)} \to 0$ as

$R \to +\infty$. Next, we show that also

$$\|\mathbb{S}_\alpha^{\mathrm{loc},q}(f - f\eta_R)\|_{L^p(\mu_\chi)} \to 0\,, \tag{46}$$

as $R \to +\infty$. Set $\zeta_R = 1 - \eta_R$ and observe that ζ_R vanishes identically on B_{R-1}. Then, we have

$$\begin{aligned}
&[\mathbb{S}_\alpha^{\mathrm{loc},q}(f - f\eta_R)(x)]^q \\
&= \int_0^1 \Big[\frac{1}{u^\alpha V(u)} \int_{|y|\le u} |\zeta_R(xy^{-1})f(xy^{-1}) - \zeta_R(x)f(x)|\, d\rho(y)\Big]^q \frac{du}{u} \\
&\lesssim \int_0^1 \Big[\frac{1}{u^\alpha V(u)} \int_{|y|\le u} \zeta_R(xy^{-1})|f(xy^{-1}) - f(x)|\, d\rho(y)\Big]^q \frac{du}{u} \\
&\qquad + |f(x)|^q \int_0^1 \Big[\frac{1}{u^\alpha V(u)} \int_{|y|\le u} |\zeta_R(xy^{-1}) - \zeta_R(x)|\, d\rho(y)\Big]^q \frac{du}{u} \\
&\lesssim \int_0^1 \Big[\frac{1}{u^\alpha V(u)} \int_{|y|\le u} \zeta_R(xy^{-1})|f(xy^{-1}) - f(x)|\, d\rho(y)\Big]^q \frac{du}{u} \\
&\qquad + |f(x)|^q \int_0^1 \Big[\frac{1}{u^\alpha V(u)} \int_{|y|\le u} |y| \sup_{z\in B(x,1)} \sum_{j=1}^{\ell} |X_j\zeta_R(z)|\, d\rho(y)\Big]^q \frac{du}{u} \\
&\lesssim \mathbf{1}_{\{|x|\ge R-2\}}(x)\Big([\mathbb{S}_\alpha^{\mathrm{loc},q} f(x)]^q + |f(x)|^q\Big)\,,
\end{aligned}$$

since $\zeta_R(z) = 0$ if $|z| \le R - 1$, and $\|\zeta_R\|_\infty = 1$. Therefore,

$$\begin{aligned}
&\|\mathbb{S}_\alpha^{\mathrm{loc},q}(f - f\eta_R)\|^p_{L^p(\mu_\chi)} \\
&\qquad \lesssim \int_{\{|x|\ge R-2\}} [\mathbb{S}_\alpha^{\mathrm{loc},q} f(x)]^p\, d\mu_\chi(x) + \int_{\{|x|\ge R-2\}} |f(x)|^p\, d\mu_\chi(x)\,.
\end{aligned}$$

This proves (46) and therefore we can approximate any element of $F_\alpha^{p,q}$ with elements with compact support.

Step 2 Using the characterization of the norm in $B_\alpha^{p,q}$ by finite differences for $\alpha \in (0, 1)$, Theorem 9, we prove that we can approximate functions in $B_\alpha^{p,q}$ with functions with compact support.

Let $f \in B_\alpha^{p,q}$ be given and let η_R and ζ_R be as in Step 1. Then, for $y \in B_1$ and $x \in G$ we write

$$\begin{aligned}
\mathrm{D}_y(f\zeta_R)(x) &= (f\zeta_R)(xy^{-1}) - (f\zeta_R)(x) \\
&= \zeta_R(xy^{-1})\big[f(xy^{-1}) - f(x)\big] + f(x)\big[\zeta_R(xy^{-1}) - \zeta_R(x)\big] \\
&=: F_R(x, y) + G_R(x, y)\,.
\end{aligned}$$

We observe that

$$\begin{aligned}&\left(\int_{|y|\le 1}\left(\frac{\|G_R(\cdot,y)\|_{L^p(\mu_\chi)}}{|y|^\alpha}\right)^q\frac{d\rho(y)}{V(|y|)}\right)^{1/q}\\&\lesssim\left(\int_{|y|\le 1}|y|^{(1-\alpha)q}\frac{d\rho(y)}{V(|y|)}\right)^{1/q}\|\mathbf{1}_{\{|x|\ge R-2\}}f\|_{L^p(\mu_\chi)}\\&\lesssim\|\mathbf{1}_{\{|x|\ge R-2\}}f\|_{L^p(\mu_\chi)},\end{aligned}$$

that tends to 0 as $R\to+\infty$. Next, we observe that

$$|F_R(x,y)|\le \mathbf{1}_{\{|x|\ge R-2\}}|\,\mathrm{D}_y f(x)|\le|\,\mathrm{D}_y f(x)|,$$

so that

- $\|F_R(\cdot,y)\|_{L^p(\mu_\chi)}\le\left(\int_{\{|x|\ge R-2\}}|\,\mathrm{D}_y f(x)|^p\,d\mu_\chi(x)\right)^{1/p}\to 0$, as $R\to+\infty$;
- $\|F_R(\cdot,y)\|_{L^p(\mu_\chi)}\le\|\,\mathrm{D}_y f\|_{L^p(\mu_\chi)}$, which is independent of R.

Lebesgue's theorem now gives that

$$\left(\int_{|y|\le 1}\left(\frac{\|F_R(\cdot,y)\|_{L^p(\mu_\chi)}}{|y|^\alpha}\right)^q\frac{d\rho(y)}{V(|y|)}\right)^{1/q}\to 0$$

as $R\to+\infty$. Hence, recalling (35), we have

$$\begin{aligned}\mathbb{A}_\alpha^{p,q}(f-f\eta_R)\le&\left(\int_{|y|\le 1}\left(\frac{\|F_R(\cdot,y)\|_{L^p(\mu_\chi)}}{|y|^\alpha}\right)^q\frac{d\rho(y)}{V(|y|)}\right)^{1/q}\\&+\left(\int_{|y|\le 1}\left(\frac{\|G_R(\cdot,y)\|_{L^p(\mu_\chi)}}{|y|^\alpha}\right)^q\frac{d\rho(y)}{V(|y|)}\right)^{1/q}\to 0,\end{aligned}$$

as $R\to+\infty$. This completes Step 2.

Step 3 We select a smooth approximation of the identity. Precisely, for $0<\kappa\le 1/2$, select $\psi_\kappa\in C_c^\infty(B_\kappa)$, $\psi_\kappa\ge 0$, and $\|\psi_\kappa\|_{L^1(\lambda)}=1$. Moreover, we require that $\psi_\kappa\lesssim V(2\kappa)^{-1}$ (where the constant does not depend on κ).[2] We then have

$$|\psi_\kappa * f(x)|\lesssim Mf(x)\,, \tag{47}$$

[2] This can achieved as follows. Let $\tilde\psi\in C_c^\infty$, $0\le\tilde\psi\le 1$, supp $\tilde\psi\subseteq B_{\kappa/2}$, $\int_G\tilde\psi\,d\rho=1$, and define $\psi_\kappa=C_\kappa V(2\kappa)^{-1}\mathbf{1}_{B_{\kappa/4}}*\tilde\psi$. By requiring that $\|\psi_\kappa\|_{L^1(\lambda)}=1$, we obtain that $C_\kappa\approx 1$.

where M is the local maximal function, defined in (28). Indeed, we estimate

$$|\psi_\kappa * f(x)| = \left|\int_G \psi_\kappa(xy^{-1}) f(y)\, d\rho(y)\right| \lesssim \frac{1}{V(2\kappa)} \int_{B(x,\kappa)} |f(y)|\, d\rho \lesssim Mf(x),$$

as claimed.

By [4, Proposition 20 of Chapter VIII, §4, No. 7] for any $g \in L^p(\mu_\chi)$, $\|\psi_\kappa * g - g\|_{L^p(\mu_\chi)}$ tends to 0, as $\kappa \to 0$. Next, notice that, by left invariance, $W_t^{(m)}(\psi_\kappa * f) = \psi_\kappa * W_t^{(m)} f$. Moreover, if $f \in \mathcal{S}'$ has compact support, then $\psi_\kappa * f \in C_c^\infty(G)$.

Step 4 We complete the proof that C_c^∞ is dense in the Triebel–Lizorkin spaces, $F_\alpha^{p,q}$ in the case $\alpha \in (0, 1)$. To this end, let $f \in F_\alpha^{p,q}$ have compact support so that $\psi_\kappa * f \in C_c^\infty$.

Given an integer $m > \alpha/2$, by Theorem 7 we have

$$\|\psi_\kappa * f - f\|_{F_\alpha^{p,q}}$$

$$\lesssim \Big\| \Big(\sum_{j=0}^{+\infty} (2^{j\alpha/2} |\psi_\kappa * W_{2^{-j}}^{(m)} f - W_{2^{-j}}^{(m)} f|)^q \Big)^{1/q} \Big\|_{L^p(\mu_\chi)} + \|\psi_\kappa * f - f\|_{L^p(\mu_\chi)}.$$

We only need to estimate the first term in the right hand side above. Observe that, since f has compact support, $W_t^{(m)} f \in \mathcal{S}$, for all $t \in (0, 1)$. Hence, [12, Proposition 2.44] gives that, for each j fixed

$$\|\psi_\kappa * W_{2^{-j}}^{(m)} f - W_{2^{-j}}^{(m)} f\|_\infty \to 0 \qquad \text{as} \quad \kappa \to 0.$$

We wish to apply Lebesgue's theorem to the inner sum. We observe that by (47) we have that

$$(2^{j\alpha/2} |\psi_\kappa * W_{2^{-j}}^{(m)} f - W_{2^{-j}}^{(m)} f|)^q \lesssim (2^{j\alpha/2} [M(W_{2^{-j}}^{(m)} f) + |W_{2^{-j}}^{(m)} f|])^q,$$

which is (independent of κ and) summable by [17, p. 481]. Thus, the inner sum tends to 0, as $\kappa \to 0$, for every $x \in G$, that is, the family of vector-valued functions S_κ,

$$S_\kappa := \big(2^{j\alpha/2} \psi_\kappa * W_{2^{-j}}^{(m)} f\big)_j : G \to \ell^q$$

as $\kappa \to 0$ converges pointwise to the vector-valued function $S : G \to \ell^q$, where $S := \big(2^{j\alpha/2} W_{2^{-j}}^{(m)} f\big)_j$. Since, as before, for $0 < \kappa \le 1/2$,

$$\|S_\kappa(x)\|_{\ell^q} \lesssim \Big(\sum_{j=0}^{+\infty} (2^{j\alpha/2} M(W_{2^{-j}}^{(m)} f(x))^q \Big)^{1/q} \in L^p\big((G, \ell^q), \mu_\chi\big).$$

We can apply Lebesgue's theorem to obtain that $S_\kappa \to S$ in $L^p\big((G, \ell^q), \mu_\chi\big)$, that is,

$$\Big\| \Big(\sum_{j=0}^{+\infty} (2^{j\alpha/2} |\psi_\kappa * W^{(m)}_{2^{-j}} f - W^{(m)}_{2^{-j}} f|)^q \Big)^{1/q} \Big\|_{L^p(\mu_\chi)} \to 0 \qquad \text{as} \quad \kappa \to 0\,,$$

as we wished to show. Hence, C_c^∞ is dense in $F_\alpha^{p,q}$, for $p, q \in (1,\infty)$, $\alpha \in (0,1)$.

Step 5 We complete the proof that C_c^∞ is dense in $B_\alpha^{p,q}$ in the case $\alpha \in (0,1)$. Let $f \in B_\alpha^{p,q}$ have compact support so that $\psi_\kappa * f \in C_c^\infty$. We then have

$$\begin{aligned} &\|\psi_\kappa * f - f\|_{B_\alpha^{p,q}} \\ &\lesssim \Big(\int_0^1 (t^{-\alpha/2} \|\psi_\kappa * W_t^{(m)} f - W_t^{(m)} f\|_{L^p(\mu_\chi)})^q \frac{dt}{t} \Big)^{1/q} + \|\psi_\kappa * f - f\|_{L^p(\mu_\chi)}\,. \end{aligned}$$

Now, $\|\psi_\kappa * f - f\|_{L^p(\mu_\chi)}$ and $\|\psi_\kappa * W_t^{(m)} f - W_t^{(m)} f\|_{L^p(\mu_\chi)} \to 0$, as $\kappa \to 0$, the latter term for each $t \in (0,1)$ fixed. Using Young's inequality (44) we see that

$$\|\psi_\kappa * W_t^{(m)} f - W_t^{(m)} f\|_{L^p(\mu_\chi)} \lesssim \|W_t^{(m)} f\|_{L^p(\mu_\chi)}\,,$$

so that we may use Lebesgue's theorem to obtain that

$$\Big(\int_0^1 (t^{-\alpha/2} \|\psi_\kappa * W_t^{(m)} f - W_t^{(m)} f\|_{L^p(\mu_\chi)})^q \frac{dt}{t} \Big)^{1/q} \to 0 \qquad \text{as} \quad \kappa \to 0\,.$$

This gives that $\|\psi_\kappa * f - f\|_{B_\alpha^{p,q}} \to 0$, as $\kappa \to 0$, and completes the proof that C_c^∞ is dense in $B_\alpha^{p,q}$, for $p, q \in (1,\infty)$, $\alpha \in (0,1)$.

Step 6 We now prove that C_c^∞ is dense in $X_{\alpha+1}^{p,q}$, for $p, q \in (1,\infty)$, $\alpha \in (0,1)$. By [6, Theorem 4.5], $\|f\|_{X_{\alpha+1}^{p,q}} \approx \|f\|_{X_\alpha^{p,q}} + \sum_{j=1}^{\ell} \|X_j f\|_{X_\alpha^{p,q}}$ for every $f \in X_{\alpha+1}^{p,q}$. Let $\varepsilon > 0$. By the arguments in Steps 1 and 2, using the same notation, there exists $R > 0$ sufficiently large such that

$$\|f - f\eta_R\|_{X_\alpha^{p,q}} < \varepsilon\,, \qquad \text{and} \quad \|X_j f - (X_j f)\eta_R\|_{X_\alpha^{p,q}} < \varepsilon/2\,,$$

for $j = 1, \ldots, \ell$. Since $X_j \eta_R$ is a C_c^∞ function vanishing in the ball B_{R-1} and $X_j \eta_R$ are uniformly bounded by (45), by the arguments involving ζ_R in Steps 1 and 2 we can also assume that $\|f X_j \eta_R\|_{X_\alpha^{p,q}} < \varepsilon/2$, $j = 1, \ldots, \ell$. Therefore,

$$\|f - f\eta_R\|_{X_\alpha^{p,q}} < \varepsilon\,, \qquad \text{and} \quad \|X_j f - X_j(f\eta_R)\|_{X_\alpha^{p,q}} < \varepsilon\,,$$

that is, $\|f - f\eta_R\|_{X_{\alpha+1}^{p,q}} \lesssim \varepsilon$.

Next, given $f \in X^{p,q}_{\alpha+1}$ having compact support, let $\{\psi_\kappa\}$, $0 < \kappa \le 1/2$ be the approximation of the identity of Step 3, and consider $\psi_\kappa * f$. Then, by Steps 4 and 5, $\psi_\kappa * f \to f$, $X_j(\psi_\kappa * f) = \psi_\kappa * X_j f \to X_j f$, $j = 1, \ldots, \ell$, in $X^{p,q}_\alpha$, as $\kappa \to 0$. This implies that $\psi_\kappa f \to f$ in $X^{p,q}_{\alpha+1}$, as $\kappa \to 0$. This shows that C^∞_c is dense in $X^{p,q}_{\alpha+1}$, for $\alpha \in (0, 1)$.

Step 7 We now finish the proof. Arguing as in the previous step, we obtain that C^∞_c is dense in $X^{p,q}_\alpha$ for all $\alpha \in (0, \infty)\setminus\mathbb{N}$. Let n be a positive integer, and $\theta \in (0, 1)$. By [3, Theorem 4.2.2] $X^{p,q}_{n+\theta} = X^{p,q}_{n-\theta} \cap X^{p,q}_{n+\theta}$ is dense in $X^{p,q}_n$. Since C^∞_c is dense in $X^{p,q}_{n+\theta}$ which embeds continuously into $X^{p,q}_n$, we deduce that C^∞_c is dense in $X^{p,q}_n$. □

6 Isomorphisms of Triebel–Lizorkin and Besov Spaces

Goal of this section is to prove that Bessel potentials provide isomorphisms in both the Triebel–Lizorkin and Besov scales and that a simplified version of local Riesz transforms is bounded on both the Triebel–Lizorkin and Besov spaces. We continue to denote by $X^{p,q}_\alpha$ either $F^{p,q}_\alpha$ or $B^{p,q}_\alpha$. We begin by showing that for all $c \ge 0$, the fractional powers of $\Delta_\chi + cI$ are bounded on the spaces $X^{p,q}_\alpha$. Precisely, we prove the following.

Lemma 12 *Let $p, q \in (1, +\infty)$, $\alpha \ge 0$ and let $\gamma > 0$. Then, for all $c \ge 0$,*

$$(\Delta_\chi + cI)^{\gamma/2} : X^{p,q}_{\alpha+\gamma} \to X^{p,q}_\alpha$$

is bounded.

Proof If $\beta > 0$, then for $\tau > 0$ we have

$$\tau^{-\beta} = \frac{1}{\Gamma(\beta)} \int_0^{+\infty} s^{\beta-1} e^{-\tau s}\, ds\,,$$

so that, if $\gamma > 0$, $c \ge 0$ and k is an integer, $k \ge [\gamma/2] + 1$, for every $f \in \mathcal{S}$

$$\begin{aligned}(\Delta_\chi + cI)^{\gamma/2} f &= (\Delta_\chi + cI)^{-(k-\gamma/2)} (\Delta_\chi + cI)^k f \\ &= \frac{1}{\Gamma(k-\gamma/2)} \int_0^{+\infty} s^{k-\gamma/2} e^{-s(\Delta_\chi + cI)} (\Delta_\chi + cI)^k f \,\frac{ds}{s} \\ &= \frac{1}{\Gamma(k-\gamma/2)} \sum_{j=0}^{k} \sigma_j \int_0^{+\infty} s^{k-\gamma/2} e^{-cs} e^{-s\Delta_\chi} \Delta_\chi^j f\, \frac{ds}{s}\end{aligned} \tag{48}$$

for suitable positive constants σ_j.

Step 1 We prove that for all $f \in X^{p,q}_\alpha$

$$\|e^{-\frac{1}{2}\Delta_X}(\Delta_X + cI)^{\gamma/2} f\|_{L^p(\mu_X)} \lesssim \|e^{-\frac{1}{4}\Delta_X} f\|_{L^p(\mu_X)}. \tag{49}$$

Using (48), [6, Lemma 3.3], the Cauchy–Schwarz inequality and the boundedness of the g-function (9), we notice that

$$\begin{aligned}
&\|e^{-\frac{1}{2}\Delta_X}(\Delta_X + cI)^{\gamma/2} f\|_{L^p(\mu_X)} \\
&\lesssim \sum_{j=0}^{k} \int_0^{3/4} s^{k-\gamma/2} \|e^{-(s+1/4)\Delta_X} \Delta_X^j e^{-\frac{1}{4}\Delta_X} f\|_{L^p(\mu_X)} \frac{ds}{s} + \|e^{-\frac{1}{4}\Delta_X} f\|_{L^p(\mu_X)} \\
&\quad + \sum_{j=1}^{k} \left\| \left(\int_{3/4}^{+\infty} |(s\Delta_X)^j e^{-s\Delta_X} e^{-\frac{1}{2}\Delta_X} f|^2 \frac{ds}{s} \right)^{1/2} \right\|_{L^p(\mu_X)} \\
&\lesssim \sum_{j=0}^{k} \int_0^{1} \frac{s^{k-\gamma/2}}{(s+1/4)^j} \|e^{-\frac{1}{4}\Delta_X} f\|_{L^p(\mu_X)} \frac{ds}{s} + \|e^{-\frac{1}{4}\Delta_X} f\|_{L^p(\mu_X)} \\
&\quad + \sum_{j=1}^{k} \|g_j(e^{-\frac{1}{2}\Delta_X} f)\|_{L^p(\mu_X)} \\
&\lesssim \|e^{-\frac{1}{4}\Delta_X} f\|_{L^p(\mu_X)},
\end{aligned}$$

and thus (49) holds true.

In order to proceed with the main part of the estimates, we need to consider, for $m > \alpha/2$,

$$\begin{aligned}
&W_t^{(m)}(\Delta_X + cI)^{\gamma/2} f \\
&= \frac{1}{\Gamma(k-\gamma/2)} \sum_{j=0}^{k} \sigma_j \left(\int_0^1 + \int_1^{+\infty} \right) s^{k-\gamma/2} e^{-s\Delta_X} e^{-cs} \Delta_X^j W_t^{(m)} f \frac{ds}{s} \\
&=: \sum_{j=0}^{k} F_1^j(t,\cdot) + F_\infty^j(t,\cdot).
\end{aligned} \tag{50}$$

Observe that F_1^j, F_∞^j are functions defined on $(0,1) \times G$.

Step 2 We prove that

$$(\Delta_X + cI)^{\gamma/2} : B^{p,q}_{\alpha+\gamma} \to B^{p,q}_\alpha$$

is bounded, by showing that for any $f \in \mathcal{S}$

$$\int_0^1 \left(t^{-\alpha/2}\|W_t^{(m)}(\Delta_\chi + cI)^{\gamma/2} f\|_{L^p(\mu_\chi)}\right)^q \frac{dt}{t} \lesssim \int_0^1 \left(t^{-(\alpha+\gamma)/2}\|W_t^{(m+k)} f\|_{L^p(\mu_\chi)}\right)^q \frac{dt}{t}. \tag{51}$$

By the norm equivalence in Theorem 6 and (49), Step 2 will follow.

We prove (51) using the decomposition (50). First, observe that

$$\|F_1^0(t,\cdot) + F_\infty^0(t,\cdot)\|_{L^p(\mu_\chi)} \lesssim \left\| \int_0^{+\infty} s^{k-\gamma/2} e^{-cs} |e^{-s\Delta_\chi} W_t^{(m)} f| \frac{ds}{s} \right\|_{L^p(\mu_\chi)} \lesssim \|W_t^{(m)} f\|_{L^p(\mu_\chi)}.$$

Thus, we may suppose $j \geq 1$. By the Cauchy–Schwarz inequality and the boundedness of the g-function (9)

$$\begin{aligned}
\|F_\infty^j(t,\cdot)\|_{L^p(\mu_\chi)} &\lesssim \left\| \int_1^{+\infty} s^{k-\gamma/2} e^{-cs} |e^{-s\Delta_\chi} \Delta_\chi^j W_t^{(m)} f| \frac{ds}{s} \right\|_{L^p(\mu_\chi)} \\
&\lesssim \left\| \left(\int_1^{+\infty} |(s\Delta_\chi)^j e^{-s\Delta_\chi}(W_t^{(m)} f)|^2 \frac{ds}{s} \right)^{1/2} \right\|_{L^p(\mu_\chi)} \\
&\lesssim \|g_j(W_t^{(m)} f)\|_{L^p(\mu_\chi)} \\
&\lesssim \|W_t^{(m)} f\|_{L^p(\mu_\chi)}.
\end{aligned}$$

Therefore,

$$\int_0^1 \left(t^{-\alpha/2}\|F_\infty^j(t,\cdot)\|_{L^p(\mu_\chi)}\right)^q \frac{dt}{t} \lesssim \int_0^1 \left(t^{-\alpha/2}\|W_t^{(m)} f\|_{L^p(\mu_\chi)}\right)^q \frac{dt}{t} \lesssim \|f\|^q_{B_\alpha^{p,q}} \lesssim \|f\|^q_{B_{\alpha+\gamma}^{p,q}}. \tag{52}$$

Next,

$$\begin{aligned}
\|F_1^j(t,\cdot)\|_{L^p(\mu_\chi)} &\lesssim \int_0^1 s^{k-\gamma/2} \|e^{-s\Delta_\chi} \Delta_\chi^j W_t^{(m)} f\|_{L^p(\mu_\chi)} \frac{ds}{s} \\
&= \left(\int_0^t + \int_t^1 \right) t^{-j} s^{k-\gamma/2} \|e^{-s\Delta_\chi} W_t^{(m+j)} f\|_{L^p(\mu_\chi)} \frac{ds}{s} \qquad (53) \\
&=: I_j(t) + II_j(t).
\end{aligned}$$

Now,

$$\begin{aligned} I_j(t) &= \int_0^t t^{-j} s^{k-\gamma/2} \| e^{-s\Delta_X} W_t^{(m+j)} f \|_{L^p(\mu_\chi)} \frac{ds}{s} \\ &\lesssim t^{-j} \| W_t^{(m+j)} f \|_{L^p(\mu_\chi)} \int_0^t s^{k-1-\gamma/2}\, ds \\ &\approx t^{k-j-\gamma/2} \| W_t^{(m+j)} f \|_{L^p(\mu_\chi)}, \end{aligned}$$

so that, since $m + j > (\alpha + \gamma + 2j - 2k)/2$,

$$\begin{aligned} \int_0^1 \left(t^{-\alpha/2} I_j(t) \right)^q \frac{dt}{t} &\lesssim \int_0^1 \left(t^{-(\alpha+\gamma+2j-2k)/2} \| W_t^{(m+j)} f \|_{L^p(\mu_\chi)} \right)^q \frac{dt}{t} \\ &\lesssim \| f \|^q_{B^{p,q}_{\alpha+\gamma+2j-2k}} \lesssim \| f \|^q_{B^{p,q}_{\alpha+\gamma}} . \end{aligned} \tag{54}$$

Next,

$$\begin{aligned} t^{-\alpha/2} II_j(t) &= \int_t^1 t^{-j-\alpha/2} s^{k-\gamma/2} \| e^{-s\Delta_X} W_t^{(m+j)} f \|_{L^p(\mu_\chi)} \frac{ds}{s} \\ &\le \int_0^1 \mathbf{1}_{\{t<s\}} \left(\frac{t}{s} \right)^{m-\alpha/2} s^{-(\alpha+\gamma+2j-2k)/2} \| W_s^{(m+j)} f \|_{L^p(\mu_\chi)} \frac{ds}{s} . \end{aligned}$$

Hence,

$$\int_0^1 \left(t^{-\alpha/2} II_j(t) \right)^q \frac{dt}{t} \le \int_0^1 \left(\int_0^1 K(s,t) g(s) \frac{ds}{s} \right)^q \frac{dt}{t},$$

where

$$K(s,t) = \mathbf{1}_{\{t<s\}} \left(\frac{t}{s} \right)^{m-\alpha/2} \quad \text{and} \quad g(s) = s^{-(\alpha+\gamma+2j-2k)/2} \| W_s^{(m+j)} f \|_{L^p(\mu_\chi)} .$$

It is easy to check that

$$\int_0^1 K(s,t) \frac{ds}{s} \lesssim 1 \quad \text{and} \quad \int_0^1 K(s,t) \frac{dt}{t} \lesssim 1$$

so that Schur's lemma (see [11] e.g.) gives that

$$\int_0^1 \left(t^{-\alpha/2} II_j(t) \right)^q \frac{dt}{t} \lesssim \int_0^1 |g(t)|^q \frac{dt}{t} . \tag{55}$$

Thus, putting together (52) to (55) we obtain (51). This completes Step 2.

Step 3 We now prove that

$$(\Delta_\chi + cI)^{\gamma/2} : F^{p,q}_{\alpha+\gamma} \to F^{p,q}_\alpha$$

is bounded, by showing that for any $f \in \mathcal{S}$

$$\begin{aligned}
&\Big\| \Big(\int_0^1 \big(t^{-\alpha/2} |W_t^{(m)} (\Delta_\chi + cI)^{\gamma/2} f| \big)^q \frac{dt}{t} \Big)^{1/q} \Big\|_{L^p(\mu_\chi)} \\
&\quad \lesssim \Big\| \Big(\int_0^1 \big(t^{-(\alpha+\gamma)/2} |W_t^{(m+k)} f| \big)^q \frac{dt}{t} \Big)^{1/q} \Big\|_{L^p(\mu_\chi)} . \qquad (56)
\end{aligned}$$

Again, this, together with Theorem 6 and (49), will give the desired boundedness.

We use decomposition (50) again.

$$\begin{aligned}
F^j_\infty(t,\cdot) &= \frac{\sigma_j}{\Gamma(k-\gamma/2)} t^m e^{-t\Delta_\chi} \int_1^{+\infty} s^{k-\gamma/2} e^{-cs} e^{-(s-1/2)\Delta_\chi} \Delta_\chi^{m+j} e^{-\frac12 \Delta_\chi} f \, \frac{ds}{s} \\
&= \frac{\sigma_j}{\Gamma(k-\gamma/2)} t^m e^{-t\Delta_\chi} \int_{\frac12}^{+\infty} (s+\tfrac12)^{k-1-\gamma/2} e^{-c(s+\frac12)} e^{-s\Delta_\chi} \Delta_\chi^{m+j} e^{-\frac12\Delta_\chi} f \, ds .
\end{aligned}$$

We notice that, since $f \in \mathcal{S}$ and $e^{-t\Delta_\chi}$ is continuous on $\mathcal{S}'$, the integral converges in $\mathcal{S}'$. By the Cauchy–Schwarz inequality and recalling (8), we have

$$\begin{aligned}
|F^j_\infty(t,\cdot)| &\lesssim t^m e^{-t\Delta_\chi} \Big(\int_0^{+\infty} |(s\Delta_\chi)^{m+j} e^{-s\Delta_\chi} e^{-\frac12\Delta_\chi} f|^2 \frac{ds}{s} \Big)^{1/2} \\
&= t^m e^{-t\Delta_\chi} g_{m+j}(e^{-\frac12\Delta_\chi} f) .
\end{aligned}$$

Now we use [6, Proposition 3.6 (ii)] and the boundedness of the g-function to obtain

$$\begin{aligned}
&\Big\| \Big(\int_0^1 \big(t^{-\alpha/2} |F^j_\infty(t,\cdot)| \big)^q \frac{dt}{t} \Big)^{1/q} \Big\|_{L^p(\mu_\chi)} \\
&\lesssim \Big\| \Big(\int_0^1 \big(t^{m-\alpha/2} e^{-t\Delta_\chi} g_{m+j}(e^{-\frac12\Delta_\chi} f) \big)^q \frac{dt}{t} \Big)^{1/q} \Big\|_{L^p(\mu_\chi)} \qquad (57) \\
&\lesssim \| g_{m+j}(e^{-\frac12\Delta_\chi} f) \|_{L^p(\mu_\chi)} \\
&\lesssim \| e^{-\frac12\Delta_\chi} f \|_{L^p(\mu_\chi)} .
\end{aligned}$$

Finally,

$$
\begin{aligned}
&\int_0^1 \left(t^{-\alpha/2} F_1^j(t,\cdot)\right)^q \frac{dt}{t} \\
&\lesssim \int_0^1 \left(t^{-\alpha/2}\Big(\int_0^t + \int_t^1\Big) s^{k-\gamma/2} |e^{-s\Delta_X} \Delta_X^j W_t^{(m)} f| \frac{ds}{s}\right)^q \frac{dt}{t} \\
&\lesssim \int_0^1 \left(t^{-\alpha/2} \int_0^t s^{k-\gamma/2} |e^{-s\Delta_X} \Delta_X^j W_t^{(m)} f| \frac{ds}{s}\right)^q \frac{dt}{t} \qquad (58)\\
&\quad + \int_0^1 \left(t^{-\alpha/2} \int_t^1 s^{k-\gamma/2} |e^{-s\Delta_X} \Delta_X^j W_t^{(m)} f| \frac{ds}{s}\right)^q \frac{dt}{t} \\
&=: I_j + II_j\,,
\end{aligned}
$$

where in this case, I_j and II_j are functions on G. Similarly to the argument in Step 1, we have

$$
\begin{aligned}
I_j &= \int_0^1 \Big(\int_0^t \frac{t^{m-\alpha/2} s^{k-\gamma/2}}{(s+t)^{m+j}} |W_{s+t}^{(m+j)} f| \frac{ds}{s}\Big)^q \frac{dt}{t} \\
&= \int_0^1 \Big(\int_t^{2t} \frac{t^{m-\alpha/2}(\tau-t)^{k-1-\gamma/2}}{\tau^{m+j-1}} |W_\tau^{(m+j)} f| \frac{d\tau}{\tau}\Big)^q \frac{dt}{t} \\
&= \int_0^1 \Big(\int_0^1 K(\tau,t) g(\tau) \frac{d\tau}{\tau}\Big)^q \frac{dt}{t}\,,
\end{aligned}
$$

where

$$
K(\tau,t) = \mathbf{1}_{\{t<\tau<2t\}} \frac{t^{m-\alpha/2}(\tau-t)^{k-1-\gamma/2}}{\tau^{m+k-1-(\alpha+\gamma)/2}} \quad \text{and} \quad g(\tau) = \tau^{-(\alpha+\gamma+2j-2k)/2} |W_\tau^{(m+k)} f|\,.
$$

Since

$$
\int_0^1 K(\tau,t) \frac{d\tau}{\tau} \lesssim 1 \quad \text{and} \quad \int_0^1 K(\tau,t) \frac{dt}{t} \lesssim 1\,,
$$

we obtain that

$$
I_j \lesssim \int_0^1 \left(t^{-(\alpha+\gamma+2j-2k)/2} |W_t^{(m+k)} f|\right)^q \frac{dt}{t}\,. \qquad (59)
$$

To estimate II_j we use [6, Proposition 3.6 (iv)] and obtain that

$$\begin{aligned}
&\|II_j^{1/q}\|_{L^p(\mu_\chi)} \\
&= \Big\|\Big(\int_0^1 \Big(t^{m-\alpha/2}\int_t^1 s^{-m+k-j-\gamma/2}|e^{-t\Delta_\chi}W_s^{(m+j)}f|\,\frac{ds}{s}\Big)^q\,\frac{dt}{t}\Big)^{1/q}\Big\|_{L^p(\mu_\chi)} \\
&\le \Big\|\Big(\int_0^1 \Big(t^{m-\alpha/2}e^{-t\Delta_\chi}\int_t^1 s^{-m+k-j-\gamma/2}|W_s^{(m+j)}f|\,\frac{ds}{s}\Big)^q\,\frac{dt}{t}\Big)^{1/q}\Big\|_{L^p(\mu_\chi)} \\
&\lesssim \Big\|\Big(\int_0^1 \Big(t^{m-\alpha/2}\int_t^1 s^{-m+k-j-\gamma/2}|W_s^{(m+j)}f|\,\frac{ds}{s}\Big)^q\,\frac{dt}{t}\Big)^{1/q}\Big\|_{L^p(\mu_\chi)} \\
&=: \Big\|\Big(\int_0^1 \Big(\int_0^1 K(s,t)g(s)\,\frac{ds}{s}\Big)^q\,\frac{dt}{t}\Big)^{1/q}\Big\|_{L^p(\mu_\chi)},
\end{aligned} \tag{60}$$

where we have set

$$K(s,t) = \mathbf{1}_{\{t<s\}}\Big(\frac{t}{s}\Big)^{m-\alpha/2} \quad \text{and} \quad g(s) = s^{-(\alpha+\gamma+2j-2k)/2}|W_s^{(m+j)}f|.$$

Again, since

$$\int_0^1 K(s,t)\,\frac{ds}{s} \lesssim 1 \quad \text{and} \quad \int_0^1 K(s,t)\,\frac{dt}{t} \lesssim 1,$$

we see that

$$\|II_j^{1/q}\|_{L^p(\mu_\chi)} \lesssim \Big\|\Big(\int_0^1 \Big(t^{-(\alpha+\gamma+2j-2k)/2}|W_t^{(m+j)}f|\Big)^q\,\frac{dt}{t}\Big)^{1/q}\Big\|_{L^p(\mu_\chi)}. \tag{61}$$

Therefore, (58), (59) and (61) give

$$\begin{aligned}
&\Big\|\Big(\int_0^1 \Big(t^{-\alpha/2}F_1^j(t,\cdot)\Big)^q\,\frac{dt}{t}\Big)^{1/q}\Big\|_{L^p(\mu_\chi)} \\
&\lesssim \|I_j^{1/q}\|_{L^p(\mu_\chi)} + \|II_j^{1/q}\|_{L^p(\mu_\chi)} \\
&\lesssim \Big\|\Big(\int_0^1 \Big(t^{-(\alpha+\gamma+2j-2k)/2}|W_t^{(m+j)}f|\Big)^q\,\frac{dt}{t}\Big)^{1/q}\Big\|_{L^p(\mu_\chi)} \\
&\lesssim \|f\|_{F^{p,q}_{\alpha+\gamma+2j-2k}} \lesssim \|f\|_{F^{p,q}_{\alpha+\gamma}}.
\end{aligned} \tag{62}$$

This, together with (57), proves (56), and finally estimates (49) and (56) complete Step 3. The proof of the lemma is complete. □

We are finally ready to prove the main result of this section.

Theorem 13 *Let $\alpha \geq 0$, $\gamma \geq 0$, $p, q \in (1, +\infty)$, and let $X^{p,q}_\alpha$ denote either space $F^{p,q}_\alpha(\mu_\chi)$ or $B^{p,q}_\alpha(\mu_\chi)$. Then, for $c > 0$ sufficiently large,*

$$(\Delta_\chi + cI)^{\gamma/2} : X^{p,q}_{\alpha+\gamma} \to X^{p,q}_\alpha$$

is a surjective isomorphism, and its inverse is $(\Delta_\chi + cI)^{-\gamma/2}$. Moreover, if $c > 0$ is sufficiently large, then, for all $\alpha, \gamma \geq 0$, the operators

$$\Delta^\gamma_\chi(\Delta_\chi + cI)^{-\gamma} : X^{p,q}_\alpha \to X^{p,q}_\alpha$$

are bounded.

We point out that, for any $\gamma > 0$, c is to be chosen so that the local Riesz transforms $X_J(\Delta_\chi + cI)^{-|J|/2}$ are bounded on $L^p(\mu_\chi)$, for $1 < p < \infty$ and $|J| \leq [\gamma/2] + 1$. Moreover, the operators $\Delta^\gamma_\chi(\Delta_\chi + cI)^{-\gamma}$ can be thought as a simplified version of the local Riesz transforms.

Proof *Step 1* We first prove that, for $n \in \mathbb{N}$, and $c > 0$ sufficiently large so that the local Riesz transforms $X_J(\Delta_\chi + cI)^{-n}$ are bounded on $L^p(\mu_\chi)$, for $1 < p < \infty$ and $|J| \leq 2n$,

$$(\Delta_\chi + cI)^{-n} : X^{p,q}_\alpha \to X^{p,q}_{\alpha+2n}$$

is bounded.

Since $(\Delta_\chi + cI)^{-\beta} : L^p(\mu_\chi) \to L^p(\mu_\chi)$ is bounded, for $p \in (1, +\infty)$ and $\beta > 0$, see [29] or also [5], we trivially have that

$$\|e^{-\frac{1}{2}\Delta_\chi}(\Delta_\chi + cI)^{-n} f\|_{L^p(\mu_\chi)} \lesssim \|e^{-\frac{1}{2}\Delta_\chi} f\|_{L^p(\mu_\chi)}. \tag{63}$$

Let $m \geq [\alpha/2] + 1$. In the case of Besov spaces, it suffices to apply Theorem 3.2 in [5] to obtain

$$\begin{aligned}
\|W^{(m+n)}_t(\Delta_\chi + cI)^{-n} f\|_{L^p(\mu_\chi)} &= \|(t\Delta_\chi)^n(\Delta_\chi + cI)^{-n} W^{(m)}_t f\|_{L^p(\mu_\chi)} \\
&\lesssim t^n \sum_{|J| \leq 2n} \|X_J(\Delta_\chi + cI)^{-n} W^{(m)}_t f\|_{L^p(\mu_\chi)} \\
&\lesssim t^n \|W^{(m)}_t f\|_{L^p(\mu_\chi)}.
\end{aligned}$$

Hence,

$$\begin{aligned}
&\Big(\int_0^1 \Big(t^{-(n+\alpha/2)}\,\|W_t^{(m+n)}(\Delta_\chi + cI)^{-n} f\|_{L^p(\mu_\chi)}\Big)^q \frac{dt}{t}\Big)^{1/q} \\
&\lesssim \Big(\int_0^1 \Big(t^{-\alpha/2}\|W_t^{(m)} f\|_{L^p(\mu_\chi)}\Big)^q \frac{dt}{t}\Big)^{1/q} \\
&\lesssim \|f\|_{B_\alpha^{p,q}}\,.
\end{aligned} \tag{64}$$

Therefore, (63) and (64) show that $(\Delta_\chi + cI)^{-n} : B_\alpha^{p,q} \to B_{\alpha+2n}^{p,q}$ is bounded, $p, q \in (1, +\infty)$, $\alpha \geq 0$, $n \in \mathbb{N}$.

Next we consider the case of the Triebel–Lizorkin spaces. Arguing as in (48) we write

$$\begin{aligned}
W_t^{(m+n)}(\Delta_\chi + cI)^{-n} f &= \frac{1}{\Gamma(n)} \int_0^{+\infty} s^n e^{-cs} e^{-s\Delta_\chi} W_t^{(m+n)} f\, \frac{ds}{s} \\
&= \frac{1}{\Gamma(n)} \Big(\int_0^1 + \int_1^{+\infty}\Big) s^n e^{-cs} e^{-s\Delta_\chi} W_t^{(m+n)} f\, \frac{ds}{s} \\
&=: \frac{1}{\Gamma(n)} \big(A_1(t,\cdot) + A_\infty(t,\cdot)\big)\,.
\end{aligned}$$

We begin with the latter term and observe that, since $f \in \mathcal{S}$ and $e^{-t\Delta_\chi}$ is continuous on $\mathcal{S}'$, we get

$$|A_\infty(t,\cdot)| = t^{m+n} \Big| e^{-t\Delta_\chi} \int_1^{+\infty} s^n e^{-cs} e^{-(s-1/2)\Delta_\chi} \Delta_\chi^{m+n} e^{-\frac{1}{2}\Delta_\chi} f\, \frac{ds}{s}\Big|$$

which by the Cauchy–Schwarz inequality is bounded by

$$t^{m+n} e^{-t\Delta_\chi} \Big(\int_0^{+\infty} |W_s^{(m+n)} e^{-\frac{1}{2}\Delta_\chi} f|^2\, \frac{ds}{s}\Big)^{1/2} = t^{m+n} e^{-t\Delta_\chi} g_{m+n}(e^{-\frac{1}{2}\Delta_\chi} f)\,,$$

where g_{m+n} is defined in (8). Therefore, by the above estimate, Proposition 3.6 (ii) in [6] and the boundedness of the g-function,

$$\begin{aligned}
&\Big\|\Big(\int_0^1 \Big(t^{-(n+\alpha/2)}|A_\infty(t,\cdot)|\Big)^q \frac{dt}{t}\Big)^{1/q}\Big\|_{L^p(\mu_\chi)} \\
&\lesssim \Big\|\Big(\int_0^1 \Big(t^{m-\alpha/2} e^{-t\Delta_\chi} g_{m+n}(e^{-\frac{1}{2}\Delta_\chi} f)\Big)^q \frac{dt}{t}\Big)^{1/q}\Big\|_{L^p(\mu_\chi)} \\
&\lesssim \|g_{m+n}(e^{-\frac{1}{2}\Delta_\chi} f)\|_{L^p(\mu_\chi)} \\
&\lesssim \|e^{-\frac{1}{2}\Delta_\chi} f\|_{L^p(\mu_\chi)}\,.
\end{aligned} \tag{65}$$

Now we turn to $A_1(t,\cdot)$, and arguing as above and applying [6, Lemma 3.2 (i)] we get

$$
\begin{aligned}
&|A_1(t,\cdot)| \\
&\leq t^{m+n}\left(\int_0^t s^n\left|e^{-(s+\frac{t}{2})\Delta_X}e^{-\frac{t}{2}\Delta_X}\Delta_X^{m+n}f\right|\frac{ds}{s}+e^{-t\Delta_X}\int_t^1 s^n\left|e^{-s\Delta_X}\Delta_X^{m+n}f\right|\frac{ds}{s}\right)\\
&\lesssim t^{m+n}\left(e^{-ct\Delta_X}\int_0^t s^n\left|e^{-\frac{t}{2}\Delta_X}\Delta_X^{m+n}f\right|\frac{ds}{s}+e^{-t\Delta_X}\int_t^1 s^n\left|e^{-s\Delta_X}\Delta_X^{m+n}f\right|\frac{ds}{s}\right)\\
&= t^{m+n}\left(t^n\, e^{-ct\Delta_X}\left|e^{-\frac{t}{2}\Delta_X}\Delta_X^{m+n}f\right|+e^{-t\Delta_X}\int_t^1 s^n\left|e^{-s\Delta_X}\Delta_X^{m+n}f\right|\frac{ds}{s}\right).
\end{aligned}
\tag{66}
$$

As for the first term, by [6, Proposition 3.6 (i)] one has

$$
\begin{aligned}
&\left\|\left(\int_0^1\left(t^{-(n+\frac{\alpha}{2})}t^{m+2n}\,e^{-ct\Delta_X}\left|e^{-\frac{t}{2}\Delta_X}\Delta_X^{m+n}f\right|\right)^q\frac{dt}{t}\right)^{1/q}\right\|_{L^p(\mu_X)}\\
&\lesssim\left\|\left(\int_0^1 t^{-q\frac{\alpha}{2}}|W_t^{(m+n)}f|^q\frac{dt}{t}\right)^{1/q}\right\|_{L^p(\mu_X)}\\
&\lesssim\|f\|_{F_\alpha^{p,q}}.
\end{aligned}
\tag{67}
$$

For the second term on the right hand side of (66) we use Schur's test. Precisely, arguing as at the end of Step 2 in the proof of Lemma 12, we have

$$
\begin{aligned}
&\int_0^1\left(t^{m-\alpha/2}\int_t^1 s^n\left|e^{-s\Delta_X}\Delta_X^{m+n}f\right|\frac{ds}{s}\right)^q\frac{dt}{t}\\
&=\int_0^1\left(\int_0^1\mathbf{1}_{\{t<s\}}\left(\frac{t}{s}\right)^m t^{-\alpha/2}|W_s^{(m+n)}f|\frac{ds}{s}\right)^q\frac{dt}{t}\\
&\lesssim\int_0^1\left(t^{-\alpha/2}|W_t^{(n+m)}f|\right)^q\frac{dt}{t}.
\end{aligned}
\tag{68}
$$

Thus, by (67) and (68) we obtain that

$$
\left\|\left(\int_0^1\left(t^{-(n+\alpha/2)}|A_1(t,\cdot)|\right)^q\frac{dt}{t}\right)^{1/q}\right\|_{L^p(\mu_X)}\lesssim\|f\|_{F_\alpha^{p,q}}.
\tag{69}
$$

Estimates (65) and (69) complete the proof of Step 1.

Step 2 We now complete the proof.

We first observe that for $\gamma \geq 0$, the operator

$$(\Delta_\chi + cI)^{-\gamma/2} : X_\alpha^{p,q} \to X_{\alpha+\gamma}^{p,q}$$

is bounded. Indeed, it suffices to notice that, choosing $n \in \mathbb{N}$, $n > \gamma/2$ we have

$$(\Delta_\chi + cI)^{-\gamma/2} = (\Delta_\chi + cI)^{n-\gamma/2}(\Delta_\chi + cI)^{-n} .$$

The conclusion follows from the boundedness of the two operators on the right hand side given by Step 1 and Lemma 12.

It now follows that

$$(\Delta_\chi + cI)^{\gamma/2} : X_{\alpha+\gamma}^{p,q} \to X_\alpha^{p,q}$$

is a surjective isomorphisms. Indeed, given any $\phi \in \mathcal{S}$, we can write

$$\phi = (\Delta_\chi + cI)^{\gamma/2}(\Delta_\chi + cI)^{-\gamma/2}\phi = (\Delta_\chi + cI)^{-\gamma/2}(\Delta_\chi + cI)^{\gamma/2}\phi .$$

Hence, given any $f \in X_\alpha^{p,q}$, we can write

$$f = (\Delta_\chi + cI)^{\gamma/2}(\Delta_\chi + cI)^{-\gamma/2} f ,$$

since $f \in \mathcal{S}'$.

Finally, it is now clear that for $\alpha, \gamma \geq 0$,

$$\Delta_\chi^{\gamma/2}(\Delta_\chi + cI)^{-\gamma/2} : X_\alpha^{p,q} \to X_\alpha^{p,q}$$

is bounded. The proof is now complete. □

Using Theorem 11 and applying the previous theorem with $\alpha = 0$ and $\gamma = 2m$, m being a positive integer, we immediately obtain the following

Corollary 14 *Let $p, q \in (1, \infty)$ and let $X_0^{p,q}$ denote either $F_0^{p,q}$ or $B_0^{p,q}$. Then, $C_c^\infty(G)$ is dense in $X_0^{p,q}$.*

7 Final Remarks and Open Problems

In this final section we discuss some directions for future work and indicate some open problems.

First of all, we stress the fact that in this work and [6] we have limited ourselves to the cases $p, q \in (1, +\infty)$ and $\alpha \geq 0$. It would be interesting to investigate whether the spaces $F_0^{p,2}(\mu_\chi)$ with $p = 1, +\infty$, correspond respectively to the local Hardy space $\mathfrak{h}^1(\mu_\chi)$ and its dual $\mathfrak{bmo}(\mu_\chi)$, introduced in [5], in analogy to the Euclidean setting. Such spaces turn out to be useful in many problems, most noticeably in the boundedness of singular integral operators. Moreover, Triebel–Lizorkin and Besov spaces with $0 < p < 1$ or $0 < q < 1$ are quasi-Banach and their treatment often requires different techniques. Finally, the spaces $X_\alpha^{p,q}(\mu_\chi)$ with $\alpha < 0$ should appear as natural duals of the spaces with positive index of regularity and are also of considerable interest.

We recall that Besov and also Triebel–Lizorkin spaces are instrumental to applications to solvability and regularity of solutions of nonlinear differential equations, as, for instance, in the spirit of the results in Section 6 in [5]. It would also be interesting to study the homogeneous versions of Sobolev, Besov and Triebel–Lizorkin spaces in the setting of this work. These spaces, in particular the homogeneous Besov spaces, appear naturally in the Strichartz estimates for the wave equation in the Euclidean space, or Lie groups of polynomial growth, see e.g. [2, 19] and [15].

Another set of natural and interesting questions concerns the generalization of some classical geometric inequalities, which have already been studied in the setting of manifolds and metric spaces under suitable geometric assumptions. In particular, we mention the Poincaré inequality, see [31] for the classical case and [27] for Carnot–Carathéodory groups, trace inequalities, see [31] for the classical case and [8] for Carnot–Carathéodory groups, isoperimetric and Sobolev inequalities [31] and [18], to name just a few. We intend to study extensions of these classical inequalities to the case of the sub-Laplacian Δ_χ on a general Lie group G and of the Sobolev, Triebel–Lizorkin and Besov spaces. We point out that in [45] the authors proved versions of Hardy, Hardy–Sobolev, Caffarelli–Nirenberg, Gagliardo–Nirenberg inequalities in the case of the Sobolev spaces $L_\alpha^p(\mu_\chi)$.

Acknowledgments We thank Hans Triebel for pointing out some references and sharing with us a preprint version of his new book [54]. We also thank the referee for several corrections and suggestions that greatly improved the presentation of the manuscript.

References

1. Agrachev, A., Boscain, U., Gauthier, J.-P., Rossi, F.: The intrinsic hypoelliptic Laplacian and its heat kernel on unimodular Lie groups. J. Funct. Anal. **256**(8), 2621–2655 (2009)
2. Bahouri, H., Gérard, P., Xu, C.-J.: Espaces de Besov et estimations de Strichartz généralisées sur le groupe de Heisenberg. J. Anal. Math. **82**, 93–118 (2000)
3. Bergh, J., Löfström, J.: Interpolation Spaces. An Introduction. Springer-Verlag, Berlin-New York (1976). Grundlehren der Mathematischen Wissenschaften, No. 223
4. Bourbaki, N.: Integration. II. Chapters 7–9. Elements of Mathematics (Berlin). Springer, Berlin (2004). Translated from the 1963 and 1969 French originals by Sterling K. Berberian
5. Bruno, T., Peloso, M.M., Tabacco, A., Vallarino, M.: Sobolev spaces on Lie groups: Embedding theorems and algebra properties. J. Funct.Anal. **276**(10), 3014–3050 (2019)
6. Bruno, T., Peloso, M.M., Vallarino, M.: Besov and Triebel–Lizorkin spaces on Lie groups. Math. Ann. **377**, 335–377 (2020)

7. Coulhon, T., Russ, E., Tardivel-Nachef, V.: Sobolev algebras on Lie groups and Riemannian manifolds. Amer. J. Math. **123**(2), 283–342 (2001)
8. Danielli, D., Garofalo, N., Nhieu, D.-M.: Trace inequalities for Carnot-Carathéodory spaces and applications. Ann. Scuola Norm. Sup. Pisa Cl. Sci. (4) **27**(2), 195–252 (1998)
9. Feneuil, J.: Algebra properties for Besov spaces on unimodular Lie groups. Colloq. Math. **154**(2), 205–240 (2018)
10. Folland, G.B.: Subelliptic estimates and function spaces on nilpotent Lie groups. Ark. Mat. **13**(2), 161–207 (1975)
11. Folland, G.B.: Real Analysis. Pure and Applied Mathematics (New York), 2nd edn. Wiley, New York (1999). Modern techniques and their applications, A Wiley-Interscience Publication
12. Folland, G.B.: A Course in Abstract Harmonic Analysis, 2nd edn. Textbooks in Mathematics. CRC Press, Boca Raton (2016)
13. Folland, G.B., Stein, E.M.: Estimates for the $\bar{\partial}_b$ complex and analysis on the Heisenberg group. Comm. Pure Appl. Math. **27**, 429–522 (1974)
14. Furioli, G., Melzi, C., Veneruso, A.: Littlewood-Paley decompositions and Besov spaces on Lie groups of polynomial growth. Math. Nachr. **279**(9–10), 1028–1040 (2006)
15. Furioli, G., Melzi, C., Veneruso, A.: Strichartz inequalities for the wave equation with the full Laplacian on the Heisenberg group. Canad. J. Math. **59**(6), 1301–1322 (2007)
16. Gallagher, I., Sire, Y.: Besov algebras on Lie groups of polynomial growth. Studia Math. **212**(2), 119–139 (2012)
17. García-Cuerva, J., Rubio de Francia, J.L.: Weighted Norm Inequalities and Related Topics. North-Holland Mathematics Studies, vol. 116. North-Holland Publishing Co., Amsterdam (1985). Notas de Matemática [Mathematical Notes], 104
18. Garofalo, N., Nhieu, D.-M.: Isoperimetric and Sobolev inequalities for Carnot-Carathéodory spaces and the existence of minimal surfaces. Commun. Pure Appl. Math. **49**(10), 1081–1144 (1996)
19. Ginibre, J., Velo, G.: Generalized Strichartz inequalities for the wave equation. J. Funct. Anal. **133**(1), 50–68 (1995)
20. Grafakos, L.: Modern Fourier Analysis. Graduate Texts in Mathematics, 3rd edn., vol. 250. Springer, New York (2014)
21. Guivarc'h, Y.: Croissance polynomiale et périodes des fonctions harmoniques. Bull. Soc. Math. France **101**, 333–379 (1973)
22. Han, Y., Müller, D., Yang, D.: A theory of Besov and Triebel-Lizorkin spaces on metric measure spaces modeled on Carnot-Carathéodory spaces. Abstr. Appl. Anal., Art. ID 893409, 250 pp. (2008)
23. Hebisch, W., Mauceri, G., Meda, S.: Spectral multipliers for sub-Laplacians with drift on Lie groups. Math. Z. **251**(4), 899–927 (2005)
24. Heikkinen, T., Koskela, P., Tuominen, H.: Approximation and quasicontinuity of Besov and Triebel-Lizorkin functions. Trans. Amer. Math. Soc. **369**(5), 3547–3573 (2017)
25. Hörmander, L.: Estimates for translation invariant operators in L^p spaces. Acta Math. **104**, 93–140 (1960)
26. Hörmander, L.: Hypoelliptic second order differential equations. Acta Math. **119**, 147–171 (1967)
27. Jerison, D.: The Poincaré inequality for vector fields satisfying Hörmander's condition. Duke Math. J. **53**(2), 503–523 (1986)
28. Kohn, J.J.: Harmonic integrals on strongly pseudo-convex manifolds. I. Ann. Math. (2) **78**, 112–148 (1963)
29. Komatsu, H.: Fractional powers of operators. Pacific J. Math. **19**, 285–346 (1966)
30. Korányi, A., Vági, S.: Singular integrals on homogeneous spaces and some problems of classical analysis. Ann. Scuola Norm. Sup. Pisa (3) **25**, 575–648 (1971)
31. Leoni, G.: A First Course in Sobolev Spaces. Graduate Studies in Mathematics, 2nd edn., vol. 181. American Mathematical Society, Providence, RI (2017)
32. Martini, A.: Analysis of joint spectral multipliers on Lie groups of polynomial growth. Ann. Inst. Fourier (Grenoble) **62**(4), 1215–1263 (2012)

33. Meda, S.: On the Littlewood-Paley-Stein g-function. Trans. Amer. Math. Soc. **347**(6), 2201–2212 (1995)
34. Müller, D., Yang, D.: A difference characterization of Besov and Triebel-Lizorkin spaces on RD-spaces. Forum Math. **21**(2), 259–298 (2009)
35. Müller, D., Ricci, F., Stein, E.M.: Marcinkiewicz multipliers and multi-parameter structure on Heisenberg (-type) groups. I. Invent. Math. **119**(2), 199–233 (1995)
36. Müller, D., Ricci, F., Stein, E.M.: Marcinkiewicz multipliers and multi-parameter structure on Heisenberg (-type) groups. II. Math. Z. **221**(2), 267–291 (1996)
37. Müller, D., Peloso, M.M., Ricci, F.: L^p-spectral multipliers for the Hodge Laplacian acting on 1-forms on the Heisenberg group. Geom. Funct. Anal. **17**(3), 852–886 (2007)
38. Müller, D., Peloso, M.M., Ricci, F.: Analysis of the Hodge Laplacian on the Heisenberg group. Mem. Amer. Math. Soc. **233**(1095), vi+91 (2015)
39. Peloso, M.M., Ricci, F.: Tangential Cauchy-Riemann equations on quadratic CR manifolds. Atti Accad. Naz. Lincei Cl. Sci. Fis. Mat. Natur. Rend. Lincei (9) Mat. Appl. **13**(3–4), 285–294 (2002). Harmonic analysis on complex homogeneous domains and Lie groups (Rome, 2001)
40. Peloso, M.M., Ricci, F.: Analysis of the Kohn Laplacian on quadratic CR manifolds. J. Funct. Anal. **203**(2), 321–355 (2003)
41. Peloso, M.M., Vallarino, M.: Sobolev algebras on nonunimodular Lie groups. Calc. Var. Partial Differential Equations **57**(6), Art. 150, 34 (2018)
42. Ricci, F., Stein, E.M.: Harmonic analysis on nilpotent groups and singular integrals. I. Oscillatory integrals. J. Funct. Anal. **73**(1), 179–194 (1987)
43. Ricci, F., Stein, E.M.: Harmonic analysis on nilpotent groups and singular integrals. II. Singular kernels supported on submanifolds. J. Funct. Anal. **78**(1), 56–84 (1988)
44. Ricci, F., Stein, E.M.: Harmonic analysis on nilpotent groups and singular integrals. III. Fractional integration along manifolds. J. Funct. Anal. **86**(2), 360–389 (1989)
45. Ruzhansky, M., Yessirkegenov, N.: Hardy, Hardy-Sobolev, Hardy-Littlewood-Sobolev and Caffarelli-Kohn-Nirenberg inequalities on general Lie groups (2018). ArXiv e-prints, 32 pp.
46. Stein, E.M.: Singular Integrals and Differentiability Properties of Functions. Princeton Mathematical Series, No. 30. Princeton University Press, Princeton (1970)
47. Stein, E.M.: Topics in Harmonic Analysis Related to the Littlewood-Paley Theory. Annals of Mathematics Studies, No. 63. Princeton University Press/University of Tokyo Press, Princeton/Tokyo (1970)
48. Strichartz, R.S.: L^p harmonic analysis and Radon transforms on the Heisenberg group. J. Funct. Anal. **96**(2), 350–406 (1991)
49. Triebel, H.: Spaces of Besov-Hardy-Sobolev type on complete Riemannian manifolds. Ark. Mat. **24**(2), 299–337 (1986)
50. Triebel, H.: Characterizations of function spaces on a complete Riemannian manifold with bounded geometry. Math. Nachr. **130**, 321–346 (1987)
51. Triebel, H.: Function spaces on Lie groups and on analytic manifolds. In: Function Spaces and Applications (Lund, 1986). Lecture Notes in Mathematics, vol. 1302, pp. 384–396. Springer, Berlin (1988)
52. Triebel, H.: Theory of Function Spaces. II. Monographs in Mathematics, vol. 84. Birkhäuser Verlag, Basel (1992)
53. Triebel, H.: Theory of Function Spaces. Modern Birkhäuser Classics. Birkhäuser/Springer Basel AG, Basel (2010)
54. Triebel, H.: Theory of Function Spaces. IV. Monographs in Mathematics. Birkhäuser Verlag, Basel (2020)
55. Varopoulos, N.T.: Analysis on Lie groups. J. Funct. Anal. **76**(2), 346–410 (1988)
56. Varopoulos, N.T., Saloff-Coste, L., Coulhon, T.: Analysis and Geometry on Groups. Cambridge Tracts in Mathematics, vol. 100. Cambridge University Press, Cambridge (1992)

On Fourier Restriction for Finite-Type Perturbations of the Hyperbolic Paraboloid

Stefan Buschenhenke, Detlef Müller, and Ana Vargas

To Fulvio at the occasion of his seventieth birthday

Abstract In this note, we continue our research on Fourier restriction for hyperbolic surfaces, by studying local perturbations of the hyperbolic paraboloid $z = xy$, which are of the form $z = xy + h(y)$, where $h(y)$ is a smooth function of finite type. Our results build on previous joint work in which we have studied the case $h(y) = y^3/3$ by means of the bilinear method. As it turns out, the understanding of that special case becomes also crucial for the treatment of arbitrary finite type perturbation terms $h(y)$.

Keywords Hyperbolic hypersurface · Fourier restriction

1 Introduction

Our aim in this note is to provide another step in our program towards gaining an understanding of Fourier restriction for general hyperbolic surfaces.

The first author was partially supported by the ERC grant 307617. The first two authors were partially supported by the DFG grant MU 761/11-2. The third author was partially supported by the grants PID2019-105599GB-I00/ AEI / 10.13039/501100011033 (Ministerio de Ciencia e Innovación) and MTM2016-76566-P (Ministerio de Ciencia, Innovación y Universidades), Spain.

S. Buschenhenke · D. Müller
Mathematisches Seminar, C.A.-Universität Kiel, Kiel, Germany
e-mail: buschenhenke@math.uni-kiel.de; http://analysis.math.uni-kiel.de/buschenhenke/; mueller@math.uni-kiel.de; http://analysis.math.uni-kiel.de/mueller/

A. Vargas (✉)
Departmento de Matemáticas, Universidad Autónoma de Madrid, Madrid, Spain
e-mail: ana.vargas@uam.es; http://matematicas.uam.es/~AFA/

P. Ciatti, A. Martini (eds.), *Geometric Aspects of Harmonic Analysis*, Springer INdAM Series 45, https://doi.org/10.1007/978-3-030-72058-2_5

Fourier restriction for hypersurfaces with non-negative principal curvatures has been studied intensively by many authors (see, e.g., [2–6, 13–17, 20–22, 25, 28, 29, 31, 32, 35]). For the case of hypersurfaces of non-vanishing Gaussian curvature but principal curvatures of different signs, besides Tomas-Stein type Fourier restriction estimates (see, e.g., [12, 15–17, 24, 27, 30]), until recently the only case which had been studied successfully was the case of the hyperbolic paraboloid (or "saddle") in $\mathbb{R}^3$: in 2015, independently S. Lee [19] and A. Vargas [34] established results analogous to Tao's theorem [29] on elliptic surfaces (such as the 2-sphere), with the exception of the end-point, by means of the bilinear method. Recently, B. Stovall [26] was able to include also the end-point case. Moreover, C. H. Cho and J. Lee [10], and J. Kim [18], improved the range by adapting ideas by Guth [13, 14] which are based on the polynomial partitioning method. For further information on the history of the restriction problem, we refer the interested reader to our previous paper [7].

We shall here study surfaces S which are local perturbations of the hyperbolic paraboloid $z = xy$, which are given as the graph of a function $\phi(x, y) := xy + h(y)$, where the function h is smooth and of finite type at the origin, i.e.,

$$S := \{(x, y, xy + h(y)) : (x, y) \in \Omega\}, \tag{1}$$

where Ω is a sufficiently small neighborhood of the origin, and $h(y) = y^{m+2}a(y)$, with $a(0, 0) \neq 0$ and $m \geq 1$. The Fourier restriction problem, introduced by E. M. Stein in the seventies (for general submanifolds), asks for the range of exponents $\tilde{p}$ and $\tilde{q}$ for which an a priori estimate of the form

$$\left(\int_S |\widehat{f}|^{\tilde{q}} \, d\sigma \right)^{1/\tilde{q}} \leq C \|f\|_{L^{\tilde{p}}(\mathbb{R}^n)}$$

holds true for every Schwartz function $f \in \mathcal{S}(\mathbb{R}^3)$, with a constant C independent of f. Here, $d\sigma$ denotes the surface measure on S.

As usual, it will be more convenient to use duality and work in the adjoint setting. If $\mathcal{R}$ denotes the Fourier restriction operator $g \mapsto \mathcal{R}g := \hat{g}|_S$ to the surface S, its adjoint operator $\mathcal{R}^*$ is given by $\mathcal{R}^* f(\xi) = \mathcal{E}f(-\xi)$, where $\mathcal{E}$ denotes the "Fourier extension" operator given by

$$\mathcal{E}f(\xi) := \widehat{f \, d\sigma}(\xi) = \int_S f(x) e^{-i\xi \cdot x} \, d\sigma(x),$$

with $f \in L^q(S, \sigma)$. The restriction problem is therefore equivalent to the question of finding the appropriate range of exponents for which the estimate

$$\|\mathcal{E}f\|_{L^r(\mathbb{R}^3)} \leq C \|f\|_{L^q(S, d\sigma)}$$

holds true with a constant C independent of the function $f \in L^q(S, d\sigma)$.

By identifying a point $(x, y) \in \Omega$ with the corresponding point $(x, y, \phi(x, y))$ on S, we may regard our Fourier extension operator $\mathcal{E}$ as well as an operator mapping functions on Ω to functions on $\mathbb{R}^3$, which in terms of our phase function $\phi(x, y) = xy + h(y)$ can be expressed more explicitly in the form

$$\mathcal{E}f(\xi) = \int_\Omega f(x, y)e^{-i(\xi_1 x+\xi_2 y+\xi_3\phi(x,y))}\eta(x, y)\, dxdy,$$

if $\xi = (\xi_1, \xi_2, \xi_3) \in \mathbb{R}^3$, with a suitable smooth density η.

Our main result, which generalizes Theorem 1.1 in [7], is the following

Theorem 1 *Assume that $r > 10/3$ and $1/q' > 2/r$, and let $\mathcal{E}$ denote the Fourier extension operator associated to the graph S in (1) of the above phase function $\phi(x, y) := xy + h(y)$, where the function h is smooth and of finite type at the origin. Then, if Ω is a sufficiently small neighborhood of the origin,*

$$\|\mathcal{E}f\|_{L^r(\mathbb{R}^3)} \leq C_{r,q}\|f\|_{L^q(\Omega)}$$

for all $f \in L^q(\Omega)$.

For the proof of this result, we shall strongly build on the approach devised for the special case where $h(y) = y^3/3$. In many arguments, we shall be able to basically follow [7]. Therefore, we shall concentrate on explaining the new ideas and modifications that are needed to handle more general finite type perturbations.

Convention Unless stated otherwise, $C > 0$ will stand for an absolute constant whose value may vary from occurrence to occurrence. We will use the notation $A \sim_C B$ to express that $\frac{1}{C}A \leq B \leq CA$. In some contexts where the size of C is irrelevant we shall drop the index C and simply write $A \sim B$. Similarly, $A \lesssim B$ will express the fact that there is a constant C (which does not depend on the relevant quantities in the estimate) such that $A \leq CB$, and we write $A \ll B$, if the constant C is sufficiently small.

Foreword by the Second-Named Author My collaboration with Fulvio started more than thirty years ago at a time when he had still been teaching at the Politecnico di Torino. Fulvio had invited me then to come to visit him, and right away we had found an interesting joint research project, which opened new horizons for me. Our ways of mathematical thinking turned out to harmonize very well, and, not least due to the personal warmth and hospitality of Sandra and Fulvio, very soon also our families became friends. I would like to take the opportunity to express my deeply felt gratitude for all these years of wonderful and intensive collaboration and friendship with Fulvio.

2 Reduction to Perturbations of Cubic Type

Recall that we are assuming that

$$\phi(x, y) = xy + y^{m+2}a(y), \quad \text{where } a(0) \neq 0, m \geq 1. \tag{1}$$

We may assume without loss of generality that Ω is a square, and then decompose the domain Ω dyadically with respect to the y-variable into rectangular boxes

$$\Omega = \bigcup_{\pm} \bigcup_{i \geq i_0} \Omega^{\pm}_{2^{-i}},$$

where for any $\kappa = 2^{-i}$ we have $\Omega^{-}_{\kappa} = -\Omega^{+}_{\kappa}$, and $\kappa \leq y \leq 2\kappa$ on Ω^{+}_{κ}. Note that we may assume that $i_0 \gg 1$ is sufficiently large, by choosing Ω sufficiently small. By

$$\mathcal{E}^{\pm}_{\kappa} f(\xi) = \int_{\Omega^{\pm}_{\kappa}} f(x, y) e^{-i(\xi_1 x + \xi_2 y + \xi_3 \phi(x,y))} \eta(x, y)\, dx dy$$

we denote the contribution of $\Omega^{\pm}_{\kappa}$ to $\mathcal{E} f$.

Let us fix one of these subsets, say Ω^{+}_{κ}. We then apply an affine change of variables to pass to the phase

$$\phi_\kappa(x, y) := \frac{1}{\kappa}\phi\big(x, \kappa(1 + y)\big) = x(1 + y) + \kappa^{m+1}(1 + y)^{m+2} a\big(\kappa(1 + y)\big),$$

where $0 \leq y \leq 1$. Actually, by taking, say, 1000 subdomains, we may even assume that $0 \leq y \leq 1/1000$. Let us put

$$H_\kappa(y) := (1 + y)^{m+2} a\big(\kappa(1 + y)\big).$$

Then

$$\phi_\kappa(x, y) = x(1 + y) + \kappa^{m+1} H_\kappa(y) = x + xy + \kappa^{m+1} P_2(\kappa, y) + \kappa^{m+1} h_\kappa(y),$$

where $P_2(\kappa, y)$ denotes the Taylor polynomial of $H_\kappa(y)$ of degree 2 centered at $y = 0$. As in our previous paper [7], we may then write

$$\begin{aligned} x + xy + \kappa^{m+1} P_2(\kappa, y) &= xy + c_\kappa y^2 + \text{affine linear terms} \\ &= (x + c_\kappa y)y + \text{affine linear terms}. \end{aligned}$$

The linear change of variables $x \mapsto x + c_\kappa y$ then allows to reduce to the phase function

$$\tilde{\phi}_\kappa(x, y) := xy + \kappa^{m+1} h_\kappa(y),$$

for (x, y) in a sufficiently small neighborhood of the origin which can be chosen independently of κ. Note that

$$h_\kappa(0) = h'_\kappa(0) = h''_\kappa(0). \tag{2}$$

Moreover, it is easy to see that for κ sufficiently small (depending on m, $a(0) \neq 0$, $\|a'\|_\infty$, $\|a''\|_\infty$ and $\|a'''\|_\infty$), we have

$$|h'''_\kappa(y)| \geq \frac{(m+2)(m+1)m}{2}|a(0)| \geq C_3 > 0. \tag{3}$$

A similar reasoning shows that

$$|h'''_\kappa(y)| \leq 4C_3, \quad \text{and} \quad |h^{(l)}_\kappa(y)| \leq C_l \ \text{ for all } |y| \leq 1/1000, l \geq 4, \tag{4}$$

with constants C_l which are independent of κ.

Similar arguments apply to Ω_κ^-. We consider next the Fourier extension operator

$$\tilde{\mathcal{E}}_\kappa^\pm f(\xi) = \int_{\tilde{\Omega}_\kappa^\pm} f(x, y)e^{-i(\xi_1 x+\xi_2 y+\xi_3(xy+\kappa^{m+1}h_\kappa(y))}\tilde{\eta}_\kappa(x, y)\,dxdy,$$

where $\tilde{\eta}_\kappa(x, y) = \eta(x, \kappa(1 + y))$, which corresponds to the operator $\mathcal{E}_\kappa^\pm$ in the new coordinates. Then an easy scaling argument shows that the following estimates for $\mathcal{E}_\kappa^\pm$ and $\tilde{\mathcal{E}}_\kappa^\pm$ are equivalent:

$$\|\tilde{\mathcal{E}}_\kappa^\pm f\|_{L^r} \leq C\|f\|_{L^q}; \tag{5}$$

$$\|\mathcal{E}_\kappa^\pm g\|_{L^r} \leq C\kappa^{1-2/r-1/q}\|g\|_{L^q} \tag{6}$$

for all g with $\operatorname{supp} g \subset \{|y - \kappa| \leq \kappa/1000\}$ (and support in x sufficiently small).

Since we work under the assumption that $1/q' > 2/r$, we thus see that by summing a geometric series it will suffice to prove the uniform estimates (5) in order to prove Theorem 1.

3 Transversality Conditions and Admissible Pairs of Sets

In the previous section, we have seen that we may reduce to proving uniform Fourier extension estimates for phases

$$\phi(x, y) = xy + \epsilon h(y),$$

defined on a small square Q which, after a further scaling, we may assume to be the square $Q = [-1, 1] \times [-1, 1]$, where $\epsilon > 0$ is assumed to be sufficiently small, and where h is a perturbation function *of cubic type* in y of the phase xy. By this, we mean that h is smooth and satisfies

$$\begin{cases} h(0) = h'(0) = h''(0) = 0, \\ \frac{C_3}{4} \leq |h'''(y)| \leq C_3 \quad \text{for all } |y| \leq 1, \\ |h^{(l)}(y)| \leq C_l \quad \text{for all } l \geq 4 \text{ and } |y| \leq 1 \end{cases} \tag{1}$$

(compare (2)–(4), where we have applied an additional scaling by a factor 1000 in y). Here, the constants C_l will be assumed to be fixed constants, with $C_3 > 0$, and our goal will be to establish uniform estimates which will depend only on these constants (in many parts actually only on C_3), but not on ϵ.

3.1 Admissible Pairs of Sets U_1, U_2 on which Transversalities Are of a Fixed Size: An Informal Discussion

Recall next that the bilinear approach is based on bilinear estimates of the form

$$\|\mathcal{E}_{U_1}(f_1)\,\mathcal{E}_{U_2}(f_2)\|_p \leq C(U_1, U_2)\|f_1\|_2\|f_2\|_2. \tag{2}$$

Here, $\mathcal{E}_{U_1}$ and $\mathcal{E}_{U_2}$ are the Fourier extension operators associated to patches of subsurfaces $S_i := \operatorname{graph} \phi|_{U_i} \subset S,\ i = 1, 2$, with $U_i \subset \Omega$. What is crucial for obtaining useful bilinear estimates is that the two patches of surface S_1 and S_2 satisfy certain *transversality conditions,* which are stronger than just assuming that S_1 and S_2 are transversal as hypersurfaces (i.e., that all normals to S_1 are transversal to all normals to S_2). Indeed, what is needed in addition is the following (cf. [7, 19, 20, 34], or [1]):

Denoting by $H\phi$ the Hessian of ϕ, we consider the following quantity

$$\tilde{\Gamma}_z^{\phi}(z_1, z_2, z_1', z_2') := \Big\langle (H\phi)^{-1}(z)(\nabla\phi(z_2) - \nabla\phi(z_1)), \nabla\phi(z_2') - \nabla\phi(z_1')\Big\rangle. \tag{3}$$

If its modulus is bounded from below by a constant $c > 0$ for all $z_i = (x_i, y_i),\ z_i' = (x_i', y_i') \in U_i, i = 1, 2, z = (x, y) \in U_1 \cup U_2$, then we have (2) for $p > 5/3$, with a constant $C(U_1, U_2)$ that depends only on this constant c and on upper bounds for the derivatives of ϕ. If U_1 and U_2 are sufficiently small (with sizes depending on upper bounds of the first and second order derivatives of ϕ and a lower bound for the determinant of $H\phi$) this condition reduces to the estimate

$$|\Gamma_z^{\phi}(z_1, z_2)| \geq c, \tag{4}$$

for $z_i = (x_i, y_i) \in U_i, i = 1, 2, z = (x, y) \in U_1 \cup U_2$, where

$$\Gamma_z^\phi(z_1, z_2) := \Big\langle (H\phi)^{-1}(z)(\nabla\phi(z_2) - \nabla\phi(z_1)), \nabla\phi(z_2) - \nabla\phi(z_1)\Big\rangle. \tag{5}$$

It is easy to check that for $\phi(x, y) = xy + \epsilon h(y)$, we have

$$\Gamma_z^\phi(z_1, z_2) =: 2(y_2 - y_1)\,\tau_z(z_1, z_2), \tag{6}$$

where

$$\tau_z(z_1, z_2) := x_2 - x_1 + \epsilon[h'(y_2) - h'(y_1) - \frac{1}{2}h''(y)(y_2 - y_1)]. \tag{7}$$

As in [7], it will be particularly important to look at the expression (7) when $z = z_1 \in U_1$, and $z = z_2 \in U_2$, so that the two "transversalities"

$$\tau_{z_1}(z_1, z_2) = x_2 - x_1 + \epsilon[(h'(y_2) - h'(y_1) - \frac{1}{2}h''(y_1)(y_2 - y_1)] \tag{8}$$

$$\tau_{z_2}(z_1, z_2) = x_2 - x_1 + \epsilon[(h'(y_2) - h'(y_1) - \frac{1}{2}h''(y_2)(y_2 - y_1)] \tag{9}$$

become relevant. Note the following relation between these quantities:

$$\begin{aligned}|\tau_{z_1}(z_1, z_2) - \tau_{z_2}(z_1, z_2)| &= \frac{\epsilon}{2}|h''(y_2) - h''(y_1)||y_2 - y_1| \sim \epsilon|h'''(\eta)|(y_2 - y_1)^2\\ &\sim \epsilon(y_2 - y_1)^2,\end{aligned} \tag{10}$$

where η is some intermediate point.

Following Section 2 in [7], we shall try to devise neighborhoods U_1 and U_2 of two given points $z_1^0 = (x_1^0, y_1^0)$ and $z_2^0 = (x_2^0, y_2^0)$ on which these quantities are roughly constant for $z_i = (x_i, y_i) \in U_i, i = 1, 2$, and which are also essentially chosen as large as possible. The corresponding pair (U_1, U_2) of neighborhoods of z_1^0 respectively z_2^0 will be called an *admissible pair.*

As in [7], we will present the basic motivating idea in this subsection, and give a precise definition of admissible pairs in the next subsection.

In a first step, we choose a large constant $C_0 \gg 1$, which will be made precise only later, and assume that $|y_2^0 - y_1^0| \sim C_0\rho$ for some $\rho > 0$. It is then natural to allow y_1 to vary on U_1 and y_2 on U_2 by at most ρ from y_1^0 and y_2^0, respectively, i.e., we shall assume that

$$|y_i - y_i^0| \lesssim \rho, \qquad \text{for} \quad z_i \in U_i,\ i = 1, 2,$$

so that indeed

$$|y_2 - y_1| \sim C_0\rho \qquad \text{for} \quad z_i \in U_i,\ i = 1, 2. \tag{11}$$

Recall next the identity (10), which in particular implies that

$$|\tau_{z_1^0}(z_1^0, z_2^0) - \tau_{z_2^0}(z_1^0, z_2^0)| \sim C_0^2\epsilon\rho^2. \tag{12}$$

We begin with

Case 1: Assume that $|\tau_{z_1^0}(z_1^0, z_2^0)| \le |\tau_{z_2^0}(z_1^0, z_2^0)|$. Let us then write

$$|\tau_{z_1^0}(z_1^0, z_2^0)| = \epsilon\rho^2\delta, \tag{13}$$

where $\delta \geq 0$. Note, however, that obviously $\epsilon\rho^2\delta \lesssim 1$. From (12) one then easily deduces that there are two subcases:

Subcase 1(a): (the "straight box" case), where $|\tau_{z_1^0}(z_1^0, z_2^0)| \sim |\tau_{z_2^0}(z_1^0, z_2^0)|$, or, equivalently, $\delta \gtrsim 1$. In this case, also $|\tau_{z_2^0}(z_1^0, z_2^0)| \sim \epsilon\rho^2\delta$.

Subcase 1(b): (the "curved box" case), where $|\tau_{z_1^0}(z_1^0, z_2^0)| \ll |\tau_{z_2^0}(z_1^0, z_2^0)|$, or, equivalently, $\delta \ll 1$. In this case, $|\tau_{z_2^0}(z_1^0, z_2^0)| \sim \epsilon\rho^2$.

Given ρ and δ, we shall then want to devise U_1 and U_2 so that the same kind of conditions hold for all $z_1 \in U_1$ and $z_2 \in U_2$, i.e.,

$$|\tau_{z_1}(z_1, z_2)| \sim \epsilon\rho^2\delta, \text{ and } \quad |\tau_{z_2}(z_1, z_2)| \sim \epsilon\rho^2(1 \vee \delta).$$

Note that in view of (10) and (11) the second condition is redundant, and so the only additional condition that needs to be satisfied is that, for all $z_1 = (x_1, y_1) \in U_1$ and $z_2 = (x_2, y_2) \in U_2$, we have

$$|\tau_{z_1}(z_1, z_2)| \sim \epsilon\rho^2\delta.$$

The choice of the sets U_1 and U_2 becomes particularly lucid if we first assume that $z_1^0 = 0$, so let us begin by examining this case. Later we shall see that a simple change of coordinates will allow to reduce to this case for general z_1^0.

The case $z_1^0 = 0$: We shall want to choose U_2 as large as possible w.r. to y_2, so we assume that on U_2 we have $|y_2 - y_2^0| \lesssim \rho$. Let

$$a^0 := \tau_0(0, z_2^0),$$

so that $|a^0| \sim \epsilon\rho^2\delta$. Then we shall assume that on U_2 we have, say, $|\tau_0(0, z_2) - a^0| \ll \epsilon\rho^2\delta$.

If $z_1^0 = 0$, this means that we shall **define** U_2 by the following conditions:

$$|y_2 - y_2^0| \lesssim \rho,$$
$$|\tau_0(0, z_2) - \tau_0(0, z_2^0)| = |x_2 + \epsilon h'(y_2) - a^0| \ll \epsilon\rho^2\delta. \tag{14}$$

As for U_1, given our choice of U_2, what we still need is that $|\tau_{z_1}(z_1, z_2) - \tau_0(0, z_2)| \ll \epsilon\rho^2\delta$ for all $z_1 \in U_1$ and $z_2 \in U_2$, for then also $|\tau_{z_1}(z_1, z_2) - \tau_0(0, z_2^0)| \ll \epsilon\rho^2\delta$ for all such z_1, z_2.

Note that, for y_2 fixed, the equation

$$0 = \tau_{z_1}(z_1, z_2) - \tau_0(0, z_2) = -\Big(x_1 + \epsilon[h'(y_1) + \tfrac{h''(y_1)}{2}(y_2 - y_1)]\Big)$$

defines a curve $x_1 = \gamma(y_1)$, so that the condition $|\tau_{z_1}(z_1, z_2) - \tau_0(0, z_2)| \ll \epsilon\rho^2\delta$ determines essentially an $\epsilon\rho^2\delta$ neighborhood of this curve, whose slope $\partial_{y_1}\gamma$ is of order $O(\epsilon)$. Moreover, since y_2 is allowed to vary within U_2 of order $O(\rho)$, and since (1) shows that $|\partial_{y_2}(\partial_{y_1}\gamma)| = |\epsilon h'''(y_1)/2| \sim \epsilon$, we see that the natural condition to impose for U_1 is that $\epsilon\rho|y_1 - y_1^0| = \epsilon\rho|y_1| \ll \epsilon\rho^2\delta$, i.e.,

$$|y_1| \leq \rho\delta \wedge \rho = \rho(1 \wedge \delta)$$

(note here that, in Subcase 1(a), we may have $\delta \geq 1$). Moreover, by the mean value theorem and (1), we have $|h'(y_1)| \sim |h'''(\eta)|y_1^2 \sim C_3 y_1^2$ and $|h''(y_1)| \sim |h'''(\tilde\eta)y_1| \sim C_3|y_1|$ whereas $|y_2 - y_1| \sim \rho$. Thus we see that $|\epsilon[h'(y_1) + \frac{h''(y_1)}{2}(y_2 - y_1)]| \ll \epsilon\rho^2\delta$.

In combination, this shows that it will be natural to **define** U_1 by the following conditions:

$$|y_1| \lesssim \rho(1 \wedge \delta),$$
$$|x_1| \ll \epsilon\rho^2\delta. \tag{15}$$

The case of arbitrary z_1^0: Let now $z_1^0 := (x_1^0, y_1^0)$ be arbitrary. In a first step we translate the point z_1^0 to the origin by writing $z = z_1^0 + \tilde z$, i.e., $x = x_1^0 + \tilde x$, $y = y_1^0 + \tilde y$. Then

$$\begin{aligned}\phi(z) &= \phi(z_1^0 + \tilde z) = (x_1^0 + \tilde x)(y_1^0 + \tilde y) + \epsilon h(y_1^0 + \tilde y)\\ &= \tilde x\tilde y + \epsilon\frac{h''(y_1^0)}{2}(\tilde y)^2 + \epsilon H(\tilde y) + \text{affine linear terms}\\ &= \Big(\tilde x + \epsilon\frac{h''(y_1^0)}{2}\tilde y\Big)\tilde y + \epsilon H(\tilde y) + \text{affine linear terms},\end{aligned}$$

with

$$H(\tilde{y}) = h(\tilde{y} + y_1^0) - h(y_1^0) - h'(y_1^0)\tilde{y} - \frac{h''(y_1^0)}{2}(\tilde{y})^2. \tag{16}$$

By our assumptions (1) on ϕ, the error term H satisfies estimates of the form

$$\begin{cases} H(0) = H'(0) = H''(0) = 0, \\ |H'''(\tilde{y})| = |h'''(y_1^0 + \tilde{y})| \sim C_3, \\ |H^{(l)}(\tilde{y})| \le C_l \quad \text{for all } l \ge 4, \end{cases} \tag{17}$$

which means that also H is of cubic type, uniformly in z_1^0, with the same constants C_l as for h.

It is thus natural to introduce a further change of coordinates

$$x'' := \tilde{x} + \epsilon h''(y_1^0)\tilde{y} = x - x_1^0 + \epsilon\frac{h''(y_1^0)}{2}(y - y_1^0),\ y'' := \tilde{y} = y - y_1^0, \tag{18}$$

so that in these coordinates

$$\phi(z) = x''y'' + \epsilon H(y'') + \text{affine linear terms}. \tag{19}$$

This shows that in these coordinates (x'', y''), the function ϕ is again a perturbation of $x''y''$ by a perturbation function $H(y'')$ of cubic type in the sense of (1) (up to an affine linear term, which is irrelevant), uniformly in the parameter z_1^0.

We can now define the sets U_1 and U_2 by choosing them in terms of the coordinates (x'', y'') as in (15) and (14), only with the function h replaced by H, and then express those sets in terms of our original coordinates (x, y). Note also that in the coordinates (x'', y''), we have

$$(x_1^0)'' = 0,\ (y_1^0)'' = 0 \quad \text{and} \quad (x_2^0)'' = x_2^0 - x_1^0 + \epsilon\frac{h''(y_1^0)}{2}(y_2^0 - y_1^0),\ (y_2^0)'' = y_2^0 - y_1^0,$$

and $\tau_{z_1^0}(z_1^0, z_2) = x_2'' + \epsilon H'(y_2'')$. In combination with (16) this then leads to the following **choices of** U_1 **and** U_2 :

We **define** U_1 by the conditions

$$\begin{aligned} |y_1 - y_1^0| &\lesssim \rho(1 \wedge \delta), \\ |x_1 - x_1^0 + \epsilon\frac{h''(y_1^0)}{2}(y_1 - y_1^0)| &\ll \epsilon\rho^2\delta, \end{aligned} \tag{20}$$

and U_2 by the conditions

$$\begin{aligned} |y_2 - y_2^0| &\lesssim \rho, \\ |x_2 - x_1^0 + \epsilon[h'(y_2) - h'(y_1^0) - \frac{h''(y_1^0)}{2}(y_2 - y_1^0)] - a^0| &\ll \epsilon\rho^2\delta, \end{aligned} \tag{21}$$

where

$$a^0 := \tau_{z_1^0}(z_1^0, z_2^0) \tag{22}$$

is assumed to be of size $|a^0| \sim \epsilon\rho^2\delta$.

Note U_1 is essentially the affine image of a rectangular box of dimension $\epsilon\rho^2\delta \times \rho(1 \wedge \delta)$. However, when $\delta \ll 1$, then U_2 is a thin curved box, namely the segment of an $\epsilon\rho^2\delta$-neighborhood of a curve of curvature $\sim \epsilon$ lying within the horizontal strip where $|y_2 - y_2^0| \lesssim \rho$. On the other hand, when $\delta \gtrsim 1$, then it is easily seen that U_2 is essentially a rectangular box of dimension $\epsilon\rho^2\delta \times \rho$. This explains why we called Subcase 1(b) where $\delta \ll 1$ the "curved box case", and Subcase 1(a) where $\delta \gtrsim 1$ the "straight box case."

Case 2: Assume that $|\tau_{z_1^0}(z_1^0, z_2^0)| \geq |\tau_{z_2^0}(z_1^0, z_2^0)|$.

This case can easily be reduced to the previous one by symmetry. By (7), we have $\tau_z(z_1, z_2) = -\tau_z(z_2, z_1)$. Hence we just need to interchange the roles of z_1 and z_2 in the previous discussion, so that it is natural here to define $\tilde{U}_1$ by the conditions

$$\begin{aligned} |y_1 - y_1^0| &\lesssim \rho, \\ \left|x_1 - x_2^0 + \epsilon[h'(y_1) - h'(y_2^0) - \tfrac{h''(y_2^0)}{2}(y_1 - y_2^0)] - a^0\right| &\ll \epsilon\rho^2\delta, \end{aligned} \tag{23}$$

where $a^0 = \tau_{z_2^0}(z_2^0, z_1^0) = -\tau_{z_2^0}(z_1^0, z_2^0)$, and $\tilde{U}_2$ by the conditions

$$\begin{aligned} |y_2 - y_2^0| &\lesssim \rho(1 \wedge \delta), \\ |x_2 - x_2^0 + \epsilon\tfrac{h''(y_2^0)}{2}(y_2 - y_2^0)| &\ll \epsilon\rho^2\delta. \end{aligned} \tag{24}$$

3.2 *Precise Definition of Admissible Pairs within $Q \times Q$*

In view of our discussion in the previous subsection, we shall here devise more precisely certain "dyadic" subsets of $Q \times Q$ which will assume the roles of the sets U_1, respectively U_2, in such a way that on every pair of such sets each of our transversality functions is essentially of some fixed dyadic size, and which will moreover lead to a kind of Whitney decomposition of $Q \times Q$ (as will be shown in Sect. 5). Again, this mimics the approach in [7], namely Section 2.2. To begin with, as before we fix a large dyadic constant $C_0 \gg 1$.

In a first step, we perform a classical **dyadic decomposition in the** y**-variable** which is a variation of the one in [33]: For a given dyadic number $0 < \rho \lesssim 1$, we denote for $j \in \mathbb{Z}$ such that $|j|\rho \leq 1$ by $I_{j,\rho}$ the dyadic interval $I_{j,\rho} := [j\rho, j\rho + \rho)$ of length ρ, and by $V_{j,\rho}$ the corresponding horizontal "strip" $V_{j,\rho} := [-1, 1] \times I_{j,\rho}$ within Q. Given two dyadic intervals $J,\ J'$ of the same size, we say that they are *related* if their parents are adjacent but they are not adjacent. We divide each dyadic interval J in a disjoint union of dyadic subintervals $\{I_J^k\}_{1\le k\le C_0/8}$, of length $8|J|/C_0$. Then, we define (I, I') to be an *admissible pair of dyadic intervals* if and only if there are J and J' related dyadic intervals and $1 \le k,\ j \le C_0/8$ such that $I = I_J^k$ and $I' = I_{J'}^j$.

We say that a pair of strips $(V_{j_1,\rho}, V_{j_2,\rho})$ is *admissible* and write $V_{j_1,\rho} \backsim V_{j_2,\rho}$, if $(I_{j_1,\rho}, I_{j_2,\rho})$ is a pair of admissible dyadic intervals. Notice that in this case,

$$C_0/8 < |j_2 - j_1| < C_0/2. \tag{25}$$

One can easily see that this leads to the following disjoint decomposition of $Q \times Q$:

$$Q \times Q = \dot{\bigcup_{\rho}}\Big(\dot{\bigcup_{V_{j_1,\rho}\backsim V_{j_2,\rho}}} V_{j_1,\rho} \times V_{j_2,\rho}\Big), \tag{26}$$

where the first union is meant to be over all such dyadic ρ's.

In a second step, we perform a non-standard **Whitney type decomposition of any given admissible pair of strips,** to obtain subregions in which the transversalities are roughly constant.

To simplify notation, we fix ρ and an admissible pair $(V_{j_1,\rho}, V_{j_2,\rho})$, and simply write $I_i := I_{j_i,\rho},\ V_i := V_{j_i,\rho},\ i = 1, 2$, so that I_i is an interval of length ρ with left endpoint $j_i\rho$, and

$$V_1 = [-1, 1] \times I_1, \qquad V_2 = [-1, 1] \times I_2, \tag{27}$$

are rectangles of dimension $2 \times \rho$, which are vertically separated at scale $C_0\rho$. More precisely, for $z_1 = (x_1, y_1) \in V_1$ and $z_2 = (x_2, y_2) \in V_2$ we have $|y_2 - y_1| \in |j_2\rho - j_1\rho| + [-\rho, \rho]$, i.e.,

$$C_0\rho/2 \le |y_2 - y_1| \le C_0\rho. \tag{28}$$

Let $0 < \delta \lesssim \epsilon^{-1}\rho^{-2}$ be a dyadic number (note that δ could be big, depending on ρ), and let $\mathcal{J}$ be the set of points which partition the interval $[-1, 1]$ into (dyadic) intervals of the same length $\epsilon\rho^2\delta$.

Similarly, for $i = 1, 2$, we choose a finite equidistant partition $\mathcal{I}_i$ of width $\rho(1 \wedge \delta)$ of the interval I_i by points $y_i^0 \in \mathcal{I}_i$. Note: if $\delta > 1$, then $\rho(1 \wedge \delta) = \rho$, and we can choose for $\mathcal{I}_i$ just the singleton $\mathcal{I}_i = \{y_i^0\}$, where y_i^0 is the left endpoint of I_i. In view of (20), (21) and in analogy with [7], we then define:

Definition 2 For any parameters $x_1^0, t_2^0 \in \mathcal{J},\ y_1^0 \in \mathcal{I}_1$ defined in the previous lines and y_2^0 the left endpoint of I_2, we define the sets

$$\begin{aligned} U_1^{x_1^0,y_1^0,\delta} := \{(x_1, y_1) : {} & 0 \le y_1 - y_1^0 < \rho(1 \wedge \delta), \\ & 0 \le x_1 - x_1^0 + \epsilon\frac{h''(y_1^0)}{2}(y_1 - y_1^0) < \epsilon\rho^2\delta\}, \end{aligned} \tag{29}$$

$$U_2^{t_2^0,y_1^0,y_2^0,\delta} := \{(x_2,y_2) : 0 \le y_2 - y_2^0 < \rho,$$
$$0 \le x_2 - t_2^0 + \epsilon[h'(y_2) - h'(y_1^0) - \tfrac{h''(y_1^0)}{2}(y_2 - y_1^0)] < \epsilon\rho^2\delta\},$$

and the points

$$z_1^0 = (x_1^0, y_1^0), \qquad z_2^0 = (x_2^0, y_2^0) \tag{30}$$

where

$$x_2^0 := t_2^0 - \epsilon[h'(y_2^0) - h'(y_1^0) - \tfrac{h''(y_1^0)}{2}(y_2^0 - y_1^0)].$$

Observe that then

$$z_1^0 \in U_1^{x_1^0,y_1^0,\delta} \subset V_1 \quad \text{and} \quad z_2^0 \in U_2^{t_2^0,y_1^0,y_2^0,\delta} \subset V_2.$$

Indeed, z_i^0 is in some sense the "lower left" vertex of U_i, and the horizontal projection of $U_2^{t_2^0,y_1^0,y_2^0,\delta}$ equals I_2. Moreover, if we define a^0 by (22), we have that $x_1^0 + a^0 = t_2^0$, so that our definitions of the sets $U_1^{x_1^0,y_1^0,\delta}$ and $U_2^{t_2^0,y_1^0,y_2^0,\delta}$ are very close to the ones for the sets U_1 and U_2 (cf. (20), (21)) in the previous subsection. Notice also that we may re-write

$$U_2^{t_2^0,y_1^0,y_2^0,\delta} = \{z_2 = (x_2,y_2) : 0 \le \tau_{z_1^0}(z_1^0, z_2) - a^0 < \epsilon\rho^2\delta,\ 0 \le y_2 - y_2^0 < \rho\}. \tag{31}$$

In particular, $U_1^{x_1^0,y_1^0,\delta}$ is again essentially a paralellepiped of sidelengths $\sim \epsilon\rho^2\delta \times \rho(1 \wedge \delta)$, containing the point (x_1^0, y_1^0), whose longer side has slope y_1^0 with respect to the y-axis. Similarly, if $\delta \ll 1$, then $U_2^{t_2^0,y_1^0,y_2^0,\delta}$ is a thin curved box of width $\sim \epsilon\rho^2\delta$ and length $\sim \rho$, contained in a rectangle of dimension $\sim \rho^2 \times \rho$ whose axes are parallel to the coordinate axes (namely the part of a $\rho^2\delta$-neighborhood of a parabola of curvature $\sim \epsilon$ containing the point (x_1^0, y_1^0) which lies within the horizontal strip V_2). If $\delta \gtrsim 1$, then $U_2^{t_2^0,y_1^0,y_2^0,\delta}$ is essentially a rectangular box of dimension $\sim \epsilon\rho^2\delta \times \rho$ lying in the same horizontal strip.

Note also that we have chosen to use the parameter t_2^0 in place of using x_2^0 here, since with this choice by (7) the identity

$$\tau_{z_1^0}(z_1^0, z_2^0) = t_2^0 - x_1^0 \tag{32}$$

holds true, which will become quite useful in the sequel. We next have to relate the parameters $x_1^0, t_2^0, y_1^0, y_2^0$ in order to give a precise definition of an admissible pair.

Here, and in the sequel, we shall always assume that the points z_1^0, z_2^0 associated to these parameters are given by (30).

Definition 3 Let us call a pair $(U_1^{x_1^0,y_1^0,\delta}, U_2^{t_2^0,y_1^0,y_2^0,\delta})$ an *admissible pair of type 1 (at scales* $\delta,\ \rho$ *and contained in* $V_1 \times V_2$), if the following two conditions hold true:

$$\frac{C_0^2}{4}\epsilon\rho^2\delta \le |\tau_{z_1^0}(z_1^0, z_2^0)| = |t_2^0 - x_1^0| < 4\,C_0^2\epsilon\rho^2\delta, \tag{33}$$

$$\frac{C_0^2}{512}\epsilon\rho^2(1 \vee \delta) \le |\tau_{z_2^0}(z_1^0, z_2^0)| < 5\,C_0^2\epsilon\rho^2(1 \vee \delta). \tag{34}$$

By $\mathcal{P}^\delta$ we shall denote the set of all admissible pairs of type 1 at scale δ (and ρ, contained in $V_1 \times V_2$), and by $\mathcal{P}$ the corresponding union over all dyadic scales δ.

Observe that, by (10), we have $\tau_{z_2^0}(z_1^0, z_2^0) - \tau_{z_1^0}(z_1^0, z_2^0) \sim \epsilon(y_2^0 - y_1^0)^2$. In view of (33) and (28) this shows that condition (34) is automatically satisfied, unless $\delta \sim 1$.

We remark that it would indeed be more appropriate to denote the sets $\mathcal{P}^\delta$ by $\mathcal{P}^\delta_{V_1\times V_2}$, but we want to simplify the notation. In all instances in the rest of the paper $\mathcal{P}^\delta$ will be associated to a fixed admissible pair of strips (V_1, V_2), so that our imprecision will not cause any ambiguity. The next lemma can be proved by closely following the arguments in the proof of the corresponding Lemma 2.1 in [7]:

Lemma 4 *If* $(U_1^{x_1^0,y_1^0,\delta}, U_2^{t_2^0,y_1^0,y_2^0,\delta})$ *is an admissible pair of type 1, then for all* $(z_1, z_2) \in (U_1^{x_1^0,y_1^0,\delta}, U_2^{t_2^0,y_1^0,y_2^0,\delta})$,

$$|\tau_{z_1}(z_1, z_2)| \sim_8 C_0^2\epsilon\rho^2\delta \text{ and } |\tau_{z_2}(z_1, z_2)| \sim_{1000} C_0^2\epsilon\rho^2(1 \vee \delta).$$

Up to now we focused on the case $|\tau_{z_1}(z_1, z_2)| \lesssim |\tau_{z_2}(z_1, z_2)|$. For the symmetric case, corresponding to the situation where $|\tau_{z_1}(z_1, z_2)| \gtrsim |\tau_{z_2}(z_1, z_2)|$, by interchanging the roles of z_1 and z_2 we define accordingly for any $t_1^0, x_2^0 \in \mathcal{J}$, y_1^0 the left endpoint of I_1 and $y_2^0 \in \mathcal{I}_2$ the sets $\tilde U_1^{t_1^0,y_1^0,y_2^0,\delta}$ and $\tilde U_2^{x_2^0,y_2^0,\delta}$ in analogy with our discussion in [7], and denote the corresponding admissible pairs $(\tilde U_1^{t_1^0,y_1^0,y_2^0,\delta}, \tilde U_2^{x_2^0,y_2^0,\delta})$ as *admissible pairs of type 2*. We shall skip the details.

By $\tilde{\mathcal{P}}^\delta$, we shall denote the set of all admissible pairs of type 2 at scale δ (and ρ, contained in $V_1 \times V_2$), and by $\tilde{\mathcal{P}}$ the corresponding unions over all dyadic scales δ.

In analogy with Lemma 4, we have

Lemma 5 *If* $(\tilde U_1, \tilde U_2) = (\tilde U_1^{t_1^0,y_1^0,y_2^0,\delta}, \tilde U_2^{x_2^0,y_2^0,\delta}) \in \tilde{\mathcal{P}}^\delta$ *is an admissible pair of type 2, then for all* $(z_1, z_2) \in (\tilde U_1, \tilde U_2)$ *we have*

$$|\tau_{z_1}(z_1, z_2)| \sim_{1000} C_0^2\epsilon\rho^2(1 \vee \delta) \text{ and } |\tau_{z_2}(z_1, z_2)| \sim_8 C_0^2\epsilon\rho^2\delta.$$

4 The Bilinear Estimates

4.1 A Prototypical Admissible Pair in the Curved Box Case and the Crucial Scaling Transformation

In this section we shall present a "prototypical" case where U_1 and U_2 will form an admissible pair of type 1 centered at $z_1^0 = 0 \in U_1$ and $z_2^0 \in U_2$, with $\epsilon \sim 1, \rho \sim 1$ and $\delta \ll 1$, i.e., $|y_1^0 - y_2^0| \sim 1$, and $|\tau_{z_2^0}(z_1^0, z_2^0)| \sim 1$ but $|\tau_{z_1^0}(z_1^0, z_2^0)| \sim \delta \ll 1$. This means that we shall be in the curved box case. As we will show in Sect. 4.2 in detail, we can always reduce to this particular situation when the two transversalities $\tau_{z_2^0}(z_1^0, z_2^0)$ and $\tau_{z_1^0}(z_1^0, z_2^0)$ are of quite different sizes.

Fix a small number $0 < c_0 \ll 1$ ($c_0 = 10^{-10}$ will, for instance, work). Assume that $0 < \delta \leq 1/10$, and put

$$U_1 :=[0, c_0^2\delta) \times [0, c_0\delta) \tag{1}$$

$$U_2 :=\{(x_2, y_2) : 0 \leq y_2 - b < c_0, 0 \leq x_2 + F'(y_2) - a < c_0^2\delta\}, \tag{2}$$

where $|b| \sim_2 1$, $|a| \sim_4 \delta$ and F is a function of cubic type in the sense of (1), i.e.,

$$\begin{cases} F(0) = F'(0) = F''(0) = 0, \\ |F'''(y')| \sim C_3, \\ |F^{(l)}(y')| \leq C_l \quad \text{for all } l \geq 4. \end{cases} \tag{3}$$

Remark 6 Note that in the case $\epsilon = 1$, if we set $C_0 = 1/c_0$, $\rho = c_0$, then any admissible pair $(U_1, U_2) = (U_1^{0,0,\delta}, U_2^{a,0,b,\delta})$, as in (29), would satisfy (1) and (2) with the above conditions on a and b and suitable F.

Our bilinear result in this prototypical case is as follows:

Theorem 7 (Prototypical Case) *Let $p > 5/3$, and let U_1, U_2 be as in (1), (2). Assume further that $\phi(x, y) = xy + F(y)$, where F is a real-valued smooth perturbation function of cubic type, i.e., satisfying estimates (3), and denote by*

$$\mathcal{E}_{U_i} f(\xi) = \int_{U_i} f(x, y) e^{-i(\xi_1 x + \xi_2 y + \xi_3 \phi(x,y))} \eta(x, y)\, dx dy, \qquad i = 1, 2,$$

the corresponding Fourier extension operators. Then, if the constants c_0 and $\delta \ll 1$ in (1), (2) are sufficiently small,

$$\|\mathcal{E}_{U_1}(f_1), \mathcal{E}_{U_2}(f_2)\|_p \leq C_p \delta^{\frac{7}{2} - \frac{6}{p}} \|f_1\|_2 \|f_2\|_2 \tag{4}$$

for every $f_1 \in L^2(U_1)$ and every $f_2 \in L^2(U_2)$, where the constant C_p will only depend on p and the constants C_l in (3).

As in [7], it turns out that one cannot directly reduce the bilinear Fourier extension estimates in (4) to Lee's Theorem 1.1 in [19], since that would not give us the optimal dependence on δ. We shall therefore have to be more precise about the required transversality conditions. However, once we have established the correct transversality conditions in Lemma 8 below (which is the direct analogue of Lemma 2.3 in [7]), we can indeed apply our arguments from the proof of Theorem 3.3 in [7] also in the present situation and arrive at the desired bilinear estimates (4).

The crucial step will again consist in the following scaling: we introduce new coordinates $(\bar{x}, \bar{y})$ be writing $x = \delta\bar{x}$, $y = \bar{y}$, and then re-scale the phase function ϕ by putting

$$\phi^s(\bar{x}, \bar{y}) := \frac{1}{\delta}\phi(\delta\bar{x}, \bar{y}) = \bar{x}\bar{y} + \frac{F(\bar{y})}{\delta}.$$

Denote by U_i^s the corresponding re-scaled domains, i.e.,

$$U_1^s = \{(\bar{x}_1, \bar{y}_1) : 0 \leq \bar{x}_1 < c_0^2,\ 0 \leq \bar{y}_1 < c_0\delta\},$$
$$U_2^s = \{(\bar{x}_2, \bar{y}_2) : 0 \leq \bar{x}_2 + \frac{F'(\bar{y}_2)}{\delta} - \bar{a} < c_0^2,\ 0 \leq \bar{y}_2 - \bar{b} < c_0\},$$

where c_0 is small and $|\bar{a}| = |a/\delta| \sim 1$ and $\bar{b} = b \sim 1$. By S_i^s, $i = 1, 2$, we denote the corresponding scaled surface patches

$$S_i^s := \{(\bar{x}, \bar{y}, \phi^s(\bar{x}, \bar{y})) : (\bar{x}, \bar{y}) \in U_i^s\}.$$

Observe that

$$\nabla\phi^s(\bar{x}, \bar{y}) = (\bar{y}, \bar{x} + F'(\bar{y})/\delta),$$

and

$$H\phi^s(\bar{x}, \bar{y}) = \begin{pmatrix} 0 & 1 \\ 1 & F''(\bar{y})/\delta \end{pmatrix},$$

so that in particular

$$|\nabla\phi^s(\bar{z})| \lesssim 1 \tag{5}$$

for all $\bar{z} \in U_1^s \cup U_2^s$.

Assume next that $\bar{z}_1 \in U_1^s$ and $\bar{z}_2 \in U_2^s$. Since $|\bar{y}_1| \leq c_0\delta$, $|\bar{y}_2| \sim 1$, we see that

$$\begin{cases} |\frac{F'(\bar{y}_1)}{\delta}| \sim \frac{|F'''(\eta_1)\bar{y}_1^2|}{\delta} \lesssim C_3 c_0^2 \frac{\delta^2}{\delta} = c_0^2 C_3 \delta, & \frac{|F''(\bar{y}_1)|}{\delta} \sim \frac{|F'''(\tilde{\eta}_1)\bar{y}_1|}{\delta} \lesssim c_0 C_3, \\ |\frac{F'(\bar{y}_2)}{\delta}| \sim \frac{|F'''(\eta_2)\bar{y}_2^2|}{\delta} \sim \frac{C_3}{\delta}, & \frac{|F''(\bar{y}_2)|}{\delta} \sim \frac{|F'''(\tilde{\eta}_2)\bar{y}_2|}{\delta} \sim \frac{C_3}{\delta} \end{cases} \tag{6}$$

(for suitable choices of intermediate points $\eta_i, \tilde{\eta}_i$). Moreover, we then also see that

$$\nabla\phi^s(\bar{z}_2)-\nabla\phi^s(\bar{z}_1) = \left(\bar{y}_2-\bar{y}_1, \bar{x}_2+\tfrac{F'(\bar{y}_2)}{\delta}-(\bar{x}_1+\tfrac{F'(\bar{y}_1)}{\delta})\right) = (\bar{b}, \bar{a})+O(c_0). \quad (7)$$

Following further on the proof of Lemma 2.3 in [7], assume that we translate the two patches of surface S_1^s and S_2^s in such a way that the two points $\bar{z}_1$ and $\bar{z}_2$ coincide after translation, and assume that the vector $\omega = (\omega_1, \omega_2)$ is tangent to the corresponding intersection curve $\gamma(t)$ at this point. Then (7) shows that we may assume without loss of generality that

$$\omega = (-\bar{a}, \bar{b}) + O(c_0).$$

In combination with (6) this implies that

$$H\phi^s(\bar{z}_i)\cdot {}^t\omega = \begin{pmatrix} 0 & 1 \\ 1 & F''(\bar{y}_i)/\delta \end{pmatrix} \begin{pmatrix} -\bar{a}+O(c_0) \\ \bar{b}+O(c_0) \end{pmatrix}.$$

Thus, if $i = 1$, then by (6),

$$H\phi^s(\bar{z}_1)\cdot {}^t\omega = \begin{pmatrix} \bar{b}+O(c_0) \\ -\bar{a}+O(c_0) \end{pmatrix} \quad \text{and} \quad |H\phi^s(\bar{z}_1)\cdot {}^t\omega| \sim 1, \quad (8)$$

and if $i = 2$, then

$$H\phi^s(\bar{z}_2)\cdot {}^t\omega = \begin{pmatrix} \bar{b}+O(c_0) \\ -\bar{a}+\bar{b}F''(\bar{y}_2)/\delta+O(c_0)/\delta \end{pmatrix} \quad \text{and} \quad |H\phi^s(\bar{z}_1)\cdot {}^t\omega| \sim 1/\delta, \quad (9)$$

if $\delta \ll 1$ is sufficiently small.

Following [7], the refined transversalities that we need to control are given by

$$\left|TV_i^s(\bar{z}_1,\bar{z}_2)\right| := \left|\frac{\det({}^t(\nabla\phi^s(\bar{z}_1)-\nabla\phi^s(\bar{z}_2)), H\phi^s(\bar{z}_i)\cdot {}^t\omega)}{\sqrt{1+|\nabla\phi^s(\bar{z}_1)|^2}\sqrt{1+|\nabla\phi^s(\bar{z}_2)|^2}\,|H\phi^s(\bar{z}_i)\cdot {}^t\omega|}\right|,$$
$$i = 1, 2. \quad (10)$$

But, if $i = 1$, then by (7), (9), (5) and (6) we see that

$$|\det({}^t(\nabla\phi^s(\bar{z}_1)-\nabla\phi^s(\bar{z}_2)), H\phi^s(\bar{z}_1)\cdot {}^t\omega)|$$
$$= \left|\det\begin{pmatrix} \bar{b}+O(c_0) & \bar{b}+O(c_0) \\ \bar{a}+O(c_0) & -\bar{a}+O(c_0) \end{pmatrix}\right| \sim 1,$$

hence $\left|TV_1^s(\bar{z}_1,\bar{z}_2)\right| \sim 1$.

And, if $i = 2$, then by (7), (8), (5) and (6) we have

$$|\det({}^t(\nabla\phi^s(\bar{z}_1) - \nabla\phi^s(\bar{z}_2)), H\phi^s(\bar{z}_2) \cdot {}^t\omega)|$$
$$= \left|\det\begin{pmatrix} \bar{b} + O(c_0) & \bar{b} + O(c_0) \\ \bar{a} + O(c_0) & -\bar{a} + \bar{b}F''(\bar{y}_2)/\delta + O(c_0)/\delta \end{pmatrix}\right|,$$

hence also $\left|TV_2^s(\bar{z}_1, \bar{z}_2)\right| \sim (1/\delta)/(1/\delta) \sim 1$, provided δ and c_0 are sufficiently small.

We have thus proved the following lemma, from which Theorem 7 can easily be derived, as explained before, by applying the arguments from [7]:

Lemma 8 *The transversalities for the scaled patches of surface S_i^s, $i = 1, 2$, satisfy*

$$\left|TV_i^s(\bar{z}_1, \bar{z}_2)\right| \sim 1, \quad i = 1, 2.$$

We should again like to mention that estimate (4) could alternatively also be deduced from Candy's Theorem 1.4 in [8], after applying the crucial scaling in x that we used in the first step of our proof.

4.2 Reduction to the Prototypical Case

Our next goal will be to establish the following analogues of the bilinear Fourier extension estimates in Theorem 3.1 of [7]:

Theorem 9 *Let $p > 5/3$, $q \geq 2$. Then, for every admissible pair $(U_1, U_2) \in \mathcal{P}^\delta$ at scale δ, the following bilinear estimates hold true: If $\delta > 1$ and $\epsilon\delta\rho^2 \leq 1$, then*

$$\|\mathcal{E}_{U_1}(f)\mathcal{E}_{U_2}(g)\|_p \leq C_{p,q}(\epsilon\delta\rho^3)^{2(1-\frac{1}{p}-\frac{1}{q})}\|f\|_q\|g\|_q.$$

If $\delta \leq 1$, then

$$\|\mathcal{E}_{U_1}(f)\mathcal{E}_{U_2}(g)\|_p \leq C_{p,q}\ (\epsilon\rho^3)^{2(1-\frac{1}{p}-\frac{1}{q})}\,\delta^{5-\frac{3}{q}-\frac{6}{p}}\|f\|_q\|g\|_q.$$

The constants in these estimates are independent of the given admissible pair, of ϵ, ρ and of δ. The same estimates are valid for admissible pairs $(\tilde{U}_1, \tilde{U}_2) \in \tilde{\mathcal{P}}^\delta$ of type 2.

Fix $p > 5/3$ and $q \geq 2$, and assume that $U_1 = U_1^{x_1^0, y_1^0, \delta}$ and $U_2 = U_2^{t_2^0, y_1^0, y_2^0, \delta}$ form an admissible pair of type 1. We shall only discuss the case of admissible pairs of type 1; the type 2 case can be handled in the same way by symmetry.

We shall see that the bilinear estimates associated to the sets U_1, U_2 can easily be reduced by means of a suitable affine-linear transformation to either the classical bilinear estimate in [19], when $\delta \geq 1/10$, or to the estimate for the special "prototype" situation given in Sect. 4.1, when $\delta \leq 1/10$.

We first change to the coordinates (x'', y'') introduced in (18), which allows to reduce to the case where $(z_1^0)'' = 0$ and $(z_2^0)'' = (x_2^0 - x_1^0 + \epsilon \frac{h''(y_1^0)}{2}(y_2^0 - y_1^0), y_2^0 - y_1^0)$. Recall, however, that we need here to replace our original perturbation $h(y)$ by the cubic type perturbation $H(y'')$ (compare (17)). In these coordinates, U_1 corresponds to the set

$$U_1'' := \{(x_1'', y_1'') : 0 \leq y_1'' < \rho(1 \wedge \delta),\ 0 \leq x_1'' < \epsilon\rho^2\delta\},$$

and U_2 to the set

$$U_2'' = \{(x_2'', y_2'') : 0 \leq x_2'' + \epsilon H'(y_2'') - a^0 < \epsilon\rho^2\delta,\ 0 \leq y_2'' - (y_2'')^0 < \rho\},$$

where $a^0 := t_2^0 - x_1^0$ and $(y_2'')^0 := y_2^0 - y_1^0 \sim C_0\rho$ (compare (31) and (28), and note that $\tau_0(0, z_2'') = x_2'' + \epsilon H'(y_2'')$ in the coordinates (x'', y'')). Recall also from (33) that $|a^0| \sim C_0^2 \epsilon\rho^2\delta$.

This suggests to apply the following **scaling**: we change to yet other coordinates $z' = (x', y')$ by writing

$$y'' = \rho y',\ x'' = \epsilon\rho^2(1 \vee \delta)x'. \tag{11}$$

Let us accordingly introduce the function

$$F(y') := \frac{H(\rho y')}{\rho^3}, \tag{12}$$

and note that the crucial phase function $x''y'' + \epsilon H(y'')$ that arose from ϕ in (19) after the change to the coordinates (x'', y'') assumes the following form in the coordinates (x', y') :

$$x''y'' + \epsilon H(y'') = \epsilon\rho^3(1 \vee \delta)\Big(x'y' + \frac{F(y')}{1 \vee \delta}\Big) =: \epsilon\rho^3(1 \vee \delta)\, \phi_\delta(x', y'). \tag{13}$$

Observe that also the function F *is a perturbation function of cubic type, uniformly also in* ϵ *and* ρ. Indeed, the following holds true:

$$\begin{cases} F(0) = F'(0) = F''(0) = 0, \\ |F'''(y')| = |H''(\rho y')| \sim C_3, \\ |F^{(l)}(y')| = |\rho^{l-3} H^{(l)}(\rho y')| \leq C_l \quad \text{for all } l \geq 4. \end{cases} \tag{14}$$

Thus, altogether we define a change of coordinates $z' = T(z)$ by

$$x' := \epsilon^{-1}(1 \vee \delta)^{-1}\rho^{-2}(x - x_1^0 + \epsilon\frac{h''(y_1^0)}{2}(y - y_1^0)),$$
$$y' := \rho^{-1}(y - y_1^0).$$

Notice that the following lemma, in the case $\delta \leq 1/10$, corresponds to the prototypical setup up to another harmless scaling $(x', y') = (C_0^2 x''', C_0 y''')$.

Lemma 10 *We have*

$$\phi(z) = \epsilon\rho^3(1 \vee \delta)\phi_\delta(Tz) + L(z), \tag{15}$$

where L is an affine-linear map. Moreover, in these new coordinates, U_1, U_2 correspond to the sets

$$U_1' := \{(x_1', y_1') : 0 \leq y_1' < 1 \wedge \delta,\ 0 \leq x_1' < 1 \wedge \delta\} = [0, 1 \wedge \delta[^2,$$
$$U_2' = \{(x_2', y_2') : 0 \leq x_2' + \frac{F'(y_2')}{1 \vee \delta} - a < 1 \wedge \delta,\ 0 \leq y_2' - b < 1\}, \tag{16}$$

where $|b| := |\rho^{-1}(y_2^0 - y_1^0)| \sim_2 C_0$ and $|a| := |\epsilon^{-1}\rho^{-2}(1 \vee \delta)^{-1}(t_2^0 - x_1^0)| \sim_4 C_0^2\frac{\delta}{1 \vee \delta} = C_0^2(1 \wedge \delta)$. Moreover, for Lee's transversality expression Γ^{ϕ_δ} in (3) for ϕ_δ, we have that

$$|\Gamma_{z_1'}^{\phi_\delta}(z_1', z_2')| \sim C_0^3(1 \wedge \delta) \quad \text{for all } z_1' \in U_1',$$
$$|\Gamma_{z_2'}^{\phi_\delta}(z_1', z_2')| \sim C_0^3 \quad \text{for all } z_2' \in U_2', \tag{17}$$

for every $z_1' \in U_1'$ and every $z_2' \in U_2'$. Also, for $\delta \geq 1/10$, the derivatives of ϕ_δ can be uniformly (independently of δ) bounded from above.

The proof, if not clear from our previous discussions, is similar to the proof of Lemma 2.4 in [7], so we will skip the details.

Reduction of Theorems 9 to 7 Consider the scaled sets U_1', U_2' from Lemma 10.

The Case $\delta > 1/10$[1] In this case, we see that, U_1' and U_2' are squares of small side length $2c_0$, separated by a distance of size 1, and moreover (17) shows that all relevant transversalities are of size 1. Therefore we see that the conditions of Lee's Theorem 1.1 in [19] are satisfied for the patches of surface S_1' and S_2' which are the graphs of ϕ_δ (defined in (13)) over the sets U_1' and U_2'. This implies that for these

[1] We don't need to distinguish precisely the two cases $\delta > 1$ and $\delta \leq 1$ from the Theorem, since the desired bounds are comparable for $\delta \sim 1$.

patches of surface, we obtain uniform bilinear Fourier extension estimates when $p > 5/3$ and $q \geq 2$, of the form

$$\|\mathcal{E}_{U_1'}(\tilde{f})\,\mathcal{E}_{U_2'}(\tilde{g})\|_p \leq C_{p,q}\|\tilde{f}\|_q\|\tilde{g}\|_q,$$

with a constant $C_{p,q}$ which is independent of the choice of $x_1^0, y_1^0, t_2^0, y_2^0, \epsilon, \rho$ and δ. By scaling back to our original coordinates, we thus arrive at the estimate in the first case of Theorem 9 (compare with the scaling argument in Sections 2.5 and 3 of [7]).

The case $\delta \leq 1/10$ By a harmless scaling $(x', y') = (C_0^2 x, C_0 y)$, the sets U_1' and U_2' given by (16) transform to

$$\begin{aligned} U_1 &= \{(x_1, y_1) : 0 \leq x_1 < c_0^2\delta,\ 0 \leq y_1 < c_0\delta\} = [0, c_0^2\delta) \times [0, c_0\delta), \\ U_2 &= \{(x_2, y_2) : 0 \leq x_2 + c_0^2 F'(\frac{y_2}{c_0}) - a < c_0^2\delta,\ 0 \leq y_2 - b < c_0\}, \end{aligned} \tag{18}$$

where $c_0 = C_0^{-1}$ is small and $|a| \sim \delta$ and $b \sim 1$. Recall also that $c_0^3 F(\frac{y_2}{c_0})$ satisfies the cubic type estimates (14). For the sake of simplicity, let us denote this perturbation of cubic type again by F, so that in this case the phase ϕ_δ, given by (13), can be written as

$$\phi_\delta(x', y') = x'y' + F(y').$$

This means that we are in the prototypical situation. The claimed estimates for Case 2 in Theorem 9 will now follow directly from Theorem 7 for the prototypical case in combination with Hölder's inequality (to pass from L^2-norms to L^q-norms), if we again scale back to our original coordinates. □

5 The Whitney-Decomposition and Passage to Linear Restriction Estimates: Proof of Theorem 1

In order to complete the proof of Theorem 1, let us finally briefly sketch how to pass from the bilinear estimates in Theorem 7 to the crucial linear estimate in (5). Again we shall closely follow our approach in [7] and only indicate the necessary changes.

Let (V_1, V_2) be an admissible pair of strips as defined in Sect. 3.2. Recall the definition of admissible pairs of sets from the same subsection, and that we had also introduced there the sets $\mathcal{P}^\delta$ respectively $\tilde{\mathcal{P}}^\delta$ of admissible pairs of type 1 respectively type 2 at scale δ, and by $\mathcal{P}$ respectively $\tilde{\mathcal{P}}$ we had denoted the corresponding unions over all dyadic scales δ. The next lemma is in direct analogy to Lemma 4.1 in [7] and can be proved in a similar fashion.

Lemma 11 *The following covering and overlapping properties hold true:*

(i) *For fixed dyadic scale* δ, *the subsets* $U_1 \times U_2$, $(U_1, U_2) \in \mathcal{P}^\delta$, *of* $V_1 \times V_2 \subset Q \times Q$ *are pairwise disjoint, as likewise are the subsets* $\tilde{U}_1 \times \tilde{U}_2$, $(\tilde{U}_1, \tilde{U}_2) \in \tilde{\mathcal{P}}^\delta$.

(ii) *If* δ *and* δ' *are dyadic scales, and if* $(U_1, U_2) \in \mathcal{P}^\delta$ *and* $(U_1', U_2') \in \mathcal{P}^{\delta'}$, *then the sets* $U_1 \times U_2$ *and* $U_1' \times U_2'$ *can only intersect if* $\delta/\delta' \sim_{2^7} 1$. *In the latter case, there is only bounded overlap. I.e., there is a constant* $M \leq 2^6$ *such that for every* $(U_1, U_2) \in \mathcal{P}^\delta$ *there are at most* M *pairs* $(U_1', U_2') \in \mathcal{P}^{\delta'}$ *such that* $(U_1 \times U_2) \cap (U_1' \times U_2') \neq \emptyset$, *and vice versa. The analogous statements apply to admissible pairs in* $\tilde{\mathcal{P}}$.

(iii) *If* $(U_1, U_2) \in \mathcal{P}^\delta$ *and* $(\tilde{U}_1, \tilde{U}_2) \in \tilde{\mathcal{P}}^{\delta'}$, *then* $U_1 \times U_2$ *and* $\tilde{U}_1 \times \tilde{U}_2$ *are disjoint too, except possibly when both* $\delta, \delta' \geq 1/800$ *and* $\delta \sim_{2^{10}} \delta'$. *In the latter case, there is only bounded overlap. I.e., there is a constant* $N = O(C_0)$ *such that for every* $(U_1, U_2) \in \mathcal{P}^\delta$ *there are at most* N *pairs* $(\tilde{U}_1, \tilde{U}_2) \in \tilde{\mathcal{P}}^{\delta'}$ *such that* $(U_1 \times U_2) \cap (\tilde{U}_1 \times \tilde{U}_2) \neq \emptyset$, *and vice versa.*

(iv) *The product sets associated to all admissible pairs cover* $V_1 \times V_2$ *up to a set of measure* 0, *i.e.,*

$$V_1 \times V_2 = \Big(\bigcup_{(U_1,U_2)\in\mathcal{P}} U_1 \times U_2 \Big) \cup \Big(\bigcup_{(\tilde{U}_1,\tilde{U}_2)\in\mathcal{P}} \tilde{U}_1 \times \tilde{U}_2 \Big)$$

in measure.

To handle the bounded overlap between the sets $U_1 \times U_2$ for pairs of admissible sets $(U_1, U_2) \in \mathcal{P}$ of type 1 in Lemma 11, we define for $\nu = 0, \ldots, 9$ the subset $\mathcal{P}_\nu := \bigcup_j \mathcal{P}^{2^{10j+\nu}}$ of $\mathcal{P}$. To these, we associate the subsets

$$A_\nu := \bigcup_{(U_1,U_2)\in\mathcal{P}_\nu} U_1 \times U_2, \qquad \nu = 0, \ldots, 9,$$

and likewise introduce the corresponding subsets $\tilde{A}_\nu$ associated to admissible pairs of type 2. Then we may argue as in [7] to show that it will suffice to prove restriction estimates over these sets A_ν, respectively $\tilde{A}_\nu$, over which we have "decoupled" the overlaps. Let us just look at the sets A_ν in the sequel.

To prove Theorem 1, assume that $r > 10/3$ and $1/q' > 2/r$, and put $p := r/2$. so that $p > 5/3$, $1/q' > 1/p$. By interpolation with the trivial estimate for $r = \infty$, $q = 1$, it is enough to prove the result for r close to $10/3$ and q close to 5/2, i.e., p close to 5/3 and q close to 5/2. Hence, we may assume that $p < 2$, $p < q < 2p$. Also, we can assume that $\operatorname{supp} f \subset \{(x, y) \in Q : y \geq 0\}$.

As in [7], we easily see that it will suffice to prove the following: assume a scale ρ is fixed, and that $V_1 \sim V_2$ is an admissible pair of strips at scale ρ (as defined in (27) of Sect. 3.2). Then the following holds true:

Lemma 12 *If $V_1 \sim V_2$ form an admissible pair of "strips" $V_i = V_{j_i,\rho} = [-1,1]\times I_{j_i,\rho}$, $i = 1, 2$, at scale ρ within Q, and if $f \in L^q(V_1)$ and $g \in L^q(V_2)$, then for $5/3 < p < 2$, $p < q < 2p$ we have*

$$\|\mathcal{E}_{V_1}(f)\mathcal{E}_{V_2}(g)\|_p \lesssim C_{p,q}\,\rho^{2(1-1/p-1/q)}\|f\|_q\,\|g\|_q \text{ for all } f \in L^q(V_1),\ g \in L^q(V_2). \tag{1}$$

We remark that, eventually, we shall choose $f = g$, but for the arguments to follow it is helpful to distinguish between f and g.

To prove this lemma, observe first that by means of an affine linear transformation we may "move the strips V_1, V_2 vertically" so that $j_1 = 0$, which means that V_1 contains the origin and, by (25), $j_2 \sim C_0$. This we shall assume throughout the proof.

As mentioned before, it will suffice to estimate $E((f \otimes g)\chi_{A_\nu})$ in place of $\mathcal{E}_{V_1}(f)\mathcal{E}_{V_2}(g)$, and the same arguments as in [7] then show that we may decompose

$$(f\otimes g)\chi_{A_\nu} = \sum_\delta \sum_{i,i',j} f^\delta_{i,j} \otimes g^\delta_{i',j},$$

where

$$f^\delta_{i,j} = f\chi_{U_1^{i\epsilon\rho^2\delta,\,j\rho(1\wedge\delta),\,\delta}},\quad g^\delta_{i',j} = g\chi_{U_2^{i'\epsilon\rho^2\delta,\,j\rho(1\wedge\delta),\,j_2\rho,\,\delta}},$$

and where each $\big(U_1^{i\epsilon\rho^2\delta,\,j\rho(1\wedge\delta),\,\delta},\ U_2^{i'\epsilon\rho^2\delta,\,j\rho(1\wedge\delta),\,j_2\rho,\,\delta}\big)$ forms an admissible pair, i.e., (33), (34) are satisfied. This means in particular that $|i - i'| \sim C_0^2$. The summation in δ is here meant as summation over all dyadic δ such that $\delta \lesssim (\epsilon\rho^2)^{-1}$.

We may and shall also assume that f and g are supported on the set $\{y \geq 0\}$. Then

$$E((f\otimes g)\chi_{A_\nu}) = \sum_{\delta\gtrsim 1}\sum_{i,i'} \widehat{f^\delta_i d\sigma}\,\widehat{g^\delta_{i'} d\sigma} + \sum_{\delta\ll 1}\sum_{i,i',j} \widehat{f^\delta_{i,j} d\sigma}\,\widehat{g^\delta_{i',j} d\sigma}. \tag{2}$$

The first sum can be treated by more classical arguments (compare, e.g., [19] or [34]), which in view of the first estimate in Proposition 9 then leads to a bound for the contribution of that sum to $\|\mathcal{E}_{V_1}(f)\mathcal{E}_{V_2}(g)\|_p$ in (1) of the order

$$\sum_{1\lesssim\delta\lesssim(\epsilon\rho^2)^{-1}} C_{p,q}\ (\delta\epsilon\rho^3)^{2(1-\frac{1}{p}-\frac{1}{q})}\|f\|_q\|g\|_q \lesssim \rho^{2(1-\frac{1}{p}-\frac{1}{q})}\|f\|_q\|g\|_q,$$

as required. We leave the details to the interested reader. Note that for this first sum, there is no gain when $\epsilon > 0$ is getting small (which is to be expected), in contrast to what will happen for the second sum.

We shall now concentrate on the second sum in (2) where $\delta \ll 1$. Here, the admissibility conditions reduce to $|i - i'| \sim C_0^2$, $j_2 \sim C_0$.

We fix δ, and simplify notation by writing $f_{i,j} := f_{i,j}^{\delta}$, $g_{i,j} := g_{i,j}^{\delta}$, and $U_{1,i,j} := U_1^{i\epsilon\rho^2\delta, j\rho(1\wedge\delta),\delta}$, $U_{2,i',j} := U_2^{i'\epsilon\rho^2\delta, j\rho(1\wedge\delta), j_2\rho,\delta}$.

As a first step in proving estimate (1), we exploit some almost orthogonality with respect to the x-coordinate, following a classical approach (compare, e.g., [21, 22]).

Lemma 13 *For $1 \le p \le 2$, we have*

$$\Big\| \sum_{i,|i-i'|\sim C_0^2,\, j} \widehat{f_{i,j}d\sigma}\,\widehat{g_{i',j}d\sigma} \Big\|_p^p \lesssim \sum_{N=0}^{(\epsilon\rho^2)^{-1}} \Big\| \sum_{\substack{i\in[N\delta^{-1},(N+1)\delta^{-1}],\\ |i-i'|\sim C_0^2,\, j}} \widehat{f_{i,j}d\sigma}\,\widehat{g_{i',j}d\sigma} \Big\|_p^p .$$

Proof Assume that $i \in [N\delta^{-1}, (N+1)\delta^{-1}]$, and that $z_1 = (x_1, y_1) \in U_{1,i,j}$ and $z_2 = (x_2, y_2) \in U_{2,i',j}$, where $|i - i'| \sim C_0^2$, which means that $(U_{1,i,j}, U_{2,i',j}) \in \mathcal{P}^{\delta}$ is an admissible pair. Then, in a similar way as in the proof of the corresponding lemma in [7], by means of Taylor expansions (where we only need to make use of the estimates for third derivatives of h) one sees that $|x_2 - x_1| \lesssim CC_0^2\epsilon\rho^2$. This implies that $x_1 + x_2 = 2N\epsilon\rho^2 + O(\epsilon\rho^2)$, where the constant in the error term is of order C_0^2, hence

$$U_{1,i,j} + U_{2,i',j} \subset [2N\epsilon\rho^2 - C\,C_0^2\epsilon\rho^2, 2N\epsilon\rho^2 + C\,C_0^2\epsilon\rho^2] \times [0, 2C_0\rho].$$

These statements become even more lucid if we first apply the scaling $y = \rho y'$, $x = \epsilon\rho^2 x'$, that we had already introduced in (11), for then we may assume that in our definition of the sets $U_{1,i,j}$, $U_{2,i',j}$ we have $\epsilon = 1$ and $\rho = 1$. We also remark that the constant C will depend here only on the constant C_3 which controls third derivatives of h in (1).

Notice that the family of intervals $\big\{[2N\epsilon\rho^2 - C\,C_0^2\epsilon\rho^2, 2N\epsilon\rho^2 + C\,C_0^2\epsilon\rho^2]\big\}_{N=0}^{(\epsilon\rho^2)^{-1}}$ is almost pairwise disjoint. Therefore we may argue as in the proof of Lemma 6.1 in [33] in order to derive the desired estimate. □

We proceed in analogy with [7]: $U_{1,i,j}$ is a rectangular box, now of dimension $\epsilon\rho^2\delta \times \rho\delta$, and we shall further decompose the curved box $U_{2,i',j}$ into essentially rectangular boxes of the same dimensions $\epsilon\rho^2\delta \times \rho\delta$, by decomposing them in the y-coordinate into $O(1/\delta)$ intervals of length $\rho\delta$. I.e., we shall put

$$U_{2,i',j}^k := \{(x, y) \in U_{2,i',j} : 0 \le y - k\rho\delta < \rho\delta\}.$$

Then

$$U_{2,i',j} = \dot{\bigcup_k} U^k_{2,i',j},$$

where the union is over a set of $O(1/\delta)$ indices k. Accordingly, we decompose $g_{i',j} = \sum_k g^k_{i',j}$, where $g^k_{i',j} := g\chi_{U^k_{2,i',j}}$. Then we have the following uniform square function estimate:

Lemma 14 *For $1 < p \leq 2$ there exists a constant $C_p > 0$ such that for every $N = 0, \ldots, (\epsilon\rho^2)^{-1}$ we have*

$$\Big\| \sum_{\substack{i\in[N\delta^{-1},(N+1)\delta^{-1}],\\ |i-i'|\sim C_0^2,\, j}} \widehat{f_{i,j}d\sigma}\, \widehat{g_{i',j}d\sigma} \Big\|_p$$

$$\leq C_p \Big\| \Big(\sum_{\substack{i\in[N\delta^{-1},(N+1)\delta^{-1}],\\ |i-i'|\sim C_0^2,\, j,k}} |\widehat{f_{i,j}d\sigma}\, \widehat{g^k_{i',j}d\sigma}|^2 \Big)^{1/2} \Big\|_p. \tag{3}$$

Proof of Lemma 14 Notice first that a translation in x by $N\rho^2$ allows to reduce to the case $N = 0$, which we shall thus assume. Then the relevant sets $U_{1,i,j}$ and $U_{2,i',j}$ will all have their x-coordinates in the interval $[0, \epsilon\rho^2]$.

For i, i', j, k as above, set $S_{1,i,j} := \{(\xi, \phi(\xi)) : \xi \in U_{1,i,j}\}$, $S^k_{2,i',j} := \{(\xi, \phi(\xi)) : \xi \in U^k_{2,i',j}\}$, and denote by $(x', y') = D_{\epsilon,\rho}(x, y) := (\epsilon\rho^2 x, \rho y)$ the scaling transformation which changes coordinates from $z = (x, y)$ to $z' = (x', y')$. The key to the square function estimate (3) is the following almost orthogonality lemma:

Lemma 15 *Assume $N = 0$, and denote by $\tilde{D}_{\epsilon,\rho}$, $\rho > 0$, the scaling transformation on the ambient space $\mathbb{R}^3$ which is given by $\tilde{D}_{\epsilon,\rho}(x, y, w) := (\epsilon\rho^2 x, \rho y, \epsilon\rho^3 w)$. Then there is a family of cubes $\{Q^k_{i,i',j}\}_{i\in[0,\delta^{-1}],|i-i'|\sim C_0^2,j,k}$ in $\mathbb{R}^3$ with bounded overlap, whose sides are parallel to the coordinate axes and of length $\sim \delta$, such that $S_{1,i,j} + S^k_{2,i',j} \subset \tilde{D}_{\epsilon,\rho}(Q^k_{i,i',j})$.*

We remark that the amount of the overlap is in fact entirely controlled by the size of the constant C_3 in (1) (and on our choice of C_0), but not on the constants C_l for $l \geq 4$ in (1).

Proof of Lemma 15 Note first that by our assumptions we have $V_1, V_2 \subset [0, 1] \times [0, 2C_0\rho]$. Since

$$\tilde{D}^{-1}_{\epsilon,\rho}\big(D_{\epsilon,\rho}(z'), \phi(D_{\epsilon,\rho}(z'))\big) = (x', y', x'y' + F(y'))$$

(compare Sect. 4.2), we may apply this scaling in order to reduce our considerations to the case where $\epsilon = \rho = 1$, if we replace the perturbation term h by the function F which, according to (14), shares the same type of estimates as h. Notice also that, after scaling, the sets corresponding to V_1, V_2 in the new coordinates then satisfy $V_1, V_2 \subset [0, (\epsilon\rho^2)^{-1}] \times [0, 2C_0]$.

Therefore, from now on we shall work under these assumptions, denoting the new coordinates again by (x, y) in place of (x', y'), in order to defray the notation.

Notice also that if $i \in [0, \delta^{-1}]$, $|i - i'| \sim C_0^2$, then the corresponding patches of surface $S_{1,i,j}$ and $S^k_{2,i',j}$ are contained in boxes of side length, say, 2δ, and sides parallel to the axes, whose projections to the x-axis lie within the unit interval $[0, 1]$. Therefore we can choose for $Q^k_{i,i',j}$ a square of side length 4δ, with sides parallel to the axes, with the property that $S_{1,i,j} + S^k_{2,i',j} \subset Q^k_{i,i',j}$. We shall prove that the overlap is bounded, with a bound depending only on C_0 and the constant C_3 in (1).

Note that, if $(x_1, y_1) \in U_{1,i,j}$ and $(x_2, y_2) \in U^k_{2,i',j}$ with $|i - i'| \sim C_0^2$, then, by Lemma 4 we have

$$\left|x_2 - x_1 + F'(y_2) - F'(y_1) - \tfrac{1}{2}F''(y_1)(y_2 - y_1)\right| \sim C_0^2\delta.$$

It suffices to prove the following: if $(x_1, y_1), (x_2, y_2)$ and $(x_1', y_1'), (x_2', y_2')$ are so that each coordinate of these points is bounded by a large multiple of C_0, the y-coordinates are positive and satisfy $y_2 - y_1 \gtrsim C_0$, $y_2' - y_1' \gtrsim C_0$ (by the y-separation (28)), and

$$\begin{aligned}
x_2 - x_1 + F'(y_2) - F'(y_1) - \tfrac{1}{2}F''(y_1)(y_2 - y_1) &\sim C_0^2\delta,\\
x_2' - x_1' + F'(y_2') - F'(y_1') - \tfrac{1}{2}F''(y_1')(y_2' - y_1') &\sim C_0^2\delta,\\
x_1 + x_2 &= x_1' + x_2' + O(\delta),\\
y_1 + y_2 &= y_1' + y_2' + O(\delta),\\
x_1y_1 + F(y_1) + x_2y_2 + F(y_2) &= x_1'y_1' + F(y_1') + x_2'y_2' + F(y_2') + O(\delta),
\end{aligned}$$

then

$$x_1' = x_1 + O(\delta),\ y_1' = y_1 + O(\delta),\ x_2' = x_2 + O(\delta),\ y_2' = y_2 + O(\delta). \tag{4}$$

To prove this, set

$$a := x_1 + x_2, \quad b := y_1 + y_2, \quad a' := x_1' + x_2', \quad b' := y_1' + y_2',$$

and

$$t_1 := x_1y_1 + F(y_1), \qquad t_2 := x_2y_2 + F(y_2).$$

The analogous quantities defined by $(x_1', y_1'), (x_2', y_2')$ are denoted by t_1' and t_2'. Notice that by our assumptions, a and b only vary of order $O(\delta)$ if we replace $(x_1, y_1), (x_2, y_2)$ by $(x_1', y_1'), (x_2', y_2')$. Then,

$$t_1 + t_2 = 2x_1y_1 - bx_1 - ay_1 + ab + F(y_1) + F(b - y_1).$$

We next choose c with $|c| \sim C_0^2$, such that $x_2 - x_1 + F'(y_2) - F'(y_1) - \frac{1}{2}F''(y_1)(y_2 - y_1) = c\delta$. Then we may re-write

$$x_1 = \big(a - c\delta + F'(b - y_1) - F'(y_1) - \tfrac{1}{2}F''(y_1)(b - 2y_1)\big)/2,$$

which implies that

$$\begin{aligned} t_1 + t_2 &= \Big(y_1 - \frac{b}{2}\Big)\Big(a - c\delta + F'(b - y_1) - F'(y_1) - F''(y_1)\Big(\frac{b}{2} - y_1\Big)\Big) \\ &\quad - ay_1 + ab + F(y_1) + F(b - y_1) \\ &= ab/2 + O(\delta) + \psi(y_1), \end{aligned}$$

where we have set

$$\psi(y) := \Big(y - \frac{b}{2}\Big)[F'(b - y) - F'(y) + \Big(y - \frac{b}{2}\Big)F''(y)] + F(y) + F(b - y).$$

We compute that the derivative of ψ is given by

$$\begin{aligned} \psi'(y) &= \Big(y - \frac{b}{2}\Big)[F''(y) - F''(b - y) + \Big(y - \frac{b}{2}\Big)F'''(y)] \\ &= \Big(y - \frac{b}{2}\Big)^2[2F'''(\eta) + F'''(y)], \end{aligned} \tag{5}$$

where η is some intermediate point between y and $b - y$.

Similarly, $t_1' + t_2' = a'b'/2 + O(\delta) + \psi(y_1')$. Since $a = a' + O(\delta)$, $b = b' + O(\delta)$, hence $ab = a'b' + O(\delta)$. By our assumption, $t_1 + t_2 = t_1' + t_2' + O(\delta)$, we conclude that

$$\psi(y_1) = \psi(y_1') + O(\delta). \tag{6}$$

Here, the implicit constant in $O(\delta)$ depends so far only on C_0. But, because of the y-separation (28), we have $|y_2 - y_1| \gtrsim C_0$, and since $b = y_2 + y_1$, we see that $|y_1 - b/2| \sim C_0$. Moreover, since $|F'''| \sim C_3$, so that F''' in particular does not change sign, we deduce from (5) that for all relevant y's we have

$$|\psi'(y)| \sim C_3|y - b/2|^2 \sim C_3C_0^2 \gg 1,$$

if we choose C_0 sufficiently large.

In combination with (6) this shows that we must have $y_1' = y_1 + O(\delta)$, where the implicit constant in $O(\delta)$ depends only on C_3 and C_0, hence also $y_2' = y_2 + O(\delta)$, and then our first three assumptions imply also the remaining assertions in (4).

This finishes the proof of the almost orthogonality Lemma 15. □

By means of the preceding lemmas and Rubio de Francia's estimate [23] (see also [9, 11]) we can now argue in almost exactly the same way as in [7] in order to estimate the contribution of the second sum

$$\sum_{\delta \ll 1} \sum_{i,i',j} \widehat{f_{i,j}^{\delta} d\sigma}\, \widehat{g_{i',j}^{\delta} d\sigma}$$

in (2) to $\|\mathcal{E}_{V_1}(f)\mathcal{E}_{V_2}(g)\|_p$ in (1). In this way, we see that it is of the order

$$\sum_{\delta \ll 1} C_{p,q}\ \epsilon^{2(1-\frac{1}{p}-\frac{1}{q})} \delta^{5-2/q-7/p} \rho^{6(1-1/p-1/q)} \|f\|_q \|g\|_q$$

$$\lesssim C_{p,q}\ \epsilon^{2(1-\frac{1}{p}-\frac{1}{q})} \rho^{6(1-1/p-1/q)} \|f\|_q \|g\|_q.$$

This estimate is even stronger than the required estimate in (1). Notice that the additional factor $\epsilon^{2(1-\frac{1}{p}-\frac{1}{q})}$ appears here, due to the estimate in Theorem 9 for Case 2, which was not present in [7] (where we had $\epsilon = 1$). Also, the power of ρ is better than needed, but these gains do not help for the total estimate of $\|\mathcal{E}_{V_1}(f)\mathcal{E}_{V_2}(g)\|_p$, because of the presence of first sum in (2), in which $\delta \gtrsim 1$. We leave the details to the interested reader.

This completes the proof of Lemma 12. □

By means of Lemma 12, we may finally argue as in the last part of the proof of Theorem 1.1 in [7] in order to sum the contributions by all admissible pairs of "horizontal strips" $V_1 \sim V_2$ and arrive at the estimate (5), thus completing the proof of Theorem 1. □

Acknowledgments The authors would like to express their sincere gratitude to the referee for many valuable suggestions which have greatly helped to improve the presentation of the material in this article.

References

1. Bejenaru, I.: Optimal bilinear restriction estimates for general hypersurfaces and the role of the shape operator. Int. Math. Res. Not. IMRN **23**, 7109–7147 (2017)
2. Bourgain, J.: Besicovitch-type maximal operators and applications to Fourier analysis. Geom. Funct. Anal. **22**, 147–187 (1991)
3. Bourgain, J.: Some new estimates on oscillatory integrals. Essays in Fourier Analysis in honor of E. M. Stein. Princeton Math. Ser. 42, pp. 83–112. Princeton University Press, Princeton, (1995)

4. Bourgain, J.: Estimates for cone multipliers. Oper. Theory Adv. Appl. **77**, 1–16 (1995)
5. Bourgain, J., Guth, L., Bounds on oscillatory integral operators based on multilinear estimates. Geom. Funct. Anal. **21**, 1239–1295 (2011)
6. Buschenhenke, S., Müller, D., Vargas, A.: A Fourier restriction theorem for a two-dimensional surface of finite type. Anal. PDE **10**(4), 817–891 (2017)
7. Buschenhenke, S., Müller, D., Vargas, A.: A Fourier restriction theorem for a perturbed hyperbolic paraboloid. Proc. London Math. Soc. **120**(3), 124–154 (2020)
8. Candy, T.: Multi-scale bilinear restriction estimates for general phases. Math. Ann. **375**(1–2), 777–843 (2019). https://doi.org/10.1007/s00208-019-01841-4
9. Carleson, L.: On the Littlewood-Paley theorem. Report, Mittag-Leffler Inst., Djursholm (1967)
10. Cho, C.-H., Lee, J.: Improved restriction estimate for hyperbolic surfaces in $\mathbb{R}^3$. J. Funct. Anal. **273**(3), 917–945 (2017)
11. Córdoba, A.: Some remarks on the Littlewood-Paley theory. Rend. Circ. Mat. Palermo Ser II **1**(Supplemento), 75–80 (1981)
12. Greenleaf, A.: Principal curvature and harmonic analysis. Indiana Univ. Math. J. **30**(4), 519–537 (1981)
13. Guth, L.: A restriction estimate using polynomial partitioning. J. Am. Math. Soc. **29**(2), 371–413 (2016)
14. Guth, L.: Restriction estimates using polynomial partitioning II. Acta Math. **221**(1), 81–142 (2016)
15. Ikromov, I.A., Kempe, M., Müller, D.: Estimates for maximal functions associated with hypersurfaces in $\mathbb{R}^3$ and related problems in harmonic analysis. Acta Math. **204**, 151–271 (2010)
16. Ikromov, I.A., Müller, D.: Uniform estimates for the Fourier transform of surface carried measures in $\mathbb{R}^3$ and an application to Fourier restriction. J. Fourier Anal. Appl. **17**(6), 1292–1332 (2011)
17. Ikromov, I.A., Müller, D.: Fourier restriction for hypersurfaces in three dimensions and Newton polyhedra. In: Annals of Mathematics Studies, vol, 194. Princeton University Press, Princeton (2016)
18. Kim, J.: Some remarks on Fourier restriction estimates (2017). Preprint. arXiv:1702.01231
19. Lee, S., Bilinear restriction estimates for surfaces with curvatures of different signs. Trans. Am. Math. Soc. **358**(8), 3511–2533 (2005)
20. Lee, S., Vargas, A., Restriction estimates for some surfaces with vanishing curvatures. J. Funct. Anal. 258 (2010), no. 9, 2884–2909.
21. Moyua, A., Vargas, A., Vega, L.: Schrödinger maximal function and restriction properties of the Fourier transform. Int. Math. Res. Notices **16**, 793–815 (1996)
22. Moyua, A., Vargas, A., Vega, L.: Restriction theorems and maximal operators related to oscillatory integrals in $\mathbb{R}^3$. Duke Math. J. **96**(3), 547–574 (1999)
23. Rubio de Francia, J.L.: Estimates for some square functions of Littlewood-Paley type. Publ. Sec. Mat. Univ. Autónoma Barcelona **27**(2), 81–108 (1983)
24. Stein, E.M.: Oscillatory Integrals in Fourier Analysis. Beijing Lectures in Harmonic Analysis. Princeton Univ. Press, Princeton (1986)
25. Stovall, B.: Linear and bilinear restriction to certain rotationally symmetric hypersurfaces. Trans. Am. Math. Soc. **369**(6), 4093–4117 (2017)
26. Stovall, B.: Scale invariant Fourier restriction to a hyperbolic surface. Anal. PDE **12**(5), 1215–1224 (2019)
27. Strichartz, R.S.: Restrictions of Fourier transforms to quadratic surfaces and decay of solutions of wave equations. Duke Math. J. **44**(3), 705–714 (1977)
28. Tao, T.: Endpoint bilinear restriction theorems for the cone, and some sharp null-form estimates. Math. Z. **238**, 215–268 (2001)
29. Tao, T.: A Sharp bilinear restriction estimate for paraboloids. Geom. Funct. Anal. **13**, 1359–1384 (2003)
30. Tomas, P.A.: A restriction theorem for the Fourier transform. Bull. Am. Math. Soc. **81**, 477–478 (1975)

31. Tao, T., Vargas, A.: A bilinear approach to cone multipliers I. Restriction estimates. Geom. Funct. Anal. **10**, 185–215 (2000)
32. Tao, T., Vargas, A.: A bilinear approach to cone multipliers II. Applications. Geom. Funct. Anal. **10**, 216–258 (2000)
33. Tao, T., Vargas, A., Vega, L.: A bilinear approach to the restriction and Kakeya conjectures. J. Am. Math. Soc. **11**(4), 967–1000 (1998)
34. Vargas, A.: Restriction theorems for a surface with negative curvature. Math. Z. **249**, 97–111 (2005)
35. Wolff, T.: A sharp bilinear cone restriction estimate. Ann. Math. Second Series **153**(3), 661–698 (2001)

On Young's Convolution Inequality for Heisenberg Groups

Michael Christ

To Fulvio Ricci with admiration, in celebration of his 70th birthday

Abstract Young's convolution inequality provides an upper bound for the convolution of functions in terms of L^p norms. It is known that for certain groups, including Heisenberg groups, the optimal constant in this inequality is equal to that for Euclidean space of the same topological dimension, yet no functions attain exact equality. We characterize ordered pairs of functions that nearly achieve equality for Heisenberg groups. The analysis relies on a characterization of approximate solutions of a certain class of functional equations. A result of this type is developed for a class of such equations.

Keywords Young's convolution inequality · Maximizer · Symmetry · Functional equation

1 Introduction

This paper characterizes ordered triples of functions that nearly saturate Young's convolution inequality for Heisenberg groups, viewed as an inequality for the trilinear form $\langle f * g, h\rangle$. We first review Young's inequality with sharp constant for Euclidean spaces, then review the corresponding inequality for Heisenberg groups, recalling observations of Klein and Russo [11] and of Beckner [2] concerning the distinction between the Euclidean and Heisenberg settings. We introduce a group

Research supported by National Science Foundation grants DMS-1363324 and DMS-1901413.

M. Christ (✉)
Department of Mathematics, University of California, Berkeley, CA, USA
e-mail: mchrist@berkeley.edu

P. Ciatti, A. Martini (eds.), *Geometric Aspects of Harmonic Analysis*, Springer INdAM Series 45, https://doi.org/10.1007/978-3-030-72058-2_6

of symmetries of the inequality for Heisenberg groups, along with a special class of ordered triples of Gaussian functions. Our main theorem states that an ordered triple of functions nearly saturates the inequality if and only if it differs by a small amount, in the relevant norm, from the image of one of these special ordered triples of Gaussians under some element of the symmetry group. Our conclusion is of "$o(1)$" type; we do not obtain an explicit upper bound on the difference of norms as a function of the discrepancy from exact equality. However, O'Neill [15] has used the result obtained here as the starting point in a proof of such a quantitative upper bound.

The main elements of the analysis are the known analogue for Euclidean groups, a characterization of approximate solutions of certain functional equations, and structural aspects of Heisenberg group structure.

1.1 Young's Inequality for Euclidean Groups

In its classical form, Young's convolution inequality for the Euclidean group $\mathbb{R}^m$ states that the convolution $f * g$ of functions f, g satisfies the upper bound

$$\|f * g\|_{L^r(\mathbb{R}^m)} \le \|f\|_{L^p(\mathbb{R}^m)}\|g\|_{L^q(\mathbb{R}^m)} \tag{1}$$

whenever $p, q, r \in [1, \infty]$ and $r^{-1} = p^{-1} + q^{-1} - 1$. In its sharp form established by Beckner [1] for the case when all three of p, q, r' are less than or equal to 2, and subsequently established independently by Brascamp and Lieb [3] and by Beckner for the full range of exponents, it states that

$$\|f * g\|_{L^r(\mathbb{R}^m)} \le \mathbf{C}^m_{p,q}\|f\|_{L^p(\mathbb{R}^m)}\|g\|_{L^q(\mathbb{R}^m)} \tag{2}$$

with

$$\mathbf{C}_{p,q} = A_p A_q A_{r'} \text{ where } A_s = s^{1/2s} t^{-1/2t} \text{ with } t = s'; \tag{3}$$

here and below s' denotes the exponent $s' = s/(s-1)$ conjugate to s. The factor $\mathbf{C}_{p,q}$ is strictly less than 1 provided that $p, q, r \in (1, \infty)$, and $\mathbf{C}^n_{p,q}$ is the optimal constant in this inequality for all exponents and all dimensions.

Write $\mathbf{p} = (p_1, p_2, p_3)$ with $p_j \in [1, \infty]$, $\mathbf{f} = (f_1, f_2, f_3)$, and $\mathbf{x} = (x_1, x_2, x_3)$ where each $x_j \in \mathbb{R}^m$. We use the notational convention

$$\|\mathbf{f}\|_{\mathbf{p}} = \prod_{j=1}^{3} \|f_j\|_{p_j}. \tag{4}$$

An ordered triple $\mathbf{p} = (p_1, p_2, p_3)$ of exponents is said to be admissible if $p_j \in [1, \infty]$ and $\sum_{j=1}^{3} p_j^{-1} = 2$.

Rather than working with the bilinear operation $(f, g) \mapsto f * g$, we will work with the trilinear form

$$\mathcal{T}(\mathbf{f}) = \mathcal{T}_{\mathbb{R}^m}(\mathbf{f}) = \int_{x_1+x_2+x_3=0} \prod_{j=1}^{3} f_j(x_j)\, d\lambda_{\mathbb{R}^m}(\mathbf{x}) \tag{5}$$

where $\lambda_{\mathbb{R}^m}$ is the natural Lebesgue measure on

$$\Lambda_{\mathbb{R}^m} = \{\mathbf{x} \in (\mathbb{R}^m)^3 : x_1 + x_2 + x_3 = 0\}. \tag{6}$$

That is,

$$\lambda_{\mathbb{R}^m}(E) = \int_{\mathbb{R}^m \times \mathbb{R}^m} \mathbf{1}_E(x_1, x_2, -x_1 - x_2)\, dx_1\, dx_2.$$

The three variables x_1, x_2, x_3 may be freely permuted in the definition of $\lambda_{\mathbb{R}^m}$.

For $\mathbf{p} \in [1, \infty]^3$ define

$$\mathbf{A_p} = \prod_{j=1}^{3} p_j^{1/2p_j} q_j^{-1/2q_j} \tag{7}$$

where q_j is the exponent conjugate to p_j, with $\infty^{\pm 1/\infty}$ interpreted as 1. Then $\mathbf{A_p}$ is strictly less than 1 whenever $\mathbf{p}$ is admissible and each p_j belongs to the open interval $(1, \infty)$. The inequality of Beckner and Brascamp-Lieb can be restated as

$$\left|\mathcal{T}_{\mathbb{R}^m}(\mathbf{f})\right| \le \mathbf{A_p}^m \|\mathbf{f}\|_{\mathbf{p}} \tag{8}$$

whenever $\mathbf{p}$ is admissible. The factor $\mathbf{A_p}^m$ is optimal for all exponents.

By a Gaussian function G with domain equal to a Euclidean space $\mathbb{R}^m$ we mean a function

$$G(x) = ce^{-|L(x-a)|^2 + ix\cdot b} \tag{9}$$

where $0 \ne c \in \mathbb{C}$, $a \in \mathbb{R}^m$, $b \in \mathbb{R}^m$, and $L : \mathbb{R}^m \to \mathbb{R}^m$ is an invertible linear endomorphism. A linear imaginary term, $ix \cdot b$, is allowed in the exponent, but the quadratic part of the exponent is real. In other contexts, the term "Gaussian" may refer to functions that are either more, or less, general.

For the Euclidean group $\mathbb{R}^m$, maximizing triples[1] $\mathbf{f}$ for Young's convolution inequality exist for all admissible exponent triples $\mathbf{p}$ with each $p_j \in (1, \infty)$. All

[1]We abuse language mildly by writing "extremizer" or "maximizer", since saturators of the inequality are maximizers of the ratio of the functional on the left-hand side of the inequality to the product of the norms on the right.

such triples were characterized by Brascamp and Lieb [3]. See also Lieb [12] for related results. Suppose that $\|f_j\|_{p_j} > 0$ for each index j. If $|\mathcal{T}_{\mathbb{R}^m}(\mathbf{f})| = \mathbf{A}_\mathbf{p}^m \|\mathbf{f}\|_\mathbf{p}$ then each function f_j is a Gaussian function $G_j = c_j e^{-\rho_j |L_j(x-a_j)|^2 + ix\cdot b_j}$. Moreover, the ordered triple (G_1, G_2, G_3) is compatible in the sense that $a_1 + a_2 + a_3 = 0$, $b_1 = b_2 = b_3$, $L_1 = L_2 = L_3$, and $(\rho_1, \rho_2, \rho_3) = (sp_1', sp_2', sp_3')$ for some $s \in \mathbb{R}^+$, where $p_j' = p_j/(p_j - 1)$ is the exponent conjugate to p_j. Conversely, if each f_j is a Gaussian and if these functions are compatible in the sense indicated, then $|\mathcal{T}_{\mathbb{R}^m}(\mathbf{f})| = \mathbf{A}_\mathbf{p}^m \|\mathbf{f}\|_\mathbf{p}$.

A qualitative stability property of extremizers of Young's inequality for $\mathbb{R}^m$ was established in [4]. If $\|f_j\|_{p_j} = 1$ for each index j and if $\mathcal{T}(\mathbf{f}) \ge \mathbf{A}_\mathbf{p}^m - \delta$ then $\mathbf{f}$ lies within distance $\varepsilon(\delta)$ of a maximizing triple of Gaussians, in the sense that $\|f_j - G_j\|_{p_j} \le \varepsilon(\delta)$, and $\varepsilon(\delta) \to 0$ as $\delta \to 0$. The present paper establishes a corresponding result for $\mathbb{H}^d$, but the formulation is more intricate.

For $\mathbb{R}^m$, a quantitative stability theorem with $\varepsilon(\delta) = C(m, \mathbf{p})\delta^{1/2}$ is established in [10] for the case in which $p_j \le 2$ for all three indices j. For a partial range of admissible exponents $\mathbf{p}$, but not for the full range, this is a corollary of a corresponding theorem for the Hausdorff–Young inequality devloped in [8]. A quantitative statement with a smaller (and optimal) exponent is also established for the case in which some exponent exceeds 2.

1.2 Young's Inequality for Heisenberg Groups

Let $d \in \mathbb{N}$, and identify $\mathbb{R}^{2d+1}$ with $\mathbb{R}^{2d} \times \mathbb{R}$. The Heisenberg group $\mathbb{H}^d$ is $\mathbb{R}^{2d+1}$ as a set, with the group law

$$z \cdot z' = (x, t) \cdot (x', t') = (x + x',\ t + t' + \sigma(x, x')) \tag{10}$$

where $z = (x, t)$, $z' = (x', t')$, and $\sigma : \mathbb{R}^{2d} \times \mathbb{R}^{2d} \to \mathbb{R}^1$ is the symplectic form

$$\sigma(x, x') = \sum_{j=1}^{d} \left(x_j x'_{j+d} - x_{j+d} x'_j \right). \tag{11}$$

Although we use multiplicative notation for the group law, we denote the the group identity element by $0 = (0, 0)$. The Heisenberg multiplicative inverse of $v = (x, t)$ is $v^{-1} = (-x, -t)$. There are alternative isomorphic formulations of this group law, some of which are in common use. By a Gaussian function $G : \mathbb{H}^d \to \mathbb{C}$ we mean a Gaussian function $G : \mathbb{R}^{2d+1} \to \mathbb{C}$, with respect to the coordinate system for $\mathbb{H}^d$ introduced above.

L^p norms on $\mathbb{H}^d$ are defined with respect to Lebesgue measure on $\mathbb{R}^{2d+1}$, and will be denoted by $\|\cdot\|_{L^p}$ and more succinctly by $\|\cdot\|_p$. Throughout this paper, integrals over $\mathbb{H}^d$, and measures of subsets of $\mathbb{H}^d$, are understood with respect to Lebesgue measure on $\mathbb{R}^{2d+1}$, unless the contrary is explicitly indicated. Convolution is defined to be $f * g(u) = \int_{\mathbb{H}^d} f(uv^{-1})g(v)\,dv$. This bilinear operation is associative, but not commutative, on the Schwartz space.

We phrase Young's inequality for $\mathbb{H}^d$ in terms of the trilinear form

$$\mathcal{T}_{\mathbb{H}^d}(\mathbf{f}) = \int_{z_1 z_2 z_3 = 0} \prod_{j=1}^{3} f_j(z_j)\,d\lambda(\mathbf{z}) \tag{12}$$

where $z_1 z_2 z_3$ is the threefold $\mathbb{H}^d$ product and $\lambda = \lambda_{\mathbb{H}^d}$ is the natural Lebesgue measure on

$$\Lambda_{\mathbb{H}^d} = \{\mathbf{z} \in (\mathbb{H}^d)^3 : z_1 z_2 z_3 = 0\}. \tag{13}$$

That is, for $E \subset \Lambda_{\mathbb{H}^d}$,

$$\lambda(E) = \int_{\mathbb{H}^d \times \mathbb{H}^d} \mathbf{1}_E(z_1, z_2, z_2^{-1} z_1^{-1})\,dz_1\,dz_2$$

and the roles of the variables z_1, z_2, z_3 can be interchanged provided that noncommutativity of the group law is taken properly into account. Just as in the Euclidean case, it is elementary that $|\mathcal{T}_{\mathbb{H}^d}(\mathbf{f})| \le \|\mathbf{f}\|_{\mathbf{p}}$ whenever $f_j \in L^{p_j}$ for all j and $\mathbf{p}$ is admissible.

Klein and Russo [11] and Beckner [2] have observed that the sharper inequality

$$\left|\mathcal{T}_{\mathbb{H}^d}(\mathbf{f})\right| \le \mathbf{A}_{\mathbf{p}}^{2d+1} \|\mathbf{f}\|_{\mathbf{p}} \tag{14}$$

holds, with the same constant factor on the right-hand side as for Euclidean space of dimension $2d+1$. We will abuse language mildly by referring to tuples of functions that maximize the ratio $|\mathcal{T}_{\mathbb{H}^d}(\mathbf{f})|/\|\mathbf{f}\|_{\mathbf{p}}$ as maximizers of the inequality (14).

The quantity $\mathbf{A}_{\mathbf{p}}^{2d+1}$ is the optimal constant in (14). Beckner has observed further that there exist no maximizing functions, that is, $|\mathcal{T}_{\mathbb{H}^d}(\mathbf{f})|$ is strictly less than $\mathbf{A}_{\mathbf{p}}^{2d+1} \|\mathbf{f}\|_{\mathbf{p}}$ whenever all three functions have positive norms and each $p_j \in (1, \infty)$.[2]

The nonexistence of maximizing functions can be viewed differently. For each $s \in \mathbb{R}$, the set $\mathbb{R}^{2d+1}$ is a group under the operation $+_s$ defined by

$$(x, t) +_s (x', t') = (x + x', t + t' + s\sigma(x, x')). \tag{15}$$

[2]Klein and Russo do not explicitly discuss existence of extrenizers for Young's inequality, but do prove a closely related result: There exist no nonzero maximizers for the Heisenberg group analogue of the $L^p \to L^{p'}$ Hausdorff–Young inequality when the conjugate exponent p' is an even integer.

This group is isomorphic to $\mathbb{H}^d$ when $s \neq 0$, and to the Euclidean group $\mathbb{R}^{2d+1}$ when $s = 0$. Haar measure is Lebesgue measure in these coordinates, for all s. The optimal constant in Young's convolution inequality is $\mathbf{A}_{\mathbf{p}}^{2d+1}$ for every s. A datum $(\mathbf{f}, s)$ realizes this optimal constant if and only if $s = 0$ and $\mathbf{f}$ is a maximizing ordered triple $\mathbf{G}$ for $\mathbb{R}^{d+1}$. Theorem 7, below, could be reformulated as an assertion that $(\mathbf{f}, s)$ nearly realizes the optimal constant only if $(\mathbf{f}, s)$ is close to such a datum $(\mathbf{G}, 0)$, in an appropriate sense. This point of view is pursued in [15].

In a series of works [4–10] we have studied various sharp inequalities for which maximizing functions (respectively ordered tuples of functions or sets) exist and have previously been characterized. We have shown that functions (respectively ordered tuples of functions or sets) that nearly saturate the inequalities are nearly equal, in appropriate norms or other measures of approximation, to maximizing functions (respectively ordered tuples of functions or sets). The present paper characterizes near-maximizers, in a setting in which no maximizers exist.

2 Definitions and Main Theorem

Our main result, Theorem 7, will state that if $\mathbf{f}$ nearly maximizes the ratio $|\mathcal{T}_{\mathbb{H}^d}(\mathbf{f})|/\|\mathbf{f}\|_{\mathbf{p}}$ then there exists an ordered triple (G_1, G_2, G_3) of Gaussians with certain properties, such that $\|f_j - G_j\|_{p_j}$ is appropriately small for each index j. In order to formulate this result precisely, several definitions are required.

2.1 *The Symplectic Group*

Denote by Sp(2d) the symplectic group of all invertible linear mappings $S : \mathbb{R}^{2d} \to \mathbb{R}^{2d}$ satisfying

$$\sigma(Sx, Sx') = \sigma(x, x') \quad \text{for all } x, x' \in \mathbb{R}^{2d}. \tag{16}$$

To $S \in \text{Sp(2d)}$ is associated the group automorphism $(x, t) \mapsto (Sx, t)$ of $\mathbb{H}^d$.

Let J denote the $2d \times 2d$ matrix

$$J = \begin{pmatrix} 0 & I \\ -I & 0 \end{pmatrix} \tag{17}$$

where I is the $d \times d$ identity matrix. Since $\sigma(x, y) = \langle x, Jy\rangle$ for $x, y \in \mathbb{R}^{2d}$, the identity $\sigma(Sx, Sy) \equiv \sigma(x, y)$ that defines Sp(2d) is equivalent to $\langle Sx, JSy\rangle \equiv \langle x, Jy\rangle$. Thus $S \in \text{Sp(2d)}$ if and only if $S^* JS = J$.

2.2 Symmetries

Let $\Psi = (\psi_1^*, \psi_2^*, \psi_3^*)$ be an ordered 3-tuple of invertible linear mappings ψ_j^* : $L^{p_j}(\mathbb{H}^d) \to L^{p_j}(\mathbb{H}^d)$. Consider the functional

$$\Phi(\mathbf{f}) = \frac{|\mathcal{T}_{\mathbb{H}^d}(\mathbf{f})|}{\|\mathbf{f}\|_{\mathbf{p}}}, \tag{18}$$

defined for all $\mathbf{f}$ satisfying $\|\mathbf{f}\|_{\mathbf{p}} \neq 0$, with $\|\mathbf{f}\|_{\mathbf{p}} = \prod_{j=1}^{3} \|f_j\|_{p_j}$ as in (4). Given $\mathbf{p}$, we say that Ψ is a symmetry of the inequality (14), or of the functional Φ, if $\Phi(\Psi\mathbf{f}) = \Phi(\mathbf{f})$ for all $\mathbf{f} \in L^{p_1} \times L^{p_2} \times L^{p_3}$ with $\|\mathbf{f}\|_{\mathbf{p}} \neq 0$. These 3-tuples form a group under componentwise composition.

Most of the symmetries of Φ relevant to our considerations are defined in terms of mappings of the underlying space $\mathbb{H}^d$. To any diffeomorphism ψ of $\mathbb{H}^d$ we associate a linear operator on functions $f : \mathbb{H}^d \to \mathbb{C}$, defined by

$$\psi^*(f) = f \circ \psi.$$

We next list four families of ordered triples (ψ_1, ψ_2, ψ_3) of diffeomorphisms of $\mathbb{H}^d$ such that $\Psi = (\psi_1^*, \psi_2^*, \psi_3^*)$ is a symmetry of Φ. The first three of these families are:

$$\begin{cases} (i) \quad \psi_j(x,t) = (rx, r^2 t) \ \text{ with } r \in \mathbb{R}^+ \\ (ii) \quad \psi_j(z) = (u_j z w_j) \ \text{ with } w_1 = u_2^{-1},\ w_2 = u_3^{-1}, \text{ and } w_3 = u_1^{-1}. \\ (iii) \ \psi_j(x,t) = (Sx, t) \ \text{with } S \in \mathrm{Sp(2d)}. \end{cases} \tag{19}$$

The fourth family is defined by

$$\psi_j(x,t) = (x, t + \varphi_j(x)) \tag{20}$$

where $(\varphi_1, \varphi_2, \varphi_3)$ is an ordered triple of affine mappings from $\mathbb{R}^{2d}$ to $\mathbb{R}^1$ that satisfies $\sum_{k=1}^{3} \varphi_k(x_k) = 0$ whenever $\sum_{k=1}^{3} x_k = 0$. In (i), r is independent of j; likewise S is independent of j in (iii). In (ii), $u_j z_j w_j$ is the $\mathbb{H}^d$ group product of these three elements.

A fifth family of symmetries is defined in terms of modulations of functions, rather than diffeomorphisms of the underlying space. For any $u \in \mathbb{R}^{2d}$ define $\Psi = (\psi_1, \psi_2, \psi_3)$ by

$$(\psi_j f)(x,t) = e^{iu \cdot x} f(x,t). \tag{21}$$

The exponent $iu \cdot x$ depends only on the coordinate x, not on t.

Each component of each element of each of these five families is an invertible bounded linear operator on $L^p(\mathbb{H}^d)$ for all $p \in [1, \infty]$. By the composition $\Psi \circ \Psi'$

of two such ordered triples we mean the ordered triple $(\psi_1 \circ \psi_1', \psi_2 \circ \psi_2', \psi_3 \circ \psi_3')$ defined by componentwise composition.

Lemma 1 *Each of the ordered triples of linear operators Ψ listed above is a symmetry of the ratio Φ for every admissible* $\mathbf{p}$.

The straightforward verifications are left to the reader. □

Definition 2 $\mathfrak{G}(\mathbb{H}^d)$ denotes the group of all ordered triples $\Psi = (\psi_1, \psi_2, \psi_3)$, with each ψ_j an invertible linear transformation of $L^{p_j}(\mathbb{H}^d)$, that can be expressed as compositions of finitely many symmetries of the inequality (14), with each factor being one of the five types introduced above.

A sixth family of symmetries of Φ is $\mathbf{f} \mapsto (c_1 f_1, c_2 f_2, c_3 f_3)$ with each $c_j \in \mathbb{C} \setminus \{0\}$. We have chosen not to include these symmetries in $\mathfrak{G}(\mathbb{H}^d)$. The definition of $\mathfrak{G}(\mathbb{H}^d)$ could be modified by including them, resulting in a corresponding modification of the statement of Theorem 7 below.

2.3 *Special Ordered Triples of Gaussians on* $\mathbb{H}^d$

Definition 3 Let $d \geq 1$ and $\varepsilon > 0$. A canonical ε-diffuse Gaussian is a function $G : \mathbb{H}^d \to \mathbb{C}$ of the form

$$G(x, t) = e^{-|Lx|^2} e^{-at^2} e^{ibt}$$

where $a > 0$, $b \in \mathbb{R}$, and $L : \mathbb{R}^{2d} \to \mathbb{R}^{2d}$ is an invertible linear endomorphism, which together satisfy

$$\max(a^{1/2}, a, |b|) \cdot \|L^{-1}\|^2 \leq \varepsilon. \tag{22}$$

Notation 4 $\gamma(\mathbf{p}) = (\gamma_1, \gamma_2, \gamma_3)$ with

$$\gamma_j = p_j', \tag{23}$$

the exponent conjugate to p_j.

Definition 5 Let $\mathbf{p}$ be admissible. An ordered triple $\mathbf{G} = (G_1, G_2, G_3)$ of canonical ε-diffuse Gaussians

$$G_j(x, t) = e^{-|L_j x|^2} e^{-a_j t^2} e^{ib_j t}$$

is said to be $\mathbf{p}$-compatible if there exist L, a, b such that $L_j = \gamma_j^{1/2} L$, $a_j = \gamma_j a$, and $b_j = b$ for all $j \in \{1, 2, 3\}$.

Definition 6 Let $d \geq 1$ and let $\varepsilon > 0$ be small. An ordered triple $\mathbf{G} = (G_1, G_2, G_3)$ of Gaussian functions $G_j : \mathbb{H}^d \to \mathbb{C}$ is ε-diffuse and $\mathbf{p}$-compatible if there exist $\Psi \in \mathfrak{S}(\mathbb{H}^d)$, scalars $c_j \in \mathbb{C} \setminus \{0\}$, and a $\mathbf{p}$-compatible ordered triple $(\tilde{G}_1, \tilde{G}_2, \tilde{G}_3)$ of canonical ε-diffuse Gaussian functions such that

$$G_j = c_j \psi_j \tilde{G}_j \quad \text{for each index } j \in \{1, 2, 3\}.$$

2.4 *Main Theorem*

Theorem 7 *For each $d \geq 1$ and each admissible ordered triple $\mathbf{p}$ of exponents there exists a function $\delta \mapsto \varepsilon(\delta)$ satisfying $\lim_{\delta \to 0} \varepsilon(\delta) = 0$ with the following property. Let $\mathbf{f} \in L^{\mathbf{p}}(\mathbb{H}^d)$ and suppose that $\|f_j\|_{p_j} \neq 0$ for each $j \in \{1, 2, 3\}$. Let $\delta \in (0, 1)$ and suppose that $|\mathcal{T}_{\mathbb{H}^d}(\mathbf{f})| \geq (1 - \delta)\mathbf{A}_{\mathbf{p}}^{2d+1} \|\mathbf{f}\|_{\mathbf{p}}$. Then there exists a $\mathbf{p}$-compatible $\varepsilon(\delta)$-diffuse ordered triple of Gaussians $\mathbf{G} = (G_1, G_2, G_3)$ such that*

$$\|f_j - G_j\|_{p_j} \leq \varepsilon(\delta) \|f_j\|_{p_j} \quad \text{for each } j \in \{1, 2, 3\}. \tag{24}$$

Thus $G_j = c_j \psi_j \tilde{G}_j$ where $c_j \in \mathbb{C} \setminus \{0\}$, $(\tilde{G}_1, \tilde{G}_2, \tilde{G}_3)$ is a canonically $\varepsilon(\delta)$-diffuse $\mathbf{p}$-compatible ordered triple of Gaussians, and $\Psi = (\psi_1, \psi_2, \psi_3) \in \mathfrak{S}(\mathbb{H}^d)$.

O'Neill [15] has built on Theorem 7 together with the analysis developed in [10] to establish a quantitative form of its conclusion, in the same way that [10] quantifies [4]. Theorem 7 remains an essential part of the analysis; its conclusion provides the starting point for the perturbative analysis in [15].

3 Approximate Solutions of Functional Equations

A principal ingredient of the analysis is a quantitative expression of the unsolvability of a variant of the functional equation

$$\varphi(x) + \psi(y) + \xi(x + y) = 0. \tag{25}$$

This variant takes the form

$$\varphi(x) + \psi(y) + \xi(x + y) + \sigma(x, y) = 0 \tag{26}$$

where the functions φ, ψ, ξ have domains equal to $\mathbb{R}^{2m}$. Its unsolvability is formulated below, in quantitative terms, as Proposition 18.

An *ad hoc* argument that relies on the antisymmetry of $\sigma(x, y)$ will enable us to deduce the information needed concerning (26) from what is already known about approximate solutions of (25). This leads naturally to analogous questions about more general functional equations, for which this *ad hoc* argument may not apply. We therefore digress to present the following general result, which is suggested and motivated by considerations in this paper, but is not actually used in the proofs of the main theorems.

Consider the difference operators

$$(\Delta_h f)(x) = f(x+h) - f(x), \tag{27}$$

where $x \in \mathbb{R}^d$ and $+$ denotes the Euclidean group operation. Let $\mathbb{B}$ and $\tilde{\mathbb{B}}$ be balls in $\mathbb{R}^d$ of positive, finite radii, with $\tilde{\mathbb{B}}$ centered at the origin.

Theorem 8 *For each dimension $d \ge 1$, each nonnegative integer D, and each $\eta > 0$ there exists a function $\delta \mapsto \varepsilon(\delta)$ satisfying $\lim_{\delta \to 0} \varepsilon(\delta) = 0$ with the following property. Suppose that $|\tilde{\mathbb{B}}| \ge \eta |\mathbb{B}|$, $0 < \delta \le 1$, and $A \in [0, \infty)$. Let $\varphi : \mathbb{B} + \tilde{\mathbb{B}} \to \mathbb{C}$ be Lebesgue measurable. Let*

$$P_h(x) = \sum_{|\alpha| \le D} a_\alpha(h) x^\alpha \tag{28}$$

be a polynomial function of x of degree $\le D$ whose coefficients a_α are Lebesgue measurable functions of h. Suppose that there exists a function $\mathbb{B} \times \tilde{\mathbb{B}} \ni (x, h) \mapsto P_h(x) \in \mathbb{C}$ such that

$$\big|\Delta_h \varphi(x) - P_h(x)\big| \le A \tag{29}$$

for all $(x, h) \in \mathbb{B} \times \tilde{\mathbb{B}}$ with the exception of a set of measure $\le \delta |\mathbb{B}| \cdot |\tilde{\mathbb{B}}|$.

Then there exists a polynomial Q of degree at most $D + 1$ such that

$$\big|\varphi(x) - Q(x)\big| \le CA \tag{30}$$

for all $x \in \mathbb{B}$ outside a set of measure $\le \varepsilon(\delta)|\mathbb{B}|$. The constant C and function ε depend only on d, D, η.

This is proved in Sect. 11. It has already been applied in [13].

We will use informal language “for nearly all $y \in E$” to indicate a Lebesgue measurable subset $A \subset E$ satisfying $|A| \le o_\delta(1)|E|$, where the quantity $o_\delta(1)$ depends on δ, $\mathbf{p}$, d alone and tends to 0 as $\delta \to 0$ while $\mathbf{p}$, d remain fixed. “Nearly all $(y_1, y_2) \in E^2$” has a corresponding meaning.

In the simplest case $D = 0$, the assumption is that $|\varphi(x+h) - \varphi(x) - a(h)| \le A$ for nearly all points of $\mathbb{B} \times \tilde{\mathbb{B}}$; this assumption is an approximate version of the fundamental functional equation (25). In that special case, Theorem 8 is proved in [4].

It is natural to also record a multiplicative analogue Theorem 8. Let $\mathbb{B}$ and $\tilde{\mathbb{B}}$ be as above.

Theorem 9 *For each dimension $d \geq 1$, each nonnegative integer D, and each $\eta > 0$ there exists a function $\delta \mapsto \varepsilon(\delta)$ satisfying $\lim_{\delta\to 0}\varepsilon(\delta) = 0$ with the following property. Suppose that $|\tilde{\mathbb{B}}| \geq \eta|\mathbb{B}|$, $0 < \delta \leq 1$, and $A \in [0, 2]$. Let $\varphi : \mathbb{B} + \tilde{\mathbb{B}} \to \mathbb{R}$ be Lebesgue measurable. Suppose that there exists a function $\mathbb{B} \times \tilde{\mathbb{B}} \ni (x, h) \mapsto P_h(x) \in \mathbb{R}$ such that*

$$|e^{i(\varphi(x+h)-\varphi(x))}e^{-iP_h(x)} - 1| \leq A \tag{31}$$

for all $(x, h) \in \mathbb{B} \times \tilde{\mathbb{B}}$ with the exception of a set of measure $\leq \delta|\mathbb{B}| \cdot |\tilde{\mathbb{B}}|$. Suppose that

$$P_h(x) = \sum_{|\alpha|\leq D} a_\alpha(h)x^\alpha \tag{32}$$

is a polynomial function of x of degree $\leq D$ whose coefficients a_α are Lebesgue measurable real-valued functions of h. Then there exists a polynomial Q of degree at most $D + 1$ such that

$$|e^{i\varphi(x)}e^{-iQ(x)} - 1| \leq CA \tag{33}$$

for all $x \in \mathbb{B}$ outside a set of measure $\leq \varepsilon(\delta)|\mathbb{B}|$. The constant C and function ε depend only on d, D, η.

4 Analogue for Twisted Convolution

Consider twisted convolution of functions with domains $\mathbb{R}^{2d}$. The associated trilinear forms are

$$\mathcal{T}_{\mathbb{R}^{2d},\rho}(\mathbf{f}) = \int_{(\mathbb{R}^{2d})^3} e^{i\rho\sigma(x_1,x_2)} \prod_{j=1}^{3} f_j(x_j)\, d\lambda_{\mathbb{R}^{2d}}(\mathbf{x}) \tag{34}$$

where $0 \neq \rho \in \mathbb{R}$ is a parameter, $\mathbf{x} = (x_1, x_2, x_3) \in (\mathbb{R}^{2d})^3$, and $\lambda_{\mathbb{R}^{2d}} = \{\mathbf{x} : x_1 + x_2 + x_3 = 0\}$. Since $|\mathcal{T}_{\mathbb{R}^{2d},\rho}(\mathbf{f})| \leq \mathcal{T}_{\mathbb{R}^{2d}}(|f_1|, |f_2|, |f_3|)$, one has

$$|\mathcal{T}_{\mathbb{R}^{2d},\rho}(\mathbf{f})| \leq \mathbf{A}_{\mathbf{p}}^{2d} \prod_{j=1}^{3} \|f_j\|_{p_j} \tag{35}$$

for admissible $\mathbf{p}$. The constant $\mathbf{A}_{\mathbf{p}}^{2d}$ is optimal [11], as one sees by considering ordered triples of Gaussians that saturate Young's inequality for $\mathbb{R}^{2d}$ and are concentrated near 0. Again, there exist no maximizing triples [11].

Theorem 10 *For each $d \geq 1$ and each admissible ordered triple* $\mathbf{p}$ *of exponents there exists a function* $\delta \mapsto \varepsilon(\delta)$ *satisfying* $\lim_{\delta\to 0}\varepsilon(\delta) = 0$ *with the following property. Let* $\mathbf{f} \in L^{\mathbf{p}}(\mathbb{R}^{2d})$ *and suppose that* $\|f_j\|_{p_j} \neq 0$ *for each* $j \in \{1,2,3\}$*. Let* $\delta \in (0,1)$ *and suppose that* $|\mathcal{T}_{\mathbb{R}^{2d},\rho}(\mathbf{f})| \geq (1-\delta)\mathbf{A}_{\mathbf{p}}^{2d}\|\mathbf{f}\|_{\mathbf{p}}$*. Then there exist* $S \in \mathrm{Sp}(2d)$ *and a* $\mathbf{p}$*-compatible ordered triple of Gaussians* $\mathbf{G} = (G_1, G_2, G_3)$ *such that* $G_j^\natural = G_j \circ S$ *satisfy*

$$\|f_j - G_j^\natural\|_{p_j} \leq \varepsilon(\delta)\|f_j\|_{p_j} \text{ for } j \in \{1,2,3\} \tag{36}$$

and G_j *take the form*

$$G_j(x) = c_j e^{-p_j'|L(x-a_j)|^2} e^{ix\cdot v} e^{-i\rho\sigma(\tilde{a}_j, x)} \tag{37}$$

where $v \in \mathbb{R}^{2d}$, $0 \neq c_j \in \mathbb{C}$, $a_1 + a_2 + a_3 = 0$, $\tilde{a}_3 = 0$, $\tilde{a}_1 = a_2$, $\tilde{a}_2 = a_1$*, and* $L: \mathbb{R}^{2d} \to \mathbb{R}^{2d}$ *is an invertible linear transformation satisfying*

$$|\rho| \cdot \|L^{-1}\|^2 \leq \varepsilon(\delta). \tag{38}$$

The expressions for the modified parameters $\tilde{a}_j$ involve an arbitrary choice; alternative expressions can equally well be used, with alterations absorbed into the parameter v.

The proof of Theorem 10 follows that of Theorem 7, with some simplifications. A brief discussion is in Sect. 13.

O'Neill [14] has formulated and proved a quantitative form of this conclusion. Again, the qualitative conclusion is needed as a starting point for the quantitative analysis of a perturbative expansion of the functional about $(\mathbf{f}, \rho) = (\mathbf{G}, 0)$.

5 Nonexistence of Maximizers, and Value of the Optimal Constant

We begin by reviewing proofs that the optimal constant in Young's inequality for $\mathbb{H}^d$ equals the optimal constant for Euclidean space of dimension $2d+1$, and that maximizing triples do not exist. To show that the constant for $\mathbb{H}^d$ is at least as large as for $\mathbb{R}^{2d+1}$, let $\varepsilon > 0$ be small, and consider the ordered triple of functions $\mathbf{f}_\varepsilon = (f_{j,\varepsilon} : 1 \leq j \leq 3)$ with $f_{j,\varepsilon}(x,t) = e^{-\gamma_j|x|^2}e^{-\varepsilon\gamma_j t^2}$ and $\gamma(\mathbf{p}) = (\gamma_1, \gamma_2, \gamma_3)$ as in (23). For each $\varepsilon > 0$, $\mathbf{f}_\varepsilon$ saturates Young's inequality for $\mathbb{R}^{2d+1}$. One finds by a simple change of variables $t = \varepsilon^{-1/2}s$ that

$$\frac{\mathcal{T}_{\mathbb{H}^d}(\mathbf{f}_\varepsilon)}{\mathcal{T}_{\mathbb{R}^{2d+1}}(\mathbf{f}_\varepsilon)} \to 1 \text{ as } \varepsilon \to 0. \tag{39}$$

To prove the reverse implication, let $f_j \in L^{p_j}(\mathbb{H}^d)$ be nonzero nonnegative functions which are otherwise arbitrary. For $x \in \mathbb{R}^{2d}$ define

$$\begin{cases} F_j(x) = \|f_j(x,\cdot)\|_{L^{p_j}(\mathbb{R})} \\ f_{j,x}(t) = f_j(x,t)/F_j(x) \ \text{ if } F_j(x) \neq 0, \end{cases} \tag{40}$$

with instead $f_{j,x}(t) \equiv 0$ if $F_j(x) = 0$. Write $\mathbf{x} = (x_1, x_2, x_3) \in (\mathbb{R}^{2d})^3$. Then

$$\mathcal{T}_{\mathbb{H}^d}(\mathbf{f}) = \int_{\Lambda_{\mathbb{R}^{2d}}} \prod_{j=1}^{3} F_j(x_j)\mathcal{T}_{\mathbb{R}^1}(f_{1,x_1}, f_{2,x_2}, f^{\dagger}_{3,\mathbf{x}})\, d\lambda_{\mathbb{R}^{2d}}(\mathbf{x}) \tag{41}$$

where

$$f^{\dagger}_{3,\mathbf{x}}(s) = f_{3,x_3}(s + \sigma(x_1, x_2)). \tag{42}$$

Straightforward calculation gives $f_{3,x_3}(s+\sigma(x_1,x_2)+\sigma(x_1+x_2,x_3))$ as the natural definition of $f^{\dagger}_{3,\mathbf{x}}(s)$, but outside of a $\lambda_{\mathbb{R}^{2d}}$-null set this simplifies to $f_{3,x_3}(s + \sigma(x_1, x_2))$ since

$$x_1 + x_2 + x_3 = 0 \Longrightarrow \sigma(x_1 + x_2, x_3) = \sigma(x_1 + x_2, -x_1 - x_2) = 0.$$

Therefore

$$|\mathcal{T}_{\mathbb{R}^1}(f_{1,x_1}, f_{2,x_2}, f^{\dagger}_{3,\mathbf{x}})| \le \mathbf{A_p} \prod_{j=1}^{3} \|f_{j,x_j}\|_{p_j} \le \mathbf{A_p}$$

with equality only if $\prod_{j=1}^{3} F_j(x_j) \neq 0$ and $(f_{1,x_1}, f_{2,x_2}, f^{\dagger}_{3,\mathbf{x}})$ is a maximizing triple for Young's inequality for $\mathbb{R}^1$. Inserting this into (41) gives

$$\begin{aligned} |\mathcal{T}_{\mathbb{H}^d}(\mathbf{f})| &\le \mathbf{A_p} \int_{x_1+x_2+x_3=0} \prod_{j=1}^{3} F_j(x_j)\, d\lambda_{\mathbb{R}^{2d}}(\mathbf{x}) \\ &= \mathbf{A_p}\mathcal{T}_{\mathbb{R}^{2d}}(F_1, F_2, F_3) \le \mathbf{A_p}\mathbf{A_p}^{2d} \prod_{j=1}^{3} \|F_j\|_{L^{p_j}(\mathbb{R}^{2d})} = \mathbf{A_p}^{2d+1}\|\mathbf{f}\|_{\mathbf{p}}. \end{aligned}$$

This proves that the optimal constant for $\mathbb{H}^d$ cannot exceed the optimal constant for $\mathbb{R}^{2d+1}$.

This analysis implicitly proves that maximizers do not exist for $\mathbb{H}^d$. For arbitrary nonnegative $f_j \in L^{p_j}(\mathbb{H}^d)$ with positive norms, we have shown that equality holds only if both (i) for λ-almost every $\mathbf{x} \in \Lambda_{\mathbb{R}^{2d}}$, $(f_{1,x_1}, f_{2,x_2}, f^{\dagger}_{3,x_3})$ is a maximizing triple for Young's inequality for $\mathbb{R}^1$, and (ii) (F_1, F_2, F_3) is a maximizing triple for Young's inequality for $\mathbb{R}^{2d}$.

By the characterization of equality in Young's inequality for $\mathbb{R}^{2d}$, each F_j must be a Gaussian; in particular, F_j is nonzero almost everywhere. Likewise, $f_{j,y}$ must be a Gaussian for almost every $y \in \mathbb{R}^{2d}$ for each index $j \in \{1, 2, 3\}$. Moreover, $(f_{1,x_1}, f_{2,x_2}, f^{\dagger}_{3,x_3})$ must be **p**-compatible. Expressing

$$f_{j,y}(s) = c_j(y)e^{-\gamma_j(y)(s-a_j(y))^2+ib_j(y)s},$$

compatibility forces the functional equation

$$a_1(y_1) + a_2(y_2) + a_3(-y_1 - y_2) + \sigma(y_1, y_2) = 0 \tag{43}$$

for almost every $(y_1, y_2) \in \mathbb{R}^{2d} \times \mathbb{R}^{2d}$.

Lemma 11 *There exists no ordered triple of measurable functions $a_j : \mathbb{R}^{2d} \to \mathbb{C}$ that satisfies the functional equation (43) for almost every $(y_1, y_2) \in (\mathbb{R}^{2d})^2$.*

Proof of Lemma 11 Write (43) with the roles of y_1, y_2 interchanged, and add the result to (43). Since σ is antisymmetric, its contributions cancel, leaving

$$a(x_1) + a(x_2) + a_3(-x_1 - x_2) = 0$$

for almost every $(x_1, x_2) \in (\mathbb{R}^{2d})^3$, where $a = \frac{1}{2}a_1 + \frac{1}{2}a_2$. As is well known, any measurable solutions of this functional equation must agree almost everywhere with affine functions. Thus a_3 is affine.

Inserting this conclusion into (43), we find that there exist functions $\tilde{a}_j$, which differ from a_j by affine functions, such that $\tilde{a}_1(x_1) + \tilde{a}_2(x_2) + \sigma(x_1, x_2) = 0$ almost everywhere. By freezing almost any value of x_2 one finds that $\tilde{a}_1$ agrees almost everywhere with an affine function. The same reasoning applies to $\tilde{a}_2$. But the original equation (43) cannot hold with all three functions a_j affine, since σ is not an affine function of (y_1, y_2). □

This paper establishes a more quantitative form of Lemma 11, and reduces Theorem 7 to this result by elaborating on the reasoning shown above. Klein and Russo [11] have shown how the same type of reasoning as that shown above can be applied to certain semidirect product Lie groups. Much of the quantitative analysis below extends straightforwardly to more general semidirect products. However, each semidirect product leads to its own analogue of the variant (43) of the classical functional equation (25). In this paper we analyze only one such variant, leaving a general investigation for future work.

Remark 12 There is no solution (a_1, a_2, a_3) of (43) in the sense of distributions.

This remark does not subsume Lemma 11, since the lack of any assumption in that lemma that the functions a_j are locally integrable prevents their being interpreted as distributions.

Proof of Remark 12 Write $y_j = (y_{j,k})_{1\le k\le 2d}$. Applying $\frac{\partial^2}{\partial y_{1,m}\partial y_{1,n}}$ gives

$$\frac{\partial^2 a_1}{\partial y_{1,m}\partial y_{1,n}}(y_1) + \frac{\partial^2 a_3}{\partial y_{1,m}\partial y_{1,n}}(y_1+y_2) \equiv 0,$$

whence $\frac{\partial^2 a_3}{\partial y_{1,m}\partial y_{1,n}}(y_1+y_2)$ is independent of y_2 as a distribution. Therefore a_3, and hence a_1, are quadratic polynomials. The same applies to a_2.

Now consider any $k \in \{1,2,\dots,d\}$ and apply $\frac{\partial^2}{\partial y_{1,k}\partial y_{2,k+d}} + \frac{\partial^2}{\partial y_{2,k}\partial y_{1,k+d}}$ to both sides of (43). This differential monomial annihilates $\sigma(y_1, y_2)$. It results that $\frac{\partial^2}{\partial y_k \partial y_{k+d}} a_3 \equiv 0$. By applying $\frac{\partial^2}{\partial y_{1,m}\partial y_{2,n}}$ for other pairs m, n one obtains $\frac{\partial^2}{\partial y_m \partial y_n} a_3 \equiv 0$ for all m, n. Thus a_3 is an affine function.

Once this is known, apply to $\frac{\partial^2}{\partial y_{1,m}\partial y_{1,n}}$ to conclude that a_1 is affine. In the same way, a_2 is affine. (43) now expresses $\sigma(y_1, y_2)$ as a sum of three affine functions, contradicting the definition of σ. □

6 Sufficiency

Proposition 13 *Let* $d \ge 1$, *and let* $\mathbf{p}$ *be admissible. For each* $\varepsilon > 0$ *there exists* $\eta(\varepsilon) > 0$ *satisfying* $\lim_{\varepsilon\to 0}\eta(\varepsilon) = 0$ *with the following property. For any* $\mathbf{p}$*-compatible* ε*-diffuse ordered triple* $\mathbf{G} = (G_1, G_2, G_3)$ *of Gaussian functions* $G_j : \mathbb{H}^d \to \mathbb{C}\setminus\{0\}$,

$$|\mathcal{T}_{\mathbb{H}^d}(\mathbf{G})| \ge (1-\eta(\varepsilon))\mathbf{A}_{\mathbf{p}}^{2d+1}\prod_{j=1}^{3}\|G_j\|_{p_j}.$$

More generally, it follows immediately from the triangle inequality that if $\mathbf{G}$ is $\mathbf{p}$-compatible and ε-diffuse, and if $\|f_j - G_j\|_{p_j} < \varepsilon\|f_j\|_{p_j}$ for all $j \in \{1,2,3\}$ then

$$|\mathcal{T}_{\mathbb{H}^d}(\mathbf{f})| \ge (1-\eta(\varepsilon))\mathbf{A}_{\mathbf{p}}^{2d+1}\prod_{j=1}^{3}\|f_j\|_{p_j}$$

where the function η is modified but is still $o_\varepsilon(1)$.

The following notation will be used throughout the analysis, here and below.

Definition 14 For any invertible linear endomorphism L of $\mathbb{R}^{2d}$,

$$\sigma_L(x,y) = \sigma(L^{-1}x, L^{-1}y) \tag{44}$$

for $x, y \in \mathbb{R}^{2d}$.

Proof of Proposition 13 Since the action of $\mathfrak{G}(\mathbb{H}^d)$ preserves the ratio $|\mathcal{T}_{\mathbb{H}^d}(\mathbf{f})|/\prod_{j=1}^{3}\|f_j\|_{p_j}$, it suffices to prove this for **p**-compatible ordered triples of canonical ε-diffuse Gaussians. Thus we may assume that

$$G_j(x,t) = e^{-\gamma_j|Lx|^2}e^{-\gamma_j a t^2}e^{ibt}$$

where L is an invertible linear endomorphism of $\mathbb{R}^{2d}$, $a > 0$, $b \in \mathbb{R}$, and $\max(a^{1/2}, |b|)\|L^{-1}\|^2 \le \varepsilon$. In this situation,

$$\mathcal{T}_{\mathbb{H}^d}(\mathbf{G}) = \int_{\mathbb{R}^{2d}\times\mathbb{R}^{2d}} e^{-\gamma_1|Lx_1|^2-\gamma_2|Lx_2|^2-\gamma_3|L(x_1+x_2)|^2}$$

$$\cdot \int_{\mathbb{R}\times\mathbb{R}} e^{-\gamma_1 a t_1^2-\gamma_2 a t_2^2-\gamma_3 a(t_1+t_2+\sigma(x_1,x_2))^2}e^{i[bt_1+bt_2-b(t_1+t_2+\sigma(x_1,x_2))]}\,dt_1\,dt_2\,dx_1\,dx_2.$$

Cancelling where possible and substituting $Lx_j = y_j$ gives $\mathcal{T}_{\mathbb{H}^d}(\mathbf{G}) = |\det(L)|^{-2}\cdot I$ with

$$I = \int_{\mathbb{R}^{4d}} e^{-\gamma_1|y_1|^2-\gamma_2|y_2|^2-\gamma_3|y_1+y_2|^2}e^{-ib\sigma_L(y_1,y_2)}$$

$$\cdot \int_{\mathbb{R}^2} e^{-\gamma_1 a t_1^2-\gamma_2 a t_2^2-\gamma_3 a(t_1+t_2+\sigma_L(y_1,y_2))^2}\,dt_1\,dt_2\,dy_1\,dy_2.$$

Define

$$J = \int_{\mathbb{R}^{4d}} e^{-\gamma_1|y_1|^2-\gamma_2|y_2|^2-\gamma_3|y_1+y_2|^2}\int_{\mathbb{R}^2} e^{-\gamma_1 a t_1^2-\gamma_2 a t_2^2-\gamma_3 a(t_1+t_2)^2}\,dt_1\,dt_2\,dy_1\,dy_2.$$

G is a maximizing ordered triple for Young's inequality with exponents **p** for $\mathbb{R}^{2d+1}$, with the same coordinates (x,t). Thus $J = |\det(L)|^2\mathbf{A}_{\mathbf{p}}^{2d+1}\prod_{j=1}^{3}\|G_j\|_{p_j}$. Thus it suffices to prove that

$$|I| \ge (1-o_\varepsilon(1))J.$$

An application of Young's inequality for $\mathbb{R}^1$ to the inner integral, followed by an application Young's inequality for $\mathbb{R}^{2d}$ to the remaining outer integral, also reveals that $|I| \le |\det(L)|^2\mathbf{A}_{\mathbf{p}}^{2d+1}\prod_{j=1}^{3}\|G_j\|_{p_j}$.

Let $\varepsilon \mapsto \rho(\varepsilon)$ be a function that tends to ∞ slowly as $\varepsilon \to 0$. The same reasoning shows that if the integrand in the integral defining I is replaced by its absolute value, then the contribution of the region $\mathcal{R} = \{(y_1,y_2) \in \mathbb{R}^{4d} : |(y_1,y_2)| > \rho(\varepsilon)\}$ to the integral is $o_\varepsilon(1)$. Since $|b|\|L^{-1}\|^2 \le \varepsilon$ by hypothesis,

$$|b\sigma_L(y_1,y_2)| \le |b|\|L^{-1}\|^2\rho(\varepsilon)^2 \le \varepsilon^{1/2} \text{ uniformly for all } (y_1,y_2) \in \mathbb{R}^{4d}\setminus\mathcal{R}$$

provided that $\rho(\varepsilon)$ is chosen to satisfy $\rho(\varepsilon) \le \varepsilon^{-1/4}$. Therefore $|e^{-ib\sigma_L(y_1,y_2)} - 1| = O(\varepsilon^{1/2})$ uniformly for all $y \in \mathbb{R}^{4d} \setminus \mathcal{R}$. Therefore

$$I = \int_{\mathbb{R}^{4d}} e^{-\gamma_1|y_1|^2-\gamma_2|y_2|^2-\gamma_3|y_1+y_2|^2} \cdot \int_{\mathbb{R}^2} e^{-\gamma_1 a t_1^2-\gamma_2 a t_2^2-\gamma_3 a(t_1+t_2+\sigma_L(y_1,y_2))^2}\, dt_1\, dt_2\, dy_1\, dy_2$$

plus $o_\varepsilon(1)$.

Define $\mathcal{R}' = \{(t_1, t_2) \in \mathbb{R}^2 : |(t_1, t_2)| > \rho(\varepsilon)\}$. By the same reasoning, to complete the proof it suffices to have

$$e^{-\gamma_3 a 2(t_1+t_2)\sigma_L(y_1,y_2)} e^{-\gamma_3 a \sigma_L(y_1,y_2)^2} = 1 + o_\varepsilon(1)$$

uniformly for all (y_1, t_1, y_2, t_2) such that $(t_1, t_2) \in \mathbb{R}^2 \setminus \mathcal{R}'$ and $(y_1, y_2) \in \mathbb{R}^{4d} \setminus \mathcal{R}$. This holds because

$$|a(t_1 + t_2)\sigma_L(y_1, y_2)| \le a\rho(\varepsilon)\|L^{-1}\|^2\rho(\varepsilon)^2$$
$$|a\sigma_L(y_1, y_2)^2| \le a\|L^{-1}\|^4\rho(\varepsilon)^4,$$

while it is given that $(a^{1/2} + a)\|L^{-1}\|^2 \le \varepsilon$. □

7 Two Ingredients

In order to prove Theorem 7, we will make the reasoning in Sect. 5 quantitative. The following result from [4], the analogue for $\mathbb{R}^m$ of our main result for $\mathbb{H}^d$, will be the first of two main ingredients in the analysis.

Theorem 15 *For each admissible* $\mathbf{p} \in (1, \infty)^3$ *and each* $m \in \mathbb{N}$ *there exists a function* $\delta \mapsto \varepsilon(\delta)$ *satisfying* $\lim_{\delta\to 0^+} \varepsilon(\delta) = 0$ *with the following property. If* $0 \ne f_j \in L^{p_j}(\mathbb{R}^m)$ *and if* $\mathbf{f} = (f_j)_{1\le j\le 3}$ *satisfies* $|\mathcal{T}_{\mathbb{R}^m}(\mathbf{f})| \ge (1 - \delta)\mathbf{A}_\mathbf{p}^m\|\mathbf{f}\|_\mathbf{p}$ *then there exists an ordered triple of Gaussian functions of the form*

$$G_j(x) = c_j e^{-\gamma_j|L(x)-a_j|^2+ix\cdot b} \tag{45}$$

where $0 \ne c_j \in \mathbb{C}$, $a_j, b \in \mathbb{R}^m$, $\sum_{j=1}^3 a_j = 0$, *and* $L : \mathbb{R}^m \to \mathbb{R}^m$ *is a linear automorphism, such that*

$$\|f_j - G_j\|_{p_j} \le \varepsilon(\delta)\|f_j\|_{p_j} \tag{46}$$

for each $j \in \{1, 2, 3\}$.

The second ingredient is a quantitative expression of the unsolvability of a functional equation. In the discussion that follows, $\mathbb{B}$ always denotes a ball of finite, positive radius centered at the origin in $\mathbb{R}^d$. $\mathbb{B}^*$ denotes the ball centered at 0 whose radius is twice that of $\mathbb{B}$. Sets of Lebesgue measure zero are negligible for all considerations that follow, so we do not distinguish between open and closed balls. The Cartesian product $\mathbb{B} \times \mathbb{B}$ is denoted by $\mathbb{B}^2$. The following two lemmas are established in [4].

Lemma 16 *For each $d \in \mathbb{N}$ there exist $\delta_0 > 0$ and a function $t \mapsto \varepsilon(t)$ satisfying $\lim_{t \to 0^+} \varepsilon(t) = 0$ such that the following conclusion holds. Let $A \in [0, \infty)$ and $\delta \in (0, \delta_0]$. Let $\varphi, \psi : \mathbb{B} \to \mathbb{C}$ and $\xi : \mathbb{B}^* \to \mathbb{C}$ be Lebesgue measurable. Suppose that*

$$|\varphi(x) + \psi(y) + \xi(x + y)| \le A$$

for all $(x, y) \in \mathbb{B}^2$ outside a set of measure $\le \delta|\mathbb{B}|^2$. Then there exists an affine function h such that

$$\big|\varphi(x) - h(x)\big| \le CA \tag{47}$$

for all $x \in \mathbb{B}$ outside a set of measure $C\delta|\mathbb{B}|$. The constant C and function ε depend only on d.

In particular, the constants in the conclusions do not depend on $\mathbb{B}$. The following multiplicative variant of Lemma 16 is also proved in [4].

Lemma 17 *For each dimension $d \ge 1$ there exists a constant $K < \infty$ with the following property. Let $B \subset \mathbb{R}^d$ be a ball with positive radius, and let $\eta \in (0, \frac{1}{2}]$. For $j \in \{1, 2, 3\}$ let $f_j : 2B \to \mathbb{C}$ be Lebesgue measurable functions that vanish only on sets of Lebesgue measure zero. Suppose that*

$$\Big|\Big\{(x, y) \in B^2 : |f_1(x) f_2(y) f_3(x + y)^{-1} - 1| > \eta\Big\}\Big| < \delta|B|^2. \tag{48}$$

Then for each index j there exists a function $L_j : \mathbb{R}^d \to \mathbb{C}$ that is affine over $\mathbb{R}$ and satisfies

$$\Big|\Big\{x \in B : |f_j(x)e^{-L_j(x)} - 1| > K\eta^{1/K}\Big\}\Big| \le K\delta|B|. \tag{49}$$

The next result is concerned with a Heisenberg variant of Lemma 16.

Proposition 18 *For each $d \in \mathbb{N}$ there exists $C < \infty$ with the following property. Let $\mathbb{B}$ be any ball of finite, positive radius centered at the origin in $\mathbb{R}^{2d}$. Let $A < \infty$ and $\eta > 0$. Let $a_j : \mathbb{B}^* \to \mathbb{R}$ be Lebesgue measurable. Let $L : \mathbb{R}^{2d} \to \mathbb{R}^{2d}$ be an invertible linear transformation. Suppose that*

$$|a_1(x) + a_2(y) + a_3(x + y) + \sigma_L(x, y)| \le A \tag{50}$$

for all $(x, y) \in \mathbb{B}^2$ *outside a Lebesgue measurable set of Lebesgue measure* $\le \eta|\mathbb{B}|^2$. *Then there exists* $S \in \mathrm{Sp(2d)}$ *such that*

$$\|SL^{-1}\| \le CA^{1/2}|\mathbb{B}|^{-1/2d}. \tag{51}$$

Moreover, there exist affine functions ψ_j *for* $j \in \{1, 2, 3\}$ *satisfying*

$$\psi_1(x_1) + \psi_2(x_2) + \psi_3(-x_1 - x_2) = 0 \text{ for all } (x_1, x_2) \in \mathbb{R}^{2d} \times \mathbb{R}^{2d}$$

such that

$$|a_j(x) - \psi_j(x)| \le CA \text{ for all } x \in \mathbb{B} \text{ outside a set of measure } o_\eta(1)|\mathbb{B}|. \tag{52}$$

Recall that $\sigma_L(x, y) = \sigma(L^{-1}x, L^{-1}y)$. By $\|T\|$ we mean in (51) the norm $\sup_{0\ne x\in\mathbb{R}^{2d}} |T(x)|/|x|$. The main conclusion is that (50) cannot hold, unless L satisfies $\inf_{S\in\mathrm{Sp(2d)}} \|SL^{-1}\| = O(|\mathbb{B}|^{-1/2d}A^{1/2})$. Moreover, if (50) does hold, then $|\sigma_L(x, y)| \le CA$ for all $(x, y) \in \mathbb{B}^2$; consequently this term can be dropped from (50) to yield $|a_1(x) + a_2(y) + a_3(x + y)| \le CA$. The conclusion (52) follows from this by Lemma 16.

Proof of Proposition 18 The hypothesis states that

$$|a_1(x) + a_2(y) + a_3(x + y) + \sigma(Lx, Ly)| \le A$$

for all $(x, y) \in \mathbb{B}^2$ outside a set of measure $\le \eta|\mathbb{B}|^2$. By interchanging the roles of x, y, adding the resulting inequality to this one, and invoking the antisymmetry of σ, we conclude that

$$\big|\tilde{a}(x) + \tilde{a}(y) + a_3(x + y)\big| \le A$$

for all $(x, y) \in \mathbb{B}$ outside a set of measure $\le 2\eta|\mathbb{B}|^2$, where $2\tilde{a} = a_1 + a_2$. By Lemma 16 this implies that there exists an affine function ψ_3 such that $|a_3(x) - \psi_3(x)| \le CA$ for all $x \in \mathbb{B}$ outside a set of measure $\le C\eta|\mathbb{B}|$.

$\psi_3(x + y)$ can be expressed as an affine function of x plus an affine function of y; these functions can be incorporated into $a_1(x)$, $a_2(y)$, respectively. Combining this information with the hypotheses therefore gives

$$\big|a_1^\sharp(x) + a_2^\sharp(y) + \sigma(L^{-1}x, L^{-1}y)\big| \le CA \tag{53}$$

for nearly all $(x, y) \in \mathbb{B} \times \mathbb{B}$, where $a_j^\sharp - a_j$ is affine. Taking first differences with respect to x gives

$$\big|\Delta_h a_1^\sharp(x) + \sigma(L^{-1}h, L^{-1}y)\big| \le CA \tag{54}$$

for nearly all $x, h, y \in \mathbb{B}$ such that $x, h, x+h, y \in \mathbb{B}$. By specializing to a typical value of y, one finds that there exists a function $h \mapsto c(h)$ such that $|\Delta_h a_1^\sharp(x) - c(h)| \le CA$ for nearly all $x, h \in \mathbb{B}$ such that $x+h \in \mathbb{B}$. Therefore by Lemma 16 there exists an affine function ψ such that $|a_1^\sharp - \psi| \le CA$ for nearly all points of $\mathbb{B}$. Since $a_1 - a_1^\sharp$ is affine, the same conclusion holds for a_1. Interchanging the roles of the variables x, y in this argument produces the same conclusion for a_2.

Combining these results for all a_j with the original hypothesis, we conclude that there exists an affine function ψ of (x, y) such that $|\psi(x, y) - \sigma(L^{-1}x, L^{-1}y)| \le CA$ for nearly every $(x, y) \in \mathbb{B}^2$. The same must then hold for every $(x, y) \in \mathbb{B}^* \times \mathbb{B}^*$, since ψ, σ_L are polynomials whose degrees do not exceed 2. By applying $\partial^2/\partial x_i \partial y_j$ for arbitrary indices i, j and exploiting the affine character of ψ together with the homogeneous quadratic nature of $\sigma(L^{-1}x, L^{-1}y)$ we conclude that $|\sigma(L^{-1}x, L^{-1}y)| \le CA$ for all $(x, y) \in \mathbb{B}^2$. According to Lemma 21, this implies the existence of $S \in \mathrm{Sp(2d)}$ such that $\|SL^{-1}\| \le CA^{1/2}|\mathbb{B}|^{-1/2d}$. □

8 Proof of Theorem 7 for Nonnegative Functions

Let $\mathbf{p}$ be an admissible ordered triple of exponents in $(1, \infty)^3$, and let $\delta > 0$ be small. Let $f_j \in L^{p_j}(\mathbb{H}^d)$ for $j \in \{1, 2, 3\}$ satisfy $\|f_j\|_{p_j} = 1$, as we may suppose without loss of generality. Set $\mathbf{f} = (f_1, f_2, f_3)$. Assume that each $f_j \ge 0$, and suppose that

$$\mathcal{T}_{\mathbb{H}^d}(\mathbf{f}) \ge (1-\delta)\mathbf{A}_{\mathbf{p}}^{2d+1}\|\mathbf{f}\|_{\mathbf{p}} = (1-\delta)\mathbf{A}_{\mathbf{p}}^{2d+1}.$$

Define $F_j : \mathbb{R}^{2d} \to [0, \infty]$ and $f_{j,x} : \mathbb{R}^1 \to [0, \infty]$ as in (40). Set $\mathbf{F} = (F_1, F_2, F_3)$. For $\mathbf{x} = (x_1, x_2, x_3) \in (\mathbb{R}^{2d})^3$ define

$$f_{3,\mathbf{x}}^\dagger(s) = f_3(x_3, s + \sigma(x_1, x_2)); \tag{55}$$

as in Sect. 5, this definition will only be relevant for $\mathbf{x} \in \Lambda_{\mathbb{R}^{2d}}$, that is, when $x_3 = -x_1 - x_2$. Define a measure $\nu_{\mathbf{F}}$ on $(\mathbb{R}^{2d})^3$, supported on $\Lambda_{\mathbb{R}^{2d}}$, by

$$d\nu_{\mathbf{F}}(\mathbf{x}) = \prod_{j=1}^{3} F_j(x_j)\, d\lambda_{\mathbb{R}^{2d}}(\mathbf{x}), \tag{56}$$

where $\lambda_{\mathbb{R}^{2d}}$ is the natural $4d$-dimensional Lebesgue measure on $\Lambda_{\mathbb{R}^{2d}}$ introduced above. Since $\|F_j\|_{p_j} = \|f_j\|_{p_j} = 1$ and $\mathbf{p}$ is admissible, Young's inequality for $\mathbb{R}^{2d}$ guarantees that $\nu_{\mathbf{F}}(\mathbb{R}^{2d} \times \mathbb{R}^{2d} \times \mathbb{R}^{2d}) \le \mathbf{A}_{\mathbf{p}}^{2d}$.

Lemma 19 *For each $d \geq 1$ and each admissible ordered triple* $\mathbf{p}$ *there exists* $C < \infty$ *with the following property. Let* $f_j \in L^{p_j}(\mathbb{H}^d)$ *be nonnegative and satisfy* $\|f_j\|_{p_j} = 1$ *for each* $j \in \{1, 2, 3\}$. *Let* $\delta > 0$. *If* $\mathcal{T}_{\mathbb{H}^d}(\mathbf{f}) \geq (1-\delta)\mathbf{A}_\mathbf{p}^{2d+1}$ *then*

$$\mathcal{T}_{\mathbb{R}^{2d}}(\mathbf{F}) \geq (1-\delta)\mathbf{A}_\mathbf{p}^{2d} \tag{57}$$

and there exists a set $E \subset \Lambda_{\mathbb{R}^{2d}}$ *satisfying*

$$\nu_\mathbf{F}(E) \leq C\delta^{1/2} \tag{58}$$

such that for every $\mathbf{x} \in \Lambda_{\mathbb{R}^{2d}} \setminus E$,

$$\begin{cases} F_j(x_j) \neq 0 \text{ for each } j \in \{1,2,3\}, \\ \mathcal{T}_{\mathbb{R}^1}\big(f_{1,x_1}, f_{2,x_2}, f_{3,\mathbf{x}}^\dagger\big) \geq (1 - o_\delta(1))\mathbf{A}_\mathbf{p}. \end{cases} \tag{59}$$

A proof of Lemma 19 is implicit in the proof in Sect. 5 that the optimal constant in Young's inequality for $\mathbb{H}^d$ does not exceed the optimal constant for $\mathbb{R}^{2d+1}$. □

According to Theorem 15, there exists an ordered triple $\mathbf{G} = (G_1, G_2, G_3)$ of Gaussians $G_j : \mathbb{R}^{2d} \to [0, \infty$ that saturates Young's convolution inequality for $\mathbb{R}^{2d}$, of the form

$$G_j(x) = c_j |\det(L)|^{1/p_j} e^{-\gamma_j |L(x-a_j)|^2},$$

where $\gamma = \gamma(\mathbf{p})$, $a_1 + a_2 + a_3 = 0$, $c_j > 0$, and L is an invertible linear endomorphism of $\mathbb{R}^{2d}$, such that $\|F_j - G_j\|_{L^{p_j}(\mathbb{R}^{2d})} = o_\delta(1)$. The constants c_j are determined by requiring that $\|G_j\|_{p_j} = 1$, as we may require with no loss of generality since $\|F_j\|_{p_j} = 1$. Exponential factors $e^{ix\cdot b_j}$ appear in the conclusion of Theorem 15 but can dropped here; since $F_j \geq 0$ by its definition, $|G_j|$ is at least as accurate an approximation to F_j in L^{p_j} norm as is G_j.

Define an ordered triple of diffeomorphisms ψ_j of $\mathbb{H}^d$ by

$$(\psi_1(z_1), \psi_2(z_2), \psi_3(z_3)) = (z_1 u, u^{-1} z_2 v, v^{-1} z_3)$$

where $u = (-a_1, 0)$ and $v = (-a_1 - a_2, 0)$. Then $v^{-1} = (a_1 + a_2, 0) = (-a_3, 0)$. The triple $\Psi = (\psi_j^*)_{1 \leq j \leq 3}$ is an element of $\mathfrak{G}(\mathbb{H}^d)$, so upon replacement of f_j by $f_j \circ \psi_j$, all of the assumptions and conclusions above are unaffected, and we gain the simplification

$$G_j(x) = c_j |\det(L)|^{1/p_j} e^{-\gamma_j |Lx|^2},$$

with our standing notation $\gamma_j = p_j'$.

Lemma 20 *Let* $\mathbf{f}, L, G_j$ *be as above. There exist* $\lambda \in \mathbb{R}^+$, $S \in \mathrm{Sp(2d)}$, *positive scalars* $\tilde{c}_j$, *a set* $E' \subset \Lambda_{\mathbb{R}^{2d}}$, *affine mappings* $\varphi_j : \mathbb{R}^{2d} \to \mathbb{R}^1$, *and Lebesgue measurable functions* $h_j : \mathbb{R}^{2d} \to [0, \infty)$ *of the form*

$$h_j(x, t) = \tilde{c}_j e^{-\lambda \gamma_j (t - \varphi_j(x))^2} \tag{60}$$

such that $h_{j,x}(t) = h_j(x, t)$ *satisfy the following conclusions:*

$$\|h_{j,x}\|_{L^{p_j}(\mathbb{R})} = 1 \text{ for every } x \in \mathbb{R}^{2d} \tag{61}$$

$$\nu_{\mathbf{F}}(E') \le o_\delta(1) \tag{62}$$

$$\|f_{j,x_j} - h_{j,x_j}\|_{p_j} \le o_\delta(1) \text{ for each } j \in \{1, 2, 3\} \text{ for all } \mathbf{x} \in \Lambda_{\mathbb{R}^{2d}} \setminus E', \tag{63}$$

$$\|SL^{-1}\| \le o_\delta(1)\lambda^{-1/4}, \tag{64}$$

$$\varphi_1(x_1) + \varphi_2(x_2) + \varphi_3(x_3) = 0 \text{ for all } \mathbf{x} \in \Lambda_{\mathbb{R}^{2d}}. \tag{65}$$

Here F_j is associated to f_j as indicated above, and $(\gamma_1, \gamma_2, \gamma_3) = \gamma(\mathbf{p})$.

Proof Temporarily make the change of variables $(x, s) \mapsto (y, t)$ in $\mathbb{H}^d$, with

$$y = L(x) \text{ and } t = s. \tag{66}$$

We make this same change of variables for each index $j \in \{1, 2, 3\}$. This diffeomorphism of $\mathbb{H}^d$ is an automorphism of its group structure only if $L \in \mathrm{Sp(2d)}$, which need not hold. Therefore we will revert to the original coordinates after exploiting these new coordinates.

Set

$$\tilde{f}_j(y_j, t) = f_j(L^{-1}y_j, t), \tag{67}$$

and $\tilde{f}_{j,y_j}(t) = \tilde{f}_j(y_j, t)$. In these modified coordinates and for these modified functions, the conclusions of Lemma 19, coupled with the approximations $\|F_j - G_j\|_{p_j} = o_\delta(1)$, can be stated as follows. Set

$$\tilde{G}_j(y) = c_j e^{-\gamma_j |y|^2}. \tag{68}$$

$$d\nu_{\tilde{\mathbf{G}}}(\mathbf{y}) = \prod_{j=1}^{3} \tilde{G}_j(y_j)\, d\lambda_{\mathbb{R}^{2d}}(\mathbf{y}) \tag{69}$$

$$\tilde{f}^{\dagger}_{3,\mathbf{y}}(s) = \tilde{f}_{3,y_3}(s + \sigma_L(y_1, y_2)). \tag{70}$$

Recall the notation $\sigma_L(y_1, y_2) = \sigma(L^{-1}y_1, L^{-1}y_2)$. By Lemma 19, since $\sum_{j=1}^3 p_j^{-1} = 2$, there is a set $E \subset \Lambda_{\mathbb{R}^{2d}}$ satisfying $\nu_{\tilde{\mathbf{G}}}(E) = o_\delta(1)$ such that

$$\mathcal{T}_{\mathbb{R}^1}(\tilde{f}_{1,y_1}, \tilde{f}_{2,y_2}, \tilde{f}_{3,\mathbf{y}}^\dagger) \geq (1 - o_\delta(1))\mathbf{A_p} \quad \text{for every } \mathbf{y} = (y_1, y_2, y_3) \in \Lambda_{\mathbb{R}^{2d}} \setminus E. \tag{71}$$

Moreover, $\|\tilde{f}_{j,y_j}\|_{p_j} = 1$ for each $j \in \{1, 2, 3\}$ whenever $\mathbf{y} \in \Lambda_{\mathbb{R}^{2d}} \setminus E$.

Let $\delta \mapsto \rho(\delta)$ be a function that tends to infinity slowly as $\delta \to 0^+$, at a rate satisfying constraints to be imposed below. This function also depends on d and on $\mathbf{p}$, but is independent of $\mathbf{f}$. Define $\mathbb{B}$ to be the closed ball of radius $\rho(\delta)$ centered at the origin in $\mathbb{R}^{2d}$, and $\mathbb{B}^*$ to be the concentric ball of radius $2\rho(\delta)$.

The L^{p_j} norm of $\tilde{G}_j$ on the complement of $\mathbb{B}$ is $o_\delta(1)$ since $\lim_{\delta\to 0}\rho(\delta) = \infty$. $\tilde{G}_j$ is bounded above uniformly in δ, and is bounded below by $ce^{-C\rho(\delta)^2}$ on $\mathbb{B}^*$. Thus by (71), under the convention that $\mathbf{y} = (y_1, y_2, y_3)$ is regarded as a function $\mathbf{y}(y_1, y_2)$ of (y_1, y_2) via the relation $y_3 = -y_1 - y_2$,

$$\mathcal{T}_{\mathbb{R}^1}(\tilde{f}_{1,y_1}, \tilde{f}_{2,y_2}, \tilde{f}_{3,\mathbf{y}}^\dagger) \geq (1 - o_\delta(1))\mathbf{A_p} \tag{72}$$

for all $(y_1, y_2) \in \mathbb{B} \times \mathbb{B}$ outside a set of Lebesgue measure $\leq \nu_{\tilde{\mathbf{G}}}(E)c^{-1}e^{C\rho(\delta)^2}$.

Choose a function $\delta \mapsto \rho_0(\delta)$ satisfying $\lim_{\delta\to 0}\rho_0(\delta) = 0$, but tending to infinity so slowly that $\nu_{\tilde{\mathbf{G}}}(E)c^{-1}e^{C\rho_0(\delta)^2} \leq o_\delta(1)$. This is possible because $\nu_{\tilde{\mathbf{G}}}(E) = o_\delta(1)$ tends to zero at a rate that depends only on $\mathbf{p}, d$. We require henceforth that $\rho(\delta) \leq \rho_0(\delta)$, but will impose further restrictions on the rate of growth of ρ below. Therefore, (72) holds for all $(y_1, y_2) \in \mathbb{B}^2$ in the complement of a set of Lebesgue measure $\leq o_\delta(1)$.

By (72) and Theorem 15, for each $j \in \{1, 2, 3\}$, for all $y_j \in \mathbb{B}$ outside a set whose Lebesgue measure is $o_\delta(1)$, there exists a positive Gaussian function $\mathbb{R}^1 \ni t \mapsto g_{j,y_j}(t)$ satisfying $\|\tilde{f}_{j,y_j} - g_{j,y_j}\|_{p_j} \leq o_\delta(1)$. These functions can be chosen to depend Lebesgue measurably on the parameters y_j.

Write $g_{j,y}(t) = c_j(y)e^{-\lambda_j(y)(t-\alpha_j(y))^2}$ where λ_j, c_j, α_j are measurable functions with domains $\mathbb{R}^{2d}$; λ_j, c_j take values in $(0, \infty)$ and α_j takes values in $\mathbb{R}^1$. For all $y_j \in \mathbb{B}$ outside a set of Lebesgue measure $\leq o_\delta(1)$, $\|\tilde{f}_{j,y_j}\|_{p_j} = 1$. Therefore defining $g_{3,\mathbf{y}}^\dagger(s) = g_{3,y_3}(s + \sigma_L(y_1, y_2))$, we find that $(g_{1,y_1}, g_{2,y_2}, g_{3,-y_1-y_2}^\dagger)$ also nearly saturates Young's inequality with exponents $\mathbf{p}$ for $\mathbb{R}^1$, for all $(y_1, y_2) \in \mathbb{B}^2$ outside a set of Lebesgue measure $\leq o_\delta(1)$.

A first consequence of this near saturation is that

$$\left|\frac{\lambda_i(y_i)}{\lambda_j(y_j)} - \frac{\gamma_i}{\gamma_j}\right| = o_\delta(1) \tag{73}$$

for all $(y_1, y_2, y_3) \in \mathbb{B}^3$ outside a set of Lebesgue measure $o_\delta(1)$ for all indices $i, j \in \{1, 2, 3\}$, where y_3 continues to be defined to be $-y_1 - y_2$. This is a consequence of the characterization of near maximizers for Young's inequality for $\mathbb{R}^1$ [4]; near maximizing triples are close in norm to exactly maximizing triples, which were shown by Brascamp and Lieb to be unique up to compatible translations, common dilations, and multiplication by scalars.

Therefore there exists $\lambda \in \mathbb{R}^+$ such that

$$\lambda_j(y) = \lambda \cdot (\gamma_j + o_\delta(1)) \text{ for each index } j \in \{1, 2, 3\}, \tag{74}$$

for all $y \in \mathbb{B}$ outside a set of Lebesgue measure $o_\delta(1)$. Thus for each $j \in \{1, 2, 3\}$,

$$\left| g_{j,y}(t) - c'_j e^{-\lambda\gamma_j(t-\alpha_j(y))^2} \right| \le o_\delta(1) \tag{75}$$

in $L^{p_j}(\mathbb{R}^1)$ norm, for every $y \in \mathbb{B}$ outside a set of Lebesgue measure $o_\delta(1)$. The coefficients c'_j are now constants, rather than functions of $y \in \mathbb{R}^{2d}$.

In order for $(g_{1,y_1}, g_{2,y_2}, g^\dagger_{3,\mathbf{y}})$, with g_{j,y_j} of the form (75) and $y_3 = y_3(y_1, y_2) = -y_1 - y_2$, to $(1 - o_\delta(1))$-nearly saturate Young's inequality for $\mathbb{R}^1$ for every $(y_1, y_2) \in \mathbb{B}^2$ outside a set of Lebesgue measure $o_\delta(1)$, it is necessary that

$$\alpha_1(y_1) + \alpha_2(y_2) + \alpha_3(-y_1 - y_2) + \sigma_L(y_1, y_2) \le \lambda^{-1/2} \cdot o_\delta(1) \tag{76}$$

for all $(y_1, y_2) \in \mathbb{B}^2$ outside a set of Lebesgue measure $o_\delta(1)$. By Proposition 18, this implies the existence of affine functions $\varphi_j : \mathbb{R}^{2d} \to \mathbb{R}$ satisfying

$$\varphi_1(x_1) + \varphi_2(x_2) + \varphi_3(-x_1 - x_2) \equiv 0$$

that well approximate α_j for each $j \in \{1, 2, 3\}$ in the sense that

$$|\alpha_j(y) - \varphi_j(y)| \le o_\delta(1) \cdot \lambda^{-1/2} \ \forall\, y \in \mathbb{B} \setminus E'', \tag{77}$$

where $|E''| \le o_\delta(1)|\mathbb{B}| = o_\delta(1)\rho(\delta)^{2d}$.

The factor denoted by $o_\delta(1)$ depends on the choice of auxiliary function ρ_0, hence on d, $\mathbf{p}$, but does not depend on ρ. Therefore we may choose a function $\rho \le \rho_0$ that satisfies both $\lim_{\delta\to 0^+} \rho(\delta) = \infty$ and

$$|E''| \le o_\delta(1), \tag{78}$$

with this quantity $o_\delta(1)$ depending only on d, $\mathbf{p}$.

Moreover, by Proposition 18 there exists $S \in \mathrm{Sp(2d)}$ such that

$$\|SL^{-1}\| \le \lambda^{-1/4} o_\delta(1)|\mathbb{B}|^{-1/2d}. \tag{79}$$

Equivalently, $L = \tilde{L} \circ S$ where $\tilde{L} : \mathbb{R}^{2d} \to \mathbb{R}^{2d}$ is a linear endomorphism satisfying

$$|\tilde{L}(v)| \ge \lambda^{1/4}|\mathbb{B}|^{1/2d}\eta(\delta)^{-1}|v| \ \forall\, v \in \mathbb{R}^{2d}, \tag{80}$$

where $\eta(\delta) \to 0$ as $\delta \to 0$. These properties of L will be exploited below.

Define Gaussian functions

$$\tilde{g}_{j,y}(t) = c'_j e^{-\lambda\gamma_j(t-\varphi_j(y))^2}. \tag{81}$$

(77) and (78) together imply that $\|\tilde{f}_{j,y_j} - \tilde{g}_{j,y_j}\|_{p_j} \le o_\delta(1) = o_\delta(1)\|\tilde{f}_{j,y_j}\|_{p_j}$ for all $\mathbf{y} \in \Lambda_{\mathbb{R}^{2d}} \setminus E'$, where the exceptional set $E' \subset \Lambda_{\mathbb{R}^{2d}}$ consists of those those $\mathbf{y}$ is small such that (y_1, y_2) belongs to $E'' \cup ((\mathbb{R}^{2d})^2 \setminus \mathbb{B}^2$. This set E' is small in the sense that $\nu_{\mathbf{F}}(E') = o_\delta(1)$.

A consequence, since $\tilde{G}_j \in L^1$, is that

$$\|\tilde{f}_{j,y}(t)F_j(y) - \tilde{g}_{j,y}(t)\tilde{G}_j(y)\|_{L^{p_j}(\mathbb{B}\times\mathbb{R},\, dy\, dt)} \le o_\delta(1) \tag{82}$$

for each $j \in \{1, 2, 3\}$. Therefore

$$\|\tilde{f}_{j,y}(t)F_j(y) - \tilde{g}_{j,y}(t)\tilde{G}_j(y)\|_{L^{p_j}(\mathbb{R}^{2d}\times\mathbb{R},\, dy\, dt)} \le o_\delta(1). \tag{83}$$

Returning to the original coordinates (x, t) for $\mathbb{H}^d$, define

$$\tilde{h}_j(x, t) = \tilde{g}_{j,y}(t) = \tilde{g}_{j,L(x)}(t) = c'_j e^{-\lambda\gamma_j(t-\varphi_j\circ L(x))^2}. \tag{84}$$

The next step is to simplify matters by exploiting symmetries. We apply in sequence two elements $\Psi \in \mathfrak{G}(\mathbb{H}^d)$. The first is $\Psi = (\psi_1^*, \psi_2^*, \psi_3^*)$, where ψ_j^* is the invertible linear endomorphism of $L^{p_j}(\mathbb{H}^d)$ associated to the diffeomorphism $\psi_j(x_j, t_j) = t_j - (\varphi_j \circ L)(x_j)$ of $\mathbb{H}^d$. The second takes the form $\psi_j(x, t) = (S(x), t)$, where $S \in \mathrm{Sp(2d)}$ is as in (79). Replace f_j by $f_j \circ \psi_j$ for each of these in turn, continuing to denote by f_j the resulting functions and by F_j the associated functions with domains $\mathbb{R}^{2d}$. Likewise compose $\tilde{h}_j$ with each of these in turn, and denote by $h_j^\sharp$ the resulting composed functions. Matters are thereby reduced to the situation in which

$$h^\sharp_{j,x}(t) = c_j e^{-\lambda\gamma_j t^2},$$

$$F_j(x) = c_j e^{-\gamma_j|\tilde{L}(x)|^2},$$

$$\|f_{j,x_j} - h^\sharp_{j,x_j}\|_{p_j} \le o_\delta(1)\ \forall \mathbf{x} \in \Lambda_{\mathbb{R}^{2d}} \setminus E''$$

where $E'' \subset \Lambda_{\mathbb{R}^{2d}}$ satisfies $\nu_{\mathbf{F}}(E'') \le o_\delta(1)$ and $\tilde{L}$, λ are related by (80).

The next reduction is an automorphic change of variables in $\mathbb{H}^d$ of the form

$$(x, t) \mapsto \psi(x, t) = (z, r) = (\eta(\delta)^{-1}\lambda^{1/4}x, \eta(\delta)^{-2}\lambda^{1/2}t),$$

where $\eta(\delta)$ is the function introduced in (80). Setting $\psi_j = \psi$ for all three indices j defines an element $\Psi \in \mathfrak{G}(\mathbb{H}^d)$. In these new coordinates, after multiplying by scalars to renormalize, the conclusion is that $\|f_j - f_j^*\|_{p_j} \le o_\delta(1)$ where

$$f_j^*(z,r) = c_j e^{-\gamma_j|L'z|^2} e^{-\gamma_j \varepsilon r^2},$$

where $L' : \mathbb{R}^{2d} \to \mathbb{R}^{2d}$ is linear and satisfies $|L'z| \ge |z|$ for all $z \in \mathbb{R}^{2d}$, and $\varepsilon \le \varepsilon(\delta)$ where $\varepsilon(\delta)$ tends to 0 as $\delta \to 0$, and depends also on $\mathbf{p}$, d as well as on δ, but not otherwise on $\mathbf{f}$. This completes the analysis of nonnegative near-maximizers $\mathbf{f}$.

□

9 The Complex-Valued Case

Let $\delta > 0$ be small, and consider an arbitrary complex-valued $\mathbf{f} = (f_1, f_2, f_3)$ satisfying $\|f_j\|_{p_j} \ne 0$ for each index j, and $|\mathcal{T}_{\mathbb{H}^d}(\mathbf{f})| \ge (1-\delta)\mathbf{A}_{\mathbf{p}}^{2d+1}\|\mathbf{f}\|_{\mathbf{p}}$. Since $\mathcal{T}_{\mathbb{H}^d}(|f_1|, |f_2|, |f_3|) \ge |\mathcal{T}_{\mathbb{H}^d}(\mathbf{f})|$, we may apply the result proved above for nonnegative near-maximizers to conclude that there exists $\Psi = (\psi_1, \psi_2, \psi_3) \in \mathfrak{G}(\mathbb{H}^d)$ such that for each $j \in \{1, 2, 3\}$,

$$\begin{cases} \||f_j \circ \psi_j| - G_j\|_{p_j} \le o_\delta(1)\|f_j \circ \psi_j\|_{p_j} \\ G_j(x,t) = c_j e^{-\gamma_j|Lx|^2} e^{-\gamma_j \varepsilon t^2} \end{cases}$$

where $c_j \asymp 1$, $|Lx| \ge |x|$ for all $x \in \mathbb{R}^{2d}$, and $\varepsilon \le o_\delta(1)$. By replacing f_j by $f_j \circ \psi_j$ multiplied by an appropriate normalizing constant factor, we may also assume that $\|f_j\|_{p_j} = 1$ and that each ψ_j is the identity transformation on L^{p_j}, and then likewise that $\|G_j\|_{p_j} = 1$.

Write $f_j = e^{i\alpha_j}|f_j|$ where $\alpha_j : \mathbb{H}^d \to \mathbb{R}$ is measurable. We seek to analyze the factors $e^{i\alpha_j}$. Since $\|f_j - e^{i\alpha_j}G_j\|_{p_j} = \||f_j| - G_j\|_{p_j} \le o_\delta(1)$,

$$|\mathcal{T}_{\mathbb{H}^d}(e^{i\alpha_1}G_1, e^{i\alpha_2}G_2, e^{i\alpha_3}G_3)| \ge (1 - o_\delta(1))\mathbf{A}_p^{2d+1}.$$

Thus it suffices to prove that $(e^{i\alpha_j}G_j : 1 \le j \le 3)$ satisfies the conclusions of Theorem 7. So we redefine f_j to be $e^{i\alpha_j}G_j$ henceforth.

By multiplying these functions by unimodular constants, we may assume without loss of generality that $\mathcal{T}_{\mathbb{H}^d}(\mathbf{f})$ is real and positive. Since then

$$\operatorname{Re}\mathcal{T}_{\mathbb{H}^d}(f_1, f_2, f_3) \ge (1 - o_\delta(1))\mathcal{T}_{\mathbb{H}^d}(|f_1|, |f_2|, |f_3|),$$

we conclude that

$$|\prod_{j=1}^{3} e^{i\alpha_j(z_j)} - 1| = o_\delta(1) \text{ for all } \mathbf{z} \in (\mathbb{H}^d)^3 \text{ outside a set satisfying } \nu_{\mathbf{G}}(E) \le o_\delta(1) \tag{85}$$

where $d\nu_{\mathbf{G}}(\mathbf{z}) = \prod_j G_j(z_j)\, d\lambda_{\mathbb{H}^d}(\mathbf{z})$.

Let $\rho = \rho(\delta)$ be a positive quantity that tends to infinity slowly as $\delta \to 0$ and is to be chosen below, and let $\mathbb{B} \subset \mathbb{R}^{2d}$ be the ball of radius 1 centered at 0. By (85),

$$\left| e^{i\alpha_1(L^{-1}y_1,t_1)} e^{i\alpha_2(L^{-1}y_2,t_2)} e^{i\alpha_3(-L^{-1}y_1-L^{-1}y_2,-t_1-t_2-\sigma_L(y_1,y_2))} - 1 \right| \le o_\delta(1) \tag{86}$$

for all $((y_1, t_1), (y_2, t_2)) \in (\mathbb{B} \times [-\rho\varepsilon^{-1/2}, \rho\varepsilon^{-1/2}])^2$ outside a set of Lebesgue measure less than or equal to $o_\delta(1) \cdot \varepsilon^{-1}$ provided that the function ρ is chosen so that $\rho(\delta) \to \infty$ sufficiently slowly as $\delta \to 0$. Therefore according to Lemma 17, for each index j, α_j takes the form

$$e^{i\alpha_j(L^{-1}y,t)} = e^{i(a_j(y)t+b_j(y)+o_\delta(1))} \tag{87}$$

for $y \in \mathbb{B}$ and $|t| \le \rho(\delta)\varepsilon^{-1/2}$ outside a set of Lebesgue measure $o_\delta(1)\varepsilon^{-1/2}$. The coefficients a_j, b_j are real-valued measurable functions.

Invoking (87) together with (86) for typical (t_1, t_2) and also for typical (t_1', t_2') satisfying $|t_j|, |t_j'| \le \rho(\delta)\varepsilon^{-1/2}$, considering products of the exponential factors, and setting $u_j = t_j' - t_j$ gives

$$\left| e^{iu_1a_1(L^{-1}y_1)} e^{iu_2a_2(L^{-1}y_2)} e^{-i(u_1+u_2)a_3(-L^{-1}y_1-L^{-1}y_2)} - 1 \right| \le o_\delta(1) \tag{88}$$

for nearly all $(y_1, y_2) \in \mathbb{B}^2$ and nearly all (u_1, u_2) satisfying $|u_j| \le \frac{1}{2}\rho(\delta)\varepsilon^{-1/2}$ outside a set of Lebesgue measure $o_\delta(1)\varepsilon^{-1}$. The advantage of (88) over (86) is that b_j and σ_L have been eliminated.

This last inequality can be equivalently written

$$\left| e^{iu_1[a_1(L^{-1}y_1)-a_3(-L^{-1}(y_1-y_2))]} e^{iu_2[a_2(L^{-1}y_1)-a_3(-L^{-1}(y_1-y_2))]} - 1 \right| \le o_\delta(1) \tag{89}$$

for nearly all $(y_1, y_2) \in \mathbb{B}^2$. The net coefficient of u_1 in the exponent is the vector $a_1(L^{-1}(y_1) - a_3(-L^{-1}(y_1 - y_2))$. By applying Lemma 25, below, to appropriate two-dimensional slices of $\mathbb{B}^2$, we conclude from (89)

$$\left| a_1(L^{-1}(y_1) - a_3(-L^{-1}(y_1 - y_2)) \right| \le o_\delta(1)\varepsilon^{1/2} \tag{90}$$

for nearly all $(y_1, y_2) \in \mathbb{B}^2$. Note that unlike the functions α_j, which are only determined up to addition of arbitrary measurable functions taking values in $2\pi\mathbb{Z}$, this linear combination of the constituent parts a_j can be pinned down as an $\mathbb{R}$-valued function.

Therefore there exists a real number $\tilde{a}$ such that $|a_j(L^{-1}y) - \tilde{a}| \le o_\delta(1)\varepsilon^{1/2}$ for nearly all $y \in \mathbb{B}$ for $j = 1, 3$. The same reasoning gives the same conclusion for $j = 2$. Thus for each $j \in \{1, 2, 3\}$,

$$e^{i\alpha_j(L^{-1}y,t)} = e^{i\tilde{a}t}e^{ib_j(y)} + o_\delta(1) \tag{91}$$

for all $(y, t) \in \mathbb{B} \times [-\rho(\delta)\varepsilon^{-1/2}, \rho(\delta)\varepsilon^{-1/2}]$ outside a set of Lebesgue measure $o_\delta(1)\varepsilon^{-1/2}$. Thus

$$\|e^{i\alpha_j(x,t)}G_j(x,t) - e^{i(\tilde{a}t+b_j(L_j(x))}G_j(x,t)\|_{L^{p_j}(\mathbb{H}^d)} \le o_\delta(1)\|f_j\|_{p_j}; \tag{92}$$

so we may replace $\alpha_j(x,t)$ by $\tilde{a}t + b_j(L(x))$.

Inserting this into (86) gives

$$\left|e^{ib_1(L^{-1}y_1)}e^{ib_2(L^{-1}y_2)}e^{ib_3(-L^{-1}y_1-L^{-1}y_2)}e^{-i\tilde{a}\sigma_L(y_1,y_2)} - 1\right| \le o_\delta(1) \tag{93}$$

for nearly all $(y_1, y_2) \in \mathbb{B}^2$. From the antisymmetry of σ_L it follows that

$$\left|e^{i(b_1+b_2)(L^{-1}y_1)}e^{i(b_1+b_2)(L^{-1}y_2)}e^{i2b_3(-L^{-1}y_1-L^{-1}y_2)} - 1\right| \le o_\delta(1) \tag{94}$$

for nearly all $(y_1, y_2) \in \mathbb{B}^2$; this can be deduced by interchanging y_1 with y_2 and considering the product of the two resulting left-hand sides of (93).

According to Lemma 17, the functions $e^{i2b_3 \circ L^{-1}}$ and $e^{i(b_1+b_2)\circ L^{-1}}$ nearly agree with exponentials of imaginary affine functions, at nearly all points of $\mathbb{B}$. Since

$$\left|e^{i2b_1(L^{-1}y_1)}e^{i2b_2(L^{-1}y_2)}e^{i2b_3(-L^{-1}y_1-L^{-1}y_2)}e^{-i2\tilde{a}\sigma_L(y_1,y_2)} - 1\right| \le o_\delta(1) \tag{95}$$

for nearly all $(y_1, y_2) \in \mathbb{B}^2$ by (93), it follows by invoking this information for b_3 that

$$e^{i2b_1(L^{-1}y_1)}e^{i2b_2(L^{-1}y_2)}e^{-i2\tilde{a}\sigma_L(y_1,y_2)}$$

is nearly equal to the exponential of an imaginary affine function of (y_1, y_2), at nearly all points of $\mathbb{B}^2$.

Next consider the ratio

$$\frac{e^{i2b_1(L^{-1}y_1)}e^{i2b_2(L^{-1}(u+y_2))}e^{-i2\tilde{a}\sigma_L(y_1,u+y_2)}}{e^{i2b_1(L^{-1}y_1)}e^{i2b_2(L^{-1}y_2)}e^{-i2\tilde{a}\sigma_L(y_1,y_2)}} = e^{i2b_2(L^{-1}(u+y_2))}e^{-i2b_2(L^{-1}(y_2))}e^{-i2\tilde{a}\sigma_L(y_1,u)}. \quad (96)$$

From the conclusion of the preceding paragraph one can deduce that the right-hand side of (96) nearly coincides with the exponential of an imaginary affine function of u alone, at nearly all points (y_1, y_2, u) with $y_1 \in \mathbb{B}$ and $y_2, u \in \frac{1}{2}\mathbb{B}$. On the right-hand side, only the last exponential factor depends on y_1, so by regarding this quantity as a function of y_1 we conclude that $|\tilde{a}| \cdot |\sigma_L(v, u)| \le o_\delta(1)$ for nearly all $(v, u) \in (\frac{1}{4}\mathbb{B})^2$. Therefore

$$|\tilde{a}| \cdot \sup_{|x|,|y|\le 1} |\sigma_L(x, y)| \le o_\delta(1). \quad (97)$$

Therefore by Lemma 21, below, there exists $S \in \mathrm{Sp(2d)}$ such that $|\tilde{a}| \cdot \|SL^{-1}\|^2 \le o_\delta(1)$.

Combining this with (93) yields

$$\left|e^{ib_1(L^{-1}y_1)}e^{ib_2(L^{-1}y_2)}e^{ib_3(-L^{-1}y_1-L^{-1}y_2)} - 1\right| \le o_\delta(1) \quad (98)$$

for nearly all $(y_1, y_2) \in \mathbb{B}^2$. By Lemma 17, for each $j \in \{1, 2, 3\}$ there exists an affine function $L_j : \mathbb{R}^{2d} \to \mathbb{R}$ such that

$$|e^{ib_j(L^{-1}y)} - e^{iL_j(y)}| \le o_\delta(1)$$

for nearly all $y \in \mathbb{B}$. Thus

$$e^{i\alpha_j(L^{-1}y,t)} = e^{i\tilde{a}t}e^{iL_j(y)} + o_\delta(1) \quad (99)$$

for $(y, t) \in \mathbb{B} \times \mathbb{R}$ satisfying $|t| \le \rho(\delta)\varepsilon^{-1/2}$ outside a set of Lebesgue measure $\le o_\delta(1)\varepsilon^{-1/2}$, where $\tilde{a}$ satisfies (97).

This concludes the proof of Theorem 7 in the general complex-valued case. □

10 Some Matrix Algebra

Lemma 21 *For any invertible linear endomorphism* $L : \mathbb{R}^{2d} \to \mathbb{R}^{2d}$,

$$\|L^* J L\|^{1/2} = \inf_{S\in \mathrm{Sp(2d)}} \|S^{-1}L\|. \quad (100)$$

Proof That $\|L^*JL\| \le \inf_{S\in \mathrm{Sp(2d)}} \|S^{-1}L\|^2$ is immediate. For any L and any $S \in$ Sp(2d),

$$\|L^*JL\| = \|(S^{-1}L)^*S^*JS(S^{-1}L)\| = \|(S^{-1}L)^*J(S^{-1}L)\|$$
$$\le \|S^{-1}L\|\|J\|\|S^{-1}L\| = \|S^{-1}L\|^2.$$

To establish the reverse inequality, note that since L^*JL is a nonsingular antisymmetric real matrix, its eigenvalues are imaginary, and come in conjugate pairs; if $i\lambda$ is an eigenvalue then $\lambda \ne 0$ and $-i\lambda$ is also an eigenvalue, and the eigenspace associated to $-i\lambda$ has the same dimension as the eigenspace associated to $i\lambda$; coordinatewise complex conjugation interchanges these two eigenspaces. Therefore L^*JL can be written in the form $O_1^*KO_1$ where $O_1 \in O(2d)$ and K takes the form

$$K = \begin{pmatrix} 0 & t_1 & 0 & 0 & \cdots & 0 \\ -t_1 & 0 & 0 & 0 & \cdots & 0 \\ 0 & 0 & 0 & t_2 & \cdots & 0 \\ 0 & 0 & -t_2 & 0 & \cdots & 0 \\ \vdots & \vdots & \vdots & \vdots & \cdots & \vdots \end{pmatrix} \tag{101}$$

with 2×2 blocks $\begin{pmatrix} 0 & t_j \\ -t_j & 0 \end{pmatrix}$ along the diagonal, where $t_j \in \mathbb{R}^+$ and the eigenvalues are $\pm it_j$. Now $t_j \le \|L^*JL\|$. Defining

$$T = \begin{pmatrix} t_1^{1/2} & 0 & 0 & 0 & 0 & \cdots \\ 0 & t_1^{1/2} & 0 & 0 & 0 & \cdots \\ 0 & 0 & t_2^{1/2} & 0 & 0 & \cdots \\ 0 & 0 & 0 & t_2^{1/2} & 0 & \cdots \\ \vdots & \vdots & \vdots & \vdots & \vdots & \cdots \end{pmatrix} \tag{102}$$

gives

$$K = T^*\tilde{J}T \tag{103}$$

where

$$\tilde{J} = \begin{pmatrix} 0 & 1 & 0 & 0 & \cdots & 0 \\ -1 & 0 & 0 & 0 & \cdots & 0 \\ 0 & 0 & 0 & 1 & \cdots & 0 \\ 0 & 0 & -1 & 0 & \cdots & 0 \\ \vdots & \vdots & \vdots & \vdots & \cdots & \vdots \end{pmatrix} \tag{104}$$

with 2×2 blocks $\begin{pmatrix} 0 & 1 \\ -1 & 0 \end{pmatrix}$ along the diagonal. Now $\tilde{J} = O_2^* J O_2$ for a certain permutation matrix $O_2 \in O(2d)$ and thus we have

$$L^* J L = M^* J M \tag{105}$$

where $M = O_2 T O_1$. Equivalently,

$$(LM^{-1})^* J (LM^{-1}) = J, \tag{106}$$

so $LM^{-1} \in \mathrm{Sp(2d)}$. That is, $L = SM$ with $S \in \mathrm{Sp(2d)}$. Equivalently, $M = S^{-1}L$ satisfies

$$\|M\| = \|O_2 T O_1\| \le \|O_2\| \|T\| \|O_1\| = \|T\| = \|K\|^{1/2} = \|L^* J L\|^{1/2},$$

as required. □

11 Integration of Difference Relations

In this section we establish Theorem 8, which is motivated by considerations that have arisen in this paper, but on which the main theorems do not rely. This is done in the hope that it will prove useful in other problems. We continue to use the expressions "nearly every" and "nearly all" in the same sense as above.

The next lemma is elementary; the proof is omitted.

Lemma 22 *For each $d, m \in \mathbb{N}$ there exists $C < \infty$ with the following property. Let $q(x, y) = \sum_{0 \le |\alpha| \le m} a_\alpha(y) x^\alpha$ where $a_\alpha : \mathbb{R}^d \to \mathbb{C}$ are Lebesgue measurable functions. Suppose that $|q(x, y)| \le 1$ for nearly every $(x, y) \in \mathbb{B} \times \tilde{\mathbb{B}}$. Then for any multi-index β satisfying $0 \le |\beta| \le m$, $|a_\beta(y)| \le C$ for all $y \in \tilde{\mathbb{B}}$.*

Before embarking on the core of the proof of Theorem 8 we introduce several simplifications. Firstly, it suffices to prove the theorem in the case in which $\mathbb{B}$ is centered at 0, for the hypotheses and conclusions are invariant under translation. Secondly, it suffices to prove this for the ball $\mathbb{B}$ centered at 0 of radius 1. For if the result holds for some ball centered at 0, then it holds uniformly for all such balls, because the hypotheses and conclusions are invariant under dilations. Thirdly, it suffices to prove the theorem for $A = 1$, since hypotheses and conclusions are invariant under multiplication of φ by positive scalars, and since the case $A = 0$ follows from the case $A > 0$ with uniform bounds by a straightforward limiting argument. Fourthly, assuming $\mathbb{B}$ to be centered at the origin, it suffices to prove that there exists $\rho > 0$, depending only on d, D, such that the conclusion holds for all $x \in \rho\mathbb{B} = \{\rho y : y \in \mathbb{B}\}$ outside a set of measure $\varepsilon\rho^d|\mathbb{B}|$. Indeed, the full conclusion for $\mathbb{B}$ itself then follows by combining this weaker conclusion with a

Whitney decomposition of $\mathbb{B}$, as in [4]. One arranges that each Whitney cube Q_k is contained in a ball B_k of comparable diameter, such that the ball B_k^* concentric with B_k with radius enlarged by a factor of ρ^{-1} is contained in $\mathbb{B}$. Invoking the weaker result in its translation and dilation invariant form gives an approximation by an affine function on B_k, provided that $|B_k|/|\mathbb{B}|$ is not too small as a function of δ. These affine functions patch together on most of $\mathbb{B}$ to yield a single globally defined affine function, up to a suitably small additive error. The same reasoning reduces the case of small parameters η to $\eta = 1$.

The proof of the theorem will involve multiple steps in which $\mathbb{B}$ is replaced by a ball $\rho'\mathbb{B}$ where $\rho' > 0$ depends only on d, D. The final constant ρ is the product of all these factors ρ'. We will simplify notation by allowing the value of ρ to change from one step to the next, so that each of these factors ρ', and products of successive factors, are denoted by ρ.

The fifth simplification is one of language. Various conclusions will hold for all $x \in \rho\mathbb{B}$ except for a set of measure at most $\tau\rho^d|\mathbb{B}|$, where $\tau > 0$ depends only on d, D, δ, and $\tau \to 0$ as $\delta \to 0$. In this circumstance we will not specify a function $\delta \mapsto \tau(\delta)$, but will simply write that the conclusions in question hold for nearly all $x \in \rho\mathbb{B}$. In the same sense we will write "for nearly all $(x, y) \in \rho\mathbb{B} \times \rho\mathbb{B}$", and so on.

In the proof we write $O(1)$ for a quantity that is bounded above by some constant depending only on D, η. The value of this quantity is permitted to change from one occurrence to the next.

We will argue by induction on the degree D. The key to this induction is the observation that Theorem 8 implies an additional conclusion.

Corollary 23 *Let D be a nonnegative integer. Under the hypotheses of Theorem 8, for each multi-index satisfying $|\alpha| = D$, there exists a linear function ξ_α such that the coefficients a_α in (28) satisfy*

$$|a_\alpha(h) - \xi_\alpha(h)| \le CA \text{ for nearly all } h \in \rho\mathbb{B}. \tag{107}$$

Proof To prove this, assuming Theorem 8 for the given degree D, let Q be a polynomial of degree $\le D + 1$ that satisfies the conclusion (30). Then assuming as we may that $\mathbb{B}$ is centered at 0 and has radius 1, $|\Delta_h Q(x) - \Delta_h\varphi(x)| \le CA$ for nearly all $(x, h) \in (\rho\mathbb{B})^2$. Expand $\Delta_h Q(x) = \sum_{|\alpha|\le D} \tilde{a}_\alpha(h)x^\alpha$ where $\tilde{a}_\alpha$ are polynomials of degrees $\le D + 1 - |\alpha|$, and $\tilde{a}_\alpha(0) = 0$. In particular, $\tilde{a}_\alpha$ is linear when $|\alpha| = D$.

Consider $\Delta_h Q - \Delta_h\varphi$. Substituting for $\Delta_h\varphi$ the expression $\sum_{|\alpha|\le D} a_\alpha(h)x^\alpha + O(A)$ given in the hypothesis yields

$$\Big|\sum_{|\alpha|\le D} (a_\alpha(h) - \tilde{a}_\alpha(h))x^\alpha\Big| \le CA$$

for nearly all $(x, h) \in (\rho\mathbb{B})^2$. Invoking Lemma 22 gives $|a_\alpha(x) - \tilde{a}_\alpha(x)| \le CA$ for nearly all $x \in \rho\mathbb{B}$, which is the desired additional conclusion for $|\alpha| = D$. □

Proof of Theorem 8 We proceed by induction on D. Since the proof of Corollary 23 for degree D relied on Theorem 8 for that same degree, in the induction it is permissible to invoke Corollary 23 only for smaller degrees.

The base case $D = 0$ is a corollary of Lemma 16. Indeed, when $D = 0$ it is given that $|\varphi(x+h) - \varphi(x) - p(h)| \le A$ for nearly all points (x, h) with $x \in \mathbb{B}$ and $h \in \tilde{\mathbb{B}}$, where $p(h)$ is a polynomial of degree zero in x that depends on h; that is, $p(h)$ depends only on h. If $\tilde{\mathbb{B}}$ were equal to $\mathbb{B}^*$ then this would be a direct application of Lemma 16. The general case is proved by combining this special case with a Whitney decomposition of $\mathbb{B}$, as in the analysis in [4] and in the reduction outlined at the beginning of Sect. 11.

In the proof for the inductive step, we operate under the following convention: For $|\alpha| \le D-2$, $b_\alpha, \tilde{b}_\alpha, c_\alpha$ denote Lebesgue measurable functions, with appropriate domains. An equation involving such functions is to be interpreted as an existence statement; the assertion is that there exist measurable functions $b_\alpha, \tilde{b}_\alpha, c_\alpha$ such that the equation holds in the indicated domain. These are permitted to change from one occurrence of each symbol to the next. However, this convention is not in force for $|\alpha| \ge D-1$; for such indices, the functions b_α do not change after they are first introduced.

Assume without loss of generality that $A = 1$. For the inductive step, let $D \ge 1$, and let φ, P satisfy the hypothesis with $A = 1$. For $x, s, t \in \rho\mathbb{B}$ consider $\Delta_s\Delta_t\varphi(x)$, which takes the form

$$\begin{aligned}
\Delta_s\Delta_t\varphi(x) &= \Delta_t\Delta_s\varphi(x) \\
&= \sum_{|\alpha|\le D} a_\alpha(s)((x+t)^\alpha - x^\alpha) + O(1) \\
&= \sum_{|\alpha|=D-1} (b_\alpha(s)\cdot t)x^\alpha + \sum_{|\alpha|\le D-2} b_\alpha(s,t)x^\alpha + O(1)
\end{aligned}$$

for nearly all $(x, s, t) \in (\rho\mathbb{B})^3$ where b_α are $\mathbb{R}^d$-valued measurable functions.

Specialize to a typical $\tau \in \rho\mathbb{B}$. With $\psi = \Delta_\tau\varphi$, this conclusion becomes

$$\Delta_s\psi(y) = \sum_{|\alpha|=D-1} b_\alpha(s)\cdot\tau y^\alpha + \sum_{|\alpha|\le D-2} c_\alpha(s,\tau)y^\alpha + O(1)$$

for nearly all $(y, s, \tau) \in (\rho\mathbb{B})^3$. Therefore by induction on the degree D and Corollary 23, for each multi-index of degree $|\alpha| = D-1$, there exists an $\mathbb{R}^d$-valued linear function that agrees with b_α to within $O(1)$ at nearly every point of $\rho\mathbb{B}$. That is, there exist $\tilde{u}_\alpha \in \mathbb{R}^d \otimes \mathbb{R}^d$ satisfying

$$|b_\alpha(s) - \tilde{u}_\alpha \cdot s| = O(1) \text{ for nearly all } s \in \rho\mathbb{B}. \tag{108}$$

For $|\alpha| = D-1$, these coefficients b_α are related to the coefficients a_α in the hypothesis (28) as follows: Writing $b_\alpha(s) = (b_{\alpha,1}(s), \dots, b_{\alpha,d}(s))$, letting $e_i \in \mathbb{R}^d$

be the coordinate vector with i-th coordinate equal to 1 and all other coordinates equal to 0, and writing $\alpha = (\alpha_1, \ldots, \alpha_d)$, one has

$$b_{\alpha,i}(s) = (\alpha_i + 1)a_{\alpha+e_i}(s) + O(1)\ \forall\, |\alpha| = D - 1.$$

This is obtained by writing $\Delta_s\Delta_\tau\varphi = \Delta_\tau\Delta_s\varphi$, substituting the right-hand side of (28) for $\Delta_s\varphi$, applying Δ_τ, expanding $(x+\tau)^\alpha$, and invoking Lemma 22 to reach a conclusion for the first order Taylor expansion with respect to τ.

It follows that for each multi-index satisfying $|\beta| = D$, a_β is approximately linear in the sense that

$$|a_\beta(s) - u_\beta \cdot s| = O(1) \text{ for nearly all } s \in \rho\mathbb{B} \tag{109}$$

for certain $u_\beta \in \mathbb{R}^d \otimes \mathbb{R}^d$. Insert this conclusion into the hypotheses (29),(28) to obtain

$$\Delta_s\varphi(x) = \sum_{|\alpha|=D} (u_\alpha \cdot s)x^\alpha + \sum_{|\alpha|\le D-1} a_\alpha(s)x^\alpha + O(1) \text{ for nearly all } (x,s) \in (\rho\mathbb{B})^2. \tag{110}$$

We will show below, in Lemma 24, that there exists a homogeneous polynomial q of degree $\le D+1$ satisfying

$$\Delta_s q(x) \equiv \sum_{|\alpha|=D} u_\alpha \cdot s x^\alpha + \sum_{|\alpha|\le D-1} c_\alpha(s)x^\alpha + O(1) \tag{111}$$

for all $(x,s) \in (\rho\mathbb{B})^2$ and for some (polynomial) coefficient functions c_α. Granting this for the present, set $\psi = \varphi - q$. Then

$$\Delta_s\psi(x) = \sum_{|\alpha|\le D-1} c_\alpha(s)x^\alpha + O(1) \text{ for nearly all } (x,s) \in (\rho\mathbb{B})^2, \tag{112}$$

where c_α are measurable functions. This is the original hypothesis, with $\mathbb{B}$ replaced by $\rho\mathbb{B}$, φ replaced by ψ, and D replaced by $D-1$. Therefore it suffices to apply the induction hypothesis to conclude that ψ, and hence $\varphi = \psi + q$, have the required form. This completes the proof of Theorem 7, modulo the proof of the next lemma. □

Lemma 24 *There exists a polynomial q of degree $\le D+1$ that satisfies* (111).

Proof Apply Δ_t to both sides of (110) to obtain

$$\Delta_t\Delta_s\varphi(x) = \Delta_t \sum_{|\alpha|=D}\sum_{j=1}^{d} u_{\alpha,j}s_j x^\alpha + \Delta_t \sum_{|\alpha|\le D-1} b_\alpha(s)x^\alpha + O(1)$$

$$= \sum_{|\alpha|=D}\sum_{j=1}^{d} u_{\alpha,j}s_j \sum_{i=1}^{d}\alpha_i x^{\alpha-e_i}t_i + \sum_{|\alpha|\le D-2} b_\alpha(s,t)x^\alpha + O(1)$$

for nearly all $(x,s,t)\in(\rho\mathbb{B})^3$ where b_α are measurable functions. Since $\Delta_t\Delta_s\varphi = \Delta_s\Delta_t\varphi$, we may write the corresponding formula for $\Delta_s\Delta_t\varphi$, equate it to the one derived above, and apply Lemma 22 to deduce that for each $i,j\in\{1,2,\ldots,d\}$,

$$\sum_{|\alpha|=D} u_{\alpha,j}\alpha_i x^{\alpha-e_i} = \sum_{|\alpha|=D} u_{\alpha,i}\alpha_j x^{\alpha-e_j} + O(1) \tag{113}$$

for all $x\in\rho\mathbb{B}$. Equivalently, for each multi-index β satisfying $|\beta| = D-1$,

$$u_{\beta+e_i,j}(\beta_i+1) = u_{\beta+e_j,i}(\beta_j+1) + O(1) \tag{114}$$

for each i,j.

On the other hand, a homogeneous polynomial Q of degree $D+1$ satisfies the exact relation $\Delta_s Q(x) = \sum_{|\alpha|=D}\sum_{j=1}^{d}\tilde{u}_{\alpha,j}s_j x^\alpha + R(x,s)$ for some R, where $x\mapsto R(x,s)$ is a polynomial of degree $\le D-1$ for each s, if and only if $\partial Q(x)/\partial x_j = \sum_{|\alpha|=D}\tilde{u}_{\alpha,j}x^\alpha$ for each $j\in\{1,2,\ldots,d\}$. This system of equations is solvable for Q if and only if

$$\sum_{|\alpha|=D}\tilde{u}_{\alpha,j}\alpha_i x^{\alpha-e_i} = \sum_{|\alpha|=D}\tilde{u}_{\alpha,i}\alpha_j x^{\alpha-e_j} \tag{115}$$

for all $i\ne j\in\{1,2,\ldots,d\}$. Equivalently, for each multi-index β satisfying $|\beta| = D-1$,

$$\tilde{u}_{\beta+e_i,j}(\beta_i+1) = \tilde{u}_{\beta+e_j,i}(\beta_j+1) \tag{116}$$

for each i,j.

The tuple $(u_{\alpha,k} : |\alpha| = D \text{ and } 1\le k\le d)$ satisfies the system of approximate Eqs. (114). By elementary linear algebra, there exists a tuple $(\tilde{u}_{\alpha,k})$ with $|\tilde{u}_{\alpha,k} - u_{\alpha,k}| = O(1)$ for all α,k that satisfies the corresponding system of exact Eqs. (116). This system of equations implies the existence of a homogeneous polynomial q of degree $D+1$ that satisfies $\partial q(x)/\partial x_j = \sum_{|\alpha|=D}\tilde{u}_{\alpha,j}x^\alpha$ for each $j\in\{1,2,\ldots,d\}$. Therefore $\Delta_s q(x) = \sum_{|\alpha|=D}\sum_{j=1}^{d}\tilde{u}_{\alpha,j}s_j x^\alpha + R(x,s)$ where R is as above. □

The proof of Theorem 9 is very similar to that of Theorem 8. Details are left to the reader. □

12 A Final Lemma

The next lemma is rather trivial. The form of its conclusion contrasts with that of Lemma 17, in which the logarithms of the factors in the hypothesis are only nearly determined up to arbitrary additive corrections in $2\pi i\mathbb{Z}$. In Lemma 25, no such arbitrary additive corrections arise.

Lemma 25 *There exists $\eta_0 > 0$ with the following property. Let $v_j \in \mathbb{R}$ for $j = 1, 2$. Let $0 < \eta \le \eta_0$. Suppose that*

$$|e^{i(u_1 v_1 - u_2 v_2)} - 1| \le \eta$$

for all $(u_1, u_2) \in [0, 1]^2$ outside a set of Lebesgue measure $\le \frac{1}{2}$. Then $|v_j| \le 4\eta$ for $j = 1, 2$.

Proof There exists $u_2 \in [0, 1]$ such that

$$|e^{iu_1 v_1} - e^{iu_2 v_2}| \le \eta \tag{117}$$

for all $u_1 \in [0, 1]$ outside a set E of measure $\le \frac{1}{2}$. We may assume without loss of generality that $v_1 \ne 0$. If $|v_1| \ge 4\eta$ then choose $N \in \mathbb{N}$ satisfying $N|v_1|^{-1}\eta \in [\frac{1}{4}, \frac{1}{2}]$. There must exist an interval $I \subset [0, 1]$ of length N^{-1} such that $|E \cap I| \le \frac{1}{2}|I|$.

Because $|E \cap I| \le |I|/2$, $I \setminus E$ has diameter $\ge |I|/2$ and therefore the image of $I \setminus E$ under the mapping $I \ni t \mapsto e^{itv_1}$ has diameter $\ge cN|I|/2 = c' > 0$. If $2\eta < c'$, this contradicts (117).

The same reasoning applies to v_2. □

13 On Twisted Convolution

The translation symmetry for the functional in Young's inequality corresponds to a hybrid translation/modulation symmetry for twisted convolution. Identify $\mathbb{R}^{2d}$ with $\mathbb{R}^d \times \mathbb{R}^d$, with coordinates $x = (x', x'')$ with $x', x'' \in \mathbb{R}^d$. For $z \in \mathbb{R}^{2d}$, denote by $x \mapsto \tau_z(x)$ the translation mapping $x \mapsto x - z$ from $\mathbb{R}^{2d}$ to $\mathbb{R}^{2d}$. For any $\mathbf{z} = (z_1, z_2, z_3) \in (\mathbb{R}^{2d})^3$ satisfying $z_1 + z_2 + z_3 = 0$,

$$\mathcal{T}_{\mathbb{R}^{2d},\rho}(\tau_{z_1} f_1, \tau_{z_2} f_2, \tau_{z_3} f_3) = e^{i\rho\sigma(z_1,z_2)} \mathcal{T}_{\mathbb{R}^{2d},\rho}(g_1, g_2, g_3) \tag{118}$$

with $g_3 = f_3$,

$$g_1(x) = e^{-i\rho\sigma(z_2,x)} f_1(x), \quad \text{and} \quad g_2(x) = e^{-i\rho\sigma(z_1,x)} f_2(x). \tag{119}$$

To prove Theorem 10, using some of the analysis developed above, is straightforward. One has

$$|\mathcal{T}_{\mathbb{R}^{2d},\rho}(\mathbf{f})| \le \mathcal{T}_{\mathbb{R}^{2d}}(|f_1|, |f_2|, |f_3|),$$

so the optimal constant in the inequality for $\mathcal{T}_{\mathbb{R}^{2d},\rho}$ is less than or equal to the optimal constant in Young's convolution inequality for $\mathbb{R}^{2d}$. On the other hand, if $\mathbf{p}$ satisfies the scaling relation $\sum_{j=1}^{3} p_j^{-1} = 2$, and if $f_{j,\varepsilon}(x) = \varepsilon^{-2d/p_j} f_j(\varepsilon^{-1}x)$, then $\|f_{j,\varepsilon}\|_{p_j} \equiv \|f_j\|_{p_j}$ and

$$\mathcal{T}_{\mathbb{R}^{2d},\rho}(f_{1,\varepsilon}, f_{2,\varepsilon}, f_{3,\varepsilon}) \to \mathcal{T}_{\mathbb{R}^{2d}}(\mathbf{f})$$

as $\varepsilon \to 0$. Therefore the optimal constant for $\mathcal{T}_{\mathbb{R}^{2d},\rho}$ equals the optimal constant for $\mathcal{T}_{\mathbb{R}^{2d}}$. Therefore if $\mathbf{f}$ nearly realizes the optimal constant for $\mathcal{T}_{\mathbb{R}^{2d},\rho}$, then $(|f_j| : 1 \le j \le 3)$ nearly realizes the optimal constant for $\mathcal{T}_{\mathbb{R}^{2d}}$.

By invoking the characterization of near-maximizers for Young's inequality for $\mathbb{R}^{2d}$ together with the hybrid translation/modulation symmetry (118), one can reduce matters to the case in which $(|f_j| : 1 \le j \le 3)$ is an ordered triple of Gaussians centered at the origin, that realizes the optimal constant for Young's inequality for convolution in $\mathbb{R}^{2d}$. The remainder of the analysis is a simplified recapitulation of the above analysis for $\mathbb{H}^d$.

Acknowledgments The author is grateful to Anthony Carbery for pointing out the question addressed here, to Detlef Müller for calling his attention to the reference [11], and to Edward Scerbo for useful comments on the exposition. He thanks Joseph Wolf, as well as Professors Carbery and Müller, for stimulating conversations.

References

1. Beckner, W.: Inequalities in Fourier analysis. Ann. Math. (2) **102**(1), 159–182 (1975)
2. Beckner, W.: Geometric inequalities in Fourier analysis, Essays on Fourier analysis in honor of Elias M. Stein (Princeton, NJ, 1991), 36–68, Princeton Math. Ser., 42, Princeton Univ. Press, Princeton (1995)
3. Brascamp, H.J., Lieb, E.H.: Best constants in Young's inequality, its converse, and its generalization to more than three functions. Adv. Math. **20**(2), 151–173 (1976)
4. Christ, M.: Near-extremizers of Young's inequality for Euclidean groups. Rev. Mat. Iberoam. **35**(7), 1925–1972 (2019)
5. Christ, M.: Near equality in the two-dimensional Brunn–Minkowski inequality (1965). Preprint, math.CA arXiv:1206.1965
6. Christ, M.: Near equality in the Brunn–Minkowski inequality (1965). Preprint, math.CA arXiv:1206.1965
7. Christ, M.: Near equality in the Riesz-Sobolev inequality. Acta Math. Sin. (Engl. Ser.) **35**(6), 783–814 (2019)
8. Christ, M.: A sharpened Hausdorff–Young inequality. Preprint, math.CA arXiv:1406.1210
9. Christ, M.: A sharpened Riesz–Sobolev inequality. Preprint, math.CA arXiv:1706.02007

10. Christ, M.: Young's inequality sharpened. In: Geometric Aspects of Harmonic Analysis. Springer, Berlin (2021)
11. Klein, A., Russo, B.: Sharp inequalities for Weyl operators and Heisenberg groups. Math. Ann. **235**(2), 175–194 (1978)
12. Lieb, E.: Gaussian kernels have only Gaussian maximizers. Invent. Math. **102**, 179–208 (1990)
13. Neuman, A.M.: Functions of nearly maximal Gowers–Host–Kra norms on Euclidean spaces. University of California, Berkeley, ProQuest Dissertations Publishing, 2018
14. O'Neill, K.: A sharpened inequality for twisted convolution (2018). math.CA arXiv:1810.02050
15. O'Neill, K.: A quantitative stability theorem for convolution on the Heisenberg group (2019). Preprint, math.CA arXiv:1907.11986

Young's Inequality Sharpened

Michael Christ

Abstract A quantitative stability result with an optimal exponent is established, concerning near-maximizers for Young's convolution inequality for Euclidean groups.

Keywords Young's convolution inequality · Maximizer · Perturbative expansion · Hermite basis

1 Statements of Theorems

The Beckner-Brascamp-Lieb-Young convolution inequality [1, 4] states that for each dimension $d \geq 1$, for complex-valued functions $f_j \in L^{p_j}(\mathbb{R}^d)$, the convolution $f_1 * f_2(x) = \int f_1(x-y) f_2(y)\, dy$ satisfies

$$\|f_1 * f_2\|_q \leq \mathbf{A}_p^d \prod_{j=1}^{2} \|f_j\|_{L^{p_j}} \tag{1}$$

provided that $p_j, q \in [1, \infty]$ and $q^{-1} = p_1^{-1} + p_2^{-1} - 1$, where $\mathbf{A}_p = \prod_{j=1}^{3} \mathbf{C}_{p_j}$ with $p_3 = q'$ and $\mathbf{C}_p^2 = p^{1/p}/r^{1/r}$, where $r = p'$ denotes the exponent conjugate to p. It is convenient, for our purpose, to put (1) into more symmetric form, in terms of the trilinear form

$$\mathcal{T}(\mathbf{f}) = \int_{x_1+x_2+x_3=0} \prod_{j=1}^{3} f_j(x_j)\, d\lambda(\mathbf{x}) \tag{2}$$

M. Christ (✉)
Department of Mathematics, University of California, Berkeley, CA, USA
e-mail: mchrist@berkeley.edu

P. Ciatti, A. Martini (eds.), *Geometric Aspects of Harmonic Analysis*,
Springer INdAM Series 45, https://doi.org/10.1007/978-3-030-72058-2_7

where $\mathbf{f} = (f_j : j \in \{1,2,3\})$, $\mathbf{x} = (x_1, x_2, x_3) \in (\mathbb{R}^d)^3$, and λ is the natural Lebesgue measure on $\Lambda = \{\mathbf{x} : x_1 + x_2 + x_3 = 0\}$; $d\lambda = dx_i dx_j$ for any $i \neq j \in \{1,2,3\}$ if the third variable is regarded as a function of the other two via the additive relation defining Λ. Inequality (1) can be equivalently stated as

$$|\mathcal{T}(\mathbf{f})| \le \mathbf{A}_p^d \prod_{j=1}^{3} \|f_j\|_{L^{p_j}} \tag{3}$$

for all tuples of functions $f_j \in L^{p_j}(\mathbb{R}^d)$ and all $\mathbf{p} = (p_1, p_2, p_3) \in [1,\infty]^3$ satisfying $\sum_j p_j^{-1} = 2$. We assume henceforth that each exponent p_j belongs to the open interval $(1,\infty)$. Throughout the paper, p' denotes the exponent conjugate to p.

The constant $\mathbf{A}_p^d$ is optimal for all $\mathbf{p}, d$. Among the extremizing tuples $\mathbf{f}$ for (3) is the Gaussian triple $(e^{-\pi p_j' |x|^2} : j = 1,2,3)$. Moreover, Brascamp and Lieb [4] showed that every complex-valued maximizing triple belongs to the orbit of this single maximizer under the symmetry group $\mathbf{G} = \mathbf{G}_{d,\mathbf{p}}$ of the inequality, generated by a translation action of $\mathbb{R}^{2d}$ on $\mathbb{R}^{3d}$, the diagonal action of the general linear group Gl(d) on $\mathbb{R}^d \times \mathbb{R}^d \times \mathbb{R}^d$, multiplication of the components f_j by arbitrary complex scalars, and the diagonal action of the group of modulation operators $f \mapsto e^{ix\cdot v} f(x)$.

A stronger form of this uniqueness was established in [7]: If $|\mathcal{T}(\mathbf{f})| \ge (1-\delta)\mathbf{A}_p^d \prod_j \|f_j\|_{p_j}$ then there exists a maximizing triple $\mathbf{g} = (g_1, g_2, g_3)$ of Gaussians satisfying $\|f_j - g_j\|_{p_j} \le \varepsilon \|f_j\|_{p_j}$, where $\varepsilon \to 0$ as $\delta \to 0$; ε may be taken to depend only on $\delta, d, \mathbf{p}$. A weakness of that result is its nonquantitative nature; the proof provides no information on the rate at which ε tends to zero. A further weakness is the relatively complicated proof, which relies in turn on a corresponding strengthened uniqueness theorem for the Riesz-Sobolev inequality in dimension 1 [6], whose proof exploited ideas from additive combinatorics.

The present paper establishes a quantitative improvement of this stability result. The analysis provides an alternative, and perhaps simpler, proof of the weaker nonquantitative result in the special case of nonnegative functions.

Fixing the dimension d, define $\mathfrak{G}_\mathbf{p}$ to be the set of all maximizing triples of Gaussians for the ratio $\mathcal{T}(\mathbf{f})/\prod_j \|f_j\|_{p_j}$. For each $\mathbf{p}$, define the projective distance from a triple $\mathbf{f}$ to $\mathfrak{G}_\mathbf{p}$ by

$$\operatorname{dist}_\mathbf{p}(\mathbf{f}, \mathfrak{G}_\mathbf{p}) = \inf_{\mathbf{g}\in\mathfrak{G}_\mathbf{p}} \max_j \frac{\|f_j - g_j\|_{p_j}}{\|f_j\|_{p_j}}, \tag{4}$$

under the assumption that for every index j, $\|f_j\|_{p_j} \neq 0$.

Theorem 1 *Let K be a compact subset of $(1,2)^3$. Let $\mathbf{p} \in K$ satisfy $\sum_{j=1}^{3} p_j^{-1} = 2$. For each $d \geq 1$ there exists $c > 0$ such that for all $\mathbf{p} \in K$ and all $\mathbf{f} \in L^{\mathbf{p}}(\mathbb{R}^d)$ with $\|f_j\|_{p_j} \neq 0$ for each $j \in \{1,2,3\}$,*

$$|\mathcal{T}(\mathbf{f})| \leq \left(\mathbf{A}_p^d - c\,\mathrm{dist}_{\mathbf{p}}(\mathbf{f}, \mathfrak{G}_{\mathbf{p}})^2\right) \prod_{j=1}^{3} \|f_j\|_{p_j}. \tag{5}$$

This sharpens Young's inequality, in the same sense that Bianchi and Egnell [3] sharpened the Sobolev inequality.

The exponent 2 in the conclusion is optimal. The proof does not provide a concrete value for the coefficient c, and provides little insight into its optimal value.

The following variant extends the range of exponents to include the case in which some exponent equals 2, but sacrifices uniform dependence on $\mathbf{p}$. This loss might possibly be circumvented through a more thorough analysis of the dependence on $\mathbf{p}$ of various intermediate quantities that arise in the proof.

Theorem 2 *Let $\mathbf{p} \in (1,2]^3$ satisfy $\sum_{j=1}^{3} p_j^{-1} = 2$. For each $d \geq 1$ there exists $c > 0$ such that for all $\mathbf{f} \in L^{\mathbf{p}}(\mathbb{R}^d)$ with $\|f_j\|_{p_j} \neq 0$ for each $j \in \{1,2,3\}$,*

$$|\mathcal{T}(\mathbf{f})| \leq \left(\mathbf{A}_p^d - c\,\mathrm{dist}_{\mathbf{p}}(\mathbf{f}, \mathfrak{G}_{\mathbf{p}})^2\right) \prod_{j=1}^{3} \|f_j\|_{p_j}. \tag{6}$$

The restrictions $p_j \leq 2$ are necessary.

Proposition 3 *Let $\mathbf{p} \in (1,\infty)^3$ satisfy $\sum_{j=1}^{3} p_j^{-1} = 2$, and let $d \geq 1$. Suppose that $p_k > 2$ for some index $k \in \{1,2,3\}$. Then there exists no $c > 0$ for which the inequality (5), with $\mathrm{dist}_{\mathbf{p}}(\mathbf{f}, \mathfrak{G}_{\mathbf{p}})$ raised to the power 2, holds uniformly for all $\mathbf{f} \in L^{\mathbf{p}}(\mathbb{R}^d)$ with $\|f_j\|_{p_j} > 0$ for each $j \in \{1,2,3\}$.*

Nonetheless, a slightly weaker variant of Theorem 1 holds for the full range of exponents. It is most naturally formulated in terms of the bilinear inequality (1). Define $\mathfrak{G}'_{p_1,p_2}$ to be the set of all ordered pairs $\mathbf{g} = (g_1, g_2)$ of Gaussian functions $g_j : \mathbb{R}^d \to [0,\infty)$ of the form $g_j = c_j e^{-p'_j Q(x-a_j)}$ where $c_j \in \mathbb{R}^+$, $a_j \in \mathbb{R}^d$, and $Q(y) = \Lambda(y,y)$ where Λ is a positive definite symmetric quadratic form on $\mathbb{R}^d$. If each $f_j \neq 0$, then equality holds if and only if $(f_1, f_2) \in \mathfrak{G}'_{\mathbf{p}}$.

For each (p_1, p_2), define the projective distance from (f_1, f_2) to $\mathfrak{G}'_{p_1,p_2}$ by

$$\mathrm{Dist}_{p_1,p_2}((f_1,f_2), \mathfrak{G}'_{p_1,p_2}) = \inf_{(g_1,g_2) \in \mathfrak{G}'_{p_1,p_2}} \max_{j=1,2} \frac{\|f_j - g_j\|_{p_j}}{\|f_j\|_{p_j}}, \tag{7}$$

under the assumption that $\|f_j\|_{p_j}$ vanishes for neither index j.

Theorem 4 *Let $d \ge 1$. Let $p_1, p_2 \in (1,2)$, and define $q \in (1,\infty)$ by $q^{-1} = p_1^{-1} + p_2^{-1} - 1$. There exists $c > 0$ such that for any nonnegative functions $f_j \in L^{p_j}(\mathbb{R}^d)$ with nonzero norms,*

$$\|f_1 * f_2\|_q \le (\mathbf{A}_p^d - c\,\mathrm{Dist}_{p_1,p_2}((f_1,f_2),\mathfrak{G}'_{p_1,p_2})^2)\|f_1\|_{p_1}\|f_2\|_{p_2}. \tag{8}$$

For simplicity we have restricted the statement to nonnegative functions. For $q \ge 2$, Theorem 4 follows directly from Theorem 1 by duality.

Theorems 1 and 4 have the following analogue in the periodic setting, with $\mathbb{R}^d$ replaced by $\mathbb{T}^d = \mathbb{R}^d/\mathbb{Z}^d$. For each $n \in \mathbb{Z}^d$, denote by e_n the character $x \mapsto e^{2\pi i n\cdot x}$.

Theorem 5 *Let $\mathbf{p} \in (1,\infty)^3$ satisfy $\sum_{j=1}^3 p_j^{-1} = 2$. There exists $c > 0$ such that for any $d \ge 1$ and for every $\mathbf{f} \in L^{\mathbf{p}}(\mathbb{T}^d)$ with nonnegative components f_j satisfying $\|f_j\|_{p_j} = 1$,*

$$|\mathcal{T}(\mathbf{f})| \le 1 - c\sum_j \|f_j - 1\|_{p_j}^{r_j} \tag{9}$$

where $r_j = \max(p_j, 2)$.

More generally, if each f_j is complex-valued and $\|f_j\|_{p_j} = 1$ then

$$|\mathcal{T}(\mathbf{f})| \le 1 - c \inf_{n\in\mathbb{Z},\ |a_j|=1} \sum_j \|f_j - a_j e_n\|_{p_j}^{r_j} \tag{10}$$

The last infimum is taken over complex numbers a_j satisfying $|a_j| = 1$. The very simple proof is sketched in Sect. 11.

An outline of the proof of Theorem 1 is as follows. The first step is to reduce to small perturbations of maximizing triples. For nonnegative functions, this can be accomplished by exploiting the monotonicity of $\mathcal{T}(\mathbf{f})$ under a nonlinear heat flow. For general functions, the reduction is justified by a compactness theorem of [7]. This is discussed in Sect. 3.

Choosing $\mathbf{g} \in \mathfrak{G}_\mathbf{p}$ to approximately minimize $\max_j \|f_j - g_j\|_{p_j}$ and writing $f_j = g_j + h_j$, the trilinear form $\mathcal{T}$ can be expanded in terms of the small quantities h_j. The central issue is the strict negativity of two symmetric quadratic forms, which act on functions taking values in $L^2(\mathbb{R}^d, \mathbb{R}^3)$ (rather than in $L^2(\mathbb{R}^d, \mathbb{C}^3)$). We diagonalize these by expanding each h_j, in turn, in terms of appropriately dilated Hermite functions, reducing a quadratic form on $L^2(\mathbb{R}^d, \mathbb{R}^3)$ to an infinite system of quadratic forms on $\mathbb{R}^3$. All terms involving Hermite polynomials of sufficiently high degree are easily seen to be uniformly negative definite. An elementary algebraic analysis handles low degrees.

The proof of Theorem 4 is a variant of that of Theorem 1. It is sketched in Sect. 10.

Theorem 1 is in the same spirit as a quantitative stability result for the Riesz-Sobolev inequality developed in [9]. The two proofs are similar in structure, but there is one notable difference. The analysis here relies on Hermite functions and on resulting explicit expressions for eigenvalues for associated quadratic forms, while that in [9] relies on spherical harmonics, and leads to eigenvalues for which less useful expressions seem to be available.

A corresponding result for Young's inequality for Heisenberg groups is established in works of the author [10] (establishing a $o_\delta(1)$-type conclusion, thus reducing matters to the perturbative regime) and of O'Neill [11] (analyzing the perturbative regime by extending the machinery developed here).

2 Negative Result

The conclusion of Theorem 1 fails for rather superficial reasons if some exponent p_j exceed 2. Suppose without loss of generality that $p_1 > 2$. Let $\mathbf{p}$ be given. Let $\mathbf{g} = (e^{-p_j'|x|^2} : j \in \{1, 2, 3\})$, which satisfies $\mathcal{T}(\mathbf{g}) = \mathbf{A}_p^d \prod_j \|g_j\|_{p_j}$. Let $\varphi : \mathbb{R}^d \to [0, \infty)$ be continuous, compactly supported, and not identically zero. Choose any $0 \neq v \in \mathbb{R}^d$, and for $t, \delta \in \mathbb{R}^+$ consider $\mathbf{f} = \mathbf{f}_{\delta,t} = (f_j : j \in \{1, 2, 3\})$ where $f_j = g_j$ for $j = 2, 3$, and $f_1(x) = g_1(x) + \delta\varphi(x - tv)$. Restrict attention to $\delta \in (0, 1]$; we will eventually let δ tend to 0. Regard t as a function of δ, satisfying $t(\delta) \geq 1$ and $\lim_{\delta \to 0} t(\delta) = \infty$.

Now

$$\prod_j \|f_j\|_{p_j} = \prod_j \|g_j\|_{p_j} + O(\delta^{p_1}),$$

provided that $t(\delta)^2 \gg \ln(1/\delta)$ as $\delta \to 0$. This holds because φ has compact support and $g_1(x) = O(e^{-c|x|^2})$. Similarly, $\mathcal{T}(\mathbf{f}) = \mathcal{T}(\mathbf{g}) + O(e^{-at^2}\delta)$ for some $a = a(\mathbf{p}) > 0$ since $g_2 * g_3$ is a Gaussian. Therefore

$$\frac{\mathcal{T}(\mathbf{f})}{\prod_j \|f_j\|_{p_j}} = \mathbf{A}_p^d + O(\delta^{p_1}).$$

On the other hand, it is elementary that there exists $c' > 0$ depending on $v, d, \mathbf{p}$, such that for each δ, $\operatorname{dist}_{\mathbf{p}}(\mathbf{f}, \mathfrak{G}_{\mathbf{p}}) \geq c'\delta$ provided that $t = t(\delta)$ is sufficiently large. Therefore

$$\frac{\mathcal{T}(\mathbf{f})}{\prod_j \|f_j\|_{p_j}} \geq \mathbf{A}_p^d - c'' \operatorname{dist}_{\mathbf{p}}(\mathbf{f}, \mathfrak{G}_{\mathbf{p}})^{p_1}.$$

Since $p_1 > 2$, this contradicts (5) in the limit $\delta \to 0$. This completes the proof of Proposition 3.

3 Reduction to the Perturbative Case

It was shown in [7] that if $\|f_j\|_{p_j} = 1$ and if $|\mathcal{T}(\mathbf{f})| \geq (\mathbf{A}_p^d - \delta)$ then there exist Gaussians satisfying $\|f_j - G_j\|_{p_j} \leq o_\delta(1)$. Therefore in order to prove Theorem 1, it suffices to show that for each $\mathbf{p}, d$ there exists $\varepsilon_0 < 0$ such that the conclusion (5) holds whenever $\text{dist}_{\mathbf{p}}(\mathbf{f}, \mathfrak{G}_{\mathbf{p}}) \leq \varepsilon_0$. Therefore our analysis is devoted to this perturbative regime.

An alternative method of reduction that avoids recourse to the lengthy analysis of [7] is available. It consists of an initial step for nonnegative functions, followed by a separate argument to extend the result from nonnegative to general complex-valued functions. For nonnegative functions f_j, matters can be reduced to small perturbations of maximizing Gaussian ordered triples via a deformation argument relying on nonlinear heat evolutions. A proof of the following result may be found for instance in [2, 5].

Lemma 6 *Let $d \geq 1$, and let $\mathbf{p} \in (1, \infty)^3$ satisfy $\sum_j p_j^{-1} = 2$. Let $f_j \in L^{p_j}$ be nonnegative. There exist an ordered triple of nonnegative Gaussian functions G_j that satisfies $\|G_j\|_{p_j} = \|f_j\|_{p_j}$ and $\mathcal{T}(\mathbf{G}) = \mathbf{A}_p^d \prod_j \|G_j\|_{p_j}$ and a continuous mapping $[0, 1] \ni t \mapsto \mathbf{f}(t) = (f_j(t) : j \in \{1, 2, 3\}) \in L^{p_1} \times L^{p_2} \times L^{p_3}$ satisfying $f_j(0) = f_j$, $f_j(1) = G_j$, and $\|f_j(t)\|_{p_j} \equiv \|f_j\|_{p_j}$, such that $\mathcal{T}(\mathbf{f}(t))$ is a continuous nondecreasing function of t.*

If (5) is known to hold for any $\mathbf{f}$ for which $\text{dist}_{\mathbf{p}}(\mathbf{f}, \mathfrak{G}_{\mathbf{p}})$ is sufficiently small, then (5) can be deduced for for general nonnegative functions f_j, as follows. Suppose that $\text{dist}_{\mathbf{p}}(\mathbf{f}, \mathfrak{G}_{\mathbf{p}}) > \varepsilon_0$ and $\|f_j\|_{p_j} = 1$. By Lemma 6, there exists $s \in (0, 1)$ satisfying $\text{dist}_{\mathbf{p}}(\mathbf{f}(s), \mathfrak{G}_{\mathbf{p}}) = \varepsilon_0$. Applying (5) to $\mathbf{f}(s)$ gives

$$\mathcal{T}(\mathbf{f}(s)) \leq \mathbf{A}_p^d - c\,\text{dist}_{\mathbf{p}}(\mathbf{f}(s), \mathfrak{G}_{\mathbf{p}})^2 = \mathbf{A}_p^d - c\varepsilon_0^2.$$

Therefore

$$\mathcal{T}(\mathbf{f}) \leq \mathcal{T}(\mathbf{f}(s)) \leq \mathbf{A}_p^d - c\varepsilon_0^2 \leq \mathbf{A}_p^d - c\varepsilon_0^2\,\text{dist}_{\mathbf{p}}(\mathbf{f}, \mathfrak{G}_{\mathbf{p}})^2$$

since $\text{dist}_{\mathbf{p}}(\mathbf{f}, \mathfrak{G}_{\mathbf{p}}) \leq 1$ by its definition.

For general complex-valued functions f_j, this author is not in possession of any corresponding version of Lemma 6. However, the complex-valued case can be deduced from the nonnegative case in the same way that this reduction was executed in [7]. Indeed, for general $\mathbf{f} = (f_j)$, consider $\mathbf{F} = (|f_j|)$, whose components have the same L^{p_j} norms and which satisfies $\mathcal{T}(\mathbf{F}) \geq |\mathcal{T}(\mathbf{f})|$. Therefore if $\mathbf{f}$ is a near-maximizer, then so is $\mathbf{F}$, so each component $F_j = |f_j|$ must be nearly a Gaussian. Write $f_j = e^{i\varphi_j} F_j$. Let $\Lambda = \{\mathbf{x} : x_1 + x_2 + x_3 = 0\}$, and let λ be Lebesgue measure on Λ. Assume without loss of generality that $\mathcal{T}(\mathbf{f}) \in \mathbb{R}^+$. Then $\prod_{j=1}^3 \varphi_j(x_j)$ must be approximately equal to 1 for most (with respect to the measure $d\mu(\mathbf{x}) = \prod_{j=1}^3 F_j(x_j)\,d\lambda$) points $\mathbf{x} \in \Lambda$. By Proposition 8.1 of [7], the ordered triple $(\varphi_j(x_j) : j \in \{1, 2, 3\})$ is well approximated, modulo $2\pi i\mathbb{Z}^3$, at most points

$\mathbf{x}$ with respect to the same measure μ, by a ordered triple of three affine functions, whose sum vanishes identically on Λ. If $\mathbf{F}$ is close to a nonnegative maximizing Gaussian triple, if each φ_j is close in this sense to an affine function, and if these three affine functions are compatible in this sense, then $\mathbf{f}$ is also close to a complex-valued maximizing Gaussian triple. A precise formulation of these statements is in [7].

4 Perturbative Expansion

We begin the proof of Theorem 1 in the perturbative regime. We change notation. Write $g_j(x) = e^{-\pi p_j'|x|^2} = G^{p_j'}(x)$ where $G(x) = e^{-\pi|x|^2}$. Consider $\mathcal{T}(g_j + f_j : j \in \{1, 2, 3\})$, where $\|f_j\|_{p_j}$ is small for each index j. Assume that

$$\int g_j^{p_j-1} f_j = 0 \text{ for each } j \in \{1, 2, 3\}. \tag{11}$$

For $p < 2$, the mapping $L^p \ni f \mapsto \|g + f\|_p$ is not twice continuously differentiable. We circumvent this via a decomposition analyzed in [8]. Assume that $p_j \in (1, 2]$ for each index $j \in \{1, 2, 3\}$. Let $\eta > 0$ be a small parameter. For each $j \in \{1, 2, 3\}$, decompose $f_j = f_{j,\sharp} + f_{j,\flat}$, defining

$$f_{j,\sharp}(x) = \begin{cases} f_j(x) \text{ if } |f_j(x)| \le \eta g_j(x), \\ 0 \text{ otherwise,} \end{cases} \tag{12}$$

and $f_{j,\flat} = f_j - f_{j,\sharp}$. Then (using the fact that g_j is real-valued) [8]

$$\begin{aligned}\frac{\|g_j + f_j\|_{p_j}}{\|g_j\|_{p_j}} &\ge 1 + \tfrac{1}{2}\|g_j\|_{p_j}^{-p_j} \int \left[(p_j - 1)\operatorname{Re}(f_{j,\sharp})^2 + \operatorname{Im}(f_{j,\sharp})^2\right] g_j^{p_j-2} \\ &\quad - C\eta\|f_{j,\sharp}\|_{p_j}^2 \|g_j\|_{p_j}^{-2} + c\eta^{2-p_j}\|f_{j,\flat}\|_{p_j}^{p_j}\|g_j\|_{p_j}^{-p_j}\end{aligned} \tag{13}$$

where $c, C \in \mathbb{R}^+$ depend only on p_j. Positive terms on the right-hand are favorable for our purpose, while negative terms are unfavorable. The unfavorable third term will be ameliorated by choosing η to be sufficiently small. Because $\|f_j\|_{p_j}$ is raised to a power p_j strictly less than 2, the final term on the right will eventually be favorable, even with η very small. This reasoning breaks down when $p_j > 2$.

Write $\|\mathbf{f}\| = \max_j \|f_j\|_{p_j}$. Expand $\mathcal{T}(\mathbf{g} + \mathbf{f})$ as $\mathcal{T}(\mathbf{g})$ plus the sum of three terms $\mathcal{T}(g_m, g_n, f_k)$, plus the sum of three terms $\mathcal{T}(f_i, f_j, g_k)$, plus the remainder term $\mathcal{T}(\mathbf{f})$, which has magnitude $O_{\mathbf{g}}(\prod_j \|f_j\|_{p_j}) = O_{\mathbf{g}}(\|\mathbf{f}\|^3)$, and consequently will be negligible in this second order analysis.

The following expression $\mathcal{Q}_{\mathbf{p}}$ governs the analysis. We will abuse notation mildly by referring to it, and to related expressions below, as quadratic forms, although they are actually quadratic forms in the real and imaginary parts of our functions.

Definition 7 For $\mathbf{h} = (h_1, h_2, h_3) \in L^{\mathbf{p}}(\mathbb{R}^d, \mathbb{C}^3)$,

$$\mathcal{Q}_{\mathbf{p}}(\mathbf{h}) = \mathcal{T}(\mathbf{g})^{-1} \sum_{(i,j,k)} \big[\mathcal{T}(\operatorname{Re} h_i, \operatorname{Re} h_j, g_k) - \mathcal{T}(\operatorname{Im} h_i, \operatorname{Im} h_j, g_k)\big]$$
$$- \tfrac{1}{2} \sum_{j=1}^{3} \|g_j\|_{p_j}^{-p_j} \int \big[(p_j - 1)\operatorname{Re}(h_j)^2 + \operatorname{Im}(h_j)^2\big] g_j^{p_j-2} \tag{14}$$

with the summation extending over the three cyclic permutations (i, j, k) of $(1, 2, 3)$.

The notation $\mathbf{h} \in L^{\mathbf{p}}(\mathbb{R}^d, \mathbb{C}^3)$ indicates that each component h_j is complex-valued, and belongs to $L^{p_j}(\mathbb{R}^d)$.

Because the functions g_n are real-valued and even, and because $\mathbf{g}$ is a maximizer, if $\{1, 2, 3\} = \{k, m, n\}$ then $g_m * g_n$ must, by duality, be a positive scalar multiple of $g_k^{p_k-1}$. Therefore by (11), each of the three terms $\mathcal{T}(g_m, g_n, f_k)$ vanishes. By expanding $|\mathcal{T}(\mathbf{g}+\mathbf{f})|^2$, substituting $\mathcal{T}(\mathbf{g}) = \mathbf{A}_p^d \prod_j \|g_j\|_{p_j}$, and invoking (13), one obtains:

Lemma 8 *Let $\mathbf{p} \in (1, 2)^3$ satisfy $\sum_j p_j^{-1} = 2$. There exists $c > 0$ such that for any sufficiently small parameter $\eta > 0$, for any $\mathbf{f} \in L^{\mathbf{p}}(\mathbb{R}^d, \mathbb{C}^3)$ satisfying (11),*

$$\frac{\mathcal{T}(\mathbf{g}+\mathbf{f})}{\prod_j \|g_j + f_j\|_{p_j}} \le \mathbf{A}_p^d + \mathbf{A}_p^d \mathcal{Q}_{\mathbf{p}}(\mathbf{f}_\sharp)$$
$$+ C\eta \sum_j \|f_{j,\sharp}\|_{p_j}^2 \|g_j\|_{p_j}^{-2} - c \sum_j \eta^{2-p_j} \|f_{j,\flat}\|_{p_j}^{p_j} \|g_j\|_{p_j}^{-p_j} + O_{\mathbf{g}}(\|\mathbf{f}\|^3) \tag{15}$$

where $\mathbf{f}_\sharp = (f_{j,\sharp} : j \in \{1, 2, 3\})$ with $f_{j,\sharp}, f_{j,\flat}$ defined as in (12), and $g_j = G^{p_j'}$.

Thus matters have been reduced to obtaining a suitably negative upper bound for $\mathcal{Q}_{\mathbf{p}}(\mathbf{f}_\sharp)$. This is complicated slightly by the relationship $h_j = f_{j,\sharp}$, which need not satisfy the essential orthogonality relation $\int h_j g_j^{p_j-1} = 0$; this complication will be dealt with at the end of the analysis. What remains is mainly the analysis of $\mathcal{Q}_{\mathbf{p}}$, taking into account the orthogonality condition and the role of symmetries.

The following information will be needed for a more explicit description of $\mathcal{Q}_{\mathbf{p}}$.

Lemma 9 *Let $d \ge 1$, let $\mathbf{p} \in (1, 2)^3$ with $\sum_j p_j^{-1} = 2$, and set $g_j = G^{p_j'}$. Then*

$$\mathcal{T}(\mathbf{g}) = \prod_{l=1}^{3} (p_l')^{-d/2}. \tag{16}$$

Proof For any $r, s > 0$,

$$r^{d/2}G^r * s^{d/2}G^s = t^{d/2}G^t \quad \text{where } t^{-1} = r^{-1} + s^{-1}. \tag{17}$$

This is a consequence of the identities $\widehat{G} = G$ and $G^t(x) = G(t^{1/2}x)$. Therefore

$$\mathcal{T}(\mathbf{g}) = \int_{\mathbb{R}^d} G^{p_3'} \cdot (G^{p_1'} * G^{p_2'}) = \int_{\mathbb{R}^d} G^{p_3'} \cdot (p_1')^{-d/2}(p_2')^{-d/2}q^{d/2}G^q$$

where $q^{-1} = p_1'^{-1} + p_2'^{-1} = p_3^{-1}$. Therefore

$$\mathcal{T}(\mathbf{g}) = (p_1')^{-d/2}(p_2')^{-d/2}p_3^{d/2}\int_{\mathbb{R}^d} G^{p_3'}G^{p_3} = (p_1')^{-d/2}(p_2')^{-d/2}p_3^{d/2}(p_3 + p_3')^{-d/2}$$

$$= (p_1')^{-d/2}(p_2')^{-d/2}p_3^{d/2}(p_3p_3')^{-d/2} = \prod_{l=1}^{3}(p_l')^{-d/2}$$

using the relation $p_3 + p_3' = p_3p_3'$. □

Introduce the exponents

$$\tau_j = \tfrac{1}{2}p_jp_j' \tag{18}$$

and the functions

$$u_j = h_jg_j^{(p_j-2)/2} = h_jG^{\tau_j - p_j'}. \tag{19}$$

Equivalently, since $g_j = G^{p_j'}$, $h_j = u_jG^{(2-p_j)p_j'/2} = u_jG^{p_j'-\tau_j}$. Because of the assumption that $p_j < 2$, $p_j' > (p_j/2)p_j' = \tau_j$. Thus the factor $G^{p_j'-\tau_j}$ is a Schwartz function.

Since the exponents satisfy $p_j < 2$, the assumption that $h_j \in L^{p_j}$ does not suffice to guarantee that $u_j \in L^2$. However, the theory developed for $\mathcal{Q}_\mathbf{p}$ will be applied only to functions h_j that are $O(g_j)$ in the pointwise sense, in which event u_j will be $O(g_j^{p_j})$ and will belong to L^2.

Definition 10 Let $\mathbf{p} \in (1,2)^3$. $T_{(i,j,k)}$ denotes the bounded linear operator on $L^2(\mathbb{R}^d)$ given by

$$T_{(i,j,k)}(u) = G^{p_j'(2-p_j)/2} \cdot (G^{p_k'} * (G^{p_i'(2-p_i)/2}u)), \tag{20}$$

with $*$ denoting convolution.

Thus for real-valued functions h_m, u_m,

$$\mathcal{T}(h_i, h_j, g_k) = \langle T_{(i,j,k)}(u_i), \tilde{u}_j \rangle$$

where $\tilde{u}_j(x) = u_j(-x)$. The adjoint operator is $T^*_{(i,j,k)} = T_{(j,i,k)}$. $T_{(i,j,k)}$ is self-adjoint if and only if $p_i = p_j$, but the operator on $L^2(\mathbb{R}^d, \mathbb{C}^3)$ represented by the operator-valued matrix

$$\begin{pmatrix} 0 & T_{(1,2,3)} & T_{(2,3,1)} \\ T_{(1,2,3)} & 0 & T_{(3,1,2)} \\ T_{(2,3,1)} & T_{(3,1,2)} & 0 \end{pmatrix},$$

whose components are the operators $T_{(i,j,k)}$, is self-adjoint for arbitrary $\mathbf{p} \in (1,2)^3$.

Since

$$\|g_j\|_{p_j}^{p_j} = \int_{\mathbb{R}^d} e^{-\pi p_j p_j'} = (p_j p_j')^{-d/2},$$

for $\mathbf{h} \in L^{\mathbf{p}}(\mathbb{R}^d, \mathbb{C}^3)$ we can write

$$Q_{\mathbf{p}}(\mathbf{h}) = Q_{\mathbf{p}}^+(\operatorname{Re}\mathbf{u}) + Q_{\mathbf{p}}^-(\operatorname{Im}\mathbf{u}) \tag{21}$$

with $\operatorname{Re}\mathbf{u} = (\operatorname{Re}u_1, \operatorname{Re}u_2, \operatorname{Re}u_3)$, with $\operatorname{Im}\mathbf{u} = (\operatorname{Im}u_1, \operatorname{Im}u_2, \operatorname{Im}u_3)$, and with

$$Q_{\mathbf{p}}^+(\mathbf{v}) = \prod_{l=1}^{3}(p_l')^{d/2} \sum_{(i,j,k)} \langle T_{(i,j,k)}v_i, \tilde{v}_j\rangle - \tfrac{1}{2}\sum_{j=1}^{3}(p_j p_j')^{d/2}(p_j - 1)\|v_j\|_{L^2}^2 \tag{22}$$

$$Q_{\mathbf{p}}^-(\mathbf{v}) = -\prod_{l=1}^{3}(p_l')^{d/2} \sum_{(i,j,k)} \langle T_{(i,j,k)}v_i, \tilde{v}_j\rangle - \tfrac{1}{2}\sum_{j=1}^{3}(p_j p_j')^{d/2}\|v_j\|_{L^2}^2 \tag{23}$$

for $\mathbb{R}$-valued functions v_j, with $\sum_{(i,j,k)}$ denoting summation over the three cyclic permutations (i,j,k) of $(1,2,3)$, and with

$$\tilde{v}(x) = v(-x). \tag{24}$$

The L^2 norm is that of $L^2(\mathbb{R}^d, \mathbb{R})$, with respect to Lebesgue measure.

Observe that $Q_{\mathbf{p}}^{\pm}$ differ in two respects; their first terms have opposite signs, and factors of $(p_j - 1) \in (0,1)$ appear in $Q_{\mathbf{p}}^+$ but not in $Q_{\mathbf{p}}^-$.

The orthogonality condition $\int h_l g_l^{p_l - 1} = 0$ for each index l is equivalent to the relation $\int u_l g_l^{(2-p_l)p_l'/2} g_l^{p_l-1} = 0$. Since

$$\tfrac{1}{2}(2 - p_l)p_l' + (p_l - 1) = p_l' + p_l - \tfrac{1}{2}p_l p_l' = p_l p_l' - \tfrac{1}{2}p_l p_l' = \tfrac{1}{2}p_l p_l' = \tau_l,$$

this relation can be equivalently written

$$\int u_l G^{\tau_l} = 0. \tag{25}$$

The proof of Theorem 1 rests on properties of $Q_{\mathbf{p}}^+$ and of $Q_{\mathbf{p}}^-$ detailed in Lemmas 12 and 13, respectively. The following generalized Hermite polynomials and Hermite functions will be used to analyze these properties.

Definition 11

(i) For each $t \in \mathbb{R}^+$ and each $n \in \{0, 1, 2, \dots\}$, $P_n^{(t)}$ denotes the unique real-valued polynomial of degree n with positive leading coefficient such that $P_n^{(t)} G^t$ has $L^2(\mathbb{R})$ norm equal to 1 and is orthogonal to QG^t in $L^2(\mathbb{R}^1)$ for every polyomial Q of degree strictly less than n.

(ii) For $d > 1, \alpha = (\alpha_1, \dots, \alpha_d) \in \{0, 1, 2, \dots\}^d$, and $x = (x_1, \dots, x_d) \in \mathbb{R}^d$,

$$P_\alpha^{(t)}(x) = \prod_{k=1}^{d} P_{\alpha_k}^{(t)}(x_k).$$

(iii) The generalized Hermite functions with exponent t are $H_\alpha^{(t)} = P_\alpha^{(t)} G^t$.

In particular,

$$H_0^{(t)} = \|G^t\|_{L^2}^{-1} G^t = (2t)^{d/4} G^t. \tag{26}$$

For each $\tau \in \mathbb{R}^+$, the family of functions $H_\alpha^{(\tau)}$ forms an orthonormal basis for $L^2(\mathbb{R}^d)$.

The following two lemmas provide the information concerning the quadratic forms $Q_{\mathbf{p}}^+, Q_{\mathbf{p}}^-$ needed for our analysis. Each lemma has as a hypothesis certain orthogonality relations, which supplement (25). These two lemmas will be proved in Sect. 5. Lemma 17 will enable us to arrange for these supplementary orthogonality relations to hold in the context of the proof of Theorem 1. That proof will be concluded in Sect. 7.

Lemma 12 *For each* $\mathbf{p} \in (1, 2)^3$ *satisfying* $\sum_{j=1}^{3} p_j^{-1} = 2$, *there exists* $c = c(\mathbf{p}) > 0$ *such that for every* $d \geq 1$ *and every* $\mathbf{u} \in L^2(\mathbb{R}^d, \mathbb{R}^3)$ *satisfying*

$$\langle u_j, H_\alpha^{(\tau_j)} \rangle = 0 \text{ whenever } \begin{cases} \alpha = 0 & \text{and } j \in \{1, 2, 3\} \\ |\alpha| = 1 & \text{and } j \in \{1, 2\} \\ |\alpha| = 2 & \text{and } j = 3, \end{cases} \tag{27}$$

there holds

$$Q_{\mathbf{p}}^{+}(\mathbf{u}) \le -c \sum_{j=1}^{3} \|u_j\|_{L^2(\mathbb{R}^d)}^2. \tag{28}$$

Lemma 13 *For each* $\mathbf{p} \in (1,2)^3$ *satisfying* $\sum_{j=1}^{3} p_j^{-1} = 2$*, there exists* $c = c(\mathbf{p}) > 0$ *such that for every* $d \ge 1$ *and every* $\mathbf{u} \in L^2(\mathbb{R}^d, \mathbb{R}^3)$ *satisfying*

$$\langle u_j, H_\alpha^{(\tau_j)} \rangle = 0 \text{ whenever } \begin{cases} \alpha = 0 & \text{and } j \in \{1,2,3\} \\ |\alpha| = 1 & \text{and } j = 3 \end{cases} \tag{29}$$

there holds

$$Q_{\mathbf{p}}^{-}(\mathbf{u}) \le -c \sum_{j=1}^{3} \|u_j\|_{L^2(\mathbb{R}^d)}^2. \tag{30}$$

The constants c depend on $\mathbf{p}$, but are independent of the dimension d. However, $Q_{\mathbf{p}}^{+}, Q_{\mathbf{p}}^{-}$ are multiplied by the dimension-dependent factor $\mathbf{A}_p^d$ in (15).

The index $j = 3$ plays a distinguished role in Lemmas 12 and 13, but this is simply a matter of choice. We will eventually arrange that these lemmas are applied in situations in which the hypotheses (27) (respectively (29)) are satisfied. The number of independent linear conditions that we will be able to arrange to be satisfied, will be exactly equal to the number of such conditions appearing in these hypotheses.

5 Analysis of Quadratic Forms

In this section we prove Lemmas 12 and 13. We introduce, for each of the three indices, an orthonormal basis for $L^2(\mathbb{R}^d)$ such that each operator $T_{(i,j,k)}$ maps each element of the basis associated to the index i to a scalar multiple of the basis associated to the index j. We calculate these scalars explicitly. Lemma 12 is thereby reduced to obtaining uniformly negative bounds for a certain infinite family of explicit quadratic forms $Q_{\mathbf{p},\kappa}^{*}$ on $\mathbb{R}^3$, indexed by $\kappa \in \mathbb{N}$. These forms fail to be strictly negative for $\kappa = 1, 2$, but the orthogonality conditions (27) restore strict negativity. For $\kappa \ge 3$, these forms are uniformly negative. The analysis for Lemma 13 is a simple variant of that for Lemma 12.

5.1 Diagonalization of Scalar Operators

The following lemma is proved in Sect. 8. In the lemma, and throughout the ensuing discussion,

$$|\alpha| = |(\alpha_1, \dots, \alpha_d)| = \sum_{j=1}^{d} \alpha_j.$$

Lemma 14 *Let $d \ge 1$. Let $\mathbf{p} \in (1,2]^3$ satisfy $\sum_{j=1}^{3} p_j^{-1} = 2$. For each permutation (i,j,k) of $(1,2,3)$ and each $\alpha \in \{0,1,2,\dots\}^d$,*

$$T_{(i,j,k)}(H_\alpha^{(\tau_i)}) = \lambda_{\alpha,(i,j,k)}\, H_\alpha^{(\tau_j)} \tag{31}$$

where

$$\lambda_{\alpha,(i,j,k)} = \Big(\frac{p_i p_j}{p_i' p_j'}\Big)^{|\alpha|/2} \big(\frac{p_i}{p_i'}\big)^{d/4} \big(\frac{p_j}{p_j'}\big)^{d/4} (p_k')^{-d/2}. \tag{32}$$

One may compare with the corresponding analysis of the Riesz-Sobolev inequality in [9], where no expressions as explicit as (32) are available for the corresponding eigenvalues.

Ratios $\frac{p_l}{p_l'}$ pervade the discussion. Since

$$p_l \,/\, p_l' = p_l - 1, \tag{33}$$

the conclusion of Lemma 14 can alternatively be written

$$\lambda_{\alpha,(i,j,k)} = (p_i - 1)^{\frac{d}{4} + \frac{|\alpha|}{2}} (p_j - 1)^{\frac{d}{4} + \frac{|\alpha|}{2}} (p_k')^{-d/2}.$$

Since $p_i, p_j \le 2$ with at most one of these equal to 2,

$$0 < (p_i - 1)(p_j - 1) < 1 \;\forall i \ne j \in \{1,2,3\}$$

and consequently the eigenvalues $\lambda_{\alpha,(i,j,k)}$ satisfy

$$\lambda_{\alpha,(i,j,k)} \to 0 \text{ as } |\alpha| \to \infty,$$

and these are strictly decreasing positive functions of $|\alpha|$ for fixed $d, \mathbf{p}, (i,j,k)$.

5.2 Diagonalizing $Q_{\mathbf{p}}^{+}$

Fix $\mathbf{p}$. For $j \in \{1, 2, 3\}$ and $u_j \in L^2(\mathbb{R}^d)$ define the coefficients[1]

$$\widehat{u_j}(\alpha) = \langle u_j, H_\alpha^{(\tau_j)} \rangle = \int_{\mathbb{R}^d} u_j P_\alpha^{(\tau_j)} G^{\tau_j}. \tag{34}$$

Note that the definition of $\widehat{u_j}$ depends on the index $j \in \{1, 2, 3\}$, not merely on the function $u_j \in L^2$. In this language, the assumption (25) states that

$$\widehat{u_l}(0) = 0 \text{ for all three indices } l. \tag{35}$$

Consequently summations below will extend only over nonzero α.

We will systematically write $\sum_{(i,j,k)} \cdot$ to denote a sum over the three cyclic permutations (i, j, k) of $(1, 2, 3)$ over some quantities.

Because $\{H_\alpha^{(t)}\}$ is an orthonormal basis for $L^2(\mathbb{R}^d)$, with respect to Lebesgue measure, for each $t = \tau_l$ and each $l \in \{1, 2, 3\}$, and because these functions are even when $|\alpha|$ is even, and odd when $|\alpha|$ is odd,

$$\langle T_{(i,j,k)} u_i, \tilde{u}_j \rangle = \sum_\alpha (-1)^{|\alpha|} \lambda_{\alpha,(i,j,k)} \widehat{u_i}(\alpha) \widehat{u_j}(\alpha). \tag{36}$$

Therefore $Q_{\mathbf{p}}^{+}(\mathbf{u})$ decomposes as the sum

$$Q_{\mathbf{p}}^{+}(\mathbf{u}) = \sum_{\alpha \neq 0} \Big(\prod_{l=1}^{3} (p_l')^{d/2} \sum_{(i,j,k)} (-1)^{|\alpha|} \lambda_{\alpha,(i,j,k)} \widehat{u_j}(\alpha) \widehat{u_i}(\alpha) - \sum_{l=l}^{3} \tfrac{p_l - 1}{2} (p_l p_l')^{d/2} |\widehat{u_l}(\alpha)|^2 \Big), \tag{37}$$

where the outer sum extends over all nonzero $\alpha \in \{0, 1, 2, \ldots\}^d$.

According to Lemma 14 and the definition of $Q_{\mathbf{p}}^{+}$, this can be written more explicitly as

$$Q_{\mathbf{p}}^{+}(\mathbf{u}) = \sum_{\alpha \neq 0} Q_{\mathbf{p},|\alpha|}(\widehat{u_1}(\alpha), \widehat{u_2}(\alpha), \widehat{u_3}(\alpha))$$

[1]These are not Fourier coefficients.

where

$$Q_{\mathbf{p},\kappa}(\mathbf{v}) = (-1)^{\kappa} \sum_{(i,j,k)} (p_i' p_j' p_k')^{d/2} \Big(\frac{p_i p_j}{p_i' p_j'}\Big)^{\kappa/2} \big(\frac{p_i}{p_i'}\big)^{d/4} \big(\frac{p_j}{p_j'}\big)^{d/4} (p_k')^{-d/2} v_j v_i$$

$$- \tfrac{1}{2} \sum_{l=l}^{3} (p_l - 1)(p_l p_l')^{d/2} v_l^2. \qquad (38)$$

This last expression can be simplified. Consider any $\kappa \neq 0$. Make the change of variables $\mathbf{v} \mapsto \mathbf{w}$ in $\mathbb{R}^3$, with $\mathbf{w}$ defined by

$$w_l = v_l \cdot ((p_l - 1)/2)^{1/2} (p_l p_l')^{d/4} = v_l \cdot (p_l / p_l')^{1/2} 2^{-1/2} (p_l p_l')^{d/4}.$$

Write $|\mathbf{w}|^2 = \sum_{l=1}^{3} w_l^2$. Then $Q_{\mathbf{p},\kappa}(\mathbf{v})$ is equal to $-|\mathbf{w}|^2$ plus

$$(-1)^{\kappa} 2 \sum_{(i,j,k)} (p_i' p_j' p_k')^{d/2} \Big(\frac{p_i p_j}{p_i' p_j'}\Big)^{\kappa/2} \big(\frac{p_i}{p_i'}\big)^{d/4} \big(\frac{p_j}{p_j'}\big)^{d/4} (p_k')^{-d/2}$$

$$\cdot \Big(\frac{p_i'}{p_i} \frac{p_j'}{p_j}\Big)^{1/2} (p_i p_i' p_j p_j')^{-d/4} w_i w_j$$

$$= (-1)^{\kappa} 2 \sum_{(i,j,k)} \Big(\frac{p_i p_j}{p_i' p_j'}\Big)^{(\kappa-1)/2} w_i w_j$$

$$= (-1)^{\kappa} 2 \sum_{(i,j,k)} [(p_i - 1)(p_j - 1)]^{(\kappa-1)/2} w_i w_j.$$

Thus

$$Q_{\mathbf{p},\kappa}(\mathbf{v}) = -|\mathbf{w}|^2 + Q^*_{\mathbf{p},\kappa}(\mathbf{w})$$

with

$$Q^*_{\mathbf{p},\kappa}(\mathbf{w}) = (-1)^{\kappa} 2 \sum_{(i,j,k)} [(p_i - 1)(p_j - 1)]^{(\kappa-1)/2} w_i w_j \ \text{ for } w \in \mathbb{R}^3, \qquad (39)$$

for $\kappa \in \mathbb{N}$. To prove Lemma 12, we need to show that there exists $\eta > 0$ such that $Q^*_{\mathbf{p},\kappa}(\mathbf{w}) \leq (1 - \eta)|\mathbf{w}|^2$ for all $w \in \mathbb{R}^3$ satisfying the specified orthogonality conditions, for all $\kappa \geq 1$.

The forms $Q^*_{\mathbf{p},\kappa}$ are independent of the dimension d. The factor $(-1)^{\kappa}$ plays a significant role for $\kappa = 1$; the quantities of interest are the maximum eigenvalues of these forms, rather than the maxima of the absolute values of these eigenvalues.

5.3 Eigenvalue Analysis for $Q_{\mathbf{p}}^{+}$

Lemma 12, our central result concerning $Q_{\mathbf{p}}^{+}$, is a direct consequence of the next lemma.

Lemma 15 *Let $\mathbf{p} \in (1,2]^3$ satisfy $\sum_{l=1}^{3} p_l^{-1} = 2$. There exists $\eta = \eta(\mathbf{p}) > 0$ such that for every $\kappa \geq 3$,*

$$Q_{\mathbf{p},\kappa}^{*}(\mathbf{w}) \leq (1-\eta)|\mathbf{w}|^2 \text{ for all } \mathbf{w} \in \mathbb{R}^3. \tag{40}$$

The same conclusion holds if $\kappa = 1$ and $w_l = 0$ for at least two indices l. It likewise holds if $\kappa = 2$ and $w_l = 0$ for at least one index l.

Proof For $\kappa = 1$, (39) specializes to $-2(w_1w_2 + w_2w_3 + w_3w_1)$. If $w_l = 0$ for two indices l, then this vanishes.

For $\kappa = 2$, if $w_k = 0$ then (39) specializes to

$$2(p_i-1)^{1/2}(p_j-1)^{1/2}w_iw_j \leq (p_i-1)^{1/2}(p_j-1)^{1/2}|\mathbf{w}|^2.$$

Now $(p_i-1)(p_j-1) < 1$ since $p_l < 2$ for each index.

In order to treat the case $\kappa \geq 3$, we first analyze the case $\kappa = 2$ more closely. Set $r_j = (p_j-1)^{1/2}$ and consider the quadratic form $2\sum_{(i,j,k)} r_ir_jw_jw_j$, which is represented by the matrix

$$M = \begin{pmatrix} 0 & r_1r_2 & r_1r_3 \\ r_1r_2 & 0 & r_2r_3 \\ r_1r_3 & r_2r_3 & 0 \end{pmatrix}.$$

This matrix has characteristic polynomial $\det(tI - M)$ equal to

$$t^3 - (r_1^2r_2^2 + r_2^2r_3^2 + r_3^2r_1^2)t - 2r_1^2r_2^2r_3^2. \tag{41}$$

□

Sublemma 16 *If $\sum_{j=1}^{3} p_j^{-1} = 2$ then the quantities $s_j = p_j - 1$ satisfy*

$$s_1s_2 + s_2s_3 + s_3s_1 + 2s_1s_2s_3 = 1. \tag{42}$$

Proof Denote by $\sigma_n(\mathbf{p})$ the n-th elementary symmetric polynomial in $\mathbf{p}$. Thus $\sigma_1(\mathbf{p}) = p_1 + p_2 + p_3$, $\sigma_2(\mathbf{p}) = p_1p_2 + p_2p_3 + p_3p_1$, and $\sigma_3(\mathbf{p}) = p_1p_2p_3$. Multiplying both sides of the relation $\sum_j p_j^{-1} = 2$ by $\sigma_3(\mathbf{p})$ gives $\sigma_2(\mathbf{p}) = 2\sigma_3(\mathbf{p})$. Therefore

$$\begin{aligned} &s_1s_2 + s_2s_3 + s_3s_1 + 2s_1s_2s_3 \\ &\quad = (p_1-1)(p_2-1) + (p_2-1)(p_3-1) + (p_3-1)(p_1-1) \end{aligned}$$

$$
\begin{aligned}
&+ 2(p_1 - 1)(p_2 - 1)(p_3 - 1) \\
&= \sigma_2(\mathbf{p}) - 2\sigma_1(\mathbf{p}) + 3 + 2\sigma_3(\mathbf{p}) - 2\sigma_2(\mathbf{p}) + 2\sigma_1(\mathbf{p}) - 2 \\
&= 2\sigma_3(\mathbf{p}) - \sigma_2(\mathbf{p}) + 1 \\
&= 1.
\end{aligned}
$$

Therefore $t = 1$ is a root of the characteristic polynomial $\det(M - tI)$. Consequently this polynomial factors as

$$\det(tI - M) = (t - 1)(t^2 + t + 2(p_1 - 1)(p_2 - 1)(p_3 - 1)).$$

Since M is real and symmetric, $\det(M - tI)$ has three real roots. Therefore the quadratic factor $t^2 + t + 2(p_1 - 1)(p_2 - 1)(p_3 - 1)$ must have two real roots. Their product is positive, and their sum is negative. Thus both are negative. Thus we have shown that for any $\mathbf{w} \in S^2 = \{\mathbf{w} \in \mathbb{R}^3 : |w| = 1\}$,

$$2 \sum_{(i,j,k)} (p_i - 1)^{1/2}(p_j - 1)^{1/2} w_i w_j \le 1.$$

Now let $\kappa \ge 3$ and consider the maximum value of

$$\max_{\mathbf{w} \in S^2} 2 \sum_{(i,j,k)} (p_i - 1)^{(\kappa-1)/2}(p_j - 1)^{(\kappa-1)/2} w_i w_j,$$

which is attained in the orthant in which $w_m \ge 0$ for each $m \in \{1, 2, 3\}$. Each factor $(p_i - 1)(p_j - 1)$ is strictly less than 1. Therefore this maximum value is strictly less than the maximum value for $\kappa = 2$, which we have found to be equal to 1. Therefore

$$\max_{\mathbf{w} \in S^2} 2 \sum_{(i,j,k)} (p_i - 1)^{(\kappa-1)/2}(p_j - 1)^{(\kappa-1)/2} w_i w_j < 1,$$

uniformly for all $\kappa \ge 3$, as well. This concludes the proof of Lemma 15, hence that of Lemma 12. □

5.4 *Analysis for* $Q_{\mathbf{p}}^-$

Like $Q_{\mathbf{p}}^+$, the form $Q_{\mathbf{p}}^-$ can be reduced to a family of real-valued quadratic forms on $\mathbb{R}^3$:

$$Q_{\mathbf{p}}^-(\mathbf{u}) = \sum_{\alpha \neq 0} Q_{\mathbf{p},|\alpha|}^{\dagger}(\widehat{u_1}(\alpha), \widehat{u_2}(\alpha), \widehat{u_3}(\alpha))$$

where for $\mathbf{v} \in \mathbb{R}^3$,

$$Q^{\dagger}_{\mathbf{p},\kappa}(\mathbf{v}) = (-1)^{\kappa+1} \sum_{(i,j,k)} (p'_i p'_j p'_k)^{d/2} \Big(\frac{p_i p_j}{p'_i p'_j}\Big)^{|\alpha|/2} \Big(\frac{p_i}{p'_i}\Big)^{d/4} \Big(\frac{p_j}{p'_j}\Big)^{d/4} (p'_k)^{-d/2} v_j v_i$$
$$- \sum_{l=l}^{3} \tfrac{1}{2} (p_l p'_l)^{d/2} v_l^2. \tag{43}$$

$Q^{\dagger}_{\mathbf{p},\kappa}$ differs from $Q_{\mathbf{p},\kappa}$ in three ways. Firstly, -1 is raised to the power $\kappa + 1$, rather than κ. Secondly, in the expression for $Q^{\dagger}_{\mathbf{p},\kappa}$, the term with the index l in the final sum has a coefficient of $\frac{1}{2}$, whereas $Q_{\mathbf{p},\kappa}$ had $\frac{p_l-1}{2}$. Thirdly, $Q^{\dagger}_{\mathbf{p},\kappa}$ is subjected to fewer orthogonality conditions through the hypotheses of Lemma 13 than is $Q_{\mathbf{p},\kappa}$ through the hypotheses of Lemma 12.

Substituting $w_l = 2^{-1/2}(p_l p'_l)^{d/4} v_l$, Lemma 13 is equivalent to the assertion that there exists $\eta > 0$ satisfying

$$-Q^{*}_{\mathbf{p},\kappa+1}(\mathbf{w}) \le (1-\eta)|\mathbf{w}|^2 \tag{44}$$

for all $\mathbf{w} \in \mathbb{R}^3$ for all $\kappa \ge 2$, and for all $\mathbf{w}$ with $w_3 = 0$ for $\kappa = 1$. Here $Q^{*}_{\mathbf{p},\lambda}$ are the forms defined in (39); note that it is $Q^{*}_{\mathbf{p},\kappa+1}$, rather than $Q_{\mathbf{p},\kappa}$, that arises in the analysis of $Q^{\dagger}_{\mathbf{p},\kappa}$. We have already shown that $|Q_{\mathbf{p},\mu}|(\mathbf{w})| \le (1-\eta)|\mathbf{w}|^2$ for all $\mu \ge 3$, which gives the result needed here for $\kappa \ge 2$. For $\kappa = 1$, under the orthogonality condition $w_3 = 0$ we have

$$-Q^{*}_{\mathbf{p},\kappa+1}(\mathbf{w}) = -Q^{*}_{\mathbf{p},2}(\mathbf{w}) = -2(p_1-1)^{1/2}(p_2-1)^{1/2} w_1 w_2,$$

which satisfies the desired conclusion if and only if $(p_1 - 1)(p_2 - 1) < 1$. This condition holds, since each exponent p_l is strictly less than 2. This completes the proof of Lemma 13.

6 Balancing

In analyzing $\mathcal{T}(\mathbf{F})$ when $\text{dist}_{\mathbf{p}}(\mathbf{F}, \mathfrak{G}_{\mathbf{p}})$ is small, the representation $\mathbf{F} = \mathbf{g} + \mathbf{f}$ with $\mathbf{g} \in \mathfrak{G}_{\mathbf{p}}$ is not unique. We aim to choose $\mathbf{g} \in \mathfrak{G}_{\mathbf{p}}$ so that $\|\mathbf{F} - \mathbf{g}\|$ is comparable to $\text{dist}_{\mathbf{p}}(\mathbf{F}, \mathfrak{G}_{\mathbf{p}})$, and at the same time, so that the orthogonality conditions in the hypotheses of Lemmas 12 and 13 are satisfied.

Lemma 17 *Let $d \ge 1$. Let $\mathbf{p} \in (1,2]^3$ satisfy $\sum_{l=1}^{3} p_l^{-1} = 2$. For each $\mathbf{p}$ there exists $\delta_0 > 0$ such that for any $\mathbf{f}$ satisfying*

$$\|f_j - g_j\|_{L^{p_j}(\mathbb{R}^d)} \le \delta_0 \text{ and } \langle f_j - g_j, g_j^{p_j-1}\rangle = 0 \tag{45}$$

for each index $j \in \{1, 2, 3\}$, there exist $\mathbf{v} \in (\mathbb{R}^d)^3$ satisfying $v_1 + v_2 + v_3 = 0$, $\mathbf{a} \in \mathbb{C}^3$, $\xi \in \mathbb{R}^d$, and a symmetric $d \times d$ real matrix ψ satisfying

$$|v_j| + \|\psi - I\| + |a_j - 1| + |\xi| \le C \sum_{l=1}^{3} \|f_l - g_l\|_{L^{p_l}(\mathbb{R}^d)}$$

such that the functions $\tilde{f}_j(x) = a_j f_j(\psi(x) + v_j)e^{ix\cdot\xi}$ satisfy

$$\langle \mathrm{Re}(\tilde{f}_j) - g_j, P_\alpha^{(\tau_j)} g_j^{p_j-1} \rangle = 0 \tag{46}$$

whenever (j, α) satisfies any of the following conditions:

$$\begin{cases} \alpha = 0 \text{ and } j \in \{1, 2, 3\}, \\ |\alpha| = 1 \text{ and } j \in \{1, 2\}, \\ |\alpha| = 2 \text{ and } j = 3, \end{cases} \tag{47}$$

and

$$\langle \mathrm{Im}(\tilde{f}_j), P_\alpha^{(\tau_j)} g_j^{p_j-1} \rangle = 0 \text{ whenever } \alpha = 0, \text{ and whenever } |\alpha| = 1 \text{ and } j = 3. \tag{48}$$

Here I denotes the identity element of $\mathrm{Gl}(d)$, $\|\psi - I\|$ refers to the Hilbert-Schmidt norm on the space of all $d \times d$ real matrices, and we identify elements of $\mathrm{Gl}(d)$ with invertible matrices in the usual way.

The symmetry group generated by translations, composition with elements of $\mathrm{Gl}(d)$, modulations, and scalar multiplications does not provide sufficiently many free parameters to ensure any more vanishing conditions.

The pairing of $P_\alpha^{(\tau_j)}$ with $g_j^{p_j-1}$, rather than with G^{τ_j}, in (46) may appear unnatural, but is the correct combination in this context. Indeed, recall that $g_j(x) = e^{-\pi p_j'|x|^2} = G^{p_j'}(x)$, so that $g_j^{p_j-1} = G^{p_j'(p_j-1)} = G^{p_j}$. Thus $\tilde{f}_j$ satisfies (46) if and only if the associated function $u_j = \tilde{f}_j g_j^{(p_j-2)/2}$ satisfies the natural condition $\langle u_j, H_\alpha^{(\tau_j)} \rangle = 0$ for the indicated pairs (j, α). This orthogonality will allow us to apply Lemma 15 below, after one auxiliary manipulation.

Proof of Lemma 17 Define $h_j \in L^{p_j}$ by $f_j = g_j + h_j$, set $\tilde{f}_j(x) = a_j f_j(\psi(x) + v_j)e^{ix\cdot\xi}$ where $\psi, \mathbf{v}, \mathbf{a}, \xi$ are to be determined, and define $\tilde{h}_j$ by $\tilde{f}_j(x) = g_j + \tilde{h}_j$. Rewritten in terms of $\tilde{h}_j$, the desired relations become

$$\begin{cases} \langle \mathrm{Re}(\tilde{h}_j), P_\alpha^{(\tau_j)} g_j^{p_j-1} \rangle = 0 \\ \langle \mathrm{Im}(\tilde{h}_j), P_\alpha^{(\tau_j)} g_j^{p_j-1} \rangle = 0 \end{cases} \tag{49}$$

for the indicated pairs of indices (j, α).

Writing $a_j = 1 + b_j$,

$$\begin{aligned}\tilde{h}_j(x) &= a_j f_j(\psi(x) + v_j)e^{i\xi\cdot x} - g_j \\ &= (1+b_j)\big(G^{p_j'}(\psi(x)+v_j)e^{ix\cdot\xi} + h_j(\psi(x)+v_j)e^{ix\cdot\xi}\big) - G^{p_j'}(x).\end{aligned}$$

Write $\psi(x) = x + \phi(x)$. Expand

$$\begin{aligned}&G^{p_j'}(\psi(x)+v_j)e^{ix\cdot\xi} - G^{p_j'}(x) \\ &\qquad = G^{p_j'}(x)\big(G^{-p_j'}(x)G^{p_j'}(x+v_j+\phi(x))e^{ix\cdot\xi} - 1\big) \\ &\qquad = G^{p_j'}(x)\big(e^{-\pi p_j'[|x+v_j+\phi(x)|^2-|x|^2]}e^{ix\cdot\xi} - 1\big) \\ &\qquad = G^{p_j'}(x)x\cdot[-2p_j'(\phi(x)+v_j)+i\xi] + O((\|\phi\|+|\mathbf{v}|+|\xi|)^2)\end{aligned}$$

where $O((\|\phi\|+|\mathbf{v}|+|\xi|)^2)$ denotes a function whose $L^{p_j}(\mathbb{R}^d)$ norm is $O((\|\phi\|+|\mathbf{v}|+|\xi|)^2)$. Combining this with corresponding expansions for other terms yields

$$\begin{aligned}&\tilde{h}_j(x) = a_j h_j(\psi(x)+v_j)e^{ix\cdot\xi} \\ &\quad + G^{p_j'}(x)\big[b_j - x\cdot[2p_j'(\phi(x)+v_j) - i\xi]\big] + O((\|\phi\|+|\mathbf{v}|+|\mathbf{b}|+|\xi|)^2).\end{aligned} \tag{50}$$

The contribution of the term $a_j h_j(\psi(x)+v_j)e^{ix\cdot\xi}$ to the quantities of interest can be evaluated:

$$\begin{aligned}&\langle a_j h_j(\psi(x)+v_j)e^{ix\cdot\xi},\ P_\alpha^{(\tau_j)} g_j^{p_j-1}\rangle \\ &= \langle h_j(\psi(x)+v_j),\ P_\alpha^{(\tau_j)} g_j^{p_j-1}\rangle + O\big((|b_j|+|\xi|)\|h_j\|_{p_j}\big) \\ &= \langle h_j(\psi(x)+v_j),\ P_\alpha^{(\tau_j)} G^{p_j},\rangle + O\big((|b_j|+|\xi|)\|h_j\|_{p_j}\big) \\ &= |\det(\psi)|^{-1}\int h_j(y),\ P_\alpha^{(\tau_j)}(\psi^{-1}(y-v_j)G^{p_j}(\psi^{-1}(y-v_j))\,dy \\ &\qquad + O\big((|b_j|+|\xi|)\|h_j\|_{p_j}\big) \\ &= \langle h_j,\ P_\alpha^{(\tau_j)} g_j^{p_j-1}\rangle\ +\ O\big((|b_j|+|\xi|+\|\phi\|+|v_j|)\ \|h_j\|_{p_j}\big).\end{aligned}$$

Therefore

$$\begin{aligned}&\langle \tilde{h}_j,\ P_\alpha^{(\tau_j)} g_j^{p_j-1}\rangle \\ &= \langle h_j,\ P_\alpha^{(\tau_j)} g_j^{p_j-1}\rangle + \langle G^{p_j'}(x)\big[b_j - x\cdot[2p_j'(\phi(x)+v_j) - i\xi]\big],\ P_\alpha^{(\tau_j)} g_j^{p_j-1}\rangle\end{aligned}$$

plus a remainder which is quadratic in the sense that it is

$$O\Big((\|\phi\| + |\mathbf{v}| + |\mathbf{b}| + |\xi|)^2 \ + \ (\|\phi\| + |\mathbf{v}| + |\mathbf{b}| + |\xi|)\, \|h_j\|_{p_j}\Big).$$

In order to apply the Implicit Function Theorem to reach the desired conclusion, it suffices to verify that the linear map

$$(\mathbf{b}, \mathbf{v}, \xi, \phi) \mapsto \langle \big[b_j - x \cdot (v_j + i\xi) - 2p_j' x \cdot \phi(x) \big],\ P_\alpha^{(\tau_j)} G^{p_j p_j'} \rangle, \tag{51}$$

with (j, α) ranging over the indicated family of indices and taking the real or imaginary part as indicated in (49), is invertible.

Let $\mathcal{S}_d$ be the vector space of all symmetric real $d \times d$ matrices. Let V_d be the real vector space of all tuples $(t_\alpha : |\alpha| = 2)$ with each $t_\alpha \in \mathbb{R}$ and $\alpha = (\alpha_1, \ldots, \alpha_d)$ with each $\alpha_j \in \{0, 1, 2\}$ and $\sum_{j=1}^d \alpha_j = 2$. Then the linear mapping from $\mathcal{S}_d$ to V_d defined by $\phi \mapsto \big(\langle x \cdot \phi(x),\ P_\alpha^{(\tau_3)} G^{p_3 p_3'} \rangle : |\alpha| = 2\big)$. is invertible. This is a consequence of the alternative expression

$$\langle x \cdot \phi(x),\ P_\alpha^{(\tau_3)} G^{p_3 p_3'} \rangle = \langle x \cdot \phi(x) G^{\tau_3},\ P_\alpha^{(\tau_3)} G^{\tau_3} \rangle = \langle x \cdot \phi(x) G^{\tau_3},\ H_\alpha^{(\tau_3)} \rangle,$$

which holds since $p_3 p_3' = 2\tau_3$. For each ϕ there exists a unique scalar c_ϕ such that with $P(x) = x \cdot \phi(x)$, $(P - c)G^{\tau_3}$ is a Hermite function relative to the parameter τ_3. Moreover,

$$\langle x \cdot \phi(x) G^{\tau_3},\ H_\alpha^{(\tau_3)} \rangle = \langle (x \cdot \phi(x) - c_\phi) G^{\tau_3},\ H_\alpha^{(\tau_3)} \rangle$$

whenever $|\alpha| = 2$ since G^{τ_3} is orthogonal to $H_\alpha^{(\tau_3)}$.

Likewise, the mapping from $(\mathbf{v}, \xi)$, with the constraint $v_1 + v_2 + v_3 = 0$, to the tuple of real and/or imaginary parts of $\langle x \cdot (v_j + i\xi)\, P_\alpha^{(\tau_j)} G^{p_j p_j'} \rangle$, indexed by the pairs (j, α) with $|\alpha| = 1$ indicated in the statement of Lemma 17, is invertible. Moreover, these inner products vanish for $\alpha \in \{0, 2\}$. Finally, the mapping from $\mathbf{b}$ to $\langle b_j,\ P_0^{(\tau_j)} G^{p_j p_j'} \rangle$, indexed by all $j \in \{1, 2, 3\}$, is invertible, while $\langle b_j,\ P_\alpha^{(\tau_j)} G^{p_j p_j'} \rangle = 0$ for $\alpha \neq 0$. Thus the required invertibility holds. □

7 Conclusion of Proof

Let $\mathbf{p} \in (1, 2)^3$ satisfy $\sum_{j=1}^3 p_j^{-1} = 2$. Let $d \geq 1$ be given. It suffices to prove Theorem 1 for tuples $\mathbf{F} = (F_j : j \in \{1, 2, 3\}) = \mathbf{g} + \mathbf{f}$ with $f_j \in L^{p_j}(\mathbb{R}^d)$ and $\|f_j\|_{L^{p_j}}$ sufficiently small for each index $j \in \{1, 2, 3\}$, satisfying $\int f_j g_j^{p_j - 1} = 0$. According to Lemma 17, by transforming $\mathbf{F}$ to an appropriately chosen nearby element of its orbit under the symmetry group of the inequality, we may also suppose that $\tilde{f}_j = F_j$ satisfy the supplementary moment conditions (46) and (48) for the multi-indices α indicated in that lemma.

Let $\eta > 0$ be a small parameter. Decompose $f_j = f_{j,\sharp} + f_{j,\flat}$ as in (12), with respect to the parameter η. Define

$$h_j = f_{j,\sharp} - \sum_{|\alpha|\le 2} c_{j,\alpha} P_\alpha^{(\tau_j)} g_j \tag{52}$$

with $c_{j,\alpha}$ chosen so that the functions h_j continue to satisfy the moment conditions (46) and (48), and with the summation for each index j extending over those α for which a corresponding moment condition appears. An equation $\langle \mathrm{Re}(h_j), P_\beta^{(\tau_j)} g_j^{p_j-1}\rangle = 0$ is equivalent to

$$\langle \mathrm{Re}(f_{j,\sharp}), P_\beta^{(\tau_j)} g_j^{p_j-1}\rangle = \sum_{|\alpha|\le 2} \mathrm{Re}(c_{j,\alpha})\langle P_\alpha^{(\tau_j)} g_j, P_\beta^{(\tau_j)} g_j^{p_j-1}\rangle = 0.$$

Since $g_j^{p_j} = G^{p_j' p_j} = G^{2\tau_j}$,

$$\sum_{|\alpha|\le 2} \mathrm{Re}(c_{j,\alpha})\langle P_\alpha^{(\tau_j)} g_j, P_\beta^{(\tau_j)} g_j^{p_j-1}\rangle = \sum_{|\alpha|\le 2} \mathrm{Re}(c_{j,\alpha})\langle H_\alpha^{(\tau_j)}, H_\beta^{(\tau_j)}\rangle = \mathrm{Re}(c_{j,\beta}).$$

The conditions (48) for $\mathrm{Im}(f_{j,\sharp})$ can be written in the same manner as equations for $\mathrm{Im}(c_{j,\beta})$ for appropriate pairs (j,β). Thus there exists a solution $(c_{j,\alpha})$ to the system of Eqs. (52), and moreover, since $f_j = f_{j,\sharp} + f_{j,\flat}$ and the moments of f_j vanish, there exists a solution satisfying

$$|c_{j,\alpha}| \le C\|f_{j,\flat}\|_{p_j} \tag{53}$$

for each j and each $|\alpha| \le 2$.

Define

$$u_j = h_j G^{p_j'(p_j-2)/2};$$

thus $h_j = u_j G^{(2-p_j)p_j'/2}$. Write $\|\mathbf{f}\|_{\mathbf{p}} = \sum_{j=1}^3 \|h_j\|_{p_j}$, and define $\|\mathbf{f}_\flat\|_{\mathbf{p}}$, $\|\mathbf{f}_\sharp\|_{\mathbf{p}}$, and $\|\mathbf{h}\|_{\mathbf{p}}$ in the same way. Then each of these quantities is small by hypothesis, since $\|\mathbf{f}_\sharp\|_{\mathbf{p}} \le \|\mathbf{f}\|_{\mathbf{p}}$ and likewise for $\mathbf{f}_\flat$.

Since $|c_{j,\alpha}| = O(\|f_{j,\flat}\|_{p_j})$,

$$\mathcal{Q}_{\mathbf{p}}(\mathbf{f}_\sharp) \le \mathcal{Q}_{\mathbf{p}}(\mathbf{h}) + O(\|\mathbf{f}\|_{\mathbf{p}}\|\mathbf{f}_\flat\|_{\mathbf{p}}).$$

Consequently there exists $c' = c'(\mathbf{p}) > 0$ such that

$$\frac{\mathcal{T}(\mathbf{g}+\mathbf{f})}{\prod_j \|g_j + f_j\|_{p_j}} \le \mathbf{A}_p^d + \mathbf{A}_p^d Q_{\mathbf{p}}(\mathbf{f}_\sharp) + C\eta\|\mathbf{f}_\sharp\|_{\mathbf{p}}^2 - c\sum_j \eta^{c'}\|f_{j,\flat}\|_{p_j}^{p_j} + O(\|\mathbf{f}\|_{\mathbf{p}}^3)$$

$$\le \mathbf{A}_p^d + \mathbf{A}_p^d Q_{\mathbf{p}}(\mathbf{h}) + C\eta\|\mathbf{f}_\sharp\|_{\mathbf{p}}^2 - c\sum_j \eta^{c'}\|f_{j,\flat}\|_{p_j}^{p_j}$$

$$+ O(\|\mathbf{f}\|_{\mathbf{p}}\|\mathbf{f}_\flat\|_{\mathbf{p}}) + O(\|\mathbf{f}\|_{\mathbf{p}}^3)$$

$$= \mathbf{A}_p^d + \mathbf{A}_p^d Q_{\mathbf{p}}^+(\mathrm{Re}(\mathbf{u})) + \mathbf{A}_p^d Q_{\mathbf{p}}^-(\mathrm{Im}(\mathbf{u})) + C\eta\|\mathbf{f}_\sharp\|_{\mathbf{p}}^2$$

$$- c\sum_j \eta^{c'}\|f_{j,\flat}\|_{p_j}^{p_j} + O(\|\mathbf{f}\|_{\mathbf{p}}\|\mathbf{f}_\flat\|_{\mathbf{p}}) + O(\|\mathbf{f}\|_{\mathbf{p}}^3)$$

$$\le \mathbf{A}_p^d - 2\gamma_{\mathbf{p}}\mathbf{A}_p^d\sum_j \|u_j\|_{L^2}^2 + C\eta\|\mathbf{f}_\sharp\|_{\mathbf{p}}^2$$

$$- c\sum_j \eta^{c'}\|f_{j,\flat}\|_{p_j}^{p_j} + O(\|\mathbf{f}\|_{\mathbf{p}}\|\mathbf{f}_\flat\|_{\mathbf{p}}) + O(\|\mathbf{f}\|_{\mathbf{p}}^3)$$

where $\gamma_{\mathbf{p}}$ is strictly positive; we have invoked Lemmas 12 and 13 along with the relation (21) between $Q_{\mathbf{p}}^+$, $Q_{\mathbf{p}}^-$, and $Q_{\mathbf{p}}$ to obtain the crucial final inequality. Now

$$\|u_j\|_{L^2} = \|G^{p_j'(p_j-2)/2}h_j\|_{L^2} \ge \|G^{p_j'(p_j-2)/2}f_{j,\sharp}\|_{L^2} - C\|\mathbf{f}_\flat\|_{\mathbf{p}}$$

so we may conclude that

$$\frac{\mathcal{T}(\mathbf{g}+\mathbf{f})}{\prod_j \|g_j + f_j\|_{p_j}} \le \mathbf{A}_p^d - \gamma_{\mathbf{p}}\mathbf{A}_p^d\sum_j \|f_{j,\sharp}g_j^{(p_j-2)/2}\|_2^2$$

$$+ C\eta\|\mathbf{f}_\sharp\|_{\mathbf{p}}^2 - c\sum_j \eta^{c'}\|f_{j,\flat}\|_{p_j}^{p_j} + O(\|\mathbf{f}_\flat\|_{\mathbf{p}}^2) + O(\|\mathbf{f}\|_{\mathbf{p}}\|\mathbf{f}_\flat\|_{\mathbf{p}}) + O(\|\mathbf{f}\|_{\mathbf{p}}^3).$$

Each exponent $(p_j - 2)/2$ is negative, so $g_j^{(p_j-2)/2}$ is a strictly positive, rapidly growing function. Therefore by Hölder's inequality,

$$\|f_{j,\sharp}\|_{p_j} \le C\|f_{j,\sharp}g_j^{(p_j-2)/2}\|_2$$

and consequently if η is chosen to be sufficiently small then the adverse term $C\eta\|\mathbf{f}_\sharp\|_{\mathbf{p}}^2$ can be absorbed, yielding

$$\frac{\mathcal{T}(\mathbf{g}+\mathbf{f})}{\prod_j \|g_j + f_j\|_{p_j}} \le \mathbf{A}_p^d - c\sum_j \|f_{j,\sharp}g_j^{(p_j-2)/2}\|_2^2 - c\sum_j \eta^{c'}\|f_{j,\flat}\|_{p_j}^{p_j}$$

$$+ O(\|\mathbf{f}_\flat\|_{\mathbf{p}}^2) + O(\|\mathbf{f}\|_{\mathbf{p}}\|\mathbf{f}_\flat\|_{\mathbf{p}}) + O(\|\mathbf{f}\|_{\mathbf{p}}^3).$$

By choosing η to be a sufficiently small positive constant, we conclude that

$$\frac{\mathcal{T}(\mathbf{g}+\mathbf{f})}{\prod_j \|g_j+f_j\|_{p_j}} \le \mathbf{A}_p^d - c'' \sum_j \|f_{j,\sharp} g_j^{(p_j-2)/2}\|_2^2 - \tilde{c} \sum_j \|f_{j,\flat}\|_{p_j}^{p_j} \tag{54}$$

for some constants $c'', \tilde{c} > 0$ that depend only on p, d. Indeed, each exponent p_j is strictly less than 2. The quantity $\|\mathbf{f}\|_{\mathbf{p}}$ is small by hypothesis, and hence both $\|\mathbf{f}_\sharp\|_{\mathbf{p}}$ and $\|\mathbf{f}_\flat\|$ are likewise small. Consequently, the remainder $O(\|\mathbf{f}_\flat\|_{\mathbf{p}}^2) + O(\|\mathbf{f}\|_{\mathbf{p}}\|\mathbf{f}_\flat\|_{\mathbf{p}}) + O(\|\mathbf{f}\|_{\mathbf{p}}^3)$ can be absorbed.

Recall that $f_{j,\sharp} + f_{j,\flat} = f_j$ and that for any x, at most one of $f_{j,\sharp}(x)$ and $f_{j,\flat}(x)$ is nonzero. A majorization $\|f_{j,\sharp}\|_{p_j} \le C\|f_{j,\sharp} g_j^{(p_j-2)/2}\|_2$ follows from Hölder's inequality since $p_j < 2$ and consequently $g_j^{(2-p_j)/2}$ is a Schwartz function. Since $\|f_j\|_{p_j}$ is assumed to be small, and thus $\|f_j\|_{p_j} \le 1$,

$$\|f_{j,\flat}\|_{p_j}^{p_j} = \|f_{j,\flat}\|_{p_j}^2 \cdot \|f_{j,\flat}\|_{p_j}^{p_j-2} \ge \|f_{j,\flat}\|_{p_j}^2.$$

Therefore

$$\|f_j\|_{p_j}^2 \le C\|f_{j,\sharp}\|_{p_j}^2 + C\|f_{j,\flat}\|_{p_j}^2 \le C\|f_{j,\sharp} g_j^{(p_j-2)/2}\|_2^2 + C\|f_{j,\flat}\|_{p_j}^{p_j}.$$

Inserting this information into (54) gives

$$\frac{\mathcal{T}(\mathbf{g}+\mathbf{f})}{\prod_j \|g_j+f_j\|_{p_j}} \le \mathbf{A}_p^d - c \sum_j \|f_j\|_{p_j}^2 \tag{55}$$

for another constant $c > 0$. This completes the proof of Theorem 1 modulo the proof of Lemma 14. □

8 Hermite Functions and Singular Values

In this section, we prove Lemma 14. Assume that $p_j \in (1, 2]$ and $\sum_j p_j^{-1} = 2$.

Let $G(x) = e^{-\pi|x|^2}$, for $x \in \mathbb{R}^d$. For $u, v, w > 0$ consider the bounded linear operator $T = T_{u,v,w}$ on $L^2(\mathbb{R}^d)$ defined by

$$Tf = G^u \cdot (G^v * (G^w f)). \tag{56}$$

Let $t \in \mathbb{R}^+$ be arbitrary. If $f = G^t$ then $G^w f = G^{t+w}$, and $G^v * (G^w f)$ is a positive scalar multiple of G^r where $r^{-1} = v^{-1} + (w+t)^{-1}$. Therefore $T(G^t)$ is a positive scalar multiple of G^s, where $s = u + r$. Define $\sigma(t, u, v, w)$ to be this quantity s. Straightforward calculation gives

$$\sigma(t, u, v, w) = \frac{uv + vw + wu + (u + v)t}{v + w + t}. \tag{57}$$

If $d = 1$ and P is a polynomial of degree equal to n then for any $a, b \in \mathbb{R}^+$, the convolution $G^a * (PG^b)$ takes the form QG^c for some $c \in \mathbb{R}^+$, where Q is a polynomial of degree equal to n, and c depends only on a, b. If P has positive leading coefficient, then so does Q. Therefore $T(u, v, w)(PG^t)$ takes the form QG^s where $s = \sigma(t, u, v, w)$ and Q is a polynomial of the same degree as P; Q has positive leading coefficient if P does.

If there does exist $r \in \mathbb{R}^+$ satisfying $t = \sigma(r, w, v, u)$, then the adjoint operator $T^* = T^*_{u,v,w} = T_{w,v,u}$ maps G^r to a scalar multiple of G^t. More generally, it maps any polynomial multiple of G^r to a polynomial of the same degree multiplied by G^t.

Lemma 18 *Let $d = 1$. Let $t, u, v, w \in \mathbb{R}^+$. Suppose that there exists $r \in \mathbb{R}^+$ satisfying $t = \sigma(r, w, v, u)$. Define s, ρ by $s = \sigma(t, u, v, w)$ and $\rho = (s + r)/2$. Then for every $n \in \{0, 1, 2, \ldots\}$, $T_{u,v,w}(P_n^{(t)} G^t)$ is a positive scalar multiple of $P_n^{(\rho)} G^s$.*

Proof Let $T = T_{u,v,w}$. The definition of s and hypothesis on r guarantee that $T(G^t)$ and $T^*(G^r)$ are scalar multiples of G^s, G^t, respectively. In particular, the stated conclusion holds for $n = 0$.

We argue by induction on n, the induction hypothesis for n being that the statement holds for arbitrary t, u, v, w satisfying the hypotheses for all smaller values of n. Let $n \geq 1$. Let Q be an arbitrary polynomial of degree strictly less than n. According to the induction hypothesis, $T^*(QG^r) = RG^t$ for some polynomial R of degree $< n$. Therefore

$$\langle T(P_n^{(t)} G^t),\ QG^r \rangle = \langle P_n^{(t)} G^t,\ T^*(QG^r) \rangle = \langle P_n^{(t)} G^t,\ RG^t \rangle,$$

where the inner products are that of $L^2(\mathbb{R}^1)$ with respect to Lebesgue measure. By the definition of $P_n^{(t)}$, $\langle P_n^{(t)} G^t,\ RG^t \rangle = 0$. Thus $\langle T(P_n^{(t)} G^t),\ QG^r \rangle = 0$.

It was observed above that $T(P_n^{(t)} G^t)$ can be expressed in the form $\tilde{P} G^s$, where $\tilde{P}$ is a polynomial of degree equal to n with positive leading coefficient. We have shown in the preceding paragraph that $\tilde{P} G^s \perp QG^r$ for every polynomial Q of degree $< n$, that is, $\int \tilde{P} Q G^{s+r} = 0$. Equivalently, $\tilde{P} G^{(\rho)} \perp QG^{(\rho)}$, for every polynomial Q of degree $< n$. Since $\tilde{P}$ has degree exactly n and positive leading coefficient, it must be a positive scalar multiple of $P_n^{(\rho)}$. □

We will apply Lemma 18 in a context in which $r = s$. Its conclusion then simplifies.

Corollary 19 *Let $s, t, u, v, w \in \mathbb{R}^+$ be arbitrary. Let $d \geq 1$. Suppose that s, t satisfy*

$$\begin{cases} s = \sigma(t, u, v, w) \\ t = \sigma(s, w, v, u). \end{cases} \tag{58}$$

Then for every $\alpha \in \{0, 1, 2, \dots\}^d$, there exists $\lambda_\alpha(t, u, v, w) \in \mathbb{R}^+$ satisfying

$$T_{u,v,w}(P_\alpha^{(t)} G^t) = \lambda_\alpha(t, u, v, w) P_\alpha^{(s)} G^s. \tag{59}$$

For $d = 1$, this is a special case of the preceding lemma. The case of general dimensions d follows from the case $d = 1$ by virtue of the product structure of higher-dimensional Hermite functions.

Let $\mathbf{p} \in (1, 2)^3$ satisfy $\sum_j p_j^{-1} = 2$. Consider any permutation (i, j, k) of $(1, 2, 3)$, not necessarily cyclic. Define $T_{(i,j,k)}$ in terms of $\mathbf{p}$ by (20). Then $T_{(i,j,k)} = T_{u,v,w}$ with

$$w = p_i'(2 - p_i)/2, \qquad v = p_k', \qquad u = p_j'(2 - p_j)/2. \tag{60}$$

Lemma 20 *For any indices $i \neq j \in \{1, 2, 3\}$,*

$$\tau_j = \sigma(\tau_i, p_j'(2 - p_j)/2, p_k', p_i'(2 - p_i)/2). \tag{61}$$

Proof

$$\begin{aligned} T_{(i,j,k)} G^{\tau_i} &= T_{(i,j,k)} G^{p_i' p_i/2} = G^{p_j'(2-p_j)/2} \cdot \left(G^{p_k'} * (G^{p_i'(2-p_i)/2} G^{p_i' p_i/2})\right) \\ &= G^{p_j'(2-p_j)/2} \cdot (G^{p_k'} * G^{p_i'}) = G^{p_j'(2-p_j)/2} (p_i' p_k')^{-d/2} p_j^{d/2} G^{p_j} \\ &= (p_i' p_k')^{-d/2} p_j^{d/2} G^{p_j' p_j/2} = (p_i' p_k')^{-d/2} p_j^{d/2} G^{\tau_j}. \end{aligned}$$

We have used the relations $\widehat{G} = G$, $\int G = 1$, $(p_k')^{-1} + (p_i')^{-1} = 2 - p_k^{-1} - p_i^{-1} = p_j^{-1}$, $p_j + p_j' = p_j p_j'$, and

$$G^a * G^b = a^{-d/2} b^{-d/2} c^{d/2} G^c$$

where $c^{-1} = a^{-1} + b^{-1}$. Here, $\widehat{G}(\xi) = \int G(x) e^{-2\pi i x \cdot \xi}\, dx$ denotes the Fourier transform of G.

□

We have shown that

$$T_{(i,j,k)} G^{\tau_i} = (p_i' p_k')^{-d/2} p_j^{d/2} G^{\tau_j}. \tag{62}$$

The same applies to $T_{(j,i,k)}$. Therefore with u, v, w defined by (60), $\tau_j = \sigma(\tau_i, u, v, w)$ and $\tau_i = \sigma(\tau_j, w, v, u)$. This is the hypothesis of Corollary 19, which now yields the following key result.

Corollary 21 *Let $d \geq 1$ Let $\mathbf{p} \in (1, 2]^3$ satisfying $\sum_{j=1}^{3} p_j^{-1} = 2$. There exist positive scalars $\lambda_{\alpha,(i,j,k)}$, which depend also on $\mathbf{p}$ and on d, satisfying*

$$T_{(i,j,k)}(H_\alpha^{(\tau_i)}) = \lambda_{\alpha,(i,j,k)}\, H_\alpha^{(\tau_j)}. \tag{63}$$

We next complete the proof of Lemma 14 by calculating $\lambda_{\alpha,(i,j,k)}$. Combining (26) with (62) yields

$$\lambda_{0,(i,j,k)} = (2\tau_i)^{d/4}(2\tau_j)^{-d/4}\Big(\frac{p_j}{p_i' p_k'}\Big)^{d/2} = \Big(\frac{p_i}{p_i'}\Big)^{d/4}\Big(\frac{p_j}{p_j'}\Big)^{d/4}(p_k')^{-d/2}. \tag{64}$$

Specialize initially to the case $d = 1$, denoting the quantities λ for $d = 1$ by $\lambda_{n,(i,j,k)}$ for $n \in \{0, 1, 2, \dots\}$ and reserving the Greek subscript α for discussion below of general d. Consider $T_{(i,j,k)} P_n^{(\tau_i)} G^{\tau_i}$. Let $Q_n^{(\tau)}$ be the unique scalar multiple of $P_n^{(\tau)}$ whose leading coefficient equals 1.

Lemma 22 *Let $d = 1$. For any $n \geq 1$,*

$$\lambda_{n,(i,j,k)} = \left(\frac{p_i p_j}{p_i' p_j'}\right)^{1/2} \lambda_{n-1,(i,j,k)}. \tag{65}$$

Proof We will exploit a version of the classical raising and lowering operators. It is convenient to observe that

$$\begin{aligned}
\frac{d}{dx} Q_{n-1}^{(\tau_i)} G^{p_i'} &= Q_{n-1}^{(\tau_i)} \frac{d}{dx} G^{p_i'} + O(x^{n-2}) G^{p_i'} \\
&= (-2\pi p_i' x) Q_{n-1}^{(\tau_i)} G^{p_i'} + O(x^{n-2}) G^{p_i'} \\
&= (-2\pi p_i') Q_n^{(\tau_i)} G^{p_i'} + O(x^{n-2}) G^{p_i'}
\end{aligned}$$

where $O(x^k)G^t$ denotes the product of G^t with a polynomial of degree $\leq k$.

Denote by $\approx$ the equivalence relation on the class of products QG^{τ_j} of polynomials with G^{τ_j}, with $f \approx g$ if $f - g = RG^t$ for some polynomial R of degree strictly less than n. Now

$$\begin{aligned}
&\| Q_n^{(\tau_i)} G^{\tau_i} \|_{L^2}\, T_{(i,j,k)} P_n^{(\tau_i)} G^{\tau_i} \\
&= T_{(i,j,k)} Q_n^{(\tau_i)} G^{\tau_i} \\
&= G^{p_j'(2-p_j)/2} \cdot \big(G^{p_k'} * (G^{p_i'(2-p_i)/2} Q_n^{(\tau_i)} G^{p_i' p_i/2})\big)
\end{aligned}$$

$$= G^{p_j'(2-p_j)/2} \cdot \big(G^{p_k'} * Q_n^{(\tau_i)} G^{p_i'}\big)$$
$$\approx (-2\pi p_i')^{-1} G^{p_j'(2-p_j)/2} \cdot \big(G^{p_k'} * \frac{d}{dx}[Q_{n-1}^{(\tau_i)} G^{p_i'}]\big)$$
$$= (-2\pi p_i')^{-1} G^{p_j'(2-p_j)/2} \cdot \frac{d}{dx}\big(G^{p_k'} * Q_{n-1}^{(\tau_i)} G^{p_i'}\big)$$
$$= (-2\pi p_i')^{-1} \frac{d}{dx}\big[G^{p_j'(2-p_j)/2} \cdot \big(G^{p_k'} * Q_{n-1}^{(\tau_i)} G^{p_i'}\big)\big]$$
$$- (-2\pi p_i')^{-1}\big[\frac{d}{dx} G^{p_j'(2-p_j)/2}\big] \cdot \big(G^{p_k'} * Q_{n-1}^{(\tau_i)} G^{p_i'}\big)$$

by Leibniz's formula. We may continue

$$\|Q_n^{(\tau_i)} G^{\tau_i}\|_{L^2}\, T_{(i,j,k)} P_n^{(\tau_i)} G^{\tau_i}$$
$$\approx (-2\pi p_i')^{-1} \frac{d}{dx} T_{(i,j,k)}(Q_{n-1}^{(\tau_i)} G^{\tau_i})$$
$$+ (2\pi p_i')^{-1}\big[(-2\pi p_j'(2-p_j)/2) x G^{p_j'(2-p_j)/2}\big] \cdot \big(G^{p_k'} * Q_{n-1}^{(\tau_i)} G^{p_i'}\big)$$
$$= (-2\pi p_i')^{-1} \frac{d}{dx} T_{(i,j,k)}(Q_{n-1}^{(\tau_i)} G^{\tau_i})$$
$$+ (2\pi p_i')^{-1}(-2\pi p_j'(2-p_j)/2) x T_{(i,j,k)}(Q_{n-1}^{(\tau_i)} G^{\tau_i})$$
$$= \|Q_{n-1}^{(\tau_i)} G^{\tau_i}\|_{L^2}(-2\pi p_i')^{-1}\Big[\frac{d}{dx} + \pi p_j'(2-p_j)x\Big] T_{(i,j,k)}(P_{n-1}^{(\tau_i)} G^{\tau_i}).$$

Using the hypothesis on n to evaluate $T_{(i,j,k)}(P_{n-1}^{(\tau_i)} G^{\tau_i})$ yields

$$\|Q_n^{(\tau_i)} G^{\tau_i}\|_{L^2}\, T_{(i,j,k)} P_n^{(\tau_i)} G^{\tau_i}$$
$$\approx \|Q_{n-1}^{(\tau_i)} G^{\tau_i}\|_{L^2}(-2\pi p_i')^{-1}\Big[\frac{d}{dx} + (2\pi p_j' - 2\pi\tau_j)x\Big]\lambda_{n-1,(i,j,k)} P_{n-1}^{(\tau_j)} G^{\tau_j}$$
$$\approx \|Q_{n-1}^{(\tau_i)} G^{\tau_i}\|_{L^2}(-2\pi p_i')^{-1}\lambda_{n-1,(i,j,k)}[(-2\pi\tau_j)x + (2\pi p_j' - 2\pi\tau_j)x] P_{n-1}^{(\tau_j)} G^{\tau_j}$$
$$= \|Q_{n-1}^{(\tau_i)} G^{\tau_i}\|_{L^2}(-2\pi p_i')^{-1}\lambda_{n-1,(i,j,k)}(-2\pi p_j) x P_{n-1}^{(\tau_j)} G^{\tau_j}$$
$$= \|Q_{n-1}^{(\tau_i)} G^{\tau_i}\|_{L^2}(p_j/p_i')\lambda_{n-1,(i,j,k)} x P_{n-1}^{(\tau_j)} G^{\tau_j}$$
$$= (p_j/p_i')\lambda_{n-1,(i,j,k)}\|Q_{n-1}^{(\tau_i)} G^{\tau_i}\|_{L^2}\|Q_{n-1}^{(\tau_j)} G^{\tau_j}\|_{L^2}^{-1} x Q_{n-1}^{(\tau_j)} G^{\tau_j}$$
$$\approx (p_j/p_i')\lambda_{n-1,(i,j,k)}\|Q_{n-1}^{(\tau_i)} G^{\tau_i}\|_{L^2}\|Q_{n-1}^{(\tau_j)} G^{\tau_j}\|_{L^2}^{-1} Q_n^{(\tau_j)} G^{\tau_j}$$
$$= (p_j/p_i')\lambda_{n-1,(i,j,k)}\|Q_{n-1}^{(\tau_i)} G^{\tau_i}\|_{L^2}\|Q_{n-1}^{(\tau_j)} G^{\tau_j}\|_{L^2}^{-1}\|Q_n^{(\tau_j)} G^{\tau_j}\|_{L^2} P_n^{(\tau_j)} G^{\tau_j}.$$

We have used the identity $p_j' - 2\tau_j = p_j' - p_j p_j' = p_j'(1-p_j) = -p_j$.

We conclude that

$$\lambda_{n,(i,j,k)} P_n^{(\tau_j)} G^{\tau_j} \approx (p_j/p_i') \| Q_n^{(\tau_i)} G^{\tau_i} \|_{L^2}^{-1} \| Q_{n-1}^{(\tau_i)} G^{\tau_i} \|_{L^2}$$
$$\cdot \| Q_{n-1}^{(\tau_j)} G^{\tau_j} \|_{L^2}^{-1} \| Q_n^{(\tau_j)} G^{\tau_j} \|_{L^2} \lambda_{n-1,(i,j,k)} P_n^{(\tau_j)} G^{\tau_j}$$

and hence, since the left-hand side has already been shown to be equal to a scalar multiple of the right-hand side, that

$$\lambda_{n,(i,j,k)} = \frac{p_j}{p_i'} \frac{\| Q_{n-1}^{(\tau_i)} G^{\tau_i} \|_{L^2} \| Q_n^{(\tau_j)} G^{\tau_j} \|_{L^2}}{\| Q_{n-1}^{(\tau_j)} G^{\tau_j} \|_{L^2} \| Q_n^{(\tau_i)} G^{\tau_i} \|_{L^2}} \lambda_{n-1,(i,j,k)}. \tag{66}$$

The polynomials $Q_n^{(\tau)}$, for different parameters τ, are identical up to dilation and multiplication by normalizing scalars:

$$Q_n^{(\tau_j)}(x) = \tau_i^{n/2} \tau_j^{-n/2} Q_n^{(\tau_i)}(\tau_i^{-1/2} \tau_j^{1/2} x) \tag{67}$$

and therefore

$$\frac{\| Q_n^{(\tau_j)} G^{\tau_j} \|_{L^2}}{\| Q_n^{(\tau_i)} G^{\tau_i} \|_{L^2}} = (\tau_i/\tau_j)^{(2n+1)/4}. \tag{68}$$

Consequently

$$\lambda_{n,(i,j,k)} = (p_j/p_i')(\tau_i/\tau_j)^{1/2} \lambda_{n-1,(i,j,k)}$$
$$= \left(\frac{p_j^2 p_i p_i'}{p_i'^2 p_j p_j'} \right)^{1/2} \lambda_{n-1,(i,j,k)} = \left(\frac{p_i p_j}{p_i' p_j'} \right)^{1/2} \lambda_{n-1,(i,j,k)}. \tag{69}$$

□

It follows from (65) that

$$\lambda_{n,(i,j,k)} = \left(\frac{p_i p_j}{p_i' p_j'} \right)^{n/2} \lambda_{0,(i,j,k)} \ \forall n \geq 0 \tag{70}$$

for $d = 1$. Inserting the expression (64) for $\lambda_{0,(i,j,k)}$ and expressing higher-dimensional Hermite functions as products involving one-dimensional Hermite functions gives (32) for arbitrary indices α and dimensions d.

9 The Case $p_i = 2$

The proof of Theorem 2 is a small modification of that of Theorem 1. Suppose that each exponent belongs to $(1, 2]$, and that some exponent equals 2. Since $\sum_j p_j^{-1} = 2$ and each p_j is assumed to be strictly greater than 1, at most one of the three exponents p_j can equal 2. Since $\mathcal{T}$ is invariant under permutations, we may assume without loss of generality that $p_3 = 2$ and $p_1, p_2 \in (1, 2)$.

Modify the definitions of $f_{3,\sharp}$, $f_{3,\flat}$ by setting $f_{3,\sharp} \equiv f_3$ and $f_{3,\flat} \equiv 0$. Allowing p_3 to equal 2 has multiple consequences in the proof. The condition $|f_{3,\sharp}| \le \eta g_3$ is lost, but will not be needed since

$$\|g_3 + f_3\|_2 = (\|g_3\|_2^2 + \|f_3\|_2^2)^{1/2} = \|g_3\|_2 + \tfrac{1}{2}\|g_3\|_2^{-1}\|f_3\|_2^2 + O(\|f_3\|_2^3).$$

Thus

$$\frac{\mathcal{T}(\mathbf{g}+\mathbf{f})}{\prod_j \|g_j + f_j\|_{p_j}} \le \mathbf{A}_p^d + \mathbf{A}_p^d \mathcal{Q}_{\mathbf{p}}(\mathbf{f}_\sharp)$$
$$+ C\eta \sum_{j=1,2} \|f_{j,\sharp}\|_{p_j}^2 \|g_j\|_{p_j}^{-2} - c \sum_{j=1,2} \eta^{2-p_j} \|f_{j,\flat}\|_{p_j}^{p_j} \|g_j\|_{p_j}^{-p_j} + O(\|\mathbf{f}\|^3)$$

where the definition of $\mathcal{Q}_{\mathbf{p}}$ is unchanged. Both an adverse term involving $\|f_{3,\sharp}\|_{p_3}$, and a favorable term involving $\|f_{3,\flat}\|_{p_3}$, on the right-hand side have been lost.

Factors $G^{(2-p_3)p_3'/2}$ that appeared throughout the analysis above are now identically equal to 1. In the above analysis, the fact that u_3 belonged to L^2 resulted from the relation $|f_3| \le \eta g_3$, but now it results instead from the relation $u_3 = g_3^{(p_3-2)/2} h_3 = h_3$.

The analysis involving Hermite functions in Sects. 8 and 5.2 is unchanged; the operators $T_{(i,j,k)}$ are still compact since each is defined by convolution with a Gaussian, either preceded or followed by multiplication by a Gaussian, or both.

The analysis in Sect. 5.3 exploited the fact that each $p_j - 1$ was strictly less than 1, but in fact these factors always arose in the form of products $(p_i - 1)(p_j - 1)$ with $i \ne j$. Such a product is strictly less than 1 since neither factor can exceed 1, and at least one factor must be strictly smaller. The rest of the analysis is essentially unaffected. □

10 Bilinear Variant

Let $p_j \in (1, 2)$ for $j = 1, 2$. Let $q^{-1} = p_1^{-1} + p_2^{-1} - 1$ and assume that $q \in (1, \infty)$. Let $d \ge 1$. Let $F_j \in L^{p_j}(\mathbb{R}^d)$ be nonnegative functions. Theorem 4 states that

$$\|F_1 * F_2\|_q \le (\mathbf{A}_p^d - c\,\mathrm{Dist}_{p_1,p_2}\big((F_1, F_2), \mathfrak{G}'_{p_1,p_2}\big)^2)\|F_1\|_{p_1}\|F_2\|_{p_2}.$$

For $q \geq 2$, Theorem 1 yields a stronger conclusion by duality, so we may assume that $q \in (1, 2)$.

The proof of Theorem 4 is a modification of that of Theorem 1. We sketch it, indicating those points at which changes are required. As in the proof of Theorem 1, it suffices to establish the conclusion under the auxiliary assumption that the two functions F_j take the forms $F_j = g_j + f_j$ with $g_j = e^{-p_j'|x|^2}$, and with $\|f_j\|_{p_j}$ small. Let

$$F = (g_1 + f_1) * (g_2 + f_2) \text{ and } g = g_1 * g_2, \tag{71}$$

write $\mathbf{f} = (f_1, f_2)$, and write

$$\|\mathbf{f}\|_{\mathbf{p}} = \max(\|f_1\|_{p_1}, \|f_2\|_{p_2}). \tag{72}$$

The notation $\mathbf{p}$ will denote either (p_1, p_2) or (p_1, p_2, p_3) where $p_3 = q'$ satisfies $\sum_{j=1}^{3} p_j^{-1} = 2$, depending on context.

Assuming that $p_1, p_2 \in (1, 2)$, decompose $f_j = f_{j,\sharp} + f_{j,\flat}$ as in the proof of Theorem 1, relative to a small auxiliary parameter $\eta \in (0, 1]$. Also decompose

$$F = F_1 * F_2 = g + F_\sharp + F_\flat \tag{73}$$

by setting

$$F_\sharp = (g_1 + f_{1,\sharp}) * (g_2 + f_{2,\sharp}) - g. \tag{74}$$

Thus

$$F_\flat (g_1 + f_{1,\sharp} + f_{1,\flat}) * (g_2 + f_{2,\sharp} + f_{2,\flat}) - (g_1 + f_{1,\sharp}) * (g_2 + f_{2,\sharp}) \tag{75}$$

represents the total contribution of all terms involving $f_{1,\flat}, f_{2,\flat}$. Since $|f_{j,\sharp}| \leq \eta g_j$ pointwise,

$$|F_\sharp| = \left|f_{1,\sharp} * g_2 + g_1 * f_{2,\sharp} + f_{1,\sharp} * f_{2,\sharp}\right| \leq (2\eta + \eta^2)\eta g \leq 3\eta g \tag{76}$$

at every point of $\mathbb{R}^d$.

Taylor expansion and simple majorizations give

$$|F|^q \leq g^q + qg^{q-1}(F-g) + \tfrac{q(q-1)}{2} g^{q-2} F_\sharp^2 + C\eta^{2-q}|F_\flat|^q + Cg^{q-3}|F_\sharp|^3. \tag{77}$$

Since $\|F_\sharp\|_q + \|F_\flat\|_q = O(\|f_1\|_{p_1} + \|f_2\|_{p_2})$ is small, binomial expansion gives

$$\begin{aligned} \|F\|_q \leq \|g\|_q \Big(1 + \|g\|_q^{-q} \int g^{q-1}(F-g) + \tfrac{q-1}{2} \|g\|_q^{-q} \int g^{q-2} F_\sharp^2 \\ + C\|g\|_q^{-q} \|F_\flat\|_q^q + C\|g\|_q^{-q} \int g^{q-3}|F_\sharp|^3\Big). \end{aligned} \tag{78}$$

Assuming that $\int g_j^{p_j-1} f_j = 0$,

$$\|g_j + f_j\|_{p_j} \ge \|g_j\|_{p_j}\Big(1 + \tfrac{p_j-1}{2}\|g_j\|_{p_j}^{-p_j}\int f_{j,\sharp}^2 g_j^{p_j-2} - C\eta\|f_{j,\sharp}\|_{p_j}^2\|g_j\|_{p_j}^{-2} + c\eta^{2-p_j}\|f_{j,\flat}\|_{p_j}^{p_j}\|g_j\|_{p_j}^{-p_j}\Big), \tag{79}$$

a bound already exploited in the proof of Theorem 1. Taking the ratio of the last two inequalities gives

$$\frac{\|F\|_q}{\prod_{j=1,2}\|g_j + f_j\|_{p_j}} \le \mathbf{A}_p^d\Big(1 + \|g\|_q^{-q}\int g^{q-1}(F-g) + \tilde{Q} + \tilde{\mathcal{R}}\Big) \tag{80}$$

where the remainder term is

$$\tilde{\mathcal{R}} \le C\|F_\flat\|_q^q + C\int g^{q-3}|F_\sharp|^3 + \sum_{j=1,2}\Big(C\eta\|f_{j,\sharp}\|_{p_j}^2 - c\eta^{2-p_j}\|f_{j,\flat}\|_{p_j}^{p_j}\Big) \tag{81}$$

where $c, C \in \mathbb{R}^+$ depend only on $\mathbf{p}$, d, and

$$\tilde{Q} = \tfrac{q-1}{2}\|g\|_q^{-q}\int g^{q-2}F_\sharp^2 - \sum_{j=1,2}\tfrac{p_j-1}{2}\|g_j\|_{p_j}^{-p_j}\int f_{j,\sharp}^2 g_j^{p_j-2}. \tag{82}$$

Moreover,

$$\int g^{q-2}F_\sharp^2 \le \int g^{q-2}(f_{1,\sharp} * g_2 + g_1 * f_{2,\sharp})^2 + C\|\mathbf{f}\|_{\mathbf{p}}^3.$$

Rearrange $\tilde{Q} + \tilde{\mathcal{R}}$ as $Q + \mathcal{R}$ by incorporating this last term $C\|\mathbf{f}\|_{\mathbf{p}}^3$ from $\tilde{Q}$ into $\mathcal{R}$. With this modification,

$$Q = \tfrac{q-1}{2}\|g\|_q^{-q}\int g^{q-2}(g_1 * f_{2,\sharp} + g_2 * f_{2,\sharp})^2 - \sum_{j=1,2}\tfrac{p_j-1}{2}\|g_j\|_{p_j}^{-p_j}\int f_{j,\sharp}^2 g_j^{p_j-2}, \tag{83}$$

while

$$\mathcal{R} \le C\|F_\flat\|_q^q + C\int g^{q-3}|F_\sharp|^3 + |C\|\mathbf{f}\|_{\mathbf{p}}^3 + \sum_{j=1,2}\Big(C\eta\|f_{j,\sharp}\|_{p_j}^2 - c\eta^{2-p_j}\|f_{j,\flat}\|_{p_j}^{p_j}\Big), \tag{84}$$

and (80) holds with $\tilde{Q} + \tilde{\mathcal{R}}$ rewritten as $Q + \mathcal{R}$.

The most significant difference between this analysis, and the proof of Theorem 1, is the presence of an unfavorable term $\|F_\flat\|_q^q$ in the upper bound (81) for the remainder $\mathcal{R}$. However, $\|F_\flat\|_q \le C\max_{j=1,2}\|f_j\|_{p_j}$, and the exponent q is strictly greater than $\max(p_1, p_2)$, so $\|F_\flat\|_q^q$ will be negligible relative to the term $-\sum_{j=1,2}\eta^{2-p_j}\|f_{j,\flat}\|_{p_j}^{p_j}$, provided that $\|\mathbf{f}\|_{\mathbf{p}}$ is sufficiently small as a function of η, as will eventually be assumed. Therefore

$$\mathcal{R} \le C\eta\int g^{q-2}|F_\sharp|^2 + \sum_{j=1,2}\Big(C\eta\|f_{j,\sharp}\|_{p_j}^2 - c'\eta^{2-p_j}\|f_{j,\flat}\|_{p_j}^{p_j}\Big) \tag{85}$$

for some constants $c, C \in \mathbb{R}^+$.

The term $\int g^{q-1}(F-g)$ is not linear in $\mathbf{f}$, and needs closer examination. Expand

$$\int g^{q-1}(F-g) = \int g^{q-1}(g_1 * f_2 + f_1 * g_2) + \int g^{q-1}\cdot(f_1 * f_2). \tag{86}$$

The linear term $\int g^{q-1}(g_1 * f_2 + f_1 * g_2)$ vanishes for any $(f_1, f_2) \in L^{p_1} \times L^{p_2}$ satisfying $\int g^{p_j-1} f_j = 0$ for both indices j. For continuous compactly supported functions, this follows from a simple first variation argument, since the functional in question is maximized when $(f_1, f_2) = 0$. A straightforward passage to the limit yields the claim for general functions in $L^{p_1}\times L^{p_2}$ satisfying the moment conditions.

The remaining term,

$$\int g^{q-1}\cdot(f_1 * f_2) = \int g^{q-1}\cdot((f_{1,\sharp} + f_{1,\flat}) * (f_{2,\sharp} + f_{2,\flat})),$$

is equal to a principal quadratic term $\int g^{q-1}\cdot(f_{1,\sharp} * f_{2,\sharp})$ plus a remainder term whose absolute value is

$$\le \eta\cdot(\|f_{1,\sharp}\|_{p_1}^2 + \|f_{2,\sharp}\|_{p_2}^2) + C_\eta\cdot(\|f_{1,\flat}\|_{p_1}^2 + \|f_{2,\flat}\|_{p_2}^2).$$

We incorporate this remainder term into the remainder, $\mathcal{R}$, already introduced above.

Defining

$$\begin{aligned} Q_{\mathbf{p}}(\mathbf{h}) &= \tfrac{q-1}{2}\|g_1 * g_2\|_q^{-q}\int (g_1 * g_2)^{q-2}(h_1 * g_2 + g_1 * h_2)^2 \\ &+ \|g_1 * g_2\|_q^{-q}\int (g_1 * g_2)^{q-1}\cdot(h_1 * h_2) - \sum_{j=1,2}\tfrac{p_j-1}{2}\|g_j\|_{p_j}^{-p_j}\int h_j^2 g_j^{p_j-2}, \end{aligned} \tag{87}$$

we conclude that

$$\frac{\|(g_1+f_1)*(g_2+f_2)\|_q}{\prod_{j=1,2}\|g_j+f_j\|_{p_j}} \le \mathbf{A}_p^d + \mathbf{A}_p^d \mathcal{Q}_{\mathbf{p}}(\mathbf{f}_\sharp) + \mathbf{A}_p^d \mathcal{R}, \tag{88}$$

where $\mathcal{R}$ has been redefined but satisfies the upper bound (85).

Assuming $\int g_j^{p_j-1} f_j = 0$ for $j = 1, 2$, substitute

$$u_j = f_j g_j^{(p_j-2)/2} \tag{89}$$

and write $\mathbf{u} = (u_1, u_2) \in L^2(\mathbb{R}^d, \mathbb{R}^2)$. Then

$$\int G^{\tau_j} u_j = 0, \tag{90}$$

where $\tau_j = p_j p_j'/2$. As in the proof of Theorem 1, it suffices to prove that

$$\mathcal{Q}_{\mathbf{p}}(\mathbf{f}) \le -c\|\mathbf{u}\|_{L^2}^2 \tag{91}$$

under the assumption that u_1, u_2 satisfy (90), and under the assumption that u_1, u_2 satisfy certain auxiliary moment conditions, and that those conditions are achievable by exploiting symmetries of the inequality. Their formulation and achievability will be discussed at the end of this section.

Now

$$\begin{aligned}&\tfrac{q-1}{2}\|g_1*g_2\|_q^{-q}\int (g_1*g_2)^{q-2}(f_1*g_2+g_1*f_2)^2\\ &= \tfrac{q-1}{2}\|g_1*g_2\|_q^{-q}\int (g_1*g_2)^{q-2}(g_1^{(2-p_1)/2}u_1*g_2+g_1*g_2^{(2-p_2)/2}u_2)^2.\end{aligned}$$

We next rewrite this expression in terms of the operators $T_{(i,j,k)}$ encountered in the proof of Theorem 1. To begin,

$$g_1 * g_2 = G^{p_1'} * G^{p_2'} = (p_1')^{-d/2}(p_2')^{-d/2} r^{d/2} G^r \tag{92}$$

where

$$r^{-1} = (p_1')^{-1} + (p_2')^{-1} = 2 - p_1^{-1} - p_2^{-1} = 1 - q^{-1} \tag{93}$$

and therefore

$$g_1 * g_2 = (p_1')^{-d/2}(p_2')^{-d/2}(q')^{d/2} G^{q'}. \tag{94}$$

Thus

$$\begin{aligned}\|g_1 * g_2\|_q^q &= [(p_1')^{-d/2}(p_2')^{-d/2}(q')^{d/2}]^q \int G^{qq'} \\ &= [(p_1')^{-d/2}(p_2')^{-d/2}(q')^{d/2}]^q (qq')^{-d/2}\end{aligned}$$

and consequently

$$\|g_1 * g_2\|_q^{-q}(g_1 * g_2)^{q-2} = [(p_1')^{-d/2}(p_2')^{-d/2}(q')^{d/2}]^{-2}(qq')^{d/2}G^{q'(q-2)}. \tag{95}$$

With the aid of the identities

$$r'(r-2) = r(2-r') \text{ for } r = q \text{ and } r' = q' = p_3, \tag{96}$$

$$q - 1 = p_3'/p_3, \text{ and } p_j - 1 = p_j/p_j' \text{ for } j = 1, 2 \tag{97}$$

$$q'(q-1) = p_3(p_3'-1) = p_3/(p_3-1) = p_3', \tag{98}$$

$Q_{\mathbf{p}}$ can be rewritten as

$$\begin{aligned}&\tfrac{1}{2}(p_3'/p_3)(p_1')^d(p_2')^d(p_3)^{-d}(p_3p_3')^{d/2} \\ &\qquad \cdot \int G^{p_3'(2-p_3)}\big((G^{p_1'(2-p_1)/2}u_1) * G^{p_2'} + G^{p_1'} * (G^{p_2'(2-p_2)/2}u_2)\big)^2 \\ &+ (p_1')^{d/2}(p_2')^{d/2}(p_3)^{-d/2}(pp_3')^{d/2} \int G^{p_3'}\big((G^{p_1'(2-p_1)/2}u_1) * G^{p_2'(2-p_2)/2}u_2)\big) \\ &- \tfrac{1}{2}\sum_{j=1,2}(p_j/p_j')(p_jp_j')^{d/2}\|u_j\|_{L^2}^2.\end{aligned} \tag{99}$$

The first line of (99) is equal to

$$\tfrac{1}{2}(p_3'/p_3)(p_1')^d(p_2')^d(p_3)^{-d}(p_3p_3')^{d/2}\left\langle\begin{pmatrix} T_{(1,3,2)}^*T_{(1,3,2)} & T_{(2,3,1)}^*T_{(1,3,2)} \\ T_{(1,3,2)}^*T_{(2,3,1)} & T_{(2,3,1)}^*T_{(2,3,1)}\end{pmatrix}\mathbf{u}, \mathbf{u}\right\rangle \tag{100}$$

where $T_{(1,3,2)}$ and $T_{(2,3,1)}$ are defined in terms of the ordered triple of exponents (p_1, p_2, p_3) by (20). The second line is equal to

$$(p_1')^{d/2}(p_2')^{d/2}(p_3)^{-d/2}(p_3p_3')^{d/2}\langle T_{(1,2,3)}u_1, \tilde{u}_2\rangle, \tag{101}$$

where $\tilde{u}_2(x) = u_2(-x)$.

By the diagonalizations of the operators $T_{(i,j,k)}$ developed above, the sum of these two lines reduces to a family of quadratic forms on $\mathcal{H}_n^{(\tau_1)} \oplus \mathcal{H}_n^{(\tau_2)}$, for $n \in \mathbb{N} = \{1, 2, 3, \dots\}$, where $\mathcal{H}_n^{(\tau)}$ is the linear span of all generalized Hermite functions $H_\alpha^{(\tau)}$ with $|\alpha| = n$, expressed by the matrices

$$
\begin{aligned}
M_n = \tfrac{1}{2}(p_3'/p_3)(p_1')^d(p_2')^d(p_3)^{-d}(p_3p_3')^{d/2} \\
\cdot \begin{pmatrix} \lambda_{n,(1,3,2)}^2 & \lambda_{n,(2,3,1)}\lambda_{n,(1,3,2)} \\ \lambda_{n,(1,3,2)}\lambda_{n,(2,3,1)} & \lambda_{n,(2,3,1)}^2 \end{pmatrix} \qquad (102) \\
+ (-1)^n \tfrac{1}{2}(p_1')^{d/2}(p_2')^{d/2}(p_3)^{-d/2}(p_3p_3')^{d/2} \begin{pmatrix} 0 & \lambda_{n,(1,2,3)} \\ \lambda_{n,(1,2,3)} & 0 \end{pmatrix}.
\end{aligned}
$$

Each entry of each two by two matrix on the right-hand side is a block diagonal operator-valued matrix, equal to an identity matrix multiplied by the indicated scalar.

We are interested in $\langle M_n(\mathbf{v}), \mathbf{v}\rangle - \frac{1}{2}\sum_{j=1,2}(p_j/p_j')(p_jp_j')^{d/2}v_j^2$ for $v \in \mathbb{R}^2$. Substituting $(p_j/p_j')^{1/2}(p_jp_j')^{d/4}v_j = w_j$, we arrive at

$$\langle \tilde{M}_n\mathbf{w}, \mathbf{w}\rangle - \tfrac{1}{2}|\mathbf{w}|^2 \qquad (103)$$

where algebraic simplifications together with the substitution

$$r_j = p_j/p_j' = p_j - 1 \qquad (104)$$

give

$$
\begin{aligned}
2\tilde{M}_n = r_3^{n-1} \begin{pmatrix} r_1^{n-1} & (r_1r_2)^{(n-1)/2} \\ (r_1r_2)^{(n-1)/2} & r_2^{n-1} \end{pmatrix} \\
+ (-1)^n \begin{pmatrix} 0 & (r_1r_2)^{(n-1)/2} \\ (r_1r_2)^{(n-1)/2} & 0 \end{pmatrix}. \qquad (105)
\end{aligned}
$$

Each entry of each of these two by two matrices is equal to an identity matrix multiplied by the indicated scalar. For our purpose, these may be equivalently regarded as two by two matrices, with scalar entries. It suffices to show that there exists $c > 0$ such that

$$2\langle \tilde{M}_n\mathbf{w}, \mathbf{w}\rangle \le (1-c)|\mathbf{w}|^2 \qquad (106)$$

for all $n \in \mathbb{N}$ and all $\mathbf{w} \in \mathbb{R}^2$ satisfying appropriate auxiliary orthogonality relations.

The product $r_k r_l$ is strictly less than 1 for any $k \neq l \in \{1, 2, 3\}$. Indeed, since $\sum_{j=1}^{3} p_j^{-1} = 2$, $p_k^{-1} + p_l^{-1} > 1$, so $p_k + p_l > p_k p_l$, so

$$(p_k - 1)(p_l - 1) = -p_k - p_l + p_k p_l + 1 < 1.$$

Consequently the entries of $\tilde{M}_n$, hence its eigenvalues, tend to 0 as $n \to \infty$ for any $\mathbf{p}$ satisfying our hypotheses. Moreover, the entries for $n = 2$ are all positive, and for $n > 2$, each entry is strictly less than the corresponding entry for $n = 2$.

For $n = 2$, one calculates using the identity $2r_1r_2r_3 + r_1r_2 + r_2r_3 + r_3r_1 = 1$ that 1 is an eigenvalue of $2\tilde{M}_2$. The sum of the eigenvalues of $2\tilde{M}_2$ equals its trace, which equals $r_3(r_1 + r_2)$, so the second eigenvalue equals $r_3(r_1 + r_2) - 1$, which is strictly less than 1 since $r_3 r_j < 1$ for $j = 1, 2$, and is strictly greater than -1. Therefore the supremum over $n \geq 3$ of the largest eigenvalue of $2\tilde{M}_n$ is strictly less than 1, as desired.

It remains to treat the contributions of $n = 0, 1, 2$. To do so, we must ensure that orthogonality conditions are satisfied, so that: (i) The larger of the two eigenvalues for $n = 2$ is eliminated, leaving a single eigenvalue, which is < 1. (ii) The contributions of $n = 1$ and of $n = 0$ are entirely eliminated. This can be achieved as in Sect. 6, of the proof of Theorem 1, by exploiting symmetries of the inequality. The available symmetries, acting on $F_j = g_j + f_j$, are: multiplication of F_1, F_2 by positive scalars; independent translations of F_1, F_2; and composition of F_1, F_2 with a common element of $\mathrm{Gl}(d)$. The proof of Lemma 17 shows that it is possible to choose an element of the symmetry group of the inequality generated by these transformations so that these moment conditions are satisfied. See in particular (51).

Theorem 4 now follows, by repeating steps of the proof of Theorem 1 in a straightforward manner. □

11 The Periodic Case

Theorem 5 is concerned with the periodic setting of $\mathbb{T}^d$. The proof is far simpler than that of Theorem 1, and we provide only an outline. First consider the case of nonnegative functions F_j. Assume without loss of generality that $\|F_j\|_{p_j} = 1$.

Maximizing triples with nonnegative component functions are constant. As in the proof of Theorem 1, there exists a continuous deformation, via a nonlinear heat flow, of arbitrary tuples of nonnegative functions to maximizing tuples. Consequently, it suffices to analyze small perturbations of constants. By normalizing, one may reduce to the situation in which each function is of the form $F_j = 1 + f_j$ with $\int f_j = 0$, and $\|f_j\|_{p_j}$ is small. The functional is

$$\mathcal{T}(1 + f_1, 1 + f_2, 1 + f_3) = 1 + \mathcal{T}(f_1, f_2, f_3) = 1 + O\Big(\prod_j \|f_j\|_{p_j}\Big),$$

by virtue of the assumption that each function satisfies $\int f_j = 0$. The vanishing of all first- and second-order terms in this expansion makes the periodic case simpler.

According to (13),

$$\prod_j \|1 + f_j\|_{p_j} \geq 1 + c \sum_j \|f_j\|_{p_j}^{r_j}$$

where $c \in \mathbb{R}^+$ depends on $\mathbf{p}$. It follows from the arithmetic-geometric mean inequality and the relation $\sum_j p_j^{-1} = 2$ that

$$\prod_j \|f_j\|_{p_j} \ll \sum_j \|f_j\|_{p_j}^{r_j}$$

if each function f_j has small norm. For nonnegative functions, Theorem 5 follows by combining these facts.

Complex-valued maximizing triples take the form $(a_1 e_n, a_2 e_n, a_3 e_n)$ with each $a_j \in \mathbb{C}$ satisfying $|a_j| = 1$, $n \in \mathbb{Z}^d$, and $e_n(x) = e^{2\pi i n \cdot x}$. If $\mathbf{F}$ is a near-maximizer with $\|F_j\|_{p_j} = 1$ for each index, then $\mathbf{F}$ is close to such a maximizer. By replacing each component F_j by $F_j / a_j e_n$, we may reduce to the case in which F_j is a small norm perturbation of 1, as above. □

References

1. Beckner, W.: Inequalities in Fourier analysis. Ann. Math. (2) **102**(1), 159–182 (1975)
2. Bennett, J., Carbery, A., Christ, M., Tao, T.: The Brascamp-Lieb inequalities: finiteness, structure and extremals. Geom. Funct. Anal. **17**(5), 1343–1415 (2008)
3. Bianchi, G., Egnell, H.: A note on the Sobolev inequality. J. Funct. Anal. **100**(1), 18–24 (1991)
4. Brascamp, H.J., Lieb, E.H.: Best constants in Young's inequality, its converse, and its generalization to more than three functions. Adv. Math. **20**(2), 151–173 (1976)
5. Carlen, E.A., Lieb, E.H., Loss, M.: A sharp analog of Young's inequality on S^N and related entropy inequalities. J. Geom. Anal. **14**(3), 487–520 (2004)
6. Christ, M.: Near equality in the Riesz-Sobolev inequality. Acta Math. Sin. (Engl. Ser.) **35**(6), 783–814 (2019)
7. Christ, M.: Near-extremizers of Young's inequality for Euclidean groups. Rev. Mat. Iberoam. **35**(7), 1925–1972 (2019)
8. Christ, M.: A sharpened Hausdorff–Young inequality (2014). Preprint, math.CA arXiv:1406.1210
9. Christ, M.: A sharpened Riesz-Sobolev inequality (2017). Preprint, math.CA arXiv:1706.02007
10. Christ, M.: On Young's inequality for Heisenberg groups. In: Geometric Aspects of Harmonic Analysis. Springer, Berlin (2021)
11. O'Neill, K.: A quantitative stability theorem for convolution on the Heisenberg group (2019). Preprint, math.CA arXiv:1907.11986

Strongly Singular Integrals on Stratified Groups

Paolo Ciatti and James Wright

In honour of Fulvio Ricci on his 70th birthday

Abstract We consider a class of spectral multipliers on stratified Lie groups which generalise the class of Hörmander multipliers and include multipliers with an oscillatory factor. Oscillating multipliers have been examined extensively in the Euclidean setting where sharp, endpoint L^p estimates are well known. In the Lie group setting, corresponding L^p bounds for oscillating spectral multipliers have been established by several authors but only in the open range of exponents. In this paper we establish the endpoint $L^p(G)$ bound when G is a stratified Lie group. More importantly we begin to address whether these estimates are sharp.

1 Introduction

The following class of strongly singular convolution operators on $\mathbb{R}^n$ given by

$$T_{a,b}f(x) = \int_{|y|\leq 1} f(x-y)\,\frac{e^{i|y|^{-a}}}{|y|^b}\,dy$$

where $a > 0$ and $b \leq n(2+a)/2$ has a rich and interesting history. In the periodic setting, they were investigated by Hardy who used them to construct a variety of counterexamples. Regarding L^p boundedness properties, Hirschman [14] considered the one dimensional case and for general $n \geq 1$, Wainger [32]

P. Ciatti (✉)
Dipartimento di Ingegneria Civile, Edile e Ambientale, Padova, Italia
e-mail: paolo.ciatti@unipd.it

J. Wright
Maxwell Institute of Mathematical Sciences and the School of Mathematics, University of Edinburgh, Edinburgh, Scotland
e-mail: J.R.Wright@ed.ac.uk

P. Ciatti, A. Martini (eds.), *Geometric Aspects of Harmonic Analysis*,
Springer INdAM Series 45, https://doi.org/10.1007/978-3-030-72058-2_8

established the sharp L^p range but left open the endpoint case which C. Fefferman and Stein [9] accomplished using interpolation by proving that $T_{a,n}$ is bounded on the Hardy space $H^1(\mathbb{R}^n)$. Earlier C. Fefferman [8] established that $T_{a,n}$ satisfies a weak-type $(1, 1)$ bound. Chanillo [2] extended these results to weighted L^p estimates. It is well known that when $b > n(2+a)/2$, there are no L^p estimates.

As a convolution operator, we can view $T = T_{a,b}$ as a multiplier operator $\widehat{Tf}(\xi) = m(\xi)\widehat{f}(\xi)$ where $m = m_{\theta,\beta}$ is essentially given by

$$m_{\theta,\beta}(\xi) = \frac{e^{i|\xi|^\theta}}{|\xi|^{\theta\beta/2}} \tag{1}$$

for $|\xi|$ large. Here $0 < \theta = a/(1+a) < 1$ and $\beta = ((2+a)n - 2b)/a$. We note that m is bounded precisely when $b \le n(2+a)/2$.

The case $b = n$, or equivalently $\beta = n$ in (1), corresponds to the singular integral operators $T_{a,n}$, treated by Fefferman and Stein, whose convolution kernels just fail to be integrable. Their multipliers $m_{\theta,n}$ are not Hörmander multipliers but furnish examples of multipliers with $S^{-m}_{\rho,\delta}$ symbols where $m \ge 0$ and $\rho < 1$. In this context these multipliers were studied by Hörmander [15].

Note that the multipliers $m_{\theta,\beta}$ in (1) with $\beta > n$ (so that $b < n$) correspond to operators $T_{a,b}$ with integrable convolution kernels and hence are bounded on L^1. For any $\delta > 0$, consider the analytic family T^δ_z, $\mathrm{Re}(z) \in [0, 1]$, of operators with multipliers

$$m^\delta_z(\xi) = \frac{e^{i|\xi|^\theta}}{|\xi|^{[\theta(n+\delta)/2]z}}\chi(\xi) \tag{2}$$

where $\chi(\xi) = 0$ when $|\xi| \le 1$ and $\chi(\xi) = 1$ for large $|\xi|$. Thus T^δ_z is bounded on L^2 when $z = iy$ with $\|T^\delta_{iy}\|_{2\to 2}$ uniformly bounded in $y \in \mathbb{R}$. Also T^δ_z is bounded on L^1 when $z = 1 + iy$, again with $\|T^\delta_{1+iy}\|_{1\to 1}$ uniformly bounded in $y \in \mathbb{R}$. By analytic interpolation, we see that $m_{\theta,\beta}$ is an L^p multiplier in the open range $|1/p - 1/2| < \beta/2n$. To establish endpoint bounds, one needs to say something about the *endpoint* multipliers $m_{\theta,n}$ (the case $z = 1$ and $\delta = 0$ in (2)). More precisely in [9], Fefferman and Stein show that multipliers m^0_{1+iy} in (2) are H^1 multipliers with an operator norm at most $(1 + |y|)^{n+1}$.

Fefferman and Stein developed a more general theory of multipliers which include the examples (1) as special cases. Let K be a distribution with compact support, which is integrable away from the origin. Its Fourier transform $\widehat{K}$ is of course a function. We make the following assumptions:

$$\begin{cases} \int_{|x|>2|y|^{1-\theta}} |K(x-y) - K(x)|\, dx \le B, & 0 < |y| \le 1, \\ |\widehat{K}(\xi)| \le B\,(1+|\xi|)^{-\theta n/2}. & \end{cases} \tag{3}$$

In [9], Fefferman and Stein show if K satisfies (3), then $|\xi|^{(n-\beta)\theta/2}\widehat{K}(\xi)$, $0 \le \beta < n$, is an $L^p(\mathbb{R}^n)$ multiplier when $|1/p - 1/2| \le \beta/2n$. See [31] where this result is established in the open range $|1/p - 1/2| < \beta/2n$.

In the papers [3] and [4] (see also [22]), Chanillo, Kurtz and Sampson considered the cases $\theta > 1$ and $\theta < 0$ (here the $|\xi|$ large restriction becomes $|\xi|$ small). Hence multipliers on $\mathbb{R}^n$ of the form

$$m_{\theta,\beta}(\xi) \;=\; \frac{e^{i|\xi|^\theta}}{|\xi|^{\theta\beta/2}}\,\chi_{\pm}(\xi) \tag{4}$$

for any $\theta \in \mathbb{R}$ and $\beta \ge 0$ have been studied. Here when $\theta > 0$, we employ $\chi_+(\xi) \equiv 0$ for $|\xi| \le 1$ and $\chi_+(\xi) \equiv 1$ when $|\xi|$ is large whereas when $\theta < 0$, we use $\chi_-(\xi) \equiv 0$ when $|\xi| \ge 1$ and $\chi_-(\xi) \equiv 1$ when $|\xi|$ is small.

The case $\theta = 1$ is special and is related to the wave operator. The sharp range of L^p bounds in this case is different from the case $\theta \ne 1$; see [27] and [23]. We will not consider the case $\theta = 1$ and assume always $\theta \ne 1$.

In this paper we will put all these oscillating multipliers into a single, general framework (much like what Fefferman and Stein do in (3) when $0 < \theta < 1$) which strictly generalises the class of Hörmander multipliers and furthermore we will give a unified, purely spectral treatment which readily extends to estimates for corresponding spectral multipliers on any stratified Lie group.

1.1 Notation

Keeping track of constants and how they depend on the various parameters will be important for us. For the most part, constants C appearing in inequalities $P \le CQ$ between positive quantities P and Q will be *absolute* or *uniform* in that they can be taken to be independent of the parameters of the underlying problem. We will use $P \lesssim Q$ to denote $P \le CQ$ and $P \sim Q$ to denote $C^{-1}Q \le P \le CQ$. Furthermore, we use $P \ll Q$ to denote $P \le \delta Q$ for a sufficiently small constant $\delta > 0$ whose smallness will depend on the context.

2 The Euclidean Setting $\mathbb{R}^n$

We start in the Euclidean setting $\mathbb{R}^n$. Let $\phi \in C_0^\infty(\mathbb{R}^n)$ be supported in $\{1/2 \le |\xi| \le 4\}$ such that $\phi(\xi) = 1$ when $1 \le |\xi| \le 2$ and let $m^j(\xi) := m(2^j\xi)\phi(\xi)$. It is natural to impose conditions on the jth pieces m^j. The classical Hörmander condition requires uniform (in j) control of some L^2 Sobolev norm $\|m^j\|_{L^2_s}$ with s derivatives. Here we want to consider not only classical Hörmander multipliers but also the oscillating multipliers $m_{\theta,\beta}$ described in (4). Special among these are the *endpoint* multipliers $m_{\theta,n}$ whose bounds we interpolate with trivial L^2 bounds to deduce sharp L^p bounds for $m_{\theta,\beta}$ for general $\beta \geq 0$. Hence our conditions will not only involve a smoothness parameter $s > 0$ but also an oscillation parameter $\theta \in \mathbb{R}$ and a decay parameter $\beta \geq 0$.

For any $\theta \in \mathbb{R}$, the condition $j\theta > 0$ identifies the frequency range of interest. In fact if $\theta > 0$, then $j\theta > 0$ corresponds to $j > 0$ or $|\xi| \geq 1$ which is the relevant frequency range indicated in (4). However if $\theta < 0$, then $j\theta > 0$ corresponds to $j < 0$ or $|\xi| \leq 1$ which is the range of interest for the oscillating multipliers in (4) with $\theta < 0$. Finally when $\theta = 0$, the condition $j\theta > 0$ is vacuous.

2.1 Our Multiplier Conditions

We consider the following conditions on a multiplier m which will depend on parameters s, θ and β. For such parameters, we introduce the following class $M_{\theta,\beta,s}$ of multipliers: when $j\theta \leq 0$, we impose the standard uniform L^2 Sobolev norm control on the m^j;

$$\sup_{j:j\theta\leq 0} \|m^j\|_{L^2_s(\mathbb{R}^n)} < \infty. \tag{5}$$

For $j\theta > 0$, we consider the conditions

$$\sup_{j:j\theta>0} 2^{j\theta\beta/2}\|m^j\|_{L^\infty(\mathbb{R}^n)} < \infty, \quad \sup_{j:j\theta>0} 2^{-j\theta(2s-\beta)/2}\|m^j\|_{L^2_s(\mathbb{R}^n)} < \infty. \tag{6}$$

When $\theta = 0$, the condition (6) is vacuous and (5) reduces to the condition $\sup_j \|m^j\|_{L^2_s} < \infty$ and if this holds for some $s > n/2$, the classical Hörmander theorem states that the multiplier operator is of weak-type $(1, 1)$ and maps $H^1(\mathbb{R}^n)$ boundedly into $L^1(\mathbb{R}^n)$. See [30].

One can easily verify that the conditions (5) and (6) are satisfied for $m_{\theta,\beta}$ in (4) and for all $s > 0$. Note that in (6), the quantity $j\theta$ is always positive and so (6) expresses a growth in the Sobolev norm L^2_s of m^j (when $s > \beta/2$) and a decay in the L^∞ norm of m^j. If the L^2 Sobolev condition in (6) is satisfied for some $s > 0$, then by complex interpolation, it is also satisfied for all $0 \leq s' \leq s$ since the $s' = 0$ case $\|m^j\|_{L^2} \lesssim 2^{-j\theta d/2}$ is implied by the L^∞ condition.

2.2 Our Multiplier Classes

Therefore $\bigcup_{s>n/2} M_{0,*,s}$ is the classical class of Hörmander multipliers and so

$$\mathcal{M}_n := \bigcup_{\theta\in\mathbb{R}\setminus\{1\},s>n/2} M_{\theta,n,s}$$

gives us a natural extension of Hörmander multipliers. We also define the multiplier class

$$\mathcal{M}_\beta := \bigcup_{\theta\in\mathbb{R}\setminus\{1\}, s>n/2} \mathcal{M}_{\theta,\beta,s}. \tag{7}$$

It is easy to verify that the conditions (5) and (6) are independent on the choice of the bump function ϕ and hence for any $\beta \geq 0$, if $m \in \mathcal{M}_n$, then the multiplier $|\xi|^{(n-\beta)\theta/2} m(\xi)$ satisfies (6) with decay parameter β and the same oscillation and smoothness parameters θ and s. This puts us in the position to employ the complex interpolation argument in [9] to deduce that $m \in \mathcal{M}_\beta$ is an L^p multiplier in the sharp range $|1/p - 1/2| \leq \beta/2n$ from H^1 bounds for multiplier operators associated to $m \in \mathcal{M}_n$.

In fact one advantage of working with $\mathcal{M}_n$ (over say, the class of multipliers arising from kernels satisfying (3) in the case $0 < \theta < 1$) is the class $\mathcal{M}_n$ has the desirable property that it is invariant under multiplication by $|\xi|^{iy}$ for any real $y \in \mathbb{R}$; that is, if $m \in \mathcal{M}_n$, then $|\xi|^{iy} m(\xi)$ lies in $\mathcal{M}_n$, satisfying the bounds (5) and (6) with polynomial growth in $|y|$. Hence for the analytic interpolation argument, we only need to establish that multipliers in $\mathcal{M}_n$ map H^1 to L^1 instead of showing they map H^1 to H^1 as needed in [9]. This will be particularly useful when we move to the setting of Lie groups.

2.3 *The Basic Decomposition*

When we analyse a multiplier $m \in \mathcal{M}_\beta$, we will decompose $m = \sum_j m_j$ where $m_j(\xi) = m(\xi)\phi(2^{-j}\xi)$ for some $\phi \in C_0^\infty(\mathbb{R}^n)$ supported away from the origin such that $\sum_j \phi(2^{-j}\xi) = 1$ for all $\xi \neq 0$. Note that $m^j(\xi) = m_j(2^j\xi)$ is the jth piece on which we impose the conditions (5) and (6). We split the multiplier $m = m_{small} + m_{large}$ into two parts where

$$m_{small}(\xi) := \sum_{j:j\theta\leq 0} m_j(\xi) \quad \text{and} \quad m_{large}(\xi) := \sum_{j:j\theta>0} m_j(\xi). \tag{8}$$

If $\theta = 0$, then $m = m_{small}$ and in general we note that m_{small} is a Hörmander multipler (since (5) holds for some $s > n/2$) and so it is an L^p multiplier for all $1 < p < \infty$ (as well as a weak-type (1, 1) and an H^1 multiplier). We introduce the notation $\mathcal{K}_F$ to denote the convolution kernel associated to a multiplier F. Hence it suffices to treat the operator

$$T^l f(x) = \sum_{j:j\theta>0} \mathcal{K}_{m_j} * f(x) =: \mathcal{K}^l * f(x)$$

corresponding to the interesting frequency range where the jth pieces m^j satisfy (6).

2.4 $\mathcal{M}_n$ Versus (3)

When $m \in M_{\theta,n,s} \subset \mathcal{M}_n$ for $0 < \theta < 1$, we claim that $\mathcal{K}^d$ satisfies the condition (3) of Fefferman and Stein in [9] (see also [31]). Hence for $0 < \theta < 1$, the class of convolution operators satisfying (3) is larger than the class $\mathcal{M}_n$. In fact the L^∞ condition on the m^j in (6) is equivalent to the bound $|\widehat{\mathcal{K}^d}(\xi)| \le B(1+|\xi|)^{-\theta n/2}$. Furthermore we bound

$$\int_{|x|\ge 2|y|^{1-\theta}} |\mathcal{K}^d(x-y) - \mathcal{K}^d(x)|\, dx \le \sum_{j>0} \int_{|x|\ge 2|y|^{1-\theta}} |\mathcal{K}_{m_j}(x-y) - \mathcal{K}_{m_j}(x)|\, dx$$

and split the sum on the right $\sum_{j\in J_1} + \sum_{j\in J_2}$, where $J_1 = \{j > 0 : 2^j \ge |y|^{-1}\}$ and $J_2 = \mathbb{N} \setminus J_1$. For the sum over J_1, we bound each

$$\int_{|x|\ge 2|y|^{1-\theta}} |\mathcal{K}_{m_j}(x-y) - \mathcal{K}_{m_j}(x)|\, dx \le 2 \int_{|x|\ge |y|^{1-\theta}} |\mathcal{K}_{m_j}(x)|\, dx$$

and note that if $s > n/2$,

$$\begin{aligned} \int_{|x|\ge |y|^{1-\theta}} |\mathcal{K}_{m_j}(x)|\, dx &= \int_{|x|\ge 2^j|y|^{1-\theta}} |\mathcal{K}_{m^j}(x)|\, dx \\ &= \int_{|x|\ge 2^j|y|^{1-\theta}} |\mathcal{K}_{m^j}(x)|\, |x|^s |x|^{-s} dx \\ &\lesssim (2^j|y|^{1-\theta})^{-(s-n/2)} \|m^j\|_{L^2_s} \\ &\lesssim (2^j|y|)^{-(1-\theta)(s-n/2)} \end{aligned} \tag{9}$$

by Cauchy–Schwarz and (6). This is summable for $j \in J_1$ leaving us to treat the sum over J_2. In this case we use the bound

$$\int_{|x|\ge 2|y|^{1-\theta}} |\mathcal{K}_{m_j}(x-y) - \mathcal{K}_{m_j}(x)|\, dx \lesssim |y| \int_{|x|\ge |y|^{1-\theta}} |\nabla \mathcal{K}_{m_j}(x)|\, dx \tag{10}$$

and note that

$$\nabla \mathcal{K}_{m^j}(x) = \int i\xi \phi(\xi) m(2^j\xi) e^{ix\cdot\xi}\, d\xi =: \int \psi(\xi) m(2^j\xi) e^{ix\cdot\xi}\, d\xi$$

for some $\psi \in C_0^\infty(\mathbb{R}^n)$ supported away from 0. Therefore $\nabla\mathcal{K}_{m^j}$ satisfies the bounds in (6). We write

$$\int_{|x|\ge |y|^{1-\theta}} |\nabla \mathcal{K}_{m_j}(x)|\, dx = 2^j \int_{|x|\ge 2^j|y|^{1-\theta}} |\nabla \mathcal{K}_{m^j}(x)|\, dx$$

$$= 2^j \int_{2^j|y|^{1-\theta} \le |x| \le 2^{j\theta}} |\nabla \mathcal{K}_{m^j}(x)|\, dx + 2^j \int_{2^{j\theta}\le |x|} |\nabla \mathcal{K}_{m^j}(x)|\, dx =: I_j + II_j.$$

We note that the integration in I_j is nonempty since $|y|^{1-\theta} \leq 2^{-j(1-\theta)}$ for $j \in J_2$. By Cauchy–Schwarz and (6) we have

$$I_j \leq 2^j 2^{j\theta n/2} \|\mathcal{K}_{m^j}\|_{L^2} = 2^j 2^{j\theta n/2} \|m^j\|_{L^2} \lesssim 2^j 2^{j\theta n/2} \|m^j\|_{L^\infty} \lesssim 2^j.$$

In precisely the same way we argued in (9) we also have $|II_j| \lesssim 2^j$. Hence $\sum_{j \in J_2} |I_j + II_j| \lesssim |y|^{-1}$ and this shows that we can sum the integrals in (10) and get a uniform bound, establishing the claim that (3) holds for $\mathcal{K}^d$.

2.5 An Interlude

At this point we would like to highlight a useful bound which is trivial in the Euclidean setting but will not be so trivial in the Lie group setting. The following bound is an immediate consequence of the Cauchy–Schwarz inequality:

For any compactly support F with $\mathrm{supp}(F) \subseteq K$ (K compact),

$$\|\mathcal{K}_F\|_{L^1(\mathbb{R}^n)} \leq C_{s,K} \|F\|_{L^2_s(\mathbb{R}^n)} \tag{11}$$

holds for any $s > n/2$.

We can use (11) to conclude that if the decay parameter $\beta > n$, then the main part of the convolution kernel $\mathcal{K}^d$ is integrable for any $m \in \mathcal{M}_\beta$. To see this, note that $m \in M_{\theta,\beta,s}$ for some $\theta \in \mathbb{R}$ and $s > n/2$, and by (11),

$$\|\mathcal{K}^d\|_{L^1} \leq \sum_{j:j\theta>0} \|\mathcal{K}_{m_j}\|_{L^1} = \sum_{j:j\theta>0} \|\mathcal{K}_{m^j}\|_{L^1} \lesssim \sum_{j:j\theta>0} \|m^j\|_{L^2_{s'}} \lesssim \sum_{j:j\theta>0} 2^{-j\theta(\beta-2s')/2}$$

for any $s' > n/2$. Since $\beta > n$ and $s > n/2$, we can find an $s' \leq s$ such that $n/2 < s' < \beta/2$. Hence the above sum is convergent and this shows that $\mathcal{K}^d \in L^1(\mathbb{R}^n)$.

By embedding a general $m \in \mathcal{M}_\beta$ with $0 \leq \beta < n$ into the analytic family of multipliers $m_z(\xi) = |\xi|^{\theta/2(\beta-(n+\delta)z)} m(\xi)$ (see (2)) and using complex interpolation, we have the following observation.

Lemma 1 *If $m \in \mathcal{M}_\beta$ and $0 \leq \beta < n$, then m is an $L^p(\mathbb{R}^n)$ multiplier if $|1/p - 1/2| < \beta/2n$.*

Lemma 1 is an extension of a result in [31] from the case $0 < \theta < 1$ to the case of general $\theta \neq 1$.

2.6 The Results

As discussed above, using (7) and the complex interpolation argument in [9], we can show that any $m \in \mathcal{M}_\beta$ with $0 \leq \beta < n$ is an L^p multiplier at the endpoint $|1/p - 1/2| = \beta/2n$ IF we can show that every endpoint multiplier $m \in \mathcal{M}_n$ is bounded from $H^1(\mathbb{R}^n)$ to $L^1(\mathbb{R}^n)$. We have the following theorem.

Theorem 2 *For every $m \in \mathcal{M}_n$, the corresponding multiplier operator T_m is weak-type $(1, 1)$ and maps $H^1(\mathbb{R}^n)$ to $L^1(\mathbb{R}^n)$.*

For the proof see next section. We do not claim that Theorem 2 is really new. For the examples in (4), Theorem 2 was established in the series of papers [3, 4, 8, 9] and [22] for various cases of $\theta \in \mathbb{R} \setminus \{1\}$. What is new is the proof which gives a unified approach and extends to the Lie group setting. We have the immediate consequence improving Lemma 1.

Corollary 3 *If $m \in \mathcal{M}_\beta$ and $0 \leq \beta < n$, then m is an $L^p(\mathbb{R}^n)$ multiplier for $|1/p - 1/2| \leq \beta/2n$.*

3 The Stratified Lie Group Setting

Let $\mathfrak{g}$ be an n-dimensional, graded nilpotent Lie algebra so that

$$\mathfrak{g} = \bigoplus_{i=1}^{s} \mathfrak{g}_i$$

as a vector space and $[\mathfrak{g}_i, \mathfrak{g}_j] \subset \mathfrak{g}_{i+j}$ for all i, j. Suppose that $\mathfrak{g}_1$ generates $\mathfrak{g}$ as a Lie algebra. We call the associated, connected, simply connected Lie group G a stratified Lie group. Associated to such a group is its so-called homogeneous dimension

$$Q = \sum_j j \, \text{dimension}(\mathfrak{g}_j)$$

which is clearly always larger then or equal to the topological dimension n, they agree when $G = \mathbb{R}^n$, that is to say for G abelian.

We fix a basis $\{X_j\}$ for $\mathfrak{g}_1$ where each X_j can be identified with a unique left-invariant vector field on G which we also denote by X_j. Consider the sublaplacian $\mathcal{L} = -\sum_k X_k^2$ on G. For any Borel measurable function m on $\mathbb{R}_+ = [0, \infty)$, we can define the spectral multiplier operator

$$m(\sqrt{\mathcal{L}}) = \int_0^\infty m(\lambda) \, dE_\lambda$$

where $\{E_\lambda\}_{\lambda\geq 0}$ is the spectral resolution of $\sqrt{\mathcal{L}}$. This is a bounded operator on $L^2(G)$ precisely when $m \in L^\infty(\mathbb{R}_+)$. The negative of the classical Laplacian Δ is the corresponding differential operator when $G = \mathbb{R}^n$ and spectral multipliers on $\mathbb{R}^n$ are simply radial multipliers, which the multipliers in (4) provide specific examples.

3.1 The Multiplier Classes

We now state the conditions corresponding to (5) and (6) for spectral multipliers m defined on $\mathbb{R}_+$. Fix a smooth bump function ϕ on $\mathbb{R}$ supported in $\{1/2 \leq \lambda \leq 4\}$ such that $\phi(\lambda) = 1$ when $1 \leq \lambda \leq 2$ and let $m^j(\lambda) := m(2^j\lambda)\phi(\lambda)$. Again the conditions will depend on an oscillation parameter $\theta \in \mathbb{R}$, a decay parameter $\beta \geq 0$ and a smoothness parameter $s > 0$. We introduce the following class $M_{\theta,\beta,s}$ of spectral multipliers: when $j\theta \leq 0$, we impose the standard uniform L^2 Sobolev norm control on the m^j;

$$\sup_{j:j\theta\leq 0} \|m^j\|_{L^2_s(\mathbb{R}_+)} < \infty. \tag{12}$$

For $j\theta > 0$, we consider the condition

$$\sup_{j:j\theta>0} 2^{j\theta\beta/2}\|m^j\|_{L^\infty(\mathbb{R}_+)} < \infty, \quad 2^{-j\theta(2s-\beta)/2}\|m^j\|_{L^2_s(\mathbb{R}_+)} < \infty. \tag{13}$$

Again when $\theta = 0$, these conditions reduce to the condition $\sup_j \|m^j\|_{L^2_s} < \infty$ and if this holds for some $s > Q/2$, the fundamental works of Christ [5] and Mauceri-Meda [21] establish that the multiplier operator is of weak-type (1, 1) and bounded on $H^1(G)$.

The examples $m_{\theta,\beta}(\lambda) = e^{i\lambda^\theta}\lambda^{-\theta\beta/2}\chi_\pm(\lambda)$ from (4) satisfy conditions (12) and (13). We redefine

$$\mathcal{M}_\beta \ (= \mathcal{M}_{\beta,Q}) := \bigcup_{\theta\in\mathbb{R}\setminus\{1\}, s>Q/2} M_{\theta,\beta,s} \tag{14}$$

and stress the dependence of these classes on the homogeneous dimension Q, which we will return to later. Again this puts us in the position to employ analytic interpolation arguments to deduce that $m \in \mathcal{M}_\beta$ is an $L^p(G)$ multiplier in the range $|1/p - 1/2| \leq \beta/2Q$ from $H^1(G)$ bounds for multiplier operators associated to $m \in \mathcal{M}_Q$. Furthermore, from the invariance of $\mathcal{M}_Q$ under multiplication by λ^{iy} for any real y (with resulting polynomial in y bounds in (12) and (13)), it suffices to show $m(\sqrt{\mathcal{L}}) : H^1(G) \to L^1(G)$ for $m \in \mathcal{M}_Q$.

3.2 The Main Result

Our main result is the following theorem.

Theorem 4 *For any* $m \in \mathcal{M}_Q$, *the operator* $m(\sqrt{\mathcal{L}}) : H^1(G) \to L^1(G)$ *and is weak-type* $(1, 1)$.

As an immediate consequence, using complex interpolation (see above), we have the following endpoint result of Mauceri and Meda in [21]. See also the work of Alexopolous [1] on general Lie groups of polynomial volume growth.

Corollary 5 *Every* $m \in \mathcal{M}_\beta$ *with* $0 \leq \beta < Q$ *is an* $L^p(G)$ *multiplier in the range* $|1/p - 1/2| \leq \beta/2Q$.

3.3 The Interlude: Revisited

We now return to the estimate (11) and examine it in the Lie group context. Again we use the notation $\mathcal{K}_F$ to denote the convolution kernel of the operator $F(\sqrt{\mathcal{L}})$.

Let G be any stratified Lie group and suppose the following holds for some dimensional parameter d: for any spectral multiplier $F(\lambda)$, supported in a compact $K \subset \mathbb{R}_+$,

$$\|\mathcal{K}_F\|_{L^1(G)} \leq C_{s,K}\|F\|_{L^2_s(\mathbb{R}_+)} \tag{15}$$

holds for any $s > d/2$.

In [5] and [21], the estimate (15) was proved for $d = Q$, the homogeneous dimension, on a general stratified Lie group G. In fact the estimate (15) is key in their work. It is known that if (15) holds for some parameter d, then standard techniques allow us to deduce that if a spectral multiplier m satsifies $\sup_j \|m^j\|_{L^2_s} < \infty$ for some $s > d/2$, then $m(\sqrt{\mathcal{L}})$ is bounded on all $L^p(G)$, $1 < p < \infty$, and corresponding endpoint results on L^1 hold. See for example, [17]. Hence to determine the minimal amount of smoothness required for Hörmander-type spectral multipliers, matters can be reduced to establishing (15).

The fact that one only needs to control a little more than half the topological dimension n number of derivatives, $s > n/2$, for certain Lie groups was first observed by Müller and Stein [25] for the Heisenberg group. The ideas in [13] can be used to establish (15) for $d = n$ on any Lie group of Heisenberg-type (alternatively, one of the main estimates in [24] imply this immediately). Furthermore (15) for $d = n$ was established by Martini and Müller [19] for step 2 stratified Lie groups with $n \leq 7$ or whose centre has dimension at most 2. In another paper [20], Martini and Müller show that (15) holds for some $d < Q$ on any step 2 stratified Lie group.

The estimate (15) also has implications for our more general multipliers satsifying (12) and (13). Instead of $\mathcal{M}_\beta = \mathcal{M}_{\beta,Q}$ defined in (14), let us consider

$$\mathcal{M}_{\beta,d} := \bigcup_{\theta \in \mathbb{R}\setminus\{1\}, s > d/2} M_{\theta,\beta,s}$$

depending now on a dimensional parameter d which could be smaller than Q. Suppose now that (15) holds for some $d \leq Q$ on G. We can use (15) to conclude that if $\beta > d$, then any $m \in \mathcal{M}_{\beta,d}$ can be written as $m = m_{small} + m_{large}$ (see (8)) where m_{small} is a Hörmander multiplier with $s > d/2$ (and hence bounded on all $L^p(G)$, $1 < p < \infty$, weak-type (1, 1), etc...) and m_{large} is an $L^1(G)$ multiplier, the convolution kernel $\mathcal{K}^l$ associated to m_{large} being integrable. This follows exactly as in the Euclidean setting.

By embedding a general $m \in \mathcal{M}_{\beta,d}$ with $0 \leq \beta < d$ into the analytic family of spectral multipliers $m_z(\lambda) = \lambda^{\theta/2(\beta-(d+\delta)z)} m(\lambda)$ and using complex interpolation, we have the following observation.

Lemma 6 *Suppose that* (15) *holds on G for some $d \leq Q$. If $m \in \mathcal{M}_{\beta,d}$ and $0 \leq \beta < d$, then m is an $L^p(G)$ multiplier for $|1/p - 1/2| < \beta/2d$.*

In particular on any step 2 stratified Lie group, the result of Martini and Müller in [20], establishing that (15) holds for some $d < Q$, shows that the convolution kernel $\mathcal{K}^l$ corresponding to the interesting frequency range of any $m \in \mathcal{M}_Q = \mathcal{M}_{Q,Q}$ is integrable and therefore convolution with $\mathcal{K}^l$ is bounded on $L^1(G)$!

We should hence view Theorem 4 and Corollary 5 as placeholders for possible endpont results. It may be the case that (15) holds for some $d < Q$ on any stratified Lie group outwith the Euclidean $G = \mathbb{R}^n$ case. If so, our results do not say anything new outside the Euclidean setting.

In a forthcoming paper, we will establish the sharp result on any Lie group of Heisenberg-type, establishing Theorem 4 and Corollary 5 with Q replaced by n. Our analysis heavily relies on Müller and Seeger's work [24] on the wave equation in Lie groups of Heisenberg-type.

Finally we note that Theorem 4 implies Theorem 2 in the case of radial multipliers but the proof of Theorem 4 below easily gives a proof of Theorem 2. We will therefore give the proof of Theorem 4 only.

4 Preliminaries

For background information about Calderón–Zygmund theory and spectral multipliers on stratified groups, we refer the reader to the book of Folland and Stein [10]. If h is a Borel measurable function on $\mathbb{R}_+$, recall that $\mathcal{K}_h$ denotes the convolution kernel of the operator $h(\sqrt{\mathcal{L}})$ so that

$$h(\sqrt{\mathcal{L}})f(x) = f * \mathcal{K}_h(x) = \int_G f(x \cdot y^{-1})\, \mathcal{K}_h(y)\, dy,$$

where dy denotes Haar measure on G. Since we are identifying the Lie group G with its Lie algebra $\mathfrak{g}$ via the exponential map, the Haar measure is identified with Lebesgue measure on the Lie algebra $\mathfrak{g} \simeq \mathbb{R}^n$.

4.1 Some Basics

The stratified group G comes equipped with a group of dilations $\delta_r : G \to G$ which are automorphisms. We fix a homogeneous norm; that is, a function $|\cdot| : G \to \mathbb{R}_+$, smooth away from 0, with $|x| = 0$ if and only if $x = 0$, where 0 denotes the group identity, and $|\delta_r x| = r|x|$ for all $r \in \mathbb{R}_+$ and $x \in G$. Also if $s > 0$, then

$$h(s\sqrt{\mathcal{L}})f(x) = f * (\mathcal{K}_h)_s(x) \ \text{ where } \ (\mathcal{K}_h)_s(x) := s^{-Q}\mathcal{K}_h(\delta_{s^{-1}}x);$$

see [10]. Another standard fact from [10] is the following mean value theorem for Schwartz functions $\mathcal{S}$ on G: if $h \in \mathcal{S}(G)$, then for any $N \geq 1$,

$$|h(x \cdot y) - h(x)| \leq C_N \frac{|y|}{(1 + |x|)^N} \tag{16}$$

holds for any $y \in G$ such that $|y| \ll |x|$. We will find this useful at times. We will also find useful the following Plancherel-type identity which can be found in [5]: for $h \in L^\infty(\mathbb{R}_+)$, there is a constant c such that

$$\|\mathcal{K}_h\|^2_{L^2(G)} = c \int_0^\infty |h(t)|^2\, t^Q \frac{dt}{t}. \tag{17}$$

holds.

4.2 A Weighted L^2 Bound

We will use the following weighted L^2 estimate which is valid on a general stratified Lie group G: if F is a compactly supported spectral multiplier, then

$$\int_G |\mathcal{K}_F(x)|^2 (1 + |x|^s)^2\, dx \lesssim \|F\|^2_{L^2_s} \tag{18}$$

holds for any $s > 0$. See [29]. Note that by Cauchy–Schwarz, the bound (18) immediately shows that the key estimate (15) holds for all $s > Q/2$ on any stratified Lie group.

For the Hardy space estimate we will use (18) but we will also use this estimate with derivatives:

$$\int_G |X_j \mathcal{K}_F(x)|^2 (1 + |x|^s)^2 \, dx \lesssim \|F\|_{L^2_s}^2 \tag{19}$$

holds for any $s > 0$, $1 \le j \le k$ and any compactly supported F. Here $k = $ dimension$(\mathfrak{g}_1)$.

4.3 Fefferman–Stein Inequality

Our argument uses the Fefferman–Stein vector-valued Hardy–Littlewood maximal function inequality in the context of stratified groups. If

$$Mf(x) = \sup_{r>0} \frac{1}{r^Q} \int_{|y| \le r} |f(x \cdot y^{-1})| \, dy$$

denotes the Hardy–Littlewood maximal function on G, then for $1 < p, q < \infty$, we have

$$\Big\| \Big(\sum_j (Mf_j)^q \Big)^{1/q} \Big\|_{L^p(G)} \le C_{p,q,G} \Big\| \Big(\sum_j |f_j|^q \Big)^{1/q} \Big\|_{L^p(G)}; \tag{20}$$

see for example [30] or [12]. We will use this inequality for $\{f_j\}$ a sequence of characteristic functions of balls $B = B(x_B, r_B) := \{y \in G : |x_B^{-1} \cdot y| \le r_B\}$. We first note that if χ_B denotes the characteristic function of a ball B, then

$$M(\chi_B)(x) \sim \frac{1}{(1 + |\delta_{2^{-L(B)}}(x_B^{-1} \cdot x)|)^Q} \tag{21}$$

where $L(B)$ is chosen so that $2^{L(B)} = r_B$. Hence $M(\chi_B)$ is a weak approximation of the characteristic function χ_B itself.

4.4 Our Basic Decomposition

Let us recall the basic decomposition (8) in the context of spectral multipliers m; we choose $\phi \in C_0^\infty(\mathbb{R}_+)$ supported in $\{1/2 \le \lambda \le 2\}$ so that $\sum_{j \in \mathbb{Z}} \phi(2^{-j}\lambda) = 1$ for all

$\lambda > 0$. Hence $m(\lambda) = \sum_{j \in \mathbb{Z}} m_j(\lambda)$ where $m_j(\lambda) := m(\lambda)\phi(2^{-j}\lambda) = m^j(2^{-j}\lambda)$ and so

$$\mathcal{K}_{m_j}(x) = \mathcal{K}_m * (2^{jQ}\mathcal{K}_\phi(\delta_{2^j}\cdot))(x) = \mathcal{K}_m * (\mathcal{K}_\phi)_{2^{-j}}(x). \tag{22}$$

Therefore

$$m(\sqrt{\mathcal{L}})f(x) = \sum_{j \in \mathbb{Z}} m_j(\sqrt{\mathcal{L}})f(x) = \sum_{j \in \mathbb{Z}} f * \mathcal{K}_{m_j}(x).$$

For $m \in \mathcal{M}_Q$, we split the multiplier

$$m(\lambda) = \sum_{j \in \mathbb{Z}} m_j(\lambda) = \sum_{j \in \mathbb{Z}} m(\lambda)\phi(2^{-j}\lambda) = m_{small}(\lambda) + m_{large}(\lambda)$$

into two parts where $m_{small}(\lambda) = \sum_{j: j\theta \le 0} m_j(\lambda)$ and $m_{large}(\lambda) = \sum_{j: j\theta > 0} m_j(\lambda)$. Since m satisfies (12) for some $s > Q/2$, the results of Christ [5] and Mauceri-Meda [21] show the multiplier m_{small} is weak-type (1, 1) and bounded on $H^1(G)$ (alternatively, the argument below in the case $\theta = 0$ can be used to treat m_{small}). Hence it suffices to treat the operator $T := \sum_{j: j\theta > 0} m_j(\sqrt{\mathcal{L}})$ and in particular it will be good to keep in mind that $j\theta > 0$ is always satisfied.

5 The Proof of Theorem 4: The Weak-Type (1, 1) Bound

We have reduced matters to bounding $T = \sum_{j: j\theta > 0} m_j(\sqrt{\mathcal{L}})$ and our aim here is to show that

$$\Big|\{x \in G : |Tf(x)| \ge \alpha\}\Big| \le \frac{C}{\alpha} \|f\|_{L^1(G)} \tag{23}$$

holds uniformly for all $\alpha > 0$ and $f \in L^1(G)$. We will denote by $|\cdot|$ the Haar measure on G as well as the homogeneous norm (as well as the usual absolute value on $\mathbb{R}$ or $\mathbb{C}$). There should be no confusion.

We employ the classical Calderón–Zygmund decomposition of f at height α on G (see [10] or [30]): there exists a sequence of essentially disjoint balls $\{B = B(x_B, 2^{L(B)})\}$ such that $|\cup B| \lesssim \|f\|_{L^1}/\alpha$. Furthermore we can decompose $f = g + b$ where $|g(x)| \lesssim \alpha$ a.e $x \in G$ and $b = \sum_B b_B$, where $\mathrm{supp}(b_B) \subseteq B^*$,

$$\int_G b_B = 0, \quad \|b_B\|_{L^1} \lesssim \alpha|B| \quad \text{and} \quad \sum_B \|b_B\|_{L^1} \lesssim \|f\|_{L^1}. \tag{24}$$

Here and from now on, B^* will denote a generic dilate of B which is understood to be the appropriate dilate depending on the context and we may also take it to be a sufficiently large dilate when there is a need to do so.

The contribution of the bounded function g to the distribution function $|\{x : |Tf(x)| \geq \alpha\}|$ follows in the usual way, only the L^2 boundedness of T is used here (that is, only the fact that m is bounded is used). To establish (23), it suffices therefore to consider the contribution from T on the function $b = \sum_B b_B$ where f is large and so we write

$$Tb(x) = \sum_{(j,B)\in N} b_B * \mathcal{K}_{m_j}(x) + \sum_{(j,B)\in P} b_B * \mathcal{K}_{m_j}(x) =: \mathcal{A}(x) + \mathcal{B}(x),$$

where

$$N = \{(j, B) : j\theta > 0,\ j(1-\theta) + L(B) \leq 0\}$$

and P is the complementary set of pairs (j, B) with $j\theta > 0$.

For the sum over the pairs $(j, B) \in N$, we use L^2 estimates, the disjoint frequency supports of the $\{\phi(2^{-j}\lambda)\}$ and the smallness of m on the support of $\phi(2^{-j}\lambda)$, $m \approx 2^{-\theta j Q/2}$. Writing $\Phi_j(x) := (\mathcal{K}_\phi)_{2^{-j}}(x)$, we have

$$\begin{aligned}|\{x : |\sum_{(j,B)\in N} m_j(\sqrt{\mathcal{L}})(b_B)(x)| \geq \alpha\}| &\leq \alpha^{-2} \| \sum_{(j,B)\in N} m_j(\sqrt{\mathcal{L}})(b_B)\|_2^2 \\ &\lesssim \alpha^{-2} \sum_{j:j\theta>0} \|(\sum_{B\in N_j} b_B * \Phi_j) * \mathcal{K}_m\|_2^2 \\ &\lesssim \alpha^{-2} \sum_{j:j\theta>0} 2^{-\theta j Q} \| \sum_{B\in N_j} b_B * \Phi_j\|_2^2\end{aligned}$$

where $N_j = \{B : (j, B) \in N\}$. We write the last term on the right above as $E + F$ where

$$E := \alpha^{-2} \sum_{j:j\theta>0} 2^{-jQ\theta} \| \sum_{B\in N_j} b_B * \Phi_j \cdot \chi_{B^*}\|_2^2 \lesssim \alpha^{-2} \sum_{(j,B)\in N} 2^{-jQ\theta} \|b_B * \Phi_j\|_2^2$$

for some appropriately large dilate B^* of B and F is defined similarly with B^* replaced by $G \setminus B^*$. Since $\|b_B * \Phi_j\|_{L^2}^2 \leq \|\Phi_j\|_{L^2}^2 \|b_B\|_{L^1}^2 \lesssim \|\Phi_j\|_{L^2(G)}^2 \alpha^2 |B|^2$ and

$$\|\Phi_j\|_{L^2(G)}^2 = c \int_0^\infty |\phi(2^{-j}t)|^2 t^{Q-1}\, dt = c_\phi 2^{jQ}$$

by the Plancherel formula (17), we have $\|b_B * \Phi_j\|_{L^2}^2 \lesssim 2^{jQ}\alpha^2|B|^2$. Hence,

$$
\begin{aligned}
E &\lesssim \sum_{(j,B)\in N} 2^{jQ(1-\theta)}|B|^2 \\
&\lesssim \sum_B |B| \sum_{j(1-\theta)+L(B)\leq 0} 2^{Q(j(1-\theta)+L(B))} \\
&\lesssim \sum_B |B| \lesssim \alpha^{-1}\|f\|_1.
\end{aligned}
$$

Note that it is important that $\theta \neq 1$ in the above argument. This leaves us with F.

Using the cancellation of b_B, we have

$$
b_B * \Phi_j(x) = \int_G [\Phi_j(y^{-1}\cdot x) - \Phi_j(x_B^{-1}\cdot x)]b_B(y)\,dy.
$$

Noting that $\Phi_j(x) = 2^{jQ}\mathcal{K}_\phi(\delta_{2^j}x)$, we have for $y \in \operatorname{supp}(b_B)$ and $x \notin B^*$,

$$
|\Phi_j(y^{-1}\cdot x) - \Phi_j(x_B^{-1}\cdot x)| \lesssim 2^{jQ}\frac{2^{(1-N)(j+L(B))}}{(1+|\delta_{2^{-L(B)}}(x_B^{-1}\cdot x)|)^N}
$$

by the mean value theorem on stratified groups (16). Therefore we see that for $x \notin B^*$,

$$
|b_B * \Phi_j(x)| \lesssim \alpha 2^{(Q+1-N)(j+L(B))}M(\chi_B)(x)^{N/Q} = \alpha 2^{\epsilon(j+L(B))}M(\chi_B)(x)^q
$$

where $\epsilon = Q+1-N$ and $q = N/Q$. By choosing $N = Q+1/2$, we can make $\epsilon > 0$ and $q > 1$. This allows us to apply the Fefferman–Stein inequality (20) which yields

$$
\begin{aligned}
F &\lesssim \alpha^{-2}\sum_{j:j\theta>0} 2^{-jQ\theta}\Big\|\sum_{B\in N_j} b_B * \Phi_j(\chi_{G\setminus B^*})\Big\|_2^2 \\
&\lesssim \sum_{j:j\theta>0} 2^{-jQ\theta}\Big\|\sum_{B\in N_j}[M(2^{\epsilon(j+L(B))/q}\chi_B)]^q\Big\|_2^2 \\
&\lesssim \sum_{(j,B)\in N} 2^{-jQ\theta}2^{2\epsilon(j+L(B))}|B| \\
&= \sum_{(j,B)\in N} 2^{j(1-\theta)+L(B)}2^{-j\theta(Q-1)}|B|
\end{aligned}
$$

$$\leq \sum_{(j,B)\in N} 2^{j(1-\theta)+L(B)}|B|$$

$$\lesssim \sum_{B} |B| \lesssim \alpha^{-1}\|f\|_1,$$

since $j\theta > 0$ and $Q \geq 1$.

This completes the estimate for F and the contribution from the pairs $(j, B) \in N$. Hence $|\{x : |\mathcal{A}(x)| \geq \alpha\}| \lesssim \alpha^{-1}\|f\|_1$. Again it was important that $\theta \neq 1$ in this argument. We now turn to the contribution from the pairs $(j, B) \in P$ where we will use only L^1 estimates and the L^2 Sobolev condition in (13).

Since $|\cup B^*| \lesssim \alpha^{-1}\|f\|_1$, we see that the desired estimate $|\{x : |\mathcal{B}(x)| \geq \alpha\}| \lesssim \alpha^{-1}\|f\|_1$ reduces matters to estimating $|\{x \notin \cup B^* : |\mathcal{B}(x)| \geq \alpha\}|$ which we see is at most

$$\alpha^{-1} \int_{x\notin\cup B^*} \Big| \sum_{(j,B)\in P} m_j(\sqrt{\mathcal{L}})(b_B)(x)\Big| dx$$

$$\leq \alpha^{-1} \sum_{(j,B)\in P} \int_{x\notin B^*} |m_j(\sqrt{\mathcal{L}})(b_B)(x)| dx$$

$$\leq \alpha^{-1} \sum_{(j,B)\in P} \int |b_B(y)| \Big[\int_{|x\cdot x_B^{-1}|\gg 2^{L(B)}} |\mathcal{K}_{m_j}(x\cdot y^{-1})| dx\Big] dy.$$

The desired estimate will follow if we can show that

$$\sup_B \sum_{j:j(1-\theta)+L(B)\geq 0} \int_{|x|\gtrsim 2^{L(B)}} |\mathcal{K}_{m_j}(x)|\, dx < \infty. \tag{25}$$

In fact,

$$\int_{|x|\geq 2^{L(B)}} |\mathcal{K}_{m_j}(x)| dx = \int_{|x|\geq 2^{j+L(B)}} |\mathcal{K}_{m^j}(x)| dx$$

$$\lesssim 2^{(Q/2-s)(j+L(B))} \sqrt{\int_G |K_{m^j}(x)|^2 (1+|x|^s)^2 dx} \lesssim 2^{-(s-Q/2)(j(1-\theta)+L(B))}$$

and this sums in j with $j(1-\theta) + L(B) \geq 0$ if $s > Q/2$, uniformly in B. Here we used (18) and the L^2 Sobolev condition in (13) in the penultimate inequality. This establishes (25) and completes the proof of the weak-type (1, 1) bound in Theorem 4.

6 The Proof of Theorem 4: The Hardy Space Bound

Elements in the Hardy space $H^1(G)$ have an atomic decomposition (see [10]) and so it suffices to fix an atom a_B supported in a ball B and prove

$$\int_G |m(\sqrt{\mathcal{L}})a_B(x)|\, dx \lesssim 1 \tag{26}$$

for our spectral multiplier $m \in \mathcal{M}_Q$.

Without loss of generality we may assume that the ball B is centred at the origin. The L^2 boundedness of $m(\sqrt{\mathcal{L}})$ implies that $\int_{|x|\le C2^L} |m(\sqrt{\mathcal{L}})a_B(x)|dx \lesssim 1$ via the Cauchy–Schwarz inequality and so it suffices to show that

$$\int_{|x|\gg 2^L} |m(\sqrt{\mathcal{L}})a_B(x)|\, dx \lesssim 1 \tag{27}$$

holds where 2^L is the radius of the ball B.

From our basic decomposition $m = m_{small} + m_{large}$, it suffices as before to treat the operator $T := \sum_{j:j\theta>0} m_j(\sqrt{\mathcal{L}})$ and show that (27) holds with $m(\sqrt{\mathcal{L}})$ replaced by T.

We bound the integral in (27) by

$$\sum_{j\in N} \int_{|x|\gg 2^L} |a_B * \mathcal{K}_{m_j}(x)|dx \;+\; \sum_{j\in P} \int_{|x|\gg 2^L} |a_B * \mathcal{K}_{m_j}(x)|dx =: I + II$$

where $N = \{j : j\theta > 0,\, j(1-\theta) + L \le 0\}$ and P denotes the complementary range.

For $j \in P$, we note that when $|x| \gg 2^L$,

$$\begin{aligned} a_B * \mathcal{K}_{m_j}(x) &= \int_{|y|\le 2^L} \mathcal{K}_{m_j}(y^{-1}\cdot x)a_B(y)\, dy \\ &= \int_G \mathcal{K}_{m_j}(y^{-1}\cdot x)\chi_{E_L}(y^{-1}\cdot x)a_B(y)\, dy \end{aligned}$$

where $E_L = \{x \in G : |x| \ge 2^L\}$. Hence if we denote by $K(x) = \mathcal{K}_{m_j}(x)\chi_{E_L}(x)$,

$$\begin{aligned} \int_{|x|\gg 2^L} |a_B * \mathcal{K}_{m_j}(x)|dx &= \int_G |a_B * K(x)|dx \lesssim \int_G |K(x)|dx \\ &= \int_{|x|\ge 2^L} |\mathcal{K}_{m_j}(x)|dx = \int_{|x|\ge 2^{j+L}} |\mathcal{K}_{m^j}(x)|\, dx \\ &= \int_{|x|\ge 2^{j+L}} |\mathcal{K}_{m^j}(x)|\, \frac{1+|x|^s}{1+|x|^s}\, dx \end{aligned}$$

$$\leq 2^{-(s-Q/2)(j+L)} \sqrt{\int_G |\mathcal{K}_{m^j}(x)|^2 (1+|x|^s)^2 dx}$$

$$\lesssim 2^{-(s-Q/2)(j(1-\theta)+L)},$$

where in the last inequality we used (18) with some $s > Q/2$ and the L^2 Sobolev condition of our multiplier m as stated in (13). Since $\theta \neq 1$, this shows that II is uniformly bounded

For I, we split $N = N_1 \cup N_2$ further such that $N_1 = \{j \in N : j + L \leq 0\}$ and $N_2 = \{j \in N : j + L > 0\}$. This splits $I = I_1 + I_2$ accordingly.

For the sum over $j \in N_1$, we will use the cancellation of the atom a_B: for $j \in N_1$,

$$\int_{|x| \gg 2^L} |a_B * \mathcal{K}_{m_j}(x)| dx$$

$$\leq \int_G |a_B(y)| \Big[\int_{C 2^{j+L} \leq |x|} |\mathcal{K}_{m^j}((\delta_{2^j} y)^{-1} \cdot x) - \mathcal{K}_{m^j}(x)| dx \Big] dy$$

and so by applying the mean value theorem on stratified groups (see (16)), we see that the inner integral on the right hand side is at most

$$2^{j+L} \int_{2^{j+L} \leq |x|} \sup_{1 \leq r \leq k} |X_r \mathcal{K}_{m^j}(x)|\, dx$$

and so

$$\int_{|x| \gg 2^L} |a_B * \mathcal{K}_{m_j}(x)| dx \ \leq \ 2^{j+L} \sum_{r=1}^{k} \int_G |X_r \mathcal{K}_{m^j}(x)|\, dx.$$

Let X denote one of the X_r's—our immediate goal is to show that the bound

$$\int_G |X \mathcal{K}_{m^j}(x)|\, dx \ \leq \ C \tag{28}$$

holds, uniformly for all j. If this is the case, then we see that

$$I_1 \ = \ \sum_{j \in N_1} \int_{|x| \gg 2^L} |a_B * \mathcal{K}_{m_j}(x)|\, dx \ \lesssim \ \sum_{j \in N_1} 2^{j+L} \ \lesssim \ 1,$$

completing the analysis for I_1.

To show (28), we will use (19) for two different values of s. We split the integral in (28) into two parts:

$$\int_{|x|\leq 2^{j+L+\Lambda}} |X\mathcal{K}_{m^j}(x)|\,dx + \int_{2^{j+L+\Lambda}\leq |x|} |X\mathcal{K}_{m^j}(x)|\,dx =: S_\Lambda + L_\Lambda$$

for some large $\Lambda > 0$ to be chosen appropriately.

For S_Λ we use (19) with some $s_* < Q/2$: by Cauchy–Schwarz,

$$S_\Lambda^2 \leq 2^{2(Q/2-s_*)(j+L+\Lambda)} \int_G |X\mathcal{K}_{m^j}(x)|^2(1+|x|^{s_*})^2\,dx$$

and so using (19) and the L^2 Sobolev condition (13) of our multiplier m,

$$S_\Lambda \lesssim 2^{(Q/2-s_*)(j(1-\theta)+L)}\,2^{(Q/2-s_*)\Lambda}.$$

In a similar way, using (19) with some $s > Q/2$, we have

$$L_\Lambda^2 \leq 2^{-2(s-Q/2)(j+L+\Lambda)} \int_G |X\mathcal{K}_{m^j}(x)|^2(1+|x|^{s})^2\,dx$$

and so using the L^2 Sobolev condition (13) of our multiplier m, we see that

$$L_\Lambda \lesssim 2^{-(s-Q/2)(j(1-\theta)+L)}\,2^{-(s-Q/2)\Lambda}.$$

Optimising the two estimates gives $\Lambda = -(j(1-\theta)+L)$ which is positive since $j \in N_1$. Hence with this choice of Λ, $S_\Lambda + L_\Lambda \lesssim 1$, establishing (28) and completing the analysis for I_1.

Finally we turn to I_2, recall that $j \in N_2$ implies $j+L>0$. Here it does not make sense to use the cancellation of the atom a_B. Instead we use our knowledge of the L^2 size of a_B; $\|a_B\|_{L^2(G)} \leq |B|^{-1/2} = 2^{-LQ/2}$. We begin by splitting the integral into two parts as above:

$$\int_{2^L\leq|x|\leq 2^{L+\Lambda}} |a_B * \mathcal{K}_{m_j}(x)|\,dx + \int_{|x|\geq 2^{L+\Lambda}} |a_B * \mathcal{K}_{m_j}(x)|\,dx := S_\Lambda + L_\Lambda$$

for some appropriate Λ. For S_Λ, we use the L^∞ condition in (13) and Cauchy–Schwarz to see that

$$S_\Lambda \leq 2^{(L+\Lambda)Q/2}\|a_B*\mathcal{K}_{m_j}\|_{L^2} \leq 2^{(L+\Lambda)Q/2}2^{-j\theta Q/2}\|a_B*\Phi_j\|_{L^2} \leq 2^{\Lambda Q/2}2^{-j\theta Q/2}.$$

On the other hand, for L_Λ, we have

$$L_\Lambda \leq \int_{|x|\geq 2^{L+\Lambda}} |\mathcal{K}_{m_j}(x)|\,dx = \int_{|x|\geq 2^{j+L+\Lambda}} |\mathcal{K}_{m^j}(x)|\,dx$$

$$\leq 2^{-(s-Q/2)(j+L+\Lambda)}\sqrt{\int_G |\mathcal{K}_{m^j}(x)|^2(1+|x|^s)^2\,dx} \lesssim 2^{-(s-Q/2)(j(1-\theta)+L+\Lambda)}$$

by (19) with $s > Q/2$ and Cauchy–Schwarz. Optimising the two estimates gives Λ with $2^{s\Lambda} = 2^{-(s-Q/2)(j+L)}2^{j\theta s}$ which is positive since $j \in N$. Hence with this choice of Λ, $S_\Lambda + L_\Lambda \lesssim 2^{-(1-Q/2s)/(j+L)Q/2}$ which is summable over $j \in N_2$ since $j + L > 0$, showing that

$$I_2 = \sum_{j\in N_2} \int_{2^L\leq|x|} |a_B * \mathcal{K}_{m_j}(x)|\,dx$$

is uniformly bounded in L and this completes the analysis for I_2, establishing (27) and hence (26).

This finishes the $H^1(G)$ bound of $m(\sqrt{\mathcal{L}})$ and hence the proof of Theorem 4.

Acknowledgments We would like to thank Alessio Martini and Steve Wainger for discussing the history of the problem as well as guiding us through the literature. We also wish to thank the referee for many helpful comments and suggestions.

References

1. Alexopoulos, G.: Oscillating multipliers on Lie groups and Riemannian manifolds. Tôhoku Math. J. **46**, 457–468 (1994)
2. Chanillo, S.: Weighted norm inequalities for strongly singular convolution operators. Trans. Amer. Math. Soc. **281**(1), 77–107 (1984)
3. Chanillo, S., Kurtz, D., Sampson, G.: Weighted L^p estimates for oscillating kernels. Ark. Mat. **21**, 233–257 (1983)
4. Chanillo, S., Kurtz, D., Sampson, G.: Weighted weak (1, 1) and weighted L^p estimates for oscillating kernels. Trans. Am. Math. Soc. **295**(1), 127–145 (1986)
5. Christ, M.: L^p bounds for spectral multipliers on nilpotent groups. Trans. Am. Maths. Soc. **328**(1), 73–81 (1991)
6. Christ, M.: Weak type (1, 1) bounds for rough operators. Ann. Math. **128**, 19–42 (1988)
7. Cowling, M., Giulini, S., Meda, S.: Oscillatory integrals related to the wave equation on noncompact symmetric spaces. J. Lond. Math. Soc. **66**, 691–702 (2002)
8. Fefferman, C.: Inequalities for strongly singular integral operators. Acta Math. **24**, 9–36 (1970)
9. Fefferman, C., Stein, E.M.: H^p spaces of several variables. Acta Math. **129**, 137–193 (1972)
10. Folland, G., Stein, E.M.: Hardy Spaces on Homogeneous Groups. Math. Notes, vol. 28. Princeton University Press, Princeton (1982)
11. Giulini, S., Meda, S.: Oscillating multipliers on noncompact symmetric spaces. J. Reine Angew. Math. **409**, 93–105 (1990)

12. Grafakos, L., Liu, L., Yang, D.: Vector-valued singular integrals and maximal functions on spaces of homogeneous type. Math. Scand. **104**, 296–310 (2009)
13. Hebisch, W.: Multiplier theorem on generalized Heisenberg groups. Colloq. Mat. **65**(2), 231–239 (1993)
14. Hirschman, I.I.: On multiplier transformations. Duke Math. J. **26**, 222–242 (1959)
15. Hörmander, L.: Pseudo-differential operators and hypoelliptic equations. Proc. Sympos. Pure Math., vol. 10, pp. 138–183. Amer. Math. Soc. (1967)
16. Hulanicki, A.: A functional calculus for Rockland operators on nilpotent Lie groups. Stud. Math. **78**(3), 253–266 (1984)
17. Martini, A.: Analysis of joint spectral multipliers on Lie groups of polynomial growth. Ann. Inst. Fourier (Grenoble) (4) **62**, 1215–1263 (2012)
18. Martini, A.: Joint functional calculi and a sharp muliplier theorem for the Kohn Laplacian on the sphere. Math. Z. **286**, 1539–1574 (2017)
19. Martini, A., Müller, D.: Spectral multipliers of Euclidean type on new classes of two-step stratified groups. Proc. Lond. Math. Soc. **109**, 1229–1263 (2014)
20. Martini, A., Müller, D.: Spectral multipliers on 2-step groups: topological versus homogeneous dimension. GAFA **26**, 680–702 (2016)
21. Mauceri, G., Meda, S.: Vector-valued multipliers on stratified groups. Rev. Ibero. Mat. **6**(3), 141–154 (1990)
22. Miyachi, A.: On the weakly strongly singular integrals. Jpn. J. Math. **4**, 221–262 (1978)
23. Miyachi, A.: On some estimates for the wave equation in L^p and H^p. J. Fac. Sci. Tokyo Sci. IA **27**, 331–354 (1980)
24. Müller, D., Seeger, A.: Sharp L^p-bounds for the wave equation on groups of Heisenberg type. Anal. PDE **8**(5), 1051–1100 (2015)
25. Müller, D., Stein, E.M.: On spectral multipliers for Heisenberg and related groups. J. Math. Pures Appl **73**, 413–440 (1994)
26. Müller, D., Stein, E.M.: L^p-estimates for the wave equation on the Heisenberg group. Revista Mat. Ibero. **15**, 297–334 (1990)
27. Peral, J.: L^p estimates for the wave equation. J. Funct. Anal. **36**, 114–145 (1980)
28. Seeger, A., Sogge, C., Stein, E.M.: Regularity properties of Fourier integral operators. Ann. Math. **103**, 231–251 (1991)
29. Sikora, A.: Multiplier theorem for sub-Laplacians on homogeneous groups. C. R. Acad. Sci. Paris Sér. I Math. **315**(4), 417–419 (1992)
30. Stein, E.M.: Harmonic Analysis. Princeton University Press, Princeton (1993)
31. Stein, E.M.: Singular integrals, harmonic functions and differentiability properties of functions of several variables. Proc. Symp. Pure Math. **10**, 316–335 (1967)
32. Wainger, S.: Special Trigonometric Series in k Dimesnions, vol. 59. Mem. Amer. Math. Soc. (1965)
33. Zygmund, A.: Trigonometric series (2 vols), 2nd edn. Cambridge Univ. Press, London and New York (1977)

Singular Brascamp–Lieb: A Survey

Polona Durcik and Christoph Thiele

Abstract We present an overview of results on multi-linear singular integrals in the broader context of Brascamp–Lieb inequalities. This elaborates a lecture given at the inspiring conference on Geometric Aspects of Harmonic Analysis at Cortona 2018 in honor of Fulvio Ricci.

Keywords Multilinear form · Multilinear inequality · Singular integral · Multi-parameter singular integral · Multiplier

1 Brascamp–Lieb Forms and Inequalities

The recently active area of Brascamp–Lieb inequalities focuses on invariant multi-linear forms in functions on Euclidean spaces. By the Schwartz kernel theorem, the multi-linear forms acting on n-tuples of Schwartz functions F_j on $\mathbb{R}^{k_j}$ continuously in each argument are exactly the ones that can be written as

$$\Lambda(F_1 \otimes F_2 \otimes \cdots \otimes F_n)$$

with a unique tempered distribution Λ on $\mathbb{R}^{k_1+\cdots+k_n}$.

Brascamp–Lieb forms arise when the distribution Λ specializes to integration over an affine subspace of $\mathbb{R}^{k_1+\cdots+k_n}$ with respect to an invariant measure,

$$\int_{\mathbb{R}^{k_1+\cdots+k_n}} \Big(\prod_{j=1}^{n} F_j(x_j)\Big)\delta(\Pi(x-z))\,dx,$$

P. Durcik
California Institute of Technology, Pasadena, CA, USA
e-mail: durcik@caltech.edu

C. Thiele (✉)
Mathematisches Institut, Universität Bonn, Bonn, Germany
e-mail: thiele@math.uni-bonn.de

P. Ciatti, A. Martini (eds.), *Geometric Aspects of Harmonic Analysis*,
Springer INdAM Series 45, https://doi.org/10.1007/978-3-030-72058-2_9

where x denotes a vector with components x_j, Π is a linear map whose ker, translated by the vector z, is the affine space of integration, and δ is the Dirac delta measure on the range of the map Π. Here we have called the zero set of a linear map the ker rather than the kernel of the map so as to distinguish it from an integral kernel such as for example in the Schwartz kernel theorem.

A change of variables equates this form with

$$\int_{\mathbb{R}^{k_1+\cdots+k_n}} \Big(\prod_{j=1}^{n} F_j(x_j+z_j)\Big)\delta(\Pi x)\, dx,$$

which is a Brascamp–Lieb form with integration over a linear space, acting on translates of the functions F_j. Using such a reduction, we shall assume throughout this survey that the space of integration is linear, unless stated otherwise:

$$\int_{\mathbb{R}^{k_1+\cdots+k_n}} \Big(\prod_{j=1}^{n} F_j(x_j)\Big)\delta(\Pi x)\, dx. \tag{1}$$

A further change of variables, replacing x by $x - z$ with a vector z in the ker of Π, shows an invariance of the Brascamp–Lieb integral under translation of the functions by amounts z_j. Similarly, one observes a homogeneity of the form under simultaneous dilations of the functions.

Using the Fourier transform, one may write for a Brascamp–Lieb form

$$\widehat{\Lambda}(\widehat{F_1}\otimes\widehat{F_2}\otimes\cdots\otimes\widehat{F_n}),$$

where $\widehat{\Lambda}$ is integration over the orthogonal complement of the subspace of integration of Λ. If Π in (1) is an orthogonal projection, we may write for the Fourier transform integral

$$\int_{\mathbb{R}^{k_1+\cdots+k_n}} \Big(\prod_{j=1}^{n} \widehat{F}_j(\xi_j)\Big)\delta((1-\Pi)(\xi))\, d\xi. \tag{2}$$

This allows to identify further invariances of the form under simultaneous translations of the Fourier transforms of the functions. A translation of the Fourier transform of a function is the same as a modulation of the function itself:

$$M_\xi F(x) = F(x)e^{2\pi i x\cdot\xi}.$$

Up to scalar multiples, the multi-linear forms of Brascamp–Lieb type are determined by their translation and modulation symmetries.

One may write the integral over the subspace also as a parameterized integral. Assume the subspace has dimension m and let

$$I : \mathbb{R}^m \to \mathbb{R}^{k_1+\cdots+k_n}$$

be a parameterization. Denote by Π_j the composition of I with the projection onto the j-th coordinate space $\mathbb{R}^{k_j}$. We may then write for (1), up to scalar multiple,

$$\int_{\mathbb{R}^m} \Big(\prod_{j=1}^n F_j(\Pi_j x)\Big)\, dx. \tag{3}$$

Writing each F_j as Fourier integral, we obtain for (3)

$$\int_{\mathbb{R}^{k_1+\cdots+k_n}} \int_{\mathbb{R}^m} \Big(\prod_{j=1}^n \widehat{F}_j(\xi_j) e^{2\pi i \xi_j \cdot (\Pi_j x)}\Big)\, dx d\xi$$

$$= \int_{\mathbb{R}^{k_1+\cdots+k_n}} \Big(\prod_{j=1}^n \widehat{F}_j(\xi_j)\Big) \delta\Big(\sum_{j=1}^n \Pi_j^T \xi_j\Big) d\xi,$$

which is of the form (2) with $1 - \Pi = \sum_{j=1}^n \Pi_j^T$.

It is natural to seek bounds for Brascamp–Lieb forms by products of norms of the functions, with a choice of norms respecting the symmetries of the form. Most common are Lebesgue norms L^p, which are invariant under translations and modulations and have a homogeneity under dilations. The corresponding bounds are called Brascamp–Lieb inequalities. With a choice of exponents p_j, these inequalities are written as

$$\int_{\mathbb{R}^m} \Big(\prod_{j=1}^n F_j(\Pi_j x)\Big)\, dx \le C \prod_{j=1}^n \|F_j\|_{p_j} \tag{4}$$

with a constant C depending on the Π_j and p_j but not on the Schwartz functions F_j. In case the left-hand side is not a real number, we interpret the inequality in the sense of absolute value. Note also that by multiplying F_1 by a phase and using multi-linearity we can make the left-hand side non-negative real without changing the right-hand side.

Given a tuple of exponents, if $p_j < \infty$ for some j, then a Brascamp–Lieb inequality can only hold if the map Π_j is surjective. To see this, assume Π_j is not surjective. Let y and z parameterize respectively the range of Π_j and the orthogonal complement of this range in $\mathbb{R}^{k_j}$. Then the left-hand side of the Brascamp–Lieb inequality does not change under replacing F_j by

$$\widetilde{F}_j(y, z) := F_j(y, \lambda z),$$

while the right-hand side scales with a power of λ that is non-trivial if $p_j < \infty$.

If $p_j = \infty$, then the map Π_j need not be surjective. For example, if $m = 0$, then the projection Π_j is not surjective except in the pathological case $k_j = 0$. Nevertheless, as the Brascamp–Lieb integral becomes evaluation at a point, the Brascamp–Lieb inequality holds with all exponents equal to ∞.

Well known cases of a Brascamp–Lieb inequality are Hölder's inequality, where all maps Π_j are the identity map, Young's convolution inequality, and the Loomis-Whitney inequality where $m = n$, $k_j = n - 1$ and the one dimensional kers of the maps Π_j span the full space $\mathbb{R}^m$.

Much research has been devoted to Brascamp–Lieb and related inequalities, we refer to [3–5, 8] and the references therein. In particular, [3] proves a necessary and sufficient dimensional condition for a Brascamp–Lieb inequality to hold, namely that

$$\dim(V) \le \sum_{j=1}^{n} \frac{1}{p_j} \dim(\Pi_j V) \tag{5}$$

for every subspace V of $\mathbb{R}^m$, with equality if $V = \mathbb{R}^m$. The easy direction of this equivalence is necessity of (5). It is seen by testing the Brascamp–Lieb inequality on suitable characteristic functions F_j, generating them as limits of Schwartz functions. The supports of these functions are such that the integrand on the left-hand side of (4) is nonzero on a disc, more precisely on a one-neighborhood in $\mathbb{R}^m$ of a large ball in V of radius R. The left-hand side of the Brascamp–Lieb inequality grows in R with the order $R^{\dim(V)}$. The suitable choice of the function F_j is the characteristic function of the projection of the disc to $\mathbb{R}^{k_j}$. Its L^{p_j} norms grow with the order $R^{\dim(\Pi_j(V))/p_j}$. Letting R tend to infinity, we obtain the lower bound of (5). The equality in case $V = \mathbb{R}^m$ is obtained by using in addition small balls in $\mathbb{R}^m$.

Since $\dim(\Pi_j \mathbb{R}^m) \le \dim(\mathbb{R}^m)$, inequality (5) for $V = \mathbb{R}^m$ in case $m > 0$ implies that

$$1 \le \sum_{j=1}^{n} \frac{1}{p_j}. \tag{6}$$

When equality holds in (6), then each map Π_j is injective on $\mathbb{R}^m$ and we obtain $\dim(\Pi_j V) = \dim(V)$ for all subspaces V of $\mathbb{R}^m$. In this case, the condition (5) for $V = \mathbb{R}^m$ automatically implies the condition for all subspaces of $\mathbb{R}^m$. Assuming all Π_j are surjective as well, which is a mild assumption given the previous discussion, all Π_j are bijective. Reparameterizing the range of each Π_j, we may assume that each Π_j is the identity map and thereby identify Hölder's inequality.

While it may be tempting to study (4) with some $0 < p_j < 1$, such estimates are easily seen to fail. This is also reflected by (5). Assume for example a Brascamp–Lieb inequality with $p_1 < 1$ and denote the ker of Π_1 by W, and assume for simplicity all maps Π_j to be surjective. Then we obtain a contradiction by applying (5) twice:

$$m = k_1 + \dim(W) \leq k_1 + \sum_{j=2}^{n} \frac{1}{p_j} \dim(\Pi_j W) < \sum_{j=1}^{n} \frac{k_j}{p_j} = m.$$

The endpoint case $p_j = 1$ reduces to Brascamp–Lieb inequalities of fewer functions. We show this in case $j = n$. By a weak limiting process, the Brascamp–Lieb inequality extends to finite Borel measures in place of the n-th Schwartz function, with the total mass of the measure instead of the L^1 norm of the function on the right-hand side. In particular, one may insert translates of the Dirac delta measure. Conversely, bounds for the Brascamp–Lieb integral with translates of the Dirac delta measure as the n-th input imply by superposition the Brascamp–Lieb inequality for arbitrary Schwartz functions as n-th input. The Brascamp–Lieb inequality with a translate of the Dirac delta measure can be written as

$$\int_{\mathbb{R}^{k_1+\cdots+k_n}} \Big(\prod_{j=1}^{n-1} F_j(x_j + y_j)\Big)\delta(x_n - y_n)\delta(\Pi x)\, dx \leq C\Big(\prod_{j=1}^{n-1} \|F_j\|_{p_j}\Big),$$

which can be further written as

$$\int_{\mathbb{R}^{k_1+\cdots+k_{n-1}}} \Big(\prod_{j=1}^{n-1} F_j(x_j + y_j)\Big)\delta(\Pi(x_1, \ldots, x_{n-1}, y_n))\, dx_1, \ldots dx_{n-1}$$

$$\leq C\Big(\prod_{j=1}^{n-1} \|F_j\|_{p_j}\Big).$$

Note that the range of the restriction of Π to fixed y_n is the same as the range of Π as a consequence of the assumption that Π_n is surjective. The last display is again a Brascamp–Lieb integral with an affine linear space of integration and one input function less. Thus we have shown the desired reduction.

This observation in reverse allows to interpret the Dirac delta measure in the general Brascamp–Lieb form (1) as coming from a Schwartz function with L^1 norm on the right-hand side. Thus (4) is equivalent to the inequality

$$\int_{\mathbb{R}^{k_1+\cdots+k_n}} \Big(\prod_{j=1}^{n} F_j(x_j)\Big)F_{n+1}(\Pi x)\, dx \leq C\Big(\prod_{j=1}^{n} \|F_j\|_{p_j}\Big)\|F_{n+1}\|_1.$$

The integral on the left hand side is again a Brascamp–Lieb form (1), if written as

$$\int_{\mathbb{R}^{k_1+\cdots+k_n+k_{n+1}}} \Big(\prod_{j=1}^{n+1} F_j(x_j)\Big)\delta(x_{n+1} - \Pi(x_1, \ldots, x_n))\, dx.$$

Note that the subspace of integration is the graph of a function in the first n variables. In general Brascamp Lieb inequalities, each component x_j of an element of the subspace is determined by all other components unless $p_j = 1$. However, it is not necessary that each combination of the remaining components can be completed by an x_j to form to a point in the subspace.

We emphasize that Brascamp Lieb inequalities are positive inequalities. There is no loss in assuming that all functions are positive. This will be different in subsequent sections, where cancellation between positive and negative part of a singular kernel will be crucial.

2 Singular Brascamp–Lieb Inequalities

Coming to the main subject of this survey, one may ask whether a variant of the Brascamp–Lieb inequality continues to hold if one inserts singular integral kernels instead of finite measures into one or several input slots with $p_j = 1$. Singular integral kernels in general fail to be finite measures, but in many situations one retains inequalities thanks to cancellation between positive and negative parts of the kernel. Examples of singular integral kernels arise from integrating a mean zero Schwartz function over the group of dilations

$$K(t) = \lim_{N\to\infty} \int_0^N \lambda^k \phi(\lambda t)\, \frac{d\lambda}{\lambda}, \quad \widehat{K}(\tau) = \int_0^\infty \widehat{\phi}\Big(\frac{\tau}{\lambda}\Big) \frac{d\lambda}{\lambda}. \tag{7}$$

Such kernels are homogeneous under dilations and smooth outside the origin. They are in general not locally integrable near the origin, yet they are tempered distributions in the sense that the limit in N has to be executed after the pairing with a Schwartz function. Tempered distributions with such limits are called principal value distributions. More generally, one may consider tempered distributions K on $\mathbb{R}^k$ whose Fourier transform $\widehat{K}$, called the multiplier associated with K, is a bounded measurable function satisfying the symbol estimates

$$|\partial^\alpha \widehat{K}(\tau)| \le C|\tau|^{-|\alpha|} \tag{8}$$

for some constant C, all $\tau \neq 0$ and all multi-indices α up to suitably large order. This condition is satisfied for the above homogeneous kernels. For much of our survey it is sufficient to consider these homogeneous kernels. The Dirac delta measure is a singular integral kernel, it can be written in the form (7) with a Schwartz function of integral zero, and its Fourier transform is a constant function. A simple way to ensure that a Schwartz function has integral zero is to make it odd. Many of the interesting features of the theory can already be seen when restricting to odd kernels.

We write singular Brascamp–Lieb inequalities as

$$\int_{\mathbb{R}^m} \Big(\prod_{j=1}^{h} F_j(\Pi_j x)\Big)\Big(\prod_{j=h+1}^{n} K_j(\Pi_j x)\Big)\, dx \le C \prod_{j=1}^{h} \|F_j\|_{p_j} \tag{9}$$

with singular integral kernels K_j on $\mathbb{R}^{k_j}$ and surjective linear maps

$$\Pi_j : \mathbb{R}^m \to \mathbb{R}^{k_j}.$$

The constant C is assumed to be independent of the functions F_j, and is assumed to depend on the kernels K_j only through the constant in (8) and the bound on the order of derivatives in (8). For smooth homogeneous kernels, the constant C is controlled by some Schwartz norm of the Schwartz function ϕ in (7).

As we ask a given singular Brascamp–Lieb inequality to hold for all choices of singular integral kernels, it needs to hold for the special choice of a Dirac delta measure. In particular, the bound (9) needs to hold when all kernels are the Dirac delta measure. Note that

$$\prod_{j=h+1}^{n} \delta(\Pi_j x) = \delta(\Pi_{h+1}x, \dots, \Pi_n x),$$

where the Dirac delta measure on the right-hand side lives in dimension $k_{h+1} + \cdots + k_n$. In order for the integral in (9) to be well defined, we need the map

$$x \mapsto (\Pi_{h+1}x, \dots, \Pi_n x)$$

to be surjective. We assume this surjectivity and choose variables

$$t = (t_{h+1}, \dots, t_n)$$

on the range of this map. Changing coordinates and choosing y as vector of coordinates for the joint ker

$$W = \bigcap_{j=h+1}^{n} \ker \Pi_j\ , \tag{10}$$

we may rewrite the integral in (9) as

$$\int_{\mathbb{R}^m} \Big(\prod_{j=1}^{h} F_j(\Pi_j(y,t))\Big)\Big(\prod_{j=h+1}^{n} K_j(t_j)\Big)\, dydt. \tag{11}$$

Thanks to these conventions, it is particularly easy to reduce a singular integral by setting one kernel K_j equal to the Dirac delta measure. One removes this kernel from (11), sets the coordinate t_j equal to zero, and removes the integration over the variable t_j.

The class of singular integral kernels is invariant under dilation symmetries but not under translation or modulation symmetries. The translation symmetries of the Brascamp–Lieb integral discussed after (1) leave the singular Brascamp–Lieb form invariant only if the components z_j in the notation after (1) are zero for $j > h$, that is those j belonging to kernels. An analogous observation holds for the modulation symmetries.

The mean zero condition on the Schwartz function in (7) is an important theme in singular integral theory. To see necessity of the cancellation, consider a kernel K_n of the form (7) generated by a non-negative Schwartz function that is not constant equal to zero, and assume there is only one kernel or reduce the complexity by replacing the other kernels by Dirac delta measures. Consider (9) with characteristic functions F_j of standard unit balls in the respective dimensions similarly to the proof of necessity of (5). The right-hand side of (9) is finite. The integrand on the left-hand side is equal to $K_n(t_n)$ for y in a small ball about the origin and t_n in a small fixed interval around the origin. Uniformly in this ball in y, the integral in t_n tends to ∞ with N, because the degree of homogeneity of the singular integral kernel is critical for integration. Thus the left-hand side of (9) is unbounded.

Singular Brascamp–Lieb inequalities have seen much development in recent years, but the level of understanding is far from establishing a general criterion mirroring the condition (5). We present some necessary and some sufficient conditions.

A necessary condition for (9) can be obtained by specifying all K_j as Dirac delta measures, yielding a reduced Brascamp–Lieb inequality of lower order with integration over the joint ker W defined in (10). We obtain that Π_j needs to map W onto $\mathbb{R}^{k_j}$ if $p_j < \infty$, and (5) for the reduced inequality gives the necessary condition

$$\dim(V) \leq \sum_{j=1}^{h} \frac{1}{p_j} \dim(\Pi_j V) \tag{12}$$

for all $V \subseteq W$, with equality if $V = W$.

Due to the importance of cancellation of the singular integral kernel, we may obtain further necessary conditions for (9), namely that

$$\ker \Pi_1 + \ker \Pi_n = \mathbb{R}^m, \tag{13}$$

and similarly for other indices by permutation of the Schwartz functions and kernels. To see necessity, assume this condition is violated. By reduction we may assume $h = n - 1$. Then there is a non-zero linear functional λ on $\mathbb{R}^m$ which vanishes on $\ker \Pi_1$ and on $\ker \Pi_n$. This functional factors as

$$\lambda(x) = \rho_1(\Pi_1 x) = \rho(\Pi_n x)$$

for some suitable maps ρ_1, ρ. Let K_n be the kernel defined by (7) with the Schwartz function $e^{-|t|^2}\rho(t)$ and define for any tuple of Schwartz functions $(F_1, \ldots, F_{n-1})$

$$\widetilde{F_1} = F_1 \times (\mathrm{sgn} \circ \rho_1).$$

We obtain

$$\int_{\mathbb{R}^m} F_1(\Pi_1 x)\Big(\prod_{j=2}^{n-1} F_j(\Pi_j x)\Big)|K_n|(\Pi_n x)\, dx$$
$$= \int_{\mathbb{R}^m} \widetilde{F_1}(\Pi_1 x)\Big(\prod_{j=2}^{n-1} F_j(\Pi_j x)\Big)K_n(\Pi_n x)\, dx.$$

Approximating $\widetilde{F_1}$ by Schwartz functions and applying a hypothetical singular Brascamp–Lieb inequality for the right-hand side, we obtain the same inequality for the left-hand side, contradicting the impossibility of the inequality for the non-negative kernel $|K_n|$.

If $p_1 = \infty$, we obtain another necessary condition for a singular Brascamp–Lieb inequality, which we adapt from [42], namely

$$\bigcap_{j=2}^{n-1} \ker \Pi_j \subseteq \ker \Pi_1 \cup \ker \Pi_n\ .$$

Assume this is not the case. Pick a vector u which is in the space on the left-hand side but not in the space on the right-hand side. There is a linear functional λ_1 that factors as $\lambda_1(x) = \rho_1(\Pi_1 x)$ and is positive on u. Let $F_1 = 1_+ \circ \rho_1$ with 1_+ the characteristic function of the positive half line. Let F_j for $2 \leq j \leq h$ be the characteristic function of the unit ball.

There is also a linear functional λ that factors as $\lambda(x) = \rho(\Pi_n x)$ and is positive on u. Let K_n be the homogeneous kernel (7) generated by $e^{-|t|^2}\rho(t)$. We split the singular Brascamp–Lieb integral (9) by first integrating along lines parallel to u:

$$\int_{\ker(\lambda)} \int_{\mathbb{R}} F_1(\Pi_1(x+su))\Big(\prod_{j=2}^{n-1} F_j(\Pi_j(x+su))\Big)K_n(\Pi_n(x+su))\, ds dx$$

The middle factor in the integrand, the product over j, is independent of s and equal to 1 for x in a small neighborhood of the origin. The first factor is bounded,

$$F_1(\Pi_1(x+su)) = 1_+(\lambda_1(x) + s(\lambda_1(u)),$$

and for some sufficiently large a it vanishes for $s < -a$ and is constant 1 for $s > a$. The third factor is positive for $s > 0$. Hence the integral over $s < -a$ vanishes, is a bounded number for $-a < x' < a$, and is plus infinity for $s > a$ and x in a small neighborhood of the origin. Hence the singular Brascamp–Lieb integral is unbounded.

We come to some sufficient conditions for singular Brascamp–Lieb inequalities to hold. If one of the exponents p_j is equal to 1, we may reduce a singular Brascamp–Lieb inequality to one of lower complexity by the use of Dirac delta measures as discussed in the non-singular case. Validity of the reduced inequalities becomes a sufficient criterion for validity of the original inequality.

If

$$1 \le p_j \le 2 \tag{14}$$

for all $1 \le j \le h$, then it is useful to pass to the integral on the Fourier transform side. If Π in (1) is an orthogonal projection, the Fourier transform integral reads as

$$\int_{\mathbb{R}^{k_1+\cdots+k_n}} \Big(\prod_{j=1}^{h} \widehat{F}_j(\xi_j)\Big)\Big(\prod_{j=h+1}^{n} \widehat{K}_j(\xi_j)\Big)\delta((1-\Pi)\xi)d\xi. \tag{15}$$

This is estimated by a non-singular Brascamp–Lieb inequality in the Fourier transforms of the functions, using that the multipliers $\widehat{K}_j$ are functions in L^∞. Aiming at the dual exponents $p_j{}' = p_j/(p_j-1)$, we need the condition (5):

$$\dim(V) \le \sum_{j=1}^{h} \frac{1}{p_j{}'} \dim(V_j),$$

where V is a subspace of $\ker(1-\Pi)$, V_j is its projection onto the j-th coordinate space, and equality holds for V equal to $\ker(1-\Pi)$. We thus estimate (15) with the Brascamp–Lieb inequality by

$$C\prod_{j=1}^{h} \|\widehat{F}_j\|_{p_j{}'} \le C\prod_{j=1}^{h} \|F_j\|_{p_j}.$$

In the second inequality we have used the Hausdorff Young inequality, which is applicable by the assumption (14). An interesting variant of this theme is to estimate a singular Brascamp–Lieb integral by a mixed product of L^p norms of the functions and L^p norms of the Fourier transforms of the functions. An instance of this has been studied in [39].

3 Inequalities with One Singular Kernel and Hölder Scaling

As seen in the previous section, when all exponents p_j are at most 2, then one has a good sufficient criterion for a singular Brascamp–Lieb inequality. At the other end of the spectrum, when the p_j are large, one finds the special case of Hölder scaling

$$\sum_{j=1}^{h} \frac{1}{p_j} = 1,$$

where in an average sense the p_j are as large as they can be. This is a heavily studied case and we shall assume it throughout the rest of the survey.

Recall that in the Hölder case the condition (12) needs only to be checked for $V = W$. Each map Π_j restricted to W needs to be injective. Neglecting some trivial extensions for $p_j = \infty$, we may also assume that this map is surjective for each j. As a consequence, all k_j, $1 \leq j \leq n-1$ are equal and in particular $k_j = k_1$ and

$$m = k_1 + k_n.$$

The singular Brascamp–Lieb integral may then be written as

$$\int_{\mathbb{R}^{k_1}} \int_{\mathbb{R}^{k_n}} \Big(\prod_{j=1}^{n-1} F_j(A_j y + B_j t)\Big) K_n(t) dt dy,$$

with matrices A_j and B_j. Each of the matrices A_j has to be regular. Changing F_j by precomposing with the matrix A_j, we may assume that all A_j are equal to the identity matrix,

$$\int_{\mathbb{R}^{k_1}} \int_{\mathbb{R}^{k_n}} \Big(\prod_{j=1}^{n-1} F_j(y + B_j t)\Big) K_n(t) dt dy. \tag{16}$$

Interchanging the order of integration so that y becomes the inner variable and replacing it by $y - B_1 t$, we may in addition assume that

$$B_1 = 0.$$

Writing each F_j as Fourier integral we obtain for (16)

$$\int_{\mathbb{R}^{(n-1)k_1}} \int_{\mathbb{R}^{k_1}} \int_{\mathbb{R}^{k_n}} \Big(\prod_{j=1}^{n-1} \widehat{F}_j(\eta_j) e^{2\pi i \eta_j \cdot (y + B_j t)}\Big) K_n(t) dt dy d\eta_1 \ldots d\eta_{n-1}$$

$$= \int_{\mathbb{R}^{(n-1)k_1} : \eta_1 + \cdots + \eta_{n-1} = 0} \int_{\mathbb{R}^{k_n}} \Big(\prod_{j=1}^{n-1} \widehat{F}_j(\eta_j)\Big) \widehat{K}_n\Big(-\sum_j B_j^T \eta_j\Big) dt d\gamma,$$

where $d\gamma$ is the Lebesgue measure on the subspace $\eta_1 + \cdots + \eta_{n-1} = 0$ in $\mathbb{R}^{(n-1)k_1}$.

We look at small values of n. For $n = 2$, the singular Brascamp–Lieb integral in the discussed variables becomes

$$\int_{\mathbb{R}^{k_1}} \int_{\mathbb{R}^{k_2}} F_1(y) K_2(t)\, dt dy.$$

Taking formally the Fourier transform, one obtains

$$\widehat{F_1}(0)\widehat{K_2}(0),$$

which is undetermined by (8) and does not lead to an interesting theory.

The case $n = 3$ describes bilinear forms which dualize to linear operators. In the above coordinates, the singular Brascamp–Lieb integral can be written as

$$\int_{\mathbb{R}^{k_1}} \int_{\mathbb{R}^{k_3}} F_1(y) F_2(y + Bt) K_3(t)\, dt dy.$$

If B is not injective, we may integrate the ker of B first. This integrates the singular integral kernel towards a lower dimensional kernel, reducing the problem to a similar problem where B is injective. If B is not surjective, we may split the integration over y into integration over the range of B and the complement of the range. The integral over the range is a similar singular Brascamp–Lieb with smaller dimension, which can be estimated first. Subsequently, one can estimate the complementary integral by Hölder's inequality. Hence we may assume without loss of generality that B is regular. By changing variables and replacing the kernel K_3 by its composition with the inverse of B, we obtain the form

$$\int_{\mathbb{R}^{k_1}} \int_{\mathbb{R}^{k_3}} F_1(y) F_2(y + t) K_3(t)\, dt dy.$$

The dual linear operator is the classical convolution with a singular integral kernel, which is well understood. As a consequence, we have the desired singular Brascamp–Lieb inequality with Hölder scaling and $1 < p_1, p_2 < \infty$. The restriction $1 < p_j$ can be understood as a condition of the type (13) after a reduction by a Dirac delta function as in the discussion after (13).

We turn to the genuinely multi-linear case $n \geq 4$. Fixing n and k_1, singular Brascamp–Lieb inequalities become easier with growing k_n. In case of odd kernels this can be made rigorous by the method of rotations, which we will discuss more thoroughly later.

The largest possible k_n and thus easiest interesting case is $k_n = (n - 2)k_1$. For any larger k_n, one would necessarily violate condition (13) or be able to integrate out some of the t variables of K to reduce to a kernel of smaller dimension. The case $k_n = (n - 2)k_1$ is the classical theory of multi-linear forms of Coifman–Meyer type [12] and can be written as

$$\int_{\mathbb{R}^{k_1}} \int_{\mathbb{R}^{(n-2)k_1}} F_1(y) \Big(\prod_{j=2}^{n-1} F_j(y + B_j t_j) \Big) K_n(t_2, \dots t_{n-1})\, d(t_2, \dots, t_{n-1}) dy.$$

Note that each B_j has to be injective or else one could again reduce the problem by integrating a trivial component of a t variable. By a dimension count, each B_j is also surjective. Changing coordinates to parameterizing the range of this map and adjusting the kernel K_n suitably, we obtain the form

$$\int_{\mathbb{R}^{k_1}} \int_{\mathbb{R}^{(n-2)k_1}} F_1(y) \Big(\prod_{j=2}^{n-1} F_j(y + t_j) \Big) K_n(t_2, \dots t_{n-1})\, d(t_2, \dots, t_{n-1}) dy.$$

With a further change of variables we may write more symmetrically

$$\int_{\mathbb{R}^{k_1}} \int_{\mathbb{R}^{(n-1)k_1} : t_1 + \dots + t_{n-1} = 0} \Big(\prod_{j=1}^{n-1} F_j(y + t_j) \Big) \widetilde{K}_n(t_1, \dots t_{n-1})\, d\gamma dy$$

with $d\gamma$ the invariant measure on the subspace of $\mathbb{R}^{(n-1)k_1}$ perpendicular to the diagonal $(1, \dots, 1)$ and $\widetilde{K}_n$ suitably defined on this subspace. As a result of the classical theory, one obtains singular Brascamp–Lieb inequalities with Hölder scaling as long as

$$1 < p_j \leq \infty$$

for all indices $1 \leq j \leq n-1$. The restriction $1 < p_j$ is again a consequence of the discussion after (13). There is no restriction at ∞. An interesting theory allows to push the inequalities of Coifman–Meyer type beyond infinity. Under certain conditions on the kernel, one obtains BMO bounds, and one may consider restricted type estimates as discussed in [55], dualizing bounds in earlier work [33, 37]. Taking the Fourier transform, the Coifman–Meyer multi-linear form becomes

$$\int_{\mathbb{R}^{(n-1)k_1} : \xi_1 + \dots + \xi_{n-1} = 0} \Big(\prod_{j=1}^{n-1} F_j(\xi_j) \Big) \widehat{\widetilde{K}}_n(\xi_1, \dots \xi_{n-1})\, d\gamma$$

where the Fourier transform of $\widetilde{K}_n$ is suitably taken in the space Γ. The subspace of integration has dimension $(n-2)k_1$, which is equal to the dimension k of the multiplier. As a consequence, there are no translations of this subspace which leave the multiplier invariant. Hence the Coifman–Meyer case does not exhibit modulation symmetries. It relies on classical Calderón-Zygmund techniques that are translation and dilation invariant.

As one lowers k from the maximal interesting $(n-2)k_1$, one may no longer uniquely determine all B_j up to change of coordinates. The discussion bifurcates depending on the geometry of the B_j, and the classification of cases leads to quite elaborate linear algebraic questions. One case in every dimension is distinguished as the generic position of these matrices. It can be obtained almost surely by picking B_j randomly with respect to suitable Gaussian probability measures. The study of this generic situation has begun in the work on the bilinear Hilbert transform [47] and [30]. In the case $k_1 = 1$, the best sufficient dimensional condition in the generic situation is obtained in [55]. In the notation

$$\int_{\mathbb{R}} \int_{\mathbb{R}^{k_n}} \Big(\prod_{j=1}^{n-1} F_j(y + B_j t) \Big) K_n(t) dt dy,$$

the generic case is when each tuple of the linear functionals B_j spans the maximal possible space. One obtains the singular Brascamp–Lieb inequality with Hölder scaling for all

$$1 < p_j \leq \infty$$

provided one has the dimensional condition

$$n - 3 < 2k_n \tag{17}$$

for any $n \geq 3$. Unlike the Coifman–Meyer case, the generic singular Brascamp–Lieb integral for $k_n < (n-2)k_1$ exhibits modulation symmetries. The proof of the above result employs a modulation invariant counterpart of Calderón-Zygmund techniques called time-frequency analysis. Time-frequency analysis consists of breaking up the Brascamp–Lieb form into pieces that can be estimated by the classical Calderón-Zygmund theory. The decomposition is done in a way that orthogonality arguments in the phase plane allow to control the number of pieces. This technique originates in the works of [10, 29] and was first applied to singular Brascamp–Lieb forms in the work [47] on the bilinear Hilbert transform. An approach to time-frequency analysis through outer measures was described in [17]. The principal value limit in (7) in the context of time-frequency analysis and in particular the bilinear Hilbert transform is studied in [18, 20, 46].

While the time-frequency analysis in [55] breaks down if the condition (17) is violated, it remains an open problem whether (17) is necessary for singular Brascamp–Lieb inequalities to hold. Even under condition (17), interesting open questions remain concerning the extension of singular Brascamp–Lieb inequalities to restricted type inequalities beyond the threshold at $p_j = \infty$. This is discussed in [55], see also [16] for a discussion near the boundary of the range of exponents with known bounds.

The extension of the above result of [55] to $k_1 > 1$ is addressed in [15], proving singular Brascamp–Lieb inequalities on the form

$$\int_{\mathbb{R}^{k_1}} \int_{\mathbb{R}^{k_n}} \Big(\prod_{j=1}^{n-1} F_j(y + B_j t)\Big) K_n(t) dt dy$$

assuming $B_j : \mathbb{R}^{k_n} \to \mathbb{R}^{k_j}$ are in generic position and

$$k_1(n-3) < 2k_n. \tag{18}$$

If k_n is an integer multiple of k_1, this follows rather quickly from the methods of [55]. For the fractional multiple case, [15] uses some additional arguments from additive combinatorics. The authors restrict attention to the range $2 < p_j \leq \infty$. It is not known whether the restriction $2 < p_j$ is necessary.

A partial explanation for the break down of modulation invariant time-frequency analysis beyond (17) and (18) is the occurrence of more general symmetries. For example, consider the case of the trilinear Hilbert transform

$$\int_{\mathbb{R}} \int_{\mathbb{R}} \Big(\prod_{j=1}^{4} F_j(y + B_j t)\Big) \frac{1}{t} dt dy$$

with generic, that is pairwise different, numbers B_j. This form exhibits a symmetry under quadratic modulation

$$Q_{\alpha_j} F_j(x) = F_j(x) e^{i\alpha_j x^2}$$

where the four numbers α_j are all non-zero and satisfy

$$\sum_j \alpha_j (y + B_j t)^2 = 0.$$

It would be interesting to find extensions of time-frequency analysis that are invariant under more general symmetries and address boundedness of the trilinear Hilbert transform. This starts with a solid understanding of the type of symmetries, we refer to related work on inverse theorems for Gowers norm [34] involving generalized quadratic phase functions possibly relevant for the trilinear Hilbert transform and the more general symmetries in [35]. A variant of time frequency analysis under polynomial symmetries was developed in [49, 50, 71]. Additional symmetries may not be the only obstruction to go beyond (17), because it is not clear that all cases beyond (17) do exhibit additional symmetries.

Shrinking k_n further, the minimal non-trivial case is $k_n = 1$. The distance to k_1 is maximized if $k_1 = n - 1 = h$. If k_1 is greater than or equal to h, then the vectors B_j, $2 \leq j \leq h$ span a space of dimension less than k_1 and one may reduce to a singular

Brascamp–Lieb integral of lower order as discussed in the case $n = 3$. By the same token, if $k_1 = h$, then these vectors have to be linearly independent and thus a basis of $\mathbb{R}^{k_1}$. Since all bases are equivalent up to change of variables, one can write the singular Brascamp–Lieb integral without loss of generality in symmetric form as

$$\int_{\mathbb{R}^h} \Big(\prod_{j=1}^{h} F_j(x_1, \ldots, x_{j-1}, x_{j+1}, \ldots, x_h)\Big) \frac{1}{x_1 + \ldots + x_h} dx. \tag{19}$$

This form is called the simplex Hilbert form. Maybe the biggest challenge in the area is to understand whether this form satisfies any singular Brascamp–Lieb inequalities. By symmetry and interpolation techniques, the easiest bound to prove should be the one with all exponents equal. We formulate this as a conjecture.

Conjecture 1 There exists a constant C such that for all tuples of Schwartz functions $(F_j)_{j=1}^h$ the form (19) is bounded by

$$C \prod_{j=1}^{h} \|F_j\|_h.$$

By the method of rotations, bounds for the simplex Hilbert form imply bounds for many singular Brascamp–Lieb integrals including the so-called multi-linear Hilbert transform, another major open problem. Moreover, bounds for the simplex Hilbert form imply bounds for the Carleson and polynomial Carleson operator

$$\int_{\mathbb{R}} f(x-t) e^{i(N_1(x)t + N_2(x)t^2 + \ldots + N_d t^d)} \frac{dt}{t},$$

which was for general d studied in [49, 50] and [71]. Partial progress on the simplex Hilbert form in the case $h = 3$ can be found in [45], which in particular establishes the above conjectured bound in a dyadic model when one of the functions takes a special form. Further results concerning truncations of the simplex Hilbert form and effective bounds in the parameter of truncation are discussed in [70] based on the approach in [66], and in [28].

Having discussed generic choices of B_j in the spectrum from large k_n to small k_n, we turn attention to some of the phenomena arising when we do not ask the B_j to be in generic positions. We begin with the simplest case which displays some of the phenomena,

$$\int_{\mathbb{R}} \int_{\mathbb{R}} \Big(\prod_{j=1}^{3} F_j(y + B_j t)\Big) K_4(t) \, dt \, dy.$$

The generic case has three different real numbers B_j, this is the classical bilinear Hilbert transform. All generic cases have the same proof of Brascamp–Lieb bounds using time-frequency analysis. If two values of B_j are equal, the form changes its nature. One identifies the pointwise product of two functions, and replacing the product by a new function we obtain a singular Brascamp–Lieb integral with $n = 3$. Applying the classical theory without time-frequency analysis and then applying Hölder's inequality to resolve the product proves L^p bounds in this degenerate situation. The case that all three values of B_j are equal is even further degenerate but of no interest, it leads to the pointwise product of three functions together with the indeterminate integral in case $n = 2$. If two of the values of B_j approach each other, the historically first proof of bounds for the bilinear Hilbert transform produced a growing constant in the singular Brascamp–Lieb inequality. It was natural to seek uniform bounds, which was achieved in a series of papers [32, 48, 62, 67, 68] in the full Hölder range of exponents with $1 < p_j \leq \infty$. Some of these results were generalized to uniform bounds on other families of singular Brascamp–Lieb integrals in [56].

A more complicated classification of cases occurs for the two dimensional bilinear Hilbert transform

$$\int_{\mathbb{R}^2}\int_{\mathbb{R}^2}\Big(\prod_{j=1}^{3} F_j(y + B_j t)\Big)K_4(t)dtdy,$$

a situation first considered by Demeter and Thiele [14] and then thoroughly discussed in the PhD thesis [69]. The thesis classifies the possiblilities for the parameters B_1, B_2, B_3 into nine cases. Most cases can be normalized such that $B_1 = 0$ and $B_2 = I$, leaving only $B = B_3$ as indetermined matrix, which may be assumed to be in Jordan canonical form. A trivial pointwise product occurs if $B = 0$ or $B = I$, this results in a reduction of the complexity of the integral as in the one dimensional case. The case that all eigenvalues of B are different from 0 and 1 is the generic case covered by previous results. The case that one eigenvalue of B is equal to 0 or 1 and the other eigenvalue is different from 0 and 1 is an interesting hybrid case discussed in [14], likewise the case of a non-trivial Jordan block with eigenvalue 0 or 1. The case when B has both 0 and 1 as eigenvalue is called the twisted paraproduct and is an instance of the forms in Theorem 2 below with $m = 2$, albeit with the fourth function set constant equal to 1.

Only in one of the nine cases it is not known whether the singular Brascamp–Lieb inequality holds at a nontrivial set of exponents. This is the case where the first columns of all three matrices B_1, B_2, B_3 vanish, while the second columns respectively are $(0, 0)$, $(0, 1)$, $(1, 0)$. This case is a simplex Hilbert form discussed in the above conjecture. All remaining cases reduce to easier objects and are of lesser interest. An abundance of questions concerning uniform bounds arise between these various cases. While the method of rotations would prove uniform bounds for odd kernels from Conjecture 1, lacking a proof of the latter it may be of interest to study these uniform questions.

We turn to a class of Brascamp–Lieb integrals with modulation symmetry group spanned by rich modulations symmetries. A rich modulation symmetry is a one parameter modulation symmetry which generalizes to arbitrary phase functions. For example, the Hölder form

$$\int_{\mathbb{R}} F_1(x)F_2(x)\,dx$$

is invariant not only under replacing F_1 and F_2 by the modulated functions $M_\xi F_1$ and $M_{-\xi} F_2$ respectively, but also under replacing them by

$$F_1(x)e^{i\phi(x)}, \quad F_2(x)e^{-i\phi(x)}$$

for arbitrary real phase functions ϕ. If we consider each input function as a function in k_1 arguments, then one way that rich modulations symmetries occur is when slots of different functions share the same argument.

We consider an example where each of the $k = k_1$ slots carries two possible variables, making it $2k$ integration variables, which we denote as

$$(x_1^0, \dots, x_k^0, x_1^1, \dots, x_k^1) = x.$$

Each possibe combination of the variables occurs in one of the functions. This requires 2^k input functions parameterized by the cube Q, the set of all

$$j : \{1, 2, \dots, k\} \to \{0, 1\}.$$

Consequently, for $j \in Q$, we have

$$\Pi_j x = (x_1^{j(1)}, x_2^{j(2)}, \dots, x_k^{j(k)}).$$

We further consider a singular integral kernel K in $\mathbb{R}^k$ and an arbitrary surjective $\Pi : \mathbb{R}^{2k} \to \mathbb{R}^k$. The Brascamp–Lieb integral in question then writes as

$$\int_{\mathbb{R}^m} \Big(\prod_{j \in Q} F_j(\Pi_j x)\Big) K(\Pi x)\,dx\,. \tag{20}$$

Theorem 2 ([25]) *Given $k \geq 1$, the form (20) satisfies a singular Brascamp–Lieb inequality with $p_j = 2^k$ for all $j \in Q$ if and only if for all j*

$$k = \dim(\Pi_j(\ker \Pi)). \tag{21}$$

The condition (21) is equivalent to (12) in this situation.

While rich symmetries are very large symmetry groups and restrict techniques to those that are invariant under these symmetries, at least they have a very generic structure and one does not need to delve into the theory of polynomial or other structured symmetries. The main technique in the context of rich symmetries was pioneered in the context of the so-called twisted paraproduct in [42] and is sometimes called twisted technology. Brascamp–Lieb integrals involving rich symmetries were also studied in [7, 21, 22, 41, 44] and also in [27, 64] with applications to quantitative convergence of ergodic averages, and in [23, 26] with applications to some problems in Euclidean Ramsey theory. An application to stochastic integrals was studied in [43]. Further higher dimensional generalizations are discussed in [65].

It would be desirable to study some natural extensions of Theorem 2. One obvious generalization would be a more general range of exponents than the symmetric exponent point. Somewhat related to that is the question what happens if the corners of the cube are not fully occupied, that is the number of functions is strictly less than 2^{k_1}. In case one has L^∞ bounds, it is trivial to omit the corresponding function by estimating the constant function in L^∞, but it is not clear that all inequalities with constant functions arise from more general L^∞ bounds.

One further extension is to allow more than two variables in one slot, that is for $k_1 \geq 1$ and $l \geq 2$ we may consider $\mathbb{R}^m$ with coordinates

$$x = ((x_1^0, \dots, x_k^0), (x_1^1, \dots, x_k^1), \dots, (x_1^{l-1}, \dots, x_k^{l-1})) \in \mathbb{R}^{kl}$$

Then for all $j : \{1, 2, \dots, k\} \to \{0, \dots, l-1\}$ we may define

$$\Pi_j x = (x_1^{j(1)}, x_2^{j(2)}, \dots, x_k^{j(k)})$$

One may then ask the analoguous result as Theorem 2.

Note that also the simplex Hilbert forms of Conjecture 1 have many rich modulation symmetries. Indeed, the group of modulation symmetries of the simplex Hilbert form is spanned by rich symmetries. The space of integration in Fourier space has dimension $n(n-2)+1$. Since the singular integral kernel is one dimensional, this gives an $n(n-2)$ dimensional group of modulation symmetries of the simplex Hilbert form. However, for each of the n variables one can find $n-2$ pairs of functions so that independent rich symmetries akin to the above shown apply between this pair of functions. The forms in Theorem 2 and the suggested generalization above have the structure that each variable has a fixed slot number in which it may occur. Note that this is not the case in the simplex Hilbert form. For example, the variable x_2 typically appears in the second slot, unless in the function F_1, where the variable x_1 is omitted and the variable x_2 appears in the first slot. This mismatch is the main obstacle to apply the known twisted technology to the simplex Hilbert form.

4 Method of Rotations and More General Kernels

The method of rotation allows to write a singular Brascamp–Lieb form with one singular integral kernel as a superposition of a family of forms with lower dimensional kernels. The family of forms is generated by rotations or more general linear transformations of the space of integration.

Turning to details, a singular Brascamp–Lieb form with a homogeneous smooth kernel can be written as

$$\int_0^\infty \int_{\mathbb{R}^{k_1+\cdots+k_n}} \Big(\prod_{j=1}^{n-1} F_j(x_j)\Big) t^{k_n} \psi(tx_n)\delta(\Pi x)\, dx \frac{dt}{t} \tag{22}$$

with a smooth and compactly supported function ψ with integral zero. Assume there is a vector v such that the inner product $v \cdot x_n$ is bounded away from zero on the support of ψ. The following display is a superposition by a weight function ϕ of a family of forms generated by rank one perturbations of Π using a further fixed vector w and a varying scalar parameter a:

$$\int_{\mathbb{R}} \int_0^\infty \int_{\mathbb{R}^{k_1+\cdots+k_n}} \Big(\prod_{j=1}^{n-1} F_j(x_j)\Big) t^{k_n} \psi(tx_n)\phi(a)\delta(\Pi x + wa(v \cdot x_n))\, dx \frac{dt}{t} da.$$

We assume ϕ is smooth and compactly supported. Rescaling the variable a and combining it with the vector x_n to a vector of dimension $k_n + 1$, we recognize a new singular Brascamp–Lieb form

$$\int_0^\infty \int_{\mathbb{R}^{k_1+\cdots+(k_n+1)}} \Big(\prod_{j=1}^{n-1} F_j(x_j)\Big) t^{k_n+1} \widetilde{\psi}(tx_n, ta)\delta(\Pi x + wa)\, dx da \frac{dt}{t} \tag{23}$$

with the compactly supported smooth function

$$\widetilde{\psi}(x_n, a) := \psi(x_n) \frac{1}{|v \cdot x_n|} \phi\Big(\frac{a}{v \cdot x_n}\Big). \tag{24}$$

One verifies that $\widetilde{\psi}$ has integral zero by first integrating in a and then in x_n. If we can prove bounds for the singular Brascamp–Lieb forms (22) uniformly for all maps Π in the perturbed family, then by superposition we obtain a bound with the same exponents for (23).

Conversely, given a Brascamp–Lieb integral as in (23), one may seek to write it as superposition of Brascamp–Lieb forms with lower dimensional kernels. A general procedure exists, when the function $\widetilde{\psi}$ is odd. In addition, we assume $\widetilde{\psi}$ is compactly supported away from the origin. After a decomposition by a finite smooth partition of unity, and a suitable rotation of the coordinate system for each piece, we can assume that there is a vector v of dimension k_n such that $\widetilde{\psi}$ is supported in the union of two small neighborhoods respectively of $(v, 0)$ and $(-v, 0)$

With suitable compactly supported functions φ and ρ we may write

$$\widetilde{\psi}(x_n, a) = \frac{1}{|v \cdot x_n|}\varphi\Big(x_n, \frac{a}{v \cdot x_n}\Big) = \frac{1}{|v \cdot x_n|}\varphi\Big(x_n, \frac{a}{v \cdot x_n}\Big)\rho\Big(\frac{a}{v \cdot x_n}\Big)$$

and note that φ is odd in the first variable for fixed second variable. Taking a Fourier integral of φ in the second variable and denoting that by $\widehat{\varphi}$, we obtain

$$\widetilde{\psi}(x_n, a) = \int_{\mathbb{R}} \widehat{\varphi}(x_n, \xi)\frac{1}{|v \cdot x_n|}e^{2\pi i \xi \frac{a}{v \cdot x_n}} \rho\Big(\frac{a}{v \cdot x_n}\Big)\, d\xi.$$

For fixed ξ, the integrand is a function of the form (24) with an odd function ψ. If we can prove bounds for the family of Brascamp–Lieb integrals of lower dimensional kernels uniformly for fixed Schwartz norm of ψ of some order, then we may integrate these bounds in ξ as the Schwarz norm of $\widehat{\varphi}$ in the first variable is rapidly decreasing as a function in the second variable.

One can iterate rank one perturbations to obtain the more general superposition

$$\int_{\mathbb{R}^l}\int_0^{\infty}\int_{\mathbb{R}^{k_1+\dots+k_n}} \Big(\prod_{j=1}^{n-1} F_j(x_j)\Big) t^{k_n}\psi(tx_n)\phi(a)\delta\Big(\Pi x + \sum_{i=1}^{l} w_i a_i (v_i \cdot x_n)\Big)\, dx \frac{dt}{t} da.$$

If the function ϕ in the above calculation is replaced by an arbitrary finite Borel measure of normalized total mass, for example a Dirac delta measure, estimates for the form (23) uniformly in the choice of such measure are equivalent to estimates for the form (22) with lower dimensional kernel uniformly over the perturbation parameters in the support of ϕ. Choosing ϕ with any intermediate regularity between smooth function and Borel measure, one can view the difficulty of estimates for the superposed operator as intermediate between the two endpoint cases. Estimates for such forms with rough singular integral kernel can be of their own interest, if estimates for the lower dimensional kernels are not known or maybe known to be false in general.

An early example of this principle is provided by the Calderón commutator [9], which later appeared in the investigation of the Cauchy integral along Lipschitz curves, see [11] and the references therein. The commutator can be viewed as a rough superposition of bilinear Hilbert transforms. Calderón proposed the study of the bilinear Hilbert transform and uniform bounds for it as a stepping stone towards the commutator. However, the bilinear Hilbert transform remained an open problem for many years after bounds for the Calderón commutator were obtained using different techniques. A recent account and approach to the Calderón commutator and higher order commutators was given in [53] and in [54]. These higher order commutators can be seen as a suitable superposition of multi-linear Hilbert transforms which by themselves are not known to be bounded.

Consider a perturbation $\widetilde{\Pi}$ of Π as in (1) by a rank one map,

$$\widetilde{\Pi} = \Pi + \Pi(u) \otimes v.$$

Assume the perturbation is small and non-trivial, and the dimension of $\ker \widetilde{\Pi}$ is equal to that of $\ker \Pi$, namely m. The intersection $V = \ker \widetilde{\Pi} \cap \ker \Pi$ has then dimension $m-1$, and we may choose a unit vector $\tilde{v}$ in $\ker \Pi$ perpendicular to V. Using that the perturbation is small, we may chose $\tilde{u}$ perpendicular to $\ker \Pi$ so that $\tilde{v} - \tilde{u}$ is in $\ker \widetilde{\Pi}$. Then $\Pi + \Pi(\tilde{u}) \otimes \tilde{v}$ has the same ker as $\widetilde{\Pi}$ and we may assume after rescaling that $u = \tilde{u}$ and $v = \tilde{v}$. The embedding map I as in (3) can then also be identified as perturbed by a rank one matrix, namely $\tilde{I} = I - u \otimes I^T v$. To verify this, one checks separately that the vectors that embed under I into V have the same image under the perturbed map, and that the vector that maps to v under I maps to $v - u$ under the perturbation.

If the perturbations are such that only one component u_j of u and only the component $(I^T v)_n$ of $I^T v$ is non-zero, we may view the averaging of the form as an averaging of the function F_j. If we iterate several perturbations like that, then the averaged function takes the form

$$F(\Pi_j x, x_n) = \int_{\mathbb{R}^l} \phi(a) F_j\Big(\Pi_j x - \Big(\sum_l a_l u_l (v_l \cdot I x_n)\Big)_j\Big)\, da.$$

If there are enough averages so that the rank one matrices add to a regular matrix, and if F is in L^∞, then the averaged function $F(y, z)$ becomes a y dependent symbol in the variable z in the sense

$$|\partial_y^\alpha \partial_z^\beta F(y, z)| \le C|z|^{-|\alpha|-|\beta|}$$

for all multi-indices up to some degree depending on the regularity of the averaging function ϕ. Multiplying this symbol with the singular integral kernel gives a "space dependent" singular integral form which is nowadays seen within in the theory of $T(1)$ theorems originating in [13]. Therefore, bounds for the averaged operator can be viewed as a Brascamp–Lieb version of a $T(1)$ theorem.

In this spirit, a multi-linear $T(1)$ theorem with a variant of the bilinear Hilbert transform with space dependent singular integral kernel was proven in [6] and applied in [63] in a singular variant of a higher Calderón commutator. $T(1)$ theorems with rich modulation symmetries were proven in [44, 65] in dyadic models, it would be interesting to extend these results to the continuous setting and extend to further averaged singular Brascamp–Lieb forms.

The paper [24] discusses averages of the simplex Hilbert forms which yield singular Brascamp–Lieb forms with rich modulation symmetries. The averaged forms are such that they can be treated by twisted technology. More precisely, [24] proves bounds in cases $n = 4$ and $n = 5$ on

$$\int_{(0,1)^{n-3}} \int_{\mathbb{R}^{n-2}} \int_{\mathbb{R}} \Big(\prod_{j=1}^{n-3} F_j(y + \alpha_j B_j t)\Big) F_{n-2}(y + B_{n-1}t) F_{n-1}(y) K_n(t) dt dy d\alpha$$

for linearly independent vectors B_j.

5 Inequalities with Two Singular Kernels and Hölder Scaling

Singular Brascamp–Lieb integrals in the case of several singular integral kernels fall into the scope of multi-parameter theory. We display some of the features of multi-parameter theory using the example of two kernels. We continue to assume Hölder scaling.

Considerations analoguous to those leading to (16) from (11) turn the singular Brascamp–Lieb integral with two kernels into the form

$$\int_{\mathbb{R}^{k_1}} \int_{\mathbb{R}^{k_{n-1}}} \int_{\mathbb{R}^{k_n}} \Big(\prod_{j=1}^{n-2} F_j(y + B_j s + C_j t)\Big) K_{n-1}(s) K_n(t) dt ds dy. \tag{25}$$

Applying the Fourier transform as after (16) we obtain the alternative expression

$$\int_{\Gamma} \Big(\prod_{j=1}^{n-2} \widehat{F}_j(\xi_j)\Big) \widehat{K}_{n-1}(-\sum_{j=1}^{n-2} B_j^T \xi_j) \widehat{K}_n(-\sum_{j=1}^{n-2} C_j^T \xi_j)\, d\gamma, \tag{26}$$

where Γ is the subspace of $\mathbb{R}^{(n-2)k_1}$ determined by $\xi_1 + \cdots + \xi_{n-2} = 0$ and $d\gamma$ is the Lebesgue measure on this subspace.

Simplifying degenerations may occur. The arguments of the two multipliers in (26) can be identical, that is each C_j is equal to B_j. As the product of two multipliers is again a multiplier with analoguous symbol bounds, this reduces to a singular Brascamp–Lieb with one kernel. Another simplifying degeneration of (25) may be separation. If for every j one of the matrices B_j or C_j is zero, then we may write the integral in s and t as a product of two integrals, one in s and one in t. Then we may apply Hölder's inequality in the variable x on this product. Resolving the resulting L^p norms by pairing with a dual function, we obtain two singular Brascamp–Lieb integrals with one kernel each. Separation in (25) may occur after replacing the variable y by $y + Bs + Ct$ for suitable matrices B and C.

A family of cases occurs with counterexamples to a singular Brascamp–Lieb inequality that show a phenomenon not possible for one kernel. Assume we have a family of quadratic forms Q_j on $\mathbb{R}^{k_1}$ such that

$$\sum_{j=1}^{n-2} Q_j(y + B_j s + C_j t) = s_1 t_1$$

where s_1 and t_1 are the first components of s and t, there being no loss in generality choosing these particular components. For n large enough compared to k_1, k_{n-1}, k_n, such quadratic forms will exist in the case of generic matrices B_j and C_j. Choose functions of the form

$$F_j(x) = \phi(x)e^{-2\pi i Q_j(x)}$$

where ϕ is a non-negative smooth approximation of the characteristic function of a very large ball about the origin. Choose the kernel

$$K_n(t) = \lim_{N\to\infty} \int_0^N \lambda^{k_n} \psi(\lambda t_1)\phi(\lambda(t_2, \dots, t_{k_n}))\frac{d\lambda}{\lambda}$$

with odd ψ which is non-negative on the positive half axis and with non-negative ϕ, and similarly for K_{n-1} with odd $\widetilde{\psi}$ such that $\widehat{\widetilde{\psi}} = \psi$. Zooming into the critical integrals in s_1 and t_1 in the expression (25), we see

$$\int_{\mathbb{R}}\int_{\mathbb{R}} e^{-2\pi i s_1 t_1}\widetilde{\psi}(\mu s_1)\psi(\lambda t_1)\, ds_1 dt_1 = \mu^{-1}\psi(\mu^{-1}t_1)\psi(\lambda t_1).$$

The right-hand side is an even function in t_1 and non-negative on the positive half axis, hence it is non-negative, and it is not identically zero as one can see considering μ^{-1} near λ. The effect is that the cancellation of the kernel K_n is destroyed, resulting in unboundedness as N tends to ∞. More details of this calculation can be found in [57] for the two examples

$$\int_{\mathbb{R}^4} F_1(x_1, x_2)F_2(x_1 - t, x_2 - s)F_3(x_1 + t, x_2 + s)\frac{ds}{s}\frac{dt}{t}dx$$

and

$$\int_{\mathbb{R}^3} F_1(x)F_2(x + t)F_3(x + s)F_4(x + t + s)\frac{ds}{s}\frac{dt}{t}dx. \tag{27}$$

Multi-parameter theory is named after the various scaling parameters occurring in a product of singular integral kernels. We call the product of the multipliers in (26) the joint multiplier and write it with scaling parameters μ and λ as

$$m(\sigma, \tau) = \widehat{K}_{n-1}(\sigma)\widehat{K}_n(\tau) = \lim_{N,M\to\infty} \int_0^N \int_0^M \widehat{\phi}_{n-1}\Big(\frac{\sigma}{\mu}\Big)\widehat{\phi}_n\Big(\frac{\tau}{\lambda}\Big) \frac{d\mu}{\mu}\frac{d\lambda}{\lambda}.$$

A typical step in multi-parameter theory is the cone decomposition, which is a sorting of an integral in several scaling parameters by the size of the scaling parameters as follows:

$$m_1(\sigma,\tau)+m_2(\sigma,\tau) = \lim_{N\to\infty}\int_{0<\mu<\lambda<N} \dots \frac{d\mu}{\mu}\frac{d\lambda}{\lambda} + \lim_{M\to\infty}\int_{0<\lambda<\mu<M} \dots \frac{d\mu}{\mu}\frac{d\lambda}{\lambda}.$$

Note that the joint multiplier m in (26) satisfies the multi-parameter symbol estimate

$$|\partial_\sigma^\alpha \partial_\tau^\beta m(\sigma,\tau)| \le C|\sigma|^{-|\alpha|}|\tau|^{-|\beta|}, \tag{28}$$

where ∂_σ and ∂_τ are any partial derivatives in the σ and τ variables respectively. The cone multipliers m_1 and m_2 satisfy

$$|\partial_\sigma^\alpha \partial_\tau^\beta m_1(\sigma,\tau)| \le C|\sigma|^{-|\alpha|-|\beta|}, \tag{29}$$

$$|\partial_\sigma^\alpha \partial_\tau^\beta m_2(\sigma,\tau)| \le C|\tau|^{-|\alpha|-|\beta|}. \tag{30}$$

In some instances, bounds for the variants of (26) with the joint multiplier replaced by the cone multipliers can be established, based on the symbol estimates (29), (30). Note that these symbol estimates, say (29), are generalizations of the single kernel case $K_{n-1} = \delta$ in that the multiplier (29) is "frequency dependent" in the variable τ, a dual concept to the "space dependent" kernels discussed in the previous section. Typically, estimates for the cones hold for generic choices of the matrices B_j and C_j provided the methods of [55] or [15] for "frequency dependent" multipliers apply, which is under the suitably adapted conditions (17) and (18). An example for a singular Brascamp–Lieb form where this cone decomposition applies and uses generalized bounds for "frequency dependent" variants of the bilinear Hilbert transform is given by

$$\int_{\mathbb{R}^3} F_1(y)F_2(y+s+t)F_3(y+B_3s+C_3t)K_4(s)K_5(t)dsdtdy$$

with generic parameters B_3 and C_3.

Somewhat opposite of the case of generic matrices B, C, one finds in the literature the case when each of these matrices is either zero or elementary, meaning it has precisely one non-zero entry, and this entry is equal to one. The flag paraproducts in [51, 52] are essentially this case for $k_1 = 1$. Estimates are shown for the case

$$\int_{\mathbb{R}^5} F_1(y)F_2(y-t_1)F_3(y-t_2-s_1)F_4(y-s_2)K_5(t_1,t_2)K_6(s_1,s_2)ds_1ds_2dt_1dt_2dy,$$

which is motivated by questions in fluid dynamics, and a rather general positive conjecture is formulated in [51, 52]. While one also does a cone decomposition in this case, it is important that the multiplier retains a product structure underneath the cone decomposition, and one does not simply rely on symbol estimates (28). Necessity of the product structure is demonstrated in [31]. While a form

$$\int_{\mathbb{R}^3} F_1(y)F_2(y+t)F_3(y+s)K_4(t)K_5(s)dsdtdy$$

is bounded by the method of separation, and the joint multiplier satisfies

$$|\partial_\sigma^\alpha \partial_\tau^\beta (\widehat{K_4}(\sigma)\widehat{K_5}(\tau))| \leq C|\sigma|^{-\alpha}|\tau|^{-\beta},$$

the form obtained by replacing the joint multiplier by a general multiplier m satisfying

$$|\partial_\sigma^\alpha \partial_\tau^\beta m(\sigma,\tau)| \leq C|\sigma|^{-\alpha}|\tau|^{-\beta}$$

need not satisfy any bounds in L^p spaces.

We may consider the case of B_j and C_j being zero or elementary for $k_1 > 1$ as well. A particular instance is discussed in [57] under the name of bi-parameter paraproduct:

$$\int_{\mathbb{R}^6} F_1(y_1,y_2)F_2(y_1+s_1,y_2+t_1)F_3(y_1+s_2,y_2+t_2)K_4(s_1,s_2)K_5(t_1,t_2)dsdtdy.$$

A generalization with more kernels is discussed in [60]. These examples are not affected by the obstruction described in [31], and one may prove bounds for multipliers satisfying (28). However, already a simple modification of the above such as interchanging s_2 and t_2 in the argument of F_3 is not addressed by the discussion in [57].

A hybrid between the generic case and the flag paraproduct case is called the biest and studied in [58, 59],

$$\int_{\mathbb{R}^3} F_1(x)F_2(x+t)F_3(x+s)F_4(x-t-s)\frac{ds}{s}\frac{dt}{t}dx.$$

It arises in the theory of iterated Fourier integrals, which occur in multi-linear expansions of certain ordinary differential equations. Singular Brascamp–Lieb inequalities for this form are known and require time frequency analysis because the bilinear Hilbert transform is embedded into this object. Compare with the similar form (27). For a study of objects related to the biest see [19, 36, 38–40, 61].

A more recent development is the theory of vector valued inequalities in the context of singular Brascamp–Lieb inequalities. The helicoidal method was introduced in [1] to study forms similar to the biest through mixed norm spaces and vector-valued inequalities. A survey of the helicoidal method can be found in [2].

Acknowledgments This survey was initiated during a delightful stay at the conference *Geometric Aspects of Harmonic Analysis* in honor of Fulvio Ricci 2018 in Cortona, Italy. The second author acknowledges support by the Deutsche Forschungsgemeinschaft through the Hausdorff Center for Mathematics, DFG-EXC 2047, and the Collaborative Research Center 1060.

References

1. Benea, C., Muscalu, C.: Multiple vector-valued inequalities via the helicoidal method. Anal. PDE **9**(8), 1931–1988 (2016)
2. Benea, C., Muscalu, C.: The Helicoidal Method. Preprint (2018), arXiv:1801.10071
3. Bennett, J., Carbery, A., Christ, F.M., Tao, T.: The Brascamp-Lieb inequalities: finiteness, structure and extremals. Geom. Funct. Anal. **17**(5), 1343–1415 (2008)
4. Bennett, J., Bez, N., Cowling, M.G., Flock, T.C.: Behaviour of the Brascamp–Lieb constant. Bull. Lond. Math. Soc. **49**(3), 512–518 (2017)
5. Bennett, J., Bez, N., Buschenhenke, S., Flock, T.C.: The Nonlinear Brascamp–Lieb Inequality for Simple Data. Preprint (2018), arxiv:1801.05214
6. Benyi, A., Demeter, C., Nahmod, A., Torres, R., Thiele, C., Villarroya, F.: Modulation invariant bilinear $T(1)$ theorem. J. Anal. Math. **109**, 279–352 (2009)
7. Bernicot, F.: Fiber-wise Calderón-Zygmund decoposition and application to a bi-dimensional paraproduct. Illinois J. Math. **56**(2), 415–422 (2012)
8. Brascamp, H.J., Lieb, E.: Best constants in Young's inequality, its converse, and its generalization to more than three functions. Adv. Math. **20**(2), 151–173 (1976)
9. Calderón, A.-P.: Commutators of singular integral operators. Proc. Nat. Acad. Sci. U.S.A. **53**, 1092–1099 (1965)
10. Carleson, L.: On convergence and growth of partial sums of Fourier series. Acta Math. **116**, 135–157 (1966)
11. Coifman, R.R., McIntosh, A., Meyer, Y.: L'intégrale de Cauchy définit un opérateur borné sur L^2 pour les courbes lipschitziennes. Ann. Math. (2) **116**(2), 361–387 (1982)
12. Coifman, R., Meyer, Y.:Calderón-Zygmund and mulilinear operators. Translated from the 1990 and 1991 French originals by David Salinger. Cambridge Studies in Advanced Mathematics, vol. 48. Cambridge University, Cambridge (1997)
13. David, G., Journé, J.-L.: A boundedness criterion for generalized Calderón-Zygmund operators. Ann. Math. (2) **120**(2), 371–397 (1984)
14. Demeter, C., Thiele, C.: On the two-dimensional bilinear Hilbert transform. Am. J. Math. **132**(1), 201–256 (2010)
15. Demeter, C., Pramanik, M., Thiele, C.: Multilinear singular operators with fractional rank. Pacific J. Math. **246**(2), 293–324 (2010)
16. Di Plinio, F., Thiele, C.: Endpoint bounds for the bilinear Hilbert transform. Trans. Am. Math. Soc. **368**(6), 3931–3972 (2016)
17. Do, Y., Thiele, C.: L^p theory for outer measures and two themes of Lennart Carleson united. Bull. Am. Math. Soc. (N.S.) **52**(2), 249–296 (2015)
18. Do, Y., Oberlin, R., Palsson, E.: Variational bounds for a dyadic model of the bilinear Hilbert transform. Illinois J. Math. **57**(1), 105–119 (2013)
19. Do, Y., Muscalu, C., Thiele, C.: Variational estimates for the bilinear iterated Fourier integral. J. Funct. Anal. **272**(5), 2176–2233 (2017)
20. Do, Y., Oberlin, R., Palsson, E.: Variation-norm and fluctuation estimates for ergodic bilinear averages. Indiana Univ. Math. J. **66**(1), 55–99 (2017)
21. Durcik, P.: An L^4 estimate for a singular entangled quadrilinear form. Math. Res. Lett. **22**(5), 1317–1332 (2015)
22. Durcik, P.: L^p estimates for a singular entangled quadrilinear form. Trans. Am. Math. Soc. **369**(10), 6935–6951 (2017)

23. Durcik, P., Kovač, V.: Boxes, extended boxes, and sets of positive upper density in the Euclidean space. Preprint (2018), To appear in Math. Proc. Cambridge Philos. Soc. arXiv:1809.08692
24. Durcik, P., Roos, J.: Averages of simplex Hilbert transforms. Proc. Amer. Math. Soc. **149**(2), 633–647 (2021)
25. Durcik, P., Thiele, C.: Singular Brascamp–Lieb inequalities with cubical structure. Bull. London Math. Soc. **52**(2), 283–298 (2020)
26. Durcik, P., Kovač, V., Rimanić, L.: On side-lengths of corners in positive density subsets of the Euclidean space. Int. Math. Res. Not. **2018**(22), 6844–6869 (2018)
27. Durcik, P., Kovač, V., Škreb, K.A., Thiele, C.: Norm-variation of ergodic averages with respect to two commuting transformations. Ergodic Theory Dynam. Syst. **39**(3), 658–688 (2019)
28. Durcik, P., Kovač, V., Thiele, C.: Power-type cancellation for the simplex Hilbert transform. J. Anal. Math. **139**, 67–82 (2019)
29. Fefferman, C.: Pointwise convergence of Fourier series. Ann. Math. (2) **98**, 551–571 (1973)
30. Gilbert, J., Nahmod, A.: Bilinear operators with non-smooth symbols. I. J. Fourier Anal. Appl. **7**(5), 435–467 (2001)
31. Grafakos, L., Kalton, N.: The Marcinkiewicz multiplier condition for bilinear operators. Studia Math. **146**(2), 115–156 (2001)
32. Grafakos, L., Li, X.: Uniform bounds for the bilinear Hilbert transforms. I. Ann. Math. (2) **159**(3), 889–933 (2004)
33. Grafakos, L., Torres, R.: On multilinear singular integrals of Calderón-Zygmund type. Publ. Mat. **2002**(Extra), 57–91 (2002)
34. Green, B., Tao, T.: An inverse theorem for the Gowers $U^3(G)$ norm. Proc. Edinb. Math. Soc. (2) **51**(1), 73–153 (2008)
35. Green, B., Tao, T., Ziegler, T.: An inverse theorem for the Gowers $U^{s+1}[N]$-norm. Ann. Math. (2) **176**(2), 1231–1372 (2012)
36. Jung, J.: Iterated trilinear Fourier integrals with arbitrary symbols. Preprint (2013), arXiv:1311.1574
37. Kenig, C., Stein, E.: Multilinear estimates and fractional integration. In: Proceedings of the 6th International Conference on Harmonic Analysis and Partial Differential Equations (El Escorial, 2000). Math. Res. Lett. **6**(1), 1–15 (1999)
38. Kessler, R.: Generic multilinear multipliers associated to degenerate simplexes. Collectanea Mathematica. **70**, 153–186 (2019). https://doi.org/10.1007/s13348-018-0224-z
39. Kessler, R.: Mixed estimates for degenerate multilinear operators associated to simplexes. J. Math. Anal. Appl. **424**, 344–360 (2015)
40. Kessler, R.: L^p Estimates for semi-degenerate simplex multipliers. Rev. Mat. Iberoam. **36**(1), 99–158 (2020)
41. Kovač, V.: Bellman function technique for multilinear estimates and an application to generalized paraproducts. Indiana Univ. Math. J. **60**(3), 813–846 (2011)
42. Kovač, V.: Boundedness of the twisted paraproduct. Rev. Mat. Iberoam. **28**(4), 1143–1164 (2012)
43. Kovač, V., Škreb, K.A.: One modification of the martingale transform and its applications to paraproducts and stochastic integrals. J. Math. Anal. Appl. **426**(2), 1143–1163 (2015)
44. Kovač, V., Thiele, C.: A $T(1)$ theorem for entangled multilinear dyadic Calderón-Zygmund operators. Illinois J. Math. **57**(3), 775–799 (2013)
45. Kovač, V., Thiele, C., Zorin-Kranich, P.: Dyadic triangular Hilbert transform of two general and one not too general function. Forum Math. Sigma **3**, e25 (2015)
46. Lacey, M.: The bilinear maximal functions map into L^p for $2/3 < p \leq 1$. Ann. Math. (2) **151**(1), 35–57 (2000)
47. Lacey, M., Thiele, C.: L^p estimates for the bilinear Hilbert transform. Proc. Nat. Acad. Sci. USA **94**(1), 33–35 (1997)
48. Li, X.: Uniform bounds for the bilinear Hilbert transforms. II. Rev. Mat. Iberoam. **22**(3), 1069–1126 (2006)

49. Lie, V.: The (weak-L^2) *boundedness of the quadratic Carleson operator*. Geom. Funct. Anal. **19**(2), 457–497 (2009)
50. Lie, V.: The polynomial Carleson operator. Lie, Victor. Ann. Math. **192**(1), 47–163 (2020)
51. Muscalu, C.: Paraproducts with flag singularities I: A case study. Rev. Mat. Iberoam. **23**(2), 705–742 (2007)
52. Muscalu, C.: Flag paraproducts. Contemp. Math **505**, 131–151 (2010)
53. Muscalu, C.: Calderón commutators and the Cauchy integral on Lipschitz curves revisited: I. First commutator and generalizations. Rev. Mat. Iberoam. **30**, 727–750 (2014)
54. Muscalu, C.: Calderón commutators and the Cauchy integral on Lipschitz curves revisited II. The Cauchy integral and its generalizations. Rev. Mat. Iberoam. **30**, 1089–1122 (2014)
55. Muscalu, C., Tao, T., Thiele, C.: Multi-linear operators given by singular multipliers. J. Am. Math. Soc. **15**(2), 469–496 (2002)
56. Muscalu, C., Tao, T., Thiele, C.: Uniform estimates on multi-linear operators with modulation symmetry. Dedicated to the memory of Tom Wolff. J. Anal. Math. **88**, 255–309 (2002)
57. Muscalu, C., Pipher, J., Tao, T., Thiele, C.: Bi-parameter paraproducts. Acta Math. **193**(2), 269–296 (2004)
58. Muscalu, C., Tao, T., Thiele, C.: L^p estimates for the biest I. The Walsh case. Math. Ann. **329**(3), 401–426 (2004)
59. Muscalu, C., Tao, T., Thiele, C.: L^p estimates for the biest II. The Fourier case. Math. Ann. **329**(3), 427–461 (2004)
60. Muscalu, C., Pipher, J., Tao, T., Thiele, C.: Multi-parameter paraproducts. Rev. Mat. Iberoam. **22**(3), 963–976 (2006)
61. Muscalu, C., Tao, T., Thiele, C.: Multilinear operators associated to simplexes of arbitrary length. In: Advances in analysis: the legacy of Elias M. Stein., Princeton Mathematical Series, vol. 50, pp. 346–401. Princeton University Press, Princeton (2014)
62. Oberlin, R., Thiele, C.: New uniform bounds for a Walsh model of the bilinear Hilbert transform. Indiana Univ. Math. J. **60**(5), 1693–1712 (2011)
63. Palsson, E.: L^p estimates for a singular integral operator motivated by Calderón's second commutator. J. Funct. Anal. **262**, 1645–1678 (2012)
64. Škreb, K.A.: Norm-variation of cubic ergodic averages. Preprint (2019), arXiv:1903.04370
65. Stipčić, M.: $T(1)$ theorem for dyadic singular integral forms associated with hypergraphs. J. Math. Anal. Appl. **481**(2), Article 123496 (2020)
66. Tao, T.: Cancellation for the multilinear Hilbert transform. Collect. Math. **67**(2), 191–206 (2016)
67. Thiele, C.: A uniform estimate. Ann. Math. (2) **156**(2), 519–563 (2002)
68. Uraltsev, G., Warchalski, M.: Uniform bounds for the bilinear Hilbert transform in local L^1. In: Chapter in the PhD Thesis of M. Warchalski, RFWU Bonn (2018)
69. Warchalski, M.: Uniform estimates in one-and two-dimensional time-frequency analysis. PhD Thesis, RFWU Bonn (2018)
70. Zorin-Kranich, P.: Cancellation for the simplex Hilbert transform. Math. Res. Lett. **24**(2), 581–592 (2017)
71. Zorin-Kranich, P.: Maximal polynomial modulations of singular integrals. Adv. Math. **386**, p. 107832 (2021),

On the Restriction of Laplace–Beltrami Eigenfunctions and Cantor-Type Sets

Suresh Eswarathasan and Malabika Pramanik

Abstract Let (M, g) denote a compact Riemannian manifold without boundary. This article is an announcement of Lebesgue norm estimates of Laplace–Beltrami eigenfunctions of M when restricted to certain fractal subsets Γ of M. The proofs in their entirety appear in Eswarathasan and Pramanik (Restriction of Laplace–Beltrami eigenfunctions to random Cantor-type sets on manifolds, 2019). The sets Γ that we consider are random and of Cantor-type. For large Lebesgue exponents p, our estimates give a natural generalization of L^p bounds previously obtained in Hörmander (Acta Math 121: 193–218, 1968; Ark Math 11:1–11, 1971; Sogge J Funct Anal 77:123–138, 1988; Burq et al. Duke Math J 138(3):445–487, 2007). The estimates are shown to be sharp in this range. The novelty of our approach is the combination of techniques from geometric measure theory with well-known tools from harmonic and microlocal analysis. Random Cantor sets have appeared in a variety of contexts before, specifically in fractal geometry, multiscale analysis, additive combinatorics and fractal percolation Kahane and Peyriere (Adv Math 22(2):131–145, 1976; Laba and Pramanik, Geom Funct Anal 19:429–456, 2009; Laba and Pramanik, Duke Math J 158(3):347–411, 2011; Shmerkin and Suomala, Birkhäuser/Springer, Cham, 2017; Shmerkin and Suomala, Mem Am Math Soc 251:1195, 2018). They play a significant role in the study of optimal decay rates of Fourier transforms of measures, and in the identification of sets with arithmetic and geometric structures. Our methods, though inspired by earlier work, are not Fourier-analytic in nature.

Keywords Riemannian manifolds · Laplace-Beltrami eigenfunctions · Fractals · Random Cantor sets

---

S. Eswarathasan
School of Mathematics, Cardiff University, Cardiff, Wales, UK

Department of Mathematics and Statistics, McGill University, Montréal, QC, Canada
e-mail: eswarathasans@cardiff.ac.uk; suresh.eswarathasan@mcgill.ca

M. Pramanik (✉)
Department of Mathematics, University of British Columbia, Vancouver, BC, Canada
e-mail: malabika@math.ubc.ca

P. Ciatti, A. Martini (eds.), *Geometric Aspects of Harmonic Analysis*,
Springer INdAM Series 45, https://doi.org/10.1007/978-3-030-72058-2_10

1 Introduction

The study of eigenfunctions of Laplacians lies at the interface of several areas of mathematics, including analysis, geometry, mathematical physics and number theory. These special functions arise in physics and in partial differential equations as modes of periodic vibration of drums and membranes. In quantum mechanics, they represent the stationary energy states of a free quantum particle on a Riemannian manifold.

Let (M, g) denote a compact, connected, n-dimensional Riemannian manifold without boundary. The ubiquitous (positive) Laplace–Beltrami operator on M, denoted $-\Delta_g$, is the primary focus of this article. It is well-known [31, Chapter 3] that the spectrum of this operator is non-negative and discrete. Let us denote its eigenvalues by $\{\lambda_j^2 : j \geq 0\}$, and the corresponding eigenspaces by $\mathbb{E}_j$. Without loss of generality, the positive square roots of the distinct eigenvalues can be arranged in increasing order, with

$$0 = \lambda_0 < \lambda_1 < \lambda_2 < \cdots \lambda_j < \cdots \to \infty.$$

It is a standard fact [31, Chapter 3] that each $\mathbb{E}_j$ is finite-dimensional. Further, the space $L^2(M, dV_g)$ (of functions on M that are square-integrable with respect to the canonical volume measure dV_g) admits an orthogonal decomposition in terms of $\mathbb{E}_j$:

$$L^2(M, dV_g) = \bigoplus_{j=0}^{\infty} \mathbb{E}_j.$$

One of the fundamental questions surrounding Laplace–Beltrami eigenfunctions targets their concentration phenomena, via high-energy asymptotics or high-frequency behaviour. There are many avenues for this study. Semiclassical Wigner measures provide one way to measure concentration, as exemplified in the seminal work of Shnirelman [28], Zelditch [35], Colin de Vèrdiere [8], Gérard and Leichtnam [11], Zelditch and Zworski [36], Helffer, Martinez and Robert [13], Rudnick and Sarnak [24, 25], Lindenstrauss [21], and Anantharaman [1]. Another direction involves growth of the L^p norms of these eigenfunctions. The contribution of this article lies in the latter category. Specifically, it describes the $L^2(M) \to L^p(\Gamma)$ mapping property of a certain spectral projector (according to the spectral decomposition above), where Γ is a fractal-type subset of M. In particular, Γ does not enjoy any smooth structure. This is a significant point of departure from prior work where this feature was heavily exploited. We begin by reviewing the current research landscape that will help place the main result Theorem 3 in context.

2 Literature Review

The Weyl law, itself a major topic in spectral theory, provides an L^∞ bound on eigenfunctions on M [14]. The first results that establish L^p eigenfunction bounds for $p < \infty$ are due to Sogge [30].

Theorem 1 ([30]) *Given any manifold M as above and $p \in [2, \infty]$, there exists a constant $C = C(M, p) > 0$ such that the following inequality holds for all $\lambda \geq 1$:*

$$\|\varphi_\lambda\|_{L^p(M)} \leq C(1+\lambda)^{\delta(n,p)} \|\varphi_\lambda\|_{L^2(M)}, \text{ with}$$

$$\delta(n,p) = \begin{cases} \dfrac{n-1}{4} - \dfrac{n-1}{2p}, & \text{if } 2 \leq p \leq \dfrac{2(n+1)}{n-1}, \\ \dfrac{n-1}{2} - \dfrac{n}{p}, & \text{if } \dfrac{2(n+1)}{n-1} \leq p \leq \infty. \end{cases} \tag{1}$$

Here φ_λ is any eigenfunction of $-\Delta_g$ corresponding to the eigenvalue λ^2. The bound is sharp for the n-dimensional unit sphere $M = \mathbb{S}^n$, equipped with the surface measure.

Historically, an important motivation and source of inspiration for this line of investigation has been the Fourier restriction problem, which explores the behaviour of the Fourier transform when restricted to curved surfaces in Euclidean spaces. In fact the Stein-Tomas L^2 restriction theorem [33], originating in Euclidean harmonic analysis, was a key ingredient in an early proof of Theorem 1 for the sphere. Indeed, Theorem 1 may be viewed as a form of discrete restriction on M where the frequencies are given by the spectrum of the manifold, see for example [29]. Conversely, it is possible to recover the L^2 restriction theorem for the sphere from a spectral projection theorem such as Theorem 1 applied to the n-dimensional flat torus. The lecture notes of Yung [34, Section 2] contain a discussion of these implications.

Finer information on eigenfunction growth may be obtained through L^p bounds on φ_λ when restricted to smooth submanifolds of M. One expects φ_λ to assume large values on small sets. Thus its L^p-norm on a Lebesgue-null set such as a submanifold, if meaningful, is typically expected to be larger in comparison with the L^p norm taken over the entire manifold M, as given by Theorem 1. The first step in this direction is due to Reznikov [23], who studied eigenfunction restriction phenomena on hyperbolic surfaces via representation theoretic tools. The most general results to date on restricted norms of Laplace eigenfunctions are by Burq, Gérard and Tzvetkov [7], and independently by Hu [16]. The work of Tacy [32], using methods from an article of Koch–Tataru–Zworski [18] that gives a semiclassical version of Theorem 1, has extended these results to the setting of a semi-classical pseudo-differential operator (not merely the Laplacian) on a Riemannian manifold, while removing logarithmic losses at a critical threshold. Another particular endpoint result is due to Chen and Sogge [9]. We have summarized below the currently known best eigenfunction restriction estimates for a general manifold, combined from this body of work and for easy referencing later.

Theorem 2 ([7, 16, 32]) *Let $\Sigma \subset M$ be a smooth d-dimensional submanifold of M, equipped with the canonical measure $d\sigma$ that is naturally obtained from the metric g. Then for each $p \in [2, \infty]$, there exists a constant $C = C(M, \Sigma, p) > 0$ such that for any $\lambda \geq 1$ and any Laplace eigenfunction φ_λ associated with the eigenvalue λ^2, the following estimate holds:*

$$\|\varphi_\lambda\|_{L^p(\Sigma, d\sigma)} \leq C\,(1+\lambda)^{\delta(n,d,p)} \|\varphi_\lambda\|_{L^2(M, dV_g)}. \tag{2}$$

The exponent $\delta(n, d, p)$ admits a multi-part description. Specifically,

$$\delta(n, n-1, p) = \begin{cases} \dfrac{n-1}{4} - \dfrac{n-2}{2p}, & \text{for } 2 \leq p \leq \dfrac{2n}{n-1}, \\[2ex] \dfrac{n-1}{2} - \dfrac{n-1}{p}, & \text{for } \dfrac{2n}{n-1} \leq p \leq \infty. \end{cases} \tag{3}$$

For $d \neq n-1$,

$$\delta(n, d, p) = \frac{n-1}{2} - \frac{d}{p}, \quad \text{for } 2 \leq p \leq \infty \text{ and } (d, p) \neq (n-2, 2). \tag{4}$$

For $(d, p) = (n-2, 2)$, the exponent $\delta(n, d, p)$ is still given by (4); however, there is an additional logarithmic factor $\log^{1/2}(\lambda)$ appearing in the right hand side of inequality (2).

The proofs in [7] and [9] use a delicate analysis of oscillatory representations of the smoothed spectral projector $\rho(\lambda - \sqrt{-\Delta_g})$ restricted to submanifolds Σ, combined with refined estimates influenced by the considered geometry. Alternatively, [16] uses general mapping properties for Fourier integral operators with prescribed degenerate canonical relations to obtain bounds for the oscillatory integral operators in question. There are several recurrent features in these proofs; namely, stationary phase methods, arguments involving integration by parts, operator-theoretic convolution inequalities. This methodology heavily relies on the fact that the underlying measures are induced by Lebesgue, which in turn is a consequence of M and Σ being smooth manifolds. The present article explores the accessibility of this machinery in the absence of smoothness, and aims to find working substitutes when such methods are unavailable. This leads to a discussion of our main results.

3 Main Results

An interesting feature of the exponents $\delta(n, p)$ and $\delta(n, d, p)$ occurring in Theorems 1 and 2 respectively is that for large p, they are both of the form $(n-1)/2 - \alpha/p$, where

$$\begin{aligned} \alpha &= \text{dimension of the space on which the } L^p \text{norm of } \varphi_\lambda \text{is measured} \\ &= \left\{ \begin{array}{ll} \dim(M) = n & \text{in Theorem 1,} \\ \dim(\Sigma) = d & \text{in Theorem 2.} \end{array} \right\} \end{aligned} \tag{5}$$

In view of this commonality in (1), (3) and (4), we pose the following question: is there a class of "sparser" sets $\Gamma \subseteq \Sigma$, or equivalently a class of measures μ that are singular relative to the canonical measure on Σ, with respect to which we can estimate the growth of our eigenfunctions φ_λ? The optimal scenario would be to obtain bounds that reflect the dimensionality of the set Γ in the same way that Theorems 1 and 2 do. We answer this by announcing the main result of [10]:

Theorem 3 ([10]) *Fix positive integers $n \geq 2$ and $1 \leq d \leq n$. Let Σ be a smooth, d-dimensional submanifold of M. For each $\epsilon \in [0, 1)$, we define the critical exponent*

$$p_0 = p_0(n, d, \epsilon) := \frac{4d(1-\epsilon)}{n-1}. \tag{6}$$

Then for each choice of n, d, Σ and ϵ, there is a probability space $(\Omega, \mathcal{B}, \mathbb{P}^*)$ depending on these parameters that obeys the properties listed below.

(a) For $\mathbb{P}^*$-almost every $\omega \in \Omega$ there exists a Cantor-type subset $\Gamma_\omega \subset \Sigma$, equipped with a natural probability measure ν_ω, such that the set Γ_ω has Hausdorff dimension $d(1-\epsilon)$. For $\epsilon = 0$, ν_ω is singular with respect to the natural surface measure on Σ induced by the Riemannian metric g.

(b) For $\mathbb{P}^*$-almost every set Γ_ω obtained in (3) there exists a finite constant $C = C(\omega, n, d, p, \epsilon) > 0$ such that for all $\lambda \geq 1$, we have the eigenfunction estimate

$$\|\varphi_\lambda\|_{L^p(\Gamma_\omega, \nu_\omega)} \leq C\lambda^{\delta_p}\, \Phi(\lambda)\, \|\varphi_\lambda\|_{L^2(M, dV_g)}. \tag{7}$$

Here φ_λ denotes any L^2-eigenfunction associated with the eigenvalue λ^2 for the Laplace–Beltrami operator $-\Delta_g$ on M. For $p_0 > 2$, the exponent δ_p is given by

$$\delta_p = \delta_p(n, d, \epsilon) := \left\{ \begin{array}{ll} \dfrac{n-1}{4}, & \text{if } 2 \leq p \leq p_0, \\ \dfrac{n-1}{2} - \dfrac{d(1-\epsilon)}{p}, & \text{if } p \geq p_0. \end{array} \right\} \tag{8}$$

For $p_0 \leq 2$, the exponent $\delta_p = (n-1)/2 - d(1-\epsilon)/p$ for $2 \leq p \leq \infty$. The quantity $\Phi(\lambda)$ appearing in (7) is an increasing function that grows slower than any positive power of λ; specifically, Φ is of the form

$$\Phi(\lambda) = \exp(C' \sqrt{\log(\lambda)}), \tag{9}$$

where $C' = C'(n, d, p, \epsilon) > 0$ is an explicit constant.

(c) *The exponent δ_p in the above estimate is sharp in general for $p \geq \max(p_0, 2)$, in the following sense. Suppose that Σ is any d-dimensional submanifold of the n-dimensional unit sphere $M = \mathbb{S}^n$, $d \leq n$. Fix $\epsilon \in (0, 1]$.*

There exists a sequence of L^2-normalized spherical harmonics $\{\varphi_{\lambda_j} : j \geq 1\}$ with $\lambda_j \nearrow \infty$ such that for $\mathbb{P}^*$-almost every set Γ_ω obtained above and for every $p \geq p_0$, one can find a constant $C = C(\omega, p) > 0$ verifying the lower bound

$$\|\varphi_{\lambda_j}\|_{L^p(\Gamma_\omega, \nu_\omega)} \geq C \lambda_j^{\delta_p} \Phi(\lambda_j)^{-1} \tag{10}$$

for all λ_j sufficiently large.

4 Remarks

Let us pause for a moment to contextualize some of the important features of our result.

1. For $p \geq p_0$, the exponent δ_p in Theorem 3 (3) is of the same form alluded to in (5), namely $\delta_p = (n-1)/2 - \alpha/p$ with $\alpha = d(1-\epsilon)$. Thus our result may be viewed as a natural interpolation between the global estimates in [30] and the smooth restriction estimates in [7], bridging the estimates across a family of sets with continuously varying Hausdorff dimensions.
2. To the best of our knowledge, Theorem 3 is the first result of its kind in several distinct categories. First, it offers, for *every manifold M and every smooth submanifold Σ therein*, eigenfunction bounds over non-smooth subsets of positive but non-integral Hausdorff dimension. Second, even for integers m, our result produces new sets of dimension m, for example with $(n, d, \epsilon) = (2, 2, 1/2)$, that are not necessarily contained in any m-dimensional submanifold, and yet capture the same eigenfunction growth bounds as smooth submanifolds of the same dimension, up to sub-polynomial losses. Third, when $\epsilon = 0$, our result provides examples of singular measures supported on submanifolds with respect to which the eigenfunctions obey the same L^p growth bounds as with the induced Lebesgue measure on the same submanifold. This is reminiscent of an earlier article by Łaba and Pramanik [20], where the authors construct a random Cantor-type measure with respect to which the maximal averaging operator has the same L^p mapping properties as the Hardy–Littlewood maximal function (where the underlying measure is Lebesgue).

3. Our estimates, though sharp for general M and Σ, can be improved in special situations. This will be the case, for instance, when $M = \mathbb{T}^n$, the n-dimensional flat torus (which admits a stronger Weyl law), or if the submanifold Σ in a general manifold M has additional geometric properties, for example if Σ is a curve of nonvanishing geodesic curvature. This is consistent with similar results of this type for smooth submanifolds, see for example [2, 3, 6, 7, 12, Theorem 2] and the bibliography therein. We pursue this direction in greater detail in upcoming work.
4. The blow-up factor $\Phi(\lambda)$, which is super-logarithmic but sub-polynomial, is an artifact of the choices of parameters needed for the random Cantor construction. Many alternative parameter choices are possible within the framework of this construction, some of which yield logarithmic blow-up in lieu of $\Phi(\lambda)$, at the cost of additional technical challenges. We have opted not to pursue these improvements here. However, all estimates of this type will be accompanied by some blow-up. It is an interesting question whether there exists a member of this class of random sets for which such losses can be avoided.
5. The random measures ν_ω that we construct and their supporting sets Γ_ω have many analytic and geometric properties that are not directly exploited in the proof. In particular, these measures have optimal Fourier decay subject to the Hausdorff dimension of their support. More precisely, for almost every ω, our measures obey

$$\left|\widehat{\nu}_\omega(\xi)\right| \leq C_\xi \big(1 + |\xi|\big)^{-d(1-\epsilon)/2}, \quad |\xi| \geq 1,$$

where C_ξ is a function that grows slower than any positive power of $|\xi|$. In other words, the sets Γ_ω in Theorem 3 have the same Fourier dimension as their Hausdorff dimension, i.e. they are almost surely Salem.

Fourier decay of measures have long been known to play an important role in eigenfunction restriction problems. For instance, it appears in the work of Bourgain and Rudnick [6], where the authors obtain significant improvements on the general estimates of [7] in the special case of $M = \mathbb{T}^n$, $n = 2, 3$. More generally, the study of harmonic-analytic principles (such as Fourier decay, fractal analogues of the uncertainty principle, study of oscillatory integrals and operators) in settings where standard techniques (such as integration by parts or stationary phase) are not viable have led to major developments in spectral theory, for instance in the work surrounding resonance gaps in infinite-area hyperbolic surfaces [4, 5, 22]. We explore the mapping properties of convolution operators on random Cantor measure spaces, and establish Young-type inequalities for such measures. However, our methods are not Fourier-analytic in nature. This is another point of similarity of our work with [20], where a similar random Cantor set was constructed, but whose Fourier-analytic properties were not directly relevant to the proof. We would like to point out that random Cantor sets have appeared in a variety of contexts before, specifically in fractal geometry, multiscale analysis, additive combinatorics and fractal percolation [15, 17, 19, 26, 27]

5 Overview of the Proof

The broad strokes of our approach in [10] follow that of [7, Theorem 1], so we briefly review the main ideas involved here.

1. One starts with a microlocal approximation $\mathbb{T}_\lambda$ of the smoothed spectral projector $\rho(\lambda - \sqrt{-\Delta_g})$. The approximation $\mathbb{T}_\lambda$ is an oscillatory integral operator, whose phase function is essentially the distance function in the ambient Riemannian metric.
2. The TT^* method applied to $\mathbb{T}_\lambda \mathbb{T}_\lambda^*$ reduces the problem to estimating the L^p of the latter operator on the restricted set γ, which for [7, Theorem 1] was a smooth curve on M.
3. The integration kernel of $\mathbb{T}_\lambda \mathbb{T}_\lambda^*$ is itself an oscillatory integral, with a nondegenerate phase function. The method of stationary phase, applied to this oscillatory integral, yields a pointwise upper bound on the kernel, leading to a pointwise bound on the operator $\mathbb{T}_\lambda \mathbb{T}_\lambda^*$. The dominating operator is a convolution, with an explicit convolving factor.
4. The proof is then completed by invoking Young's convolution inequality for the Lebesgue measure on $\mathbb{R}$. The admissible exponents of the inequality are precisely those for which the convolving factor is integrable.

A careful analysis of [7, Theorem 1] shows that steps 1, 2 and 3 above extend with minor revisions to the setting of an arbitrary measure space, with γ replaced by Γ_ω. A noteworthy point of departure is the following. Whereas the natural measure on the curve γ used in [7] is absolutely continuous with respect to the translation-invariant Lebesgue measure on $\mathbb{R}$, the measure ν_ω accompanying our Cantor set Γ_ω is no longer translation invariant. The proof thus fails critically at the last step, since Young's convolution inequality is unavailable, indeed known to be false, in general measure spaces. The main contribution of this article is in deriving an analogue of Young's inequality for the convolution kernel $\mathcal{K}_\lambda$ that appears in the pointwise upper bound in step 3, and for the special class of random Cantor measures constructed earlier in the paper. Specifically, this involves estimation of the quantity $\sup \left\{ \|\mathcal{K}_\lambda(u - \cdot)\|_{L^p(\nu_\omega)} : u \in \Gamma_\omega \right\}$ for almost every $\omega \in \Omega$. The transition from the desired operator norm of $\mathbb{T}_\lambda \mathbb{T}_\lambda^*$ to the quantity above has been formalized thanks to a generalized Schur-type inequality proved in [10]. A substantial portion [10] is devoted to the estimation of this last quantity, through a series of successive reduction to various random sums.

References

1. Anantharaman, N.: Entropy and the localization of eigenfunctions. Ann. Math. (2) **168**(2), 435–475 (2008)
2. Blair, M., Sogge, C.: Logarithmic improvements in L^p-norms for eigenfunctions at the critical exponent in the presence of nonpositive curvature (2018). Preprint. arxiv.org/abs/1706.06704
3. Bourgain, J.: Geodesic restriction and L^p-estimates for eigenfunctions on Riemannian surfaces. Linear and Complex Analysis, pp. 27–35. Amer. Math. Soc. Transl. Ser. 2, 226, Adv. Math. Sci., 63 (2009)
4. Bourgain, J., Dyatlov, S.: Fourier dimension and spectral gaps for hyperbolic surfaces. Geom. Funct. Anal. **27**, 744–771 (2017)
5. Bourgain, J., Dyatlov, S.: Spectral gaps without the pressure condition. Ann. Math. **187**(3), 824–867 (2018)
6. Bourgain, J., Rudnick, Z.: Restriction of toral eigenfunctions to hypersurfaces and nodal sets. Geom. Funct. Anal. **22**, 878–937 (2012)
7. Burq, N., Gérard, P., Tzetkov, N.: Restrictions of the Laplace–Beltrami eigenfunctions to submanifolds. Duke Math. J. **138**(3), 445–487 (2007)
8. Colin de Verdière, Y.: Ergodicité et fonctions propres du Laplacien. Commun. Math. Phys. **102**, 497–502 (1985)
9. Chen, X., Sogge, C.: A few endpoint geodesic restriction estimates for eigenfunctions. Commun. Math. Phys. **329**, 435–459 (2014)
10. Eswarathasan, S., Pramanik, M.: Restriction of Laplace–Beltrami eigenfunctions to random Cantor-type sets on manifolds (2019). Pre-print
11. Gérard, P., Leichtnam, É.: Ergodic properties of eigenfunctions for the Dirichlet problem. Duke Math. J. **71**, 559–607 (1993)
12. Hassell, A., Tacy, M.: Semiclassical L^p estimates of quasimodes on curved hypersurfaces. J. Geom. Anal. **22**, 74–89 (2012)
13. Helffer, B., Martinez, A., Robert, D.: Ergodicité et limite semi-classique. Commun. Math. Phys. **109**, 313–326 (1987)
14. Hörmander, L.: Spectral function of an elliptic operator. Acta Math. **121**, 193–218 (1968)
15. Hörmander, L.: Oscillatory integrals and multipliers on FL^p. Ark. Math. **11**, 1–11 (1971)
16. Hu, R.: L^p norm estimates of eigenfunctions restricted to submanifolds. Forum Math. **21**, 1021–1052 (2009)
17. Kahane, J.-P., Peyriere, J.: Sur certaines martingales de Benoît Mandelbrot. Adv. Math. **22**(2), 131–145 (1976)
18. Koch, H., Tataru, D., Zworski, M.: Semiclassical L^p estimates. Ann. Henri Poincare **5**, 885–916 (2007)
19. Laba, I., Pramanik, M.: Arithmetic progressions in sets of fractal dimensions. Geom. Funct. Anal. **19**, 429–456 (2009)
20. Laba, I., Pramanik, M.: Maximal operators and differentiation theorems for sparse sets. Duke Math. J. **158**(3), 347–411 (2011)
21. Lindenstrauss, E.: Invariant measures and arithmetic quantum unique ergodicity. Ann. Math. (2) **163**, 165–219 (2006)
22. Naud, F.: Expanding maps on Cantor sets and analytic continuation of zeta functions. Ann. l'ENS **4**(38), 116–153 (2005)
23. Reznikov, A.: Norms of geodesic restrictions for eigenfunctions on hyperbolic surfaces and representation theory (2010). unpublished update to a work from 2005, arxiv.org/abs/math/0403437v3
24. Rudnick, Z., Sarnak, P.: The behaviour of eigenstates of arithmetic hyperbolic manifolds. Commun. Math. Phys. **161**(1), 195–213 (1994)
25. Sarnak, P.: Arithmetic quantum chaos. In: The Schur lectures (Tel Aviv, 1992), Israel Math. Conf. Proc., vol. 8, pp. 183–236. Bar-Ilan Univ., Ramat Gan (1995)

26. Shmerkin, P., Suomala, V.: A Class of Random Cantor Measures, with Applications. Recent Developments in Fractals and Related Fields. Trends Math., pp. 233–260. Birkhäuser/Springer, Cham (2017)
27. Shmerkin, P., Suomala, V.: Spatially independent martingales, intersections, and applications. Mem. Am. Math. Soc. **251**, 1195 (2018)
28. Shnirelman, A.I.: Ergodic properties of eigenfunctions (in Russian). Uspekhi Mat. Nauk **29**(6), 181–182 (1974)
29. Sogge, C.: Oscillatory integrals and spherical harmonics. Duke Math. J. **53**, 43–65 (1986)
30. Sogge, C.: Concerning the L^p norms of spectral clusters of second-order elliptic operators on compact manifolds. J. Funct. Anal. **77**, 123–138 (1988)
31. Sogge, C.: Fourier Integrals in Classical Analysis. Cambridge Tracts in Math., Cambridge Univ. Press, Cambridge (1993)
32. Tacy, M.: Semiclassical L^p estimates of quasimodes on submanifolds. Commun. PDE **35**(8), 1538–1562 (2010)
33. Tomas, P.: A restriction theorem for the Fourier transform. Bull. Am. Math. Soc. **81**, 477–478 (1975)
34. Yung, P.: Spectral projection theorems on compact manifolds. Lecture notes, available under "Notes/Slides". https://www.math.cuhk.edu.hk/~plyung/Sogge.pdf
35. Zelditch, S.: Uniform distribution of eigenfunctions on compact hyperbolic surfaces. Duke Math. J. **55**, 919–941 (1987)
36. Zelditch, S., Zworski, M.: Ergodicity of eigenfunctions for ergodic billiards. Commun. Math. Phys. **175**, 673–682 (1996)

Basis Properties of the Haar System in Limiting Besov Spaces

Gustavo Garrigós, Andreas Seeger, and Tino Ullrich

Abstract We study Schauder basis properties for the Haar system in Besov spaces $B^s_{p,q}(\mathbb{R}^d)$. We give a complete description of the limiting cases, obtaining various positive results for $q \le \min\{1, p\}$, and providing new counterexamples in other situations. The study is based on suitable estimates of the dyadic averaging operators $\mathbb{E}_N$; in particular we find asymptotically optimal growth rates for the norms of these operators in global and local situations.

Keywords Schauder basis · Basic sequence · Unconditional basis · Local Schauder basis · Dyadic averaging operators · Haar system · Besov and Triebel-Lizorkin spaces

1 Introduction

The purpose of this paper is to complete the study of the basis properties of the (inhomogeneous) Haar system in the scale of Besov spaces $B^s_{p,q}(\mathbb{R}^d)$. In view of previous results, only the endpoint cases are of interest. This is a companion to the paper [6], in which the authors consider the same endpoint questions for the Triebel-Lizorkin spaces. The outcomes for Besov and for Triebel-Lizorkin spaces, in both non-endpoint situations [5, 11, 12, 14, 19] and endpoint situations, are markedly different.

G. Garrigós
Department of Mathematics, University of Murcia, Murcia, Spain
e-mail: gustavo.garrigos@um.es

A. Seeger (✉)
Department of Mathematics, University of Wisconsin, Madison, WI, USA
e-mail: seeger@math.wisc.edu

T. Ullrich
Faculty of Mathematics, Technical University of Chemnitz, Chemnitz, Germany
e-mail: tino.ullrich@mathematik.tu-chemnitz.de

P. Ciatti, A. Martini (eds.), *Geometric Aspects of Harmonic Analysis*,
Springer INdAM Series 45, https://doi.org/10.1007/978-3-030-72058-2_11

To state the results, we first set some basic notation. Consider the one variable functions $h^{(0)} = \mathbb{1}_{[0,1)}$ and $h^{(1)} = \mathbb{1}_{[0,1/2)} - \mathbb{1}_{[1/2,1)}$. For every $\epsilon = (\epsilon_1, \dots, \epsilon_d) \in \{0,1\}^d$, for $k \in \mathbb{N}_0$ and $\mu = (\mu_1, \dots, \mu_d) \in \mathbb{Z}^d$ one defines

$$h^{\epsilon}_{k,\mu}(x) = \prod_{i=1}^{d} h^{(\epsilon_i)}(2^k x_i - \mu_i), \quad x = (x_1, \dots, x_d) \in \mathbb{R}^d.$$

Denoting $\Upsilon = \{0,1\}^d \setminus \{\mathbf{0}\}$, the Haar system is given as

$$\mathscr{H}_d = \left\{h^{\mathbf{0}}_{0,\mu}\right\}_{\mu \in \mathbb{Z}^d} \cup \left\{h^{\epsilon}_{k,\mu} : k \in \mathbb{N}_0,\ \mu \in \mathbb{Z}^d,\ \epsilon \in \Upsilon\right\}.$$

We refer to 2^k as the Haar frequency of $h^{\epsilon}_{k,\mu}$. We consider an enumeration $\mathcal{U} = \{u_n\}_{n=1}^{\infty}$ of the Haar system, and write $u_n = h^{\epsilon(n)}_{k(n),\mu(n)}$ for the corresponding frequency and position parameters $k(n)$, $\mu(n)$.

Given $R \in \mathbb{N}$, the partial sum operator $S_R \equiv S_R^{\mathcal{U}}$ is defined as the projection onto $\operatorname{span}\{u_1, \dots, u_R\}$, that is

$$S_R^{\mathcal{U}} f = \sum_{n=1}^{R} u_n^*(f) u_n \,, \tag{1}$$

where for $u_n = h^{\epsilon(n)}_{k(n),\mu(n)}$ the linear functional u_n^* is defined by

$$u_n^*(f) = 2^{k(n)d} \langle f, h^{\epsilon(n)}_{k(n),\mu(n)} \rangle \,, \tag{2}$$

at least when $f \in L^1_{loc}(\mathbb{R}^d)$. Below we shall only consider Besov spaces so that $u_n \in B^s_{p,q}$ and u_n^* extends to an element of $(B^s_{p,q})^*$ for all $n \in \mathbb{N}$, so that (2) will actually have a meaning for all $f \in B^s_{p,q}$.

We say that $\mathcal{U}$ is a *Schauder basis* of $B^s_{p,q}(\mathbb{R}^d)$ if

$$\lim_{R \to \infty} \|S_R^{\mathcal{U}} f - f\|_{B^s_{p,q}} = 0 \tag{3}$$

holds for every $f \in B^s_{p,q}$. We say that $\mathcal{U}$ is a *basic sequence* if (3) holds for every f in the $B^s_{p,q}$-closure of span $\mathscr{H}_d$. Finally, we say that $\mathscr{H}_d$ is an *unconditional basis* of $B^s_{p,q}$ if every enumeration $\mathcal{U}$ is a Schauder basis.

The above basis properties are related with the uniform bound

$$C_{\mathcal{U}} := \sup_{R \geq 1} \|S_R^{\mathcal{U}}\|_{B^s_{p,q} \to B^s_{p,q}} < \infty. \tag{4}$$

Indeed, one has

$$\|S_R^{\mathcal{U}} f - f\|_{B^s_{p,q}} \lesssim (C_{\mathcal{U}} + 1)\|f - h\|_{B^s_{p,q}} + \|S_R^{\mathcal{U}} h - h\|_{B^s_{p,q}}, \qquad h \in \operatorname{span} \mathscr{H}_d.$$

Thus, (4) implies that $\mathcal{U}$ is a basic sequence in $B^s_{p,q}$. If span($\mathscr{H}_d$) is dense in $B^s_{p,q}$, then $\mathcal{U}$ is a Schauder basis if and only if (4) holds. If in addition the bound in (4) does not depend on the enumeration $\mathcal{U}$ then $\mathscr{H}_d$ is an unconditional basis of $B^s_{p,q}$. By the uniform boundedness principle such a uniform estimate is also necessary for unconditionality. This is well-known for Banach spaces, and a proof for quasi-Banach spaces can be found in [1].

We consider the full range of indices $s \in \mathbb{R}$ and $0 < p, q \leq \infty$. When $p = \infty$ or $q = \infty$ the space $B^s_{p,q}$ is not separable, but in those cases one may consider the Schauder basis property in the $B^s_{p,q}$-closure of the Schwartz class $\mathcal{S}$, which we will denote $b^s_{p,q}$ (as in [10, Def 1.1.3]).

The pentagon $\mathfrak{P}$ depicted in Fig. 1 shows the natural index region for these problems. More precisely, Triebel showed in [14, 19] that, for all $q < \infty$, the Haar system $\mathscr{H}_d$ is an unconditional basis of $B^s_{p,q}(\mathbb{R}^d)$ for the $(1/p, s)$ parameters in the interior of the pentagon $\mathfrak{P}$; i.e. those satisfying one of the conditions (i), (ii) in Theorem 1. He also showed that for the parameters in the complement of the closure of $\mathfrak{P}$ the Haar system does not form a basis [20]. Except for a few trivial cases, the behavior at the points $(1/p, s)$ lying in the boundary of $\mathfrak{P}$ seems to be unexplored; see however the separate work [8] and Remark 7 below.

In this paper we attempt to fill these gaps, by giving an answer, positive or negative, depending on the secondary index q. Moreover, in some cases, the negative answer is replaced by slightly weaker properties, such as the local Schauder basis, or basic sequence properties.

We begin by stating complete results about unconditionality, which contain new negative cases compared to [14, 19]. We remark that the corresponding results in

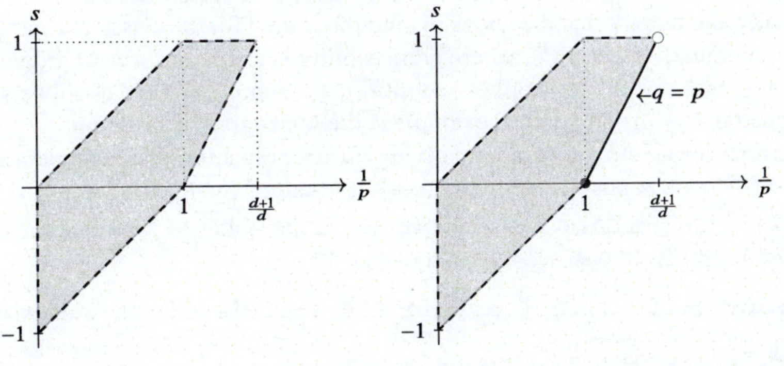

Fig. 1 Parameter domain $\mathfrak{P}$ for $\mathscr{H}_d$ in $B^s_{p,q}(\mathbb{R}^d)$. The left figure shows the region of unconditionality, and right figure the region for the Schauder basis property

Triebel-Lizorkin spaces are much more restrictive, see the discussion in [19, Remark 2.2.10] and the counterexamples in [11].

Theorem 1 *Let $0 < p, q \le \infty$ and $s \in \mathbb{R}$. Then, $\mathscr{H}_d$ is an unconditional basis of $B^s_{p,q}(\mathbb{R}^d)$ if and only if one of the following two conditions is satisfied.*

(i) $1 \le p < \infty, \quad -1 + \frac{1}{p} < s < \frac{1}{p}, \quad 0 < q < \infty.$
(ii) $\frac{d}{d+1} < p < 1, \quad d(\frac{1}{p} - 1) < s < 1, \quad 0 < q < \infty.$

The region (i)–(ii) is shown in the left of Fig. 1. In the next results we shall be concerned with the endpoint behavior when we drop unconditionality. To do so we must single out specific enumerations of the Haar system, labeled 'admissible', or 'strongly admissible'.

Definition 2

(i) An enumeration $\mathcal{U}$ is said to be *admissible* if there is a constant $b \in \mathbb{N}$ with the following property: for each cube $I_\nu = \nu + [0, 1]^d$, $\nu \in \mathbb{Z}^d$, if u_n and $u_{n'}$ are both supported in I_ν and $|\operatorname{supp}(u_n)| \ge 2^{bd} |\operatorname{supp}(u_{n'})|$, then necessarily $n < n'$.

(ii) An enumeration $\mathcal{U}$ is *strongly admissible* if there is a constant $b \in \mathbb{N}$ with the following property: for each cube I_ν, $\nu \in \mathbb{Z}^d$, if I^{**}_ν denotes the five-fold dilated cube with respect to its center, and if u_n and $u_{n'}$ are supported in I^{**}_ν with $|\operatorname{supp}(u_n)| \ge 2^{bd} |\operatorname{supp}(u_{n'})|$ then necessarily $n < n'$.

The notion in (i) was used in [5] for the case $b = 1$, but the results stated in that paper continue to hold with this slightly more general definition.

The stronger notion in (ii) turns out to be more appropriate in the endpoint cases, for which the characteristic functions of cubes may not be pointwise multipliers; cf. [10]. Loosely speaking, in a strongly admissible enumeration if a Haar frequency 2^k shows up at step n (i.e. if $u_n = h^\varepsilon_{k,\mu}$ for some $k \in \mathbb{N}_0$) then all Haar functions with Haar frequency $\le 2^{k-b}$ which are 'nearby' (in a well defined sense) have already been counted before step n. We refer to Sect. 11 for concrete examples.

Finally, we remark that the above distinction is void for the classical Haar system in the unit cube, $\mathscr{H}([0, 1)^d)$, where admissibility is a straightforward property; for $b = 1$ it means that $n < n'$ implies $|\operatorname{supp}(u_n)| \ge |\operatorname{supp}(u_{n'})|$ (and could be slightly weakened if $b \ge 2$). The typical example is the lexicographic ordering.

We now formulate a theorem involving *all* strongly admissible enumerations of the Haar system. A positive endpoint result is obtained for $B^s_{p,p}$ when $s = d/p - d$ and $\frac{d}{d+1} < p \le 1$. Also new negative results are obtained for suitable strongly admissible enumerations; see the right of Fig. 1.

Theorem 3 *Let $0 < p, q \le \infty$ and $s \in \mathbb{R}$. Then, the following statements are equivalent, i.e. (a) $\Longleftrightarrow$ (b):*

(a) Every strongly admissible enumeration of $\mathscr{H}_d$ is a Schauder basis of $B^s_{p,q}(\mathbb{R}^d)$.

(b) One of the following three conditions is satisfied:

(i) $1 \le p < \infty, \quad \frac{1}{p} - 1 < s < \frac{1}{p}, \quad 0 < q < \infty,$
(ii) $\frac{d}{d+1} < p < 1, \quad \frac{d}{p} - d < s < 1, \quad 0 < q < \infty,$
(iii) $\frac{d}{d+1} < p \le 1, \quad s = \frac{d}{p} - d, \quad q = p.$

Next we explore various weaker properties at the boundary of $\mathfrak{P}$. We say that an enumeration $\mathcal{U}$ satisfies the *local Schauder basis property* for $B^s_{p,q}$ if

$$\left\|(S_R^{\mathcal{U}} f - f)\,\chi\right\|_{B^s_{p,q}} \to 0 \tag{5}$$

for all $f \in B^s_{p,q}(\mathbb{R}^d)$ and all $\chi \in C_c^\infty(\mathbb{R}^d)$. This implies that the basis expansion holds, $g = \sum_{n=1}^\infty u_n^*(g) u_n$ in $B^s_{p,q}$, for all compactly supported $g \in B^s_{p,q}(\mathbb{R}^d)$. Similarly we say that $\mathcal{U}$ satisfies the *local basic sequence property* in $B^s_{p,q}(\mathbb{R}^d)$ when (5) holds for all $f \in \overline{\operatorname{span} \mathscr{H}_d}^{B^s_{p,q}}$ and all $\chi \in C_c^\infty(\mathbb{R}^d)$. The next theorem and Fig. 2 show the region of validity for the first of these properties.

Theorem 4 *Let $0 < p, q \le \infty$ and $s \in \mathbb{R}$. Then, the following statements are equivalent, i.e. (a) $\Longleftrightarrow$ (b):*

(a) Every strongly admissible enumeration of the Haar system $\mathscr{H}_d$ satisfies the local Schauder basis property for $B^s_{p,q}(\mathbb{R}^d)$.
(b) One of the following four conditions is satisfied:

(i) $1 \le p < \infty, \quad -1 + \frac{1}{p} < s < \frac{1}{p}, \quad 0 < q < \infty,$
(ii) $1 \le p < \infty, \quad s = -1 + \frac{1}{p}, \quad 0 < q \le 1,$
(iii) $\frac{d}{d+1} < p < 1, \quad \frac{d}{p} - d < s < 1, \quad 0 < q < \infty,$
(iv) $\frac{d}{d+1} < p < 1, \quad s = \frac{d}{p} - d, \quad 0 < q \le p.$

We remark that these local properties can be given slightly stronger statements using the Bourdaud definitions $(B^s_{p,q})_{\ell^p}$ of localized Besov spaces; see Sect. 9.2 below.

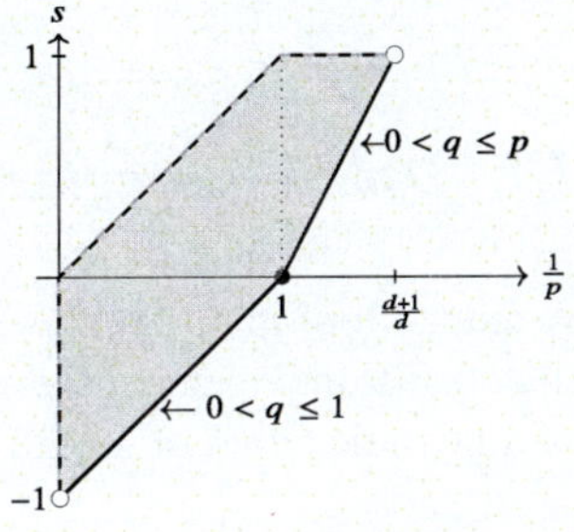

Fig. 2 Region for the local Schauder basis property in the spaces $B^s_{p,q}(\mathbb{R}^d)$, depending on the value of $q \in (0, \infty)$

Remark 5 Strong admissibility may be replaced by admissibility (for the positive results in Theorems 3 and 4) in the cases where $\mathbb{1}_{[0,1]^d}$ is a pointwise multiplier of $B^s_{p,q}(\mathbb{R}^d)$. By Triebel [16, §2.8.7] or [10, §4.6.3], the latter holds when

$$\max\left\{d(\tfrac{1}{p}-1), \tfrac{1}{p}-1\right\} < s < \tfrac{1}{p}, \tag{6}$$

so it applies to the interior points of $\mathfrak{P}$. It also applies in other positive results (such as the local basic sequence property, which will follow from Theorem 8) in the case $s = 1$, $\frac{d}{d+1} < p < 1$ corresponding to the interior of the horizontal edge of $\mathfrak{P}$.

Remark 6 A similar statement to Theorem 4 holds for $\mathscr{H}(\mathbb{T}^d)$, the Haar system in the torus in the standard lexicographic enumeration. Namely, it is a Schauder basis on $B^s_{p,q}(\mathbb{T}^d)$ if and only if one of the conditions (i), (ii), (iii), (iv) in Theorem 4 are satisfied. Moreover, in the range (6) the class of C^∞-functions with compact support in $(0,1)^d$ are dense in $B^s_{p,q}((0,1)^d)$ (see [18, §3.2]) and thus it is easy to see that the Schauder basis problem for the Haar systems on $\mathbb{T}^d$ and on $(0,1)^d$ are equivalent in this range. So far this observation does not apply to the cases corresponding to the non-horizontal edges of $\mathfrak{P}$ in higher dimensions, however see Franke's better result [3, §4.6] for the interval $(0,1)$.

Remark 7

(i) In a classical work [7], P. Oswald considered, for $0 < p < 1$, the Schauder basis property (including some endpoint results) for a class of Besov spaces on the interval, $\mathscr{B}^s_{p,q,(1)}(I)$, defined by first order differences. In these classes, which in general differ from $B^s_{p,q}$, one has a positive answer in the larger region $1/p - 1 < s < 1/p$ (in particular, for some $s \geq 1$); see [7, Theorem 3].

(ii) In a very recent separate study [8], Oswald pursued further these questions for both, the class $\mathscr{B}^s_{p,q,(1)}(I^d)$ and the standard Besov spaces $B^s_{p,q}((0,1)^d)$. He obtained analogs of the positive results in (ii)–(iv) of Theorem 4, and presented similar counterexamples as ours for the case $s = d/p - d$. Contrary to what is stated in that paper, these local results do not transfer to the spaces on $\mathbb{R}^d$ by simply enumerating the Haar system, as one may see from Theorem 3 above and the specific example in Proposition 49.

1.1 Dyadic Averaging Operators

A crucial tool in our analysis will be the dyadic averaging operator $\mathbb{E}_N$, defined as the conditional expectation with respect to the σ-algebra generated by the set $\mathscr{D}_N$ of all dyadic cubes of length 2^{-N}. That is, setting

$$I_{N,\mu} = 2^{-N}(\mu + [0,1)^d), \quad \mu \in \mathbb{Z}^d,$$

we have

$$\mathbb{E}_N f(x) = \sum_{\mu \in \mathbb{Z}^d} \mathbb{1}_{I_{N,\mu}}(x)\, 2^{Nd} \int_{I_{N,\mu}} f(y)dy\,, \tag{7}$$

at least for $f \in L^1_{loc}(\mathbb{R}^d)$.

The relation with the Haar system is given via the martingale difference operator $\mathbb{E}_{N+1} - \mathbb{E}_N$ which is the orthogonal projection onto the space generated by the Haar functions with frequency 2^N, i.e.

$$\mathbb{E}_{N+1} f - \mathbb{E}_N f = \sum_{\epsilon \in \Upsilon} \sum_{\mu \in \mathbb{Z}^d} 2^{Nd} \langle f, h^{\epsilon}_{N,\mu} \rangle h^{\epsilon}_{N,\mu}. \tag{8}$$

In addition to $\mathbb{E}_N$ we shall need another operator which involves Haar functions of a fixed frequency level. For $N \in \mathbb{N}$ and any $\mathfrak{a} \in \ell^\infty(\mathbb{Z}^d \times \Upsilon)$ we set

$$T_N[f, \mathfrak{a}] = \sum_{\epsilon \in \Upsilon} \sum_{\mu \in \mathbb{Z}^d} a_{\mu,\epsilon} 2^{Nd} \langle f, h^{\epsilon}_{N,\mu} \rangle h^{\epsilon}_{N,\mu}. \tag{9}$$

One aims for estimates of the operators $f \mapsto T_N[f, \mathfrak{a}]$ that are uniform in $\|\mathfrak{a}\|_\infty \leq 1$. The relation between the partial sum operators $S^{\mathcal{U}}_R$ and the operators $\mathbb{E}_N$ and $T_N[\cdot, \mathfrak{a}]$ is explained in Sect. 9. In particular, the uniform boundedness of these operators in $B^s_{p,q}$ implies the *local basic sequence* property for all strongly admissible enumerations $\mathcal{U}$. The region of uniform boundedness for these operators is given in the next theorems, and depicted in Fig. 3.

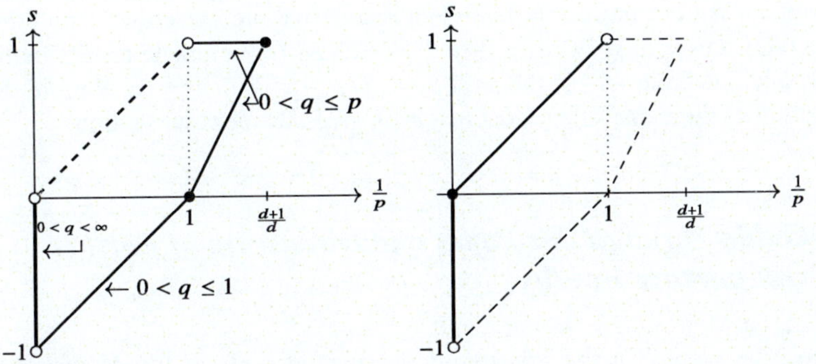

The cases $0 < q < \infty$ The case $q = \infty$

Fig. 3 Regions for uniform boundedness of $\mathbb{E}_N$ (hence for the *local* basic sequence property) in the spaces $B^s_{p,q}(\mathbb{R}^d)$

Theorem 8 *Let* $0 < p < \infty$, $0 < q \le \infty$ *and* $s \in \mathbb{R}$.

(a) The operators $\mathbb{E}_N$ *admit an extension from* $\mathcal{S}(\mathbb{R}^d)$ *to* $B^s_{p,q}(\mathbb{R}^d)$ *such that*

$$\sup_{N\ge 0} \|\mathbb{E}_N\|_{B^s_{p,q}\to B^s_{p,q}} < \infty$$

if and only if one of the following six conditions is satisfied:

(i) $p > 1, \quad s = \frac{1}{p}, \quad q = \infty,$
(ii) $p \ge 1, \quad -1 + \frac{1}{p} < s < \frac{1}{p}, \quad 0 < q \le \infty,$
(iii) $p \ge 1, \quad s = -1 + \frac{1}{p}, \quad 0 < q \le 1,$
(iv) $\frac{d}{d+1} < p < 1, \quad s = 1, \quad 0 < q \le p,$
(v) $\frac{d}{d+1} < p < 1, \quad d(\frac{1}{p} - 1) < s < 1, \quad 0 < q \le \infty,$
(vi) $\frac{d}{d+1} \le p < 1, \quad s = d(\frac{1}{p} - 1), \quad 0 < q \le p.$

(b) If one of the conditions (i)–(vi) is satisfied and if $\|\mathfrak{a}\|_{\ell^\infty(\mathbb{Z}^d\times\Upsilon)} \le 1$ *then the operators* $T_N[\cdot, \mathfrak{a}]$ *are uniformly bounded on* $B^s_{p,q}(\mathbb{R}^d)$.

Finally we state a result for $p = \infty$.

Theorem 9

(i) If $-1 < s < 0$ *then the operators* $\mathbb{E}_N$ *have uniformly bounded extensions to* $B^s_{\infty,q}(\mathbb{R}^d)$, *for all* $0 < q \le \infty$.
(ii) If $s = 0$ *then* $\mathbb{E}_N$ *admits a bounded extension to* $B^0_{\infty,q}(\mathbb{R}^d)$ *if and only if* $q = \infty$. *Moreover, we have* $\sup_N \|\mathbb{E}_N\|_{B^0_{\infty,\infty}\to B^0_{\infty,\infty}} < \infty$.
(iii) If $s = -1$ *then* $\sup_N \|\mathbb{E}_N\|_{B^{-1}_{\infty,q}\to B^{-1}_{\infty,q}} = \infty$, *for all* $0 < q \le \infty$.

Moreover, below we also investigate situations when the individual operators $\mathbb{E}_N$ are bounded but not uniformly bounded, and derive precise growth conditions for the operator norms in such cases. See Theorem 27 for complete results in the case $s = 1$, and Theorem 46 for the case $s = d/p - d$ and $p \le 1$. A more detailed description of these and other local results is given the next subsection.

1.2 *Guide Through this Paper and Discussion of Further Quantitative Results*

The positive results in the interior of the pentagon $\mathfrak{P}$ in Fig. 1, including the unconditionality property, are classical and due to Triebel [13, 14, 19]. Moreover, unboundedness results outside the closure of $\mathfrak{P}$ are discussed in those references and [5].

The new positive results in Theorems 8 and 9 at the boundary of $\mathfrak{P}$ rely on L^p bounds for the operators $L_k\mathbb{E}_N L_j$, where L_k are suitable local means and the operators act on functions with compactly supported Fourier transforms. These bounds were already contained in our previous paper [5] (see also [4] for a proof of such results using wavelets). We review these estimates in Sect. 2, see in particular Corollary 14. For both ranges $p \geq 1$ and $p \leq 1$ further straightforward estimates imply four key propositions with different outcomes in the four cases depending on the signs of $j - N$ and $k - N$. These propositions are stated for $p \leq 1$ in Sect. 3 and for $p \geq 1$ in Sect. 4.

Concerning the negative results in Theorems 8 and 9, these are presented as follows. First, when $s = 1/p$, characteristic functions of cubes (and also Haar functions) do not belong to $B^{1/p}_{p,q}$ when $q < \infty$ which rules out these cases. In Sect. 5 we shall further show that the space $b^{1/p}_{p,\infty}$ (the closure of the Schwartz class under the $B^{1/p}_{p,\infty}$ norm) intersects the algebraic span of $\mathscr{H}_d$ only in $\{0\}$. This is in contrast with the fact, shown in Sect. 8.2, that $b^{1/p}_{p,\infty}$ is actually contained in $\overline{\operatorname{span} \mathscr{H}_d}^{B^{1/p}_{p,\infty}}$ if $1 < p < \infty$, so some positive result will hold in this case; see Proposition 41.

In Sect. 6 we consider the cases $s = 1$. At the endpoint space $B^1_{1,\infty}$ we show that the operators $\mathbb{E}_N$ are individually bounded, but not uniformly bounded, and for large N we have $\|\mathbb{E}_N\|_{B^1_{1,\infty} \to B^1_{1,\infty}} \approx N$, see Theorem 27.

When $s = 1$ and $\frac{d}{d+1} < p < 1$ the operators $\mathbb{E}_N$ are also individually bounded on $B^1_{p,q}$ but not uniformly bounded if $q > p$. In these cases Theorem 27 implies that for large N we have $\|\mathbb{E}_N\|_{B^1_{p,q} \to B^1_{p,q}} \approx N^{1/p - 1/q}$. The situation is worse at the endpoint $p = d/(d+1)$, that is the vertex of $\mathfrak{P}$ where $s = 1 = d/p - d$. In this case Theorem 46 gives an exponential lower bound even for compactly supported functions, while the $\mathbb{E}_N$ fail to be individually bounded in the whole $B^1_{d/(d+1),q}(\mathbb{R}^d)$ when $q > p = \frac{d}{d+1}$.

In Sect. 7 we discuss the simpler situation on the line $s = 1/p - 1$ with $1 < p \leq \infty$. The cases $q > 1$ are easily ruled out because Haar functions do not belong to the dual space $(B^{1/p-1}_{p,q})^* = B^{1/p'}_{p',q'}$. In the cases $0 < q \leq 1$, a lower bound $\|\mathbb{E}_N\|_{B^{-1}_{\infty,q} \to B^{-1}_{\infty,q}} \gtrsim N$ is obtained by duality in Sect. 7.2.

In Sect. 10 we gather the negative results for Theorem 8 at the edge $s = d/p - d$ with $p \leq 1$. Again, we rule out the cases $q > 1$, as Haar functions do not belong to the dual space $(B^{d/p-d}_{p,q})^* = B^0_{\infty,q'}$. Moreover, we show in Theorem 45 that the individual operators $\mathbb{E}_N$ are unbounded on $B^{d/p-d}_{p,q}$ when $q > p$. We shall actually prove sharp results if one restricts to compactly supported functions. To quantify these we use the following definition.

Definition 10 Let Q be an open dyadic cube in $\mathbb{R}^d$ of side length $\geq 1/2$ and X be a (quasi-)Banach space of tempered distributions $\mathcal{S}'(\mathbb{R}^d)$. For a linear operator T defined on those $f \in X$ which are supported in Q we set

$$\operatorname{Op}(T, X, Q) = \sup\big\{\|Tf\|_X : \ \|f\|_X \leq 1, \ \operatorname{supp}(f) \subset Q\big\}. \tag{10}$$

In Theorem 46 the precise growth of $\mathrm{Op}(\mathbb{E}_N, B^{d/p-d}_{p,q}, Q)$ is obtained for the range $\frac{d}{d+1} \le p \le q \le 1$ where it is shown that

$$\mathrm{Op}\left(\mathbb{E}_N, B^{d/p-d}_{p,q}, Q\right) \approx \left(2^{Nd}|Q|\right)^{\frac{1}{p}-\frac{1}{q}}.$$

We also show that the above-mentioned lower bounds for $s = 1$ have a sharp local analogue, namely for every cube Q of sidelength ≥ 1 one has

$$\mathrm{Op}\left(\mathbb{E}_N, B^1_{p,q}, Q\right) \approx N^{\frac{1}{p}-\frac{1}{q}}, \quad \tfrac{d}{d+1} < p < 1, \quad q \in [p, \infty].$$

We now turn to the uniform boundedness of the operators $S^{\mathcal{U}}_R$, associated with strongly admissible enumerations $\mathcal{U}$. We prove in Sect. 9.1 that if $\mathbb{E}_N$ (and $T_N[\cdot, \mathfrak{a}]$) are uniformly bounded in $B^s_{p,q}$ then we also have, for each fixed Q,

$$\sup_{R\in\mathbb{N}} \mathrm{Op}\left(S^{\mathcal{U}}_R, B^s_{p,q}, Q\right) < \infty. \tag{11}$$

Assuming (11), one has the local Schauder basis property if and only if the span of the Haar system is dense in $B^s_{p,q}$.

The density of span $\mathscr{H}_d$ in $B^s_{p,q}$ is studied separately in Sect. 8. It clearly fails when $p = \infty$ or $q = \infty$ because $B^s_{p,q}$ is not separable in those cases. When $s = 1$, we also show that density fails in $B^1_{p,q}$ when $\frac{d}{d+1} \le p < 1$ and $q \le p$. This gives the negative results in Theorem 4 for those cases. We do not know, however, whether density should also fail in the remaining cases $q > p$; see our discussion in Sect. 8.1.

The positive Schauder basis results in Theorem 3 are obtained in Sect. 9.2. They follow from Sect. 9.1 and the fact that the $B^s_{p,q}(\mathbb{R}^d)$-norms can be 'localized' if and only if $p = q$. Moreover, we actually prove the Schauder basis (resp. basic sequence) property in the Bourdaud spaces $(B^s_{p,q})_{\ell^p}$, in the range of Theorem 4 (resp. 8 and 9). We remark that these spaces coincide with $B^s_{p,q}$ if and only if $p = q$. Alternatively, the positive statement in (iii) of Theorem 3 is also a special case of a more general result for Triebel-Lizorkin spaces $F^s_{p,q}(\mathbb{R}^d)$ in [6].

In Sect. 11 we construct an explicit strongly admissible enumeration $\mathcal{U}$ for which

$$S^{\mathcal{U}}_{R(m)} f = \mathbb{E}_m f \quad \text{if } \mathrm{supp}(f) \subset (-5, 5)^d,$$

for a suitable sequence $R(m)$. One can then apply the examples on unboundedness of $\mathbb{E}_N$ when restricted to functions on cubes, alluded to above, to see that (11) fails. This gives the negative results in Theorems 4 and 3 at the edge $s = d/p - d$, for $\frac{d}{d+1} < p \le 1$ and all $q > p$.

This same enumeration $\mathcal{U}$ is used in Sect. 12 to show that, when $q \in (0, p)$, the operators $S^{\mathcal{U}}_R$ are not uniformly bounded in the whole spaces $B^{d/p-d}_{p,q}(\mathbb{R}^d)$ if

$\frac{d}{d+1} < p \le 1$, or $B^{1/p-1}_{p,q}(\mathbb{R}^d)$ if $1 < p < \infty$. Hence $\mathcal{U}$ is not a Schauder basis in these cases.

Finally, regarding the negative results in Theorem 1, examples showing the failure of unconditionality for parameters $(1/p, s)$ on the boundary of $\mathfrak{P}$ are given in Sect. 13 for the case $B^{d/p-d}_{p,q}$ with $\frac{d}{d+1} < p \le 1$, and in Sect. 14 for the case $B^{1/p-1}_{p,q}$ with $1 < p < \infty$. Since the argument in Sect. 13 also applies to a similar result for Triebel-Lizorkin spaces, we include lower bounds for those as well.

2 Preparatory Results

2.1 *Besov Quasi-norms*

Let $s \in \mathbb{R}$ and $0 < p \le \infty$ be given. Throughout the paper we fix a number $A > d/p$ and an integer

$$M > A + |s| + 2. \tag{12}$$

Consider two functions $\beta_0, \beta \in C_c^\infty(\mathbb{R}^d)$, supported in $(-1/2, 1/2)^d$, with the properties $|\widehat{\beta_0}(\xi)| > 0$ if $|\xi| \le 1$, $|\widehat{\beta}(\xi)| > 0$ if $1/8 \le |\xi| \le 1$ and β has vanishing moments up to order M, that is

$$\int_{\mathbb{R}^d} \beta(x)\, x_1^{m_1} \cdots x_d^{m_d}\, dx = 0, \quad \forall\, m_i \in \mathbb{N}_0 \text{ with } m_1 + \ldots + m_d \le M. \tag{13}$$

The optimal value of M is irrelevant for the purposes of this paper, and (12) suffices for our results. We let $\beta_k := 2^{kd}\beta(2^k\cdot)$ for each $k \ge 1$, and denote

$$L_k(f) = \beta_k * f$$

whenever $f \in \mathcal{S}'(\mathbb{R}^d)$. It is then known, see e.g. [17, 2.5.3], that an equivalent quasi-norm in the Besov spaces $B^s_{p,q}(\mathbb{R}^d)$, $0 < q \le \infty$, is given by

$$\|g\|_{B^s_{p,q}} \approx \left\| \{2^{ks} L_k g\}_{k=0}^\infty \right\|_{\ell^q(L^p)}. \tag{14}$$

Recall also that $b^s_{p,q}$ denotes the closure of $\mathcal{S}(\mathbb{R}^d)$ in the $B^s_{p,q}$ norm. When $p < \infty$ and $q = \infty$, it not difficult to see that $b^s_{p,\infty}$ coincides with the set of all $g \in \mathcal{S}'(\mathbb{R}^d)$ such that

$$\lim_{k\to\infty} 2^{ks}\|L_k g\|_p = 0. \tag{15}$$

The space $b^s_{\infty,\infty}$ coincides with the set of all $g \in \mathcal{S}'(\mathbb{R}^d)$ such that $L_k g \in C_0$ for all $k \in \mathbb{N}$ and such that $\lim_{k\to\infty} 2^{ks}\|L_k g\|_\infty = 0$.

Next, let $\eta_0 \in C_c^\infty(\mathbb{R}^d)$ be supported on $\{\xi : |\xi| < 3/8\}$ and such that $\eta_0(\xi) = 1$ if $|\xi| \le 1/4$. We consider the following frequency localization operators

$$\widehat{\Lambda_0 f}(\xi) = \frac{\eta_0(\xi)}{\widehat{\beta}_0(\xi)}\widehat{f}(\xi), \tag{16a}$$

$$\widehat{\Lambda_k f}(\xi) = \frac{\eta_0(2^{-k}\xi) - \eta_0(2^{-k+1}\xi)}{\widehat{\beta}(2^{-k}\xi)}\widehat{f}(\xi), \quad k \ge 1, \tag{16b}$$

so that $f = \sum_{j=0}^\infty L_j \Lambda_j f$ with convergence in $\mathcal{S}'$. It is also well-known that

$$\|f\|_{B^s_{p,q}} \approx \Big\|\{2^{ks}\Lambda_k f\}_{k=0}^\infty\Big\|_{\ell^q(L^p)}. \tag{17}$$

In particular, if we let $\Pi_N = \sum_{j=0}^N L_j \Lambda_j$, then

$$\sup_N \|\Pi_N f\|_{B^s_{p,q}} \lesssim \|f\|_{B^s_{p,q}}. \tag{18}$$

Below we shall be interested in uniformly bounded extensions of the dyadic averaging operators $\mathbb{E}_N$ defined in (7). Observe that

$$\mathbb{E}_N - \Pi_N = \mathbb{E}_N (I - \Pi_N) - (I - \mathbb{E}_N)\Pi_N,$$

so if we denote

$$\mathbb{E}_N^\perp = I - \mathbb{E}_N,$$

then, using (14), we have

$$\begin{aligned} \|\mathbb{E}_N f - \Pi_N f\|_{B^s_{p,q}} \lesssim\ & \Big\|\{2^{ks}\sum_{j=N+1}^{\infty} L_k \mathbb{E}_N L_j \Lambda_j f\}_{k=0}^\infty\Big\|_{\ell^q(L^p)} \\ & + \Big\|\{2^{ks}\sum_{j=0}^{N} L_k \mathbb{E}_N^\perp L_j \Lambda_j f\}_{k=0}^\infty\Big\|_{\ell^q(L^p)}. \end{aligned} \tag{19}$$

Thus, as in [5], the uniform bounds of $\mathbb{E}_N$ will be reduced to suitable estimates for the compositions $L_k \mathbb{E}_N L_j$ and $L_k \mathbb{E}_N^\perp L_j$, for each $j, k \ge 0$.

2.2 Local Estimates

We consider the following Peetre maximal operators: if $j \geq 0$ and g is continuous, then

$$\mathcal{M}_j g(x) = \sup_{|h|_\infty \leq 2^{-j+5}} |g(x+h)|,$$

$$\mathcal{M}^*_{A,j} g(x) = \sup_{|h|_\infty \leq 2^5} \frac{|g(x+h)|}{(1+2^j|h|)^A},$$

$$\mathfrak{M}^{**}_{A,j} g(x) = \sup_{h \in \mathbb{R}^d} \frac{|g(x+h)|}{(1+2^j|h|)^A}.$$

Clearly, we have the pointwise relations $\mathcal{M}_j g \lesssim \mathcal{M}^*_{A,j} g \leq \mathfrak{M}^{**}_{A,j} g$.

The following lemma was proved in [5, Lemma 2.2] using the cancellation properties of $L_{\max\{j,k\}}$.

Lemma 11 *For $j, k \geq 0$ we have*

$$|L_k L_j g(x)| \lesssim 2^{-|k-j|(M-A)} \mathcal{M}^*_{A,\max\{j,k\}} g(x). \tag{20}$$

We remark that the larger maximal function $\mathfrak{M}^{**}_{A,\max\{j,k\}} f$ was used in [5, Lemma 2.2], in place of $\mathcal{M}^*_{A,\max\{j,k\}} f$. However, since the convolution kernel of $L_j L_k$ is supported on a cube of sidelength $2^{-j} + 2^{-k}$, it is clear that (20) holds as well.

From our previous work [5] we have the following crucial estimates.

Proposition 12 *Let $0 < p \leq \infty$ and*

$$B_p(j,k,N) = \begin{cases} 2^{N-j}\, 2^{\frac{j-k}{p}}\, 2^{(j-N)(d-1)(\frac{1}{p}-1)_+} & \text{if } j,k > N, \\ 2^{\frac{N-k}{p}}\, 2^{j-N} & \text{if } j \leq N < k, \\ 2^{k-N} 2^{j-N} 2^{(N-k)d(\frac{1}{p}-1)_+} & \text{if } 0 \leq j,k \leq N, \\ 2^{k-j+\frac{j-N}{p}+[N-k+(j-k)(d-1)](\frac{1}{p}-1)_+} & \text{if } k \leq N < j. \end{cases} \tag{21}$$

Then the following inequalities hold for all continuous functions g:

(i) For $j \geq N+1$,

$$\|L_k \mathbb{E}_N[L_j g]\|_p \lesssim \begin{cases} B_p(j,k,N) \|\mathcal{M}_j g\|_p & \text{if } k \geq N+1, \\ B_p(j,k,N) \|\mathcal{M}_j g\|_p + 2^{-|j-k|(M-A)} \|\mathcal{M}^*_{A,j} g\|_p & \text{if } 0 \leq k \leq N. \end{cases} \tag{22}$$

(ii) For $0 \le j \le N$,

$$\|L_k \mathbb{E}_N^{\perp}[L_j g]\|_p$$
$$\lesssim \begin{cases} B_p(j,k,N)\|\mathcal{M}_j g\|_p + 2^{-|j-k|(M-A)}\|\mathcal{M}^*_{A,j} g\|_p & \text{if } k \ge N+1, \\ B_p(j,k,N)\|\mathcal{M}_j g\|_p & \text{if } 0 \le k \le N. \end{cases} \tag{23}$$

(iii) The same bounds hold if the operators $\mathbb{E}_N$ *in (i) and* $\mathbb{E}_N^{\perp}$ *in (ii) are replaced by* $T_N[\cdot, \mathfrak{a}]$ *(as defined in (9)), uniformly in* $\|\mathfrak{a}\|_\infty \le 1$.

Remark 13 These bounds are contained in [5, §2], although the statement of [5, Proposition 2.1] is slightly less general. Namely, applying these bounds to $g \in \mathcal{S}'(\mathbb{R}^d)$ such that $\operatorname{supp} \widehat{g} \subset \{|\xi| \le 2^{j+1}\}$, and using additionally the Peetre inequality $\|\mathfrak{M}^{**}_{A,j} g\|_p \lesssim \|g\|_p$, for $A > d/p$, one obtains [5, Proposition 2.1]. The formulation here will be applied later to functions of the form $g = \zeta\, \Lambda_j f$ with $f \in \mathcal{S}'(\mathbb{R}^d)$ and $\zeta \in C_c^\infty$.

The statement of Proposition 12 can be put into a more convenient form. First, when $g = \Lambda_j f$, the Peetre maximal inequality [9] gives $\|\mathfrak{M}^{**}_{A,j}(\Lambda_j f)\|_p \lesssim \|\Lambda_j f\|_p$ provided that $A > d/p$. Next, if $M \ge A + 1$ then in the cases $k \le N < j$ and $j \le N < k$ the term $2^{-|j-k|(M-A)}$ is dominated by $B_p(j,k,N)$ and thus the statement can be simplified. Finally, we shall use the quantities

$$U_{p,s}(j,k,N) := 2^{(k-j)s} B_p(j,k,N)\,; \tag{24}$$

see also (27) and (33) below. We then obtain

Corollary 14 *Let* $U_{p,s}(j,k,N)$ *be as in (24). Then for all* $f \in \mathcal{S}'(\mathbb{R}^d)$

$$2^{ks}\|L_k \mathbb{E}_N L_j \Lambda_j f\|_p \lesssim U_{p,s}(j,k,N)\, 2^{js}\, \|\Lambda_j f\|_p, \quad \text{if } j \ge N+1, \tag{25}$$

and

$$2^{ks}\|L_k \mathbb{E}_N^{\perp} L_j \Lambda_j f\|_p \lesssim U_{p,s}(j,k,N)\, 2^{js}\, \|\Lambda_j f\|_p, \quad \text{if } j \le N. \tag{26}$$

The same holds with $\mathbb{E}_N$ *and* $\mathbb{E}_N^{\perp}$ *replaced by* $T_N[\cdot, \mathfrak{a}]$ *if* $\|\mathfrak{a}\|_\infty \le 1$.

3 Upper Bounds for $p \le 1$

In the range $p \le 1$, the constants in (24) take the following explicit form

$$U_{p,s}(j,k,N) = \begin{cases} 2^{k(s-\frac{1}{p})} 2^{j(\frac{d}{p}-d-s)} 2^{N(d-\frac{d-1}{p})} & \text{if } j,k > N \\ 2^{k(s-\frac{1}{p})} 2^{j(1-s)} 2^{N(\frac{1}{p}-1)} & \text{if } j \le N < k \\ 2^{k(s+d+1-\frac{d}{p})} 2^{j(1-s)} 2^{N(\frac{d}{p}-d-2)} & \text{if } 0 \le j,k \le N \\ 2^{k(s+d+1-\frac{d}{p})} 2^{j(\frac{d}{p}-d-s)} 2^{-N} & \text{if } k \le N < j. \end{cases} \tag{27}$$

We now state four propositions corresponding to the four cases of (27). We then sketch the straightforward proofs.

Proposition 15 *For $\frac{d-1}{d} < p \le 1$ and $r > 0$,*

$$\Big(\sum_{k>N} 2^{ksr} \Big\| \sum_{j>N} L_k \mathbb{E}_N L_j \Lambda_j f \Big\|_p^r \Big)^{1/r}$$
$$\lesssim \begin{cases} \sup_{j>N} 2^{js} \|\Lambda_j f\|_p & \text{if } d(\frac{1}{p}-1) < s < \frac{1}{p}, \\ \Big(\sum_{j>N} 2^{jsp} \|\Lambda_j f\|_p^p\Big)^{1/p} & \text{if } s = d(\frac{1}{p}-1) < \frac{1}{p}. \end{cases} \tag{28a}$$

For $p = 1$ and $s = 1$ we have

$$\sup_{k>N} 2^k \Big\| \sum_{j>N} L_k \mathbb{E}_N L_j \Lambda_j f \Big\|_1 \lesssim \sup_{j>N} 2^j \|\Lambda_j f\|_1. \tag{28b}$$

The same inequalities hold when $\mathbb{E}_N$ is replaced with $T_N[\cdot, \mathfrak{a}]$ if $\|\mathfrak{a}\|_\infty \le 1$.

Proposition 16 *For $0 < p < 1$ and $r > 0$,*

$$\Big(\sum_{k>N} 2^{ksr} \Big\| \sum_{j\le N} L_k \mathbb{E}_N^{\perp} L_j \Lambda_j f \Big\|_p^r \Big)^{1/r}$$
$$\lesssim \begin{cases} \sup_{j\le N} 2^{js} \|\Lambda_j f\|_p & \text{if } s < 1, \\ \Big(\sum_{j=0}^{N} 2^{jsp} \|\Lambda_j f\|_p^p\Big)^{1/p} & \text{if } s = 1 \\ 2^{(s-1)N} \sup_{j\le N} 2^{js} \|\Lambda_j f\|_p & \text{if } 1 < s < 1/p. \end{cases} \tag{29a}$$

Inequality (29a) also holds for $p = 1$ and $s < 1$. When $p = s = 1$ we have

$$\sup_{k>N} 2^k \Big\| \sum_{j\le N} L_k \mathbb{E}_N^{\perp} L_j \Lambda_j f \Big\|_1 \lesssim \sum_{j=0}^{N} 2^j \|\Lambda_j f\|_1 \tag{29b}$$

The same statements hold with $\mathbb{E}_N^{\perp}$ replaced by $T_N[\cdot, \mathfrak{a}]$ if $\|\mathfrak{a}\|_\infty \le 1$.

Proposition 17 *For $\frac{d}{d+2} < p \le 1$ and $r > 0$,*

$$\Big(\sum_{k\le N} 2^{ksr}\Big\|\sum_{j\le N} L_k\mathbb{E}_N^{\perp}L_j\Lambda_j f\Big\|_p^r\Big)^{1/r}$$

$$\lesssim \begin{cases} \sup_{j\le N} 2^{js}\|\Lambda_j f\|_p & \text{if } \frac{d}{p}-d-1<s<1,\\ \Big(\sum_{j=0}^N 2^{jsp}\|\Lambda_j f\|_p^p\Big)^{1/p} & \text{if } s=1\\ 2^{(s-1)N}\sup_{j\le N} 2^{js}\|\Lambda_j f\|_p & \text{if } 1<s<1/p. \end{cases} \tag{30}$$

The same inequality holds with $\mathbb{E}_N^{\perp}$ replaced by $T_N[\cdot,\mathfrak{a}]$ if $\|\mathfrak{a}\|_\infty \le 1$.

Proposition 18 *For $0 < p \le 1$ and $r > 0$,*

$$\Big(\sum_{k\le N} 2^{ksr}\Big\|\sum_{j> N} L_k\mathbb{E}_N L_j\Lambda_j f\Big\|_p^r\Big)^{1/r}$$

$$\lesssim \begin{cases} \sup_{j> N} 2^{js}\|\Lambda_j f\|_p & \text{if } s>\frac{d}{p}-d,\\ \Big(\sum_{j>N} 2^{jsp}\|\Lambda_j f\|_p^p\Big)^{1/p} & \text{if } s=\frac{d}{p}-d. \end{cases} \tag{31}$$

The same inequality holds with $\mathbb{E}_N$ replaced by $T_N[\cdot,\mathfrak{a}]$ if $\|\mathfrak{a}\|_\infty \le 1$.

3.1 Proofs

The proofs of the four propositions involve Corollary 14 and an application of the p-triangle inequality for $p \le 1$.

Proof of Proposition 15 First observe that the range of s in (28a) is nontrivial if and only if $p > d/(d-1)$. Let Σ_r denote the left hand side of (28a). Then the p-triangle inequality and Corollary 14 give

$$\Sigma_r \le \Big(\sum_{k>N} 2^{ksr}\Big[\sum_{j>N}\|L_k\mathbb{E}_N L_j\Lambda_j f\|_p^p\Big]^{\frac{r}{p}}\Big)^{1/r}$$

$$\lesssim \Big(\sum_{k>N}\Big[\sum_{j>N} U_{p,s}(j,k,N)^p\, 2^{jsp}\|\Lambda_j f\|_p^p\Big]^{\frac{r}{p}}\Big)^{1/r}.$$

When $d(\frac{1}{p}-1) < s < \frac{1}{p}$ this implies

$$\Sigma_r \lesssim \Big(\sum_{k>N}\Big[\sum_{j>N} 2^{k(s-\frac{1}{p})p}2^{j(\frac{d}{p}-d-s)p}2^{N(d-\frac{d-1}{p})p}\Big]^{\frac{r}{p}}\Big)^{1/r} \sup_{\ell>N} 2^{\ell s}\|\Lambda_\ell f\|_p,$$

which gives the asserted expression because the series above is bounded (uniformly in N). At the endpoint $s = d(\frac{1}{p} - 1) < \frac{1}{p}$ we have

$$\Sigma_r \lesssim \Big(\sum_{k>N} 2^{(k-N)(s-\frac{1}{p})r}\Big)^{1/r} \Big(\sum_{j>N} 2^{jsp} \|\Lambda_j f\|_p^p\Big)^{\frac{1}{p}},$$

which also leads to the asserted expression in (28a). Finally, if $s = p = 1$, using that $U_{p,s}(j,k,N) = 2^{N-j}$ we obtain

$$\Sigma_\infty \lesssim \sum_{j>N} 2^{N-j}\, 2^j \|\Lambda_j f\|_1 \le \sup_{\ell>N} 2^\ell \|\Lambda_\ell f\|_1.$$

The proof is complete. □

Proof of Proposition 16 The left hand side of (29a) is controlled by

$$\Big(\sum_{k>N}\Big[\sum_{j\le N} U_{p,s}(j,k,N)^p\, 2^{jsp}\|\Lambda_j f\|_p^p\Big]^{r/p}\Big)^{1/r}$$
$$\lesssim \Big(\sum_{k\ge N} 2^{k(s-\frac{1}{p})r} 2^{N(\frac{1}{p}-1)r}\Big)^{\frac{1}{r}} \Big(\sum_{j\le N} 2^{j(1-s)p}[2^{js}\|\Lambda_j f\|_p]^p\Big)^{\frac{1}{p}}.$$

If $s < 1/p$ the first sum can be evaluated as $C_2(p,s,r)2^{N(s-1)}$ and the above expression is dominated by a constant times

$$\Big(\sum_{j\le N} 2^{(j-N)(1-s)p}[2^{js}\|\Lambda_j f\|_p]^p\Big)^{\frac{1}{p}}.$$

(29a) follows immediately. The proof of (29b) is similar. □

Proof of Proposition 17 The left hand side of (30) is controlled by

$$\Big(\sum_{k\le N}\Big[\sum_{j\le N} U_{p,s}(j,k,N)^p\, 2^{jsp}\|\Lambda_j f\|_p^p\Big]^{r/p}\Big)^{1/r}$$
$$\lesssim \Big(\sum_{k\le N} 2^{k(s+d+1-\frac{d}{p})r} 2^{N(\frac{d}{p}-d-2)r}\Big)^{\frac{1}{r}} \Big(\sum_{j\le N} 2^{j(1-s)p}[2^{js}\|\Lambda_j f\|_p]^p\Big)^{\frac{1}{p}}.$$

If $s > \frac{d}{p} - d - 1$ the first factor can be evaluated to be $C_3(p,s,r)2^{N(s-1)}$ and the above expression is again dominated by a constant times

$$\Big(\sum_{j\le N} 2^{(j-N)(1-s)p}[2^{js}\|\Lambda_j f\|_p]^p\Big)^{\frac{1}{p}}.$$

Note that for the s-range in (30) to be nontrivial we want $\frac{d}{p} - d - 1 < 1$, i.e. $p > \frac{d}{d+2}$. Now (30) follows easily. □

Proof of Proposition 18 The left hand side of (31) is controlled by

$$\Big(\sum_{k\le N}\Big[\sum_{j>N} U_{p,s}(j,k,N)^p\, 2^{jsp}\|\Lambda_j f\|_p^p\Big]^{r/p}\Big)^{1/r}$$

$$\lesssim \Big(\sum_{k\le N} 2^{k(s+d+1-\frac{d}{p})r}2^{-Nr}\Big)^{\frac{1}{r}}\Big(\sum_{j>N} 2^{j(\frac{d}{p}-d-s)p}[2^{js}\|\Lambda_j f\|_p]^p\Big)^{\frac{1}{p}}.$$

In the range $s \ge \frac{d}{p} - d$ under consideration the first factor can be evaluated to be $C_4(p,s,r)2^{N(s+d-\frac{d}{p})}$ and the above expression is dominated by a constant times

$$\Big(\sum_{j>N} 2^{-(j-N)(s-\frac{d}{p}+d)p}[2^{js}\|\Lambda_j f\|_p]^p\Big)^{\frac{1}{p}}.$$

This yields (31). □

Remark 19 The proofs of Propositions 15 and 18 show that each operator $\mathbb{E}_N$ admits an extension to $B^s_{p,q}(\mathbb{R}^d)$ in the ranges of indices (iv), (v), and (vi) of Theorem 8, namely

$$\mathbb{E}_N(f) := \sum_{j=0}^{\infty}\mathbb{E}_N[L_j\Lambda_j f], \quad \text{in } B^s_{p,q}. \tag{32}$$

Indeed, for all $r > 0$ and for $J_2 > J_1 > N$ one has, in cases (iv) and (v),

$$\Big\|\mathbb{E}_N\Big(\sum_{j=J_1}^{J_2} L_j\Lambda_j f\Big)\Big\|_{B^s_{p,r}} \lesssim_N 2^{-J_1\varepsilon}\|f\|_{B^s_{p,\infty}},$$

with $\varepsilon = s - d(1/p - 1) > 0$, and in case (vi)

$$\Big\|\mathbb{E}_N\Big(\sum_{j=J_1}^{J_2} L_j\Lambda_j f\Big)\Big\|_{B^s_{p,r}} \lesssim_N \Big(\sum_{j=J_1}^{J_2} 2^{jsp}\|\Lambda_j f\|_p^p\Big)^{\frac{1}{p}}.$$

Proof of Theorem 8: Sufficiency for $\frac{d}{d+1} \le p \le 1$ In view of (18), (19) and trivial embeddings of Besov spaces, the uniform boundedness of $\mathbb{E}_N$ follows immediately from the above four propositions. □

4 Upper Bounds for $1 \le p \le \infty$

When $p \ge 1$ the constants in (24) take the form

$$U_{p,s}(j,k,N) = \begin{cases} 2^{k(s-\frac{1}{p})}2^{j(\frac{1}{p}-1-s)}2^N & \text{if } j,k > N, \\ 2^{k(s-\frac{1}{p})}2^{j(1-s)}2^{N(\frac{1}{p}-1)} & \text{if } j \le N < k \\ 2^{k(1+s)}2^{j(1-s)}2^{-2N} & \text{if } 0 \le j,k \le N, \\ 2^{k(1+s)}2^{j(\frac{1}{p}-1-s)}2^{-\frac{N}{p}} & \text{if } k \le N < j. \end{cases} \tag{33}$$

Again we state four propositions corresponding to the four cases of (33).

Proposition 20 *Suppose* $1 \le p \le \infty$. *Then*

$$\sup_{k>N} 2^{ks}\Big\| \sum_{j>N} L_k \mathbb{E}_N L_j \Lambda_j f \Big\|_p \lesssim \sup_{j>N} 2^{js}\|\Lambda_j f\|_p, \quad \text{if } \tfrac{1}{p}-1 < s \le \tfrac{1}{p}. \tag{34a}$$

Moreover, for all $r > 0$

$$\Big(\sum_{k>N} 2^{ksr}\Big\| \sum_{j>N} L_k \mathbb{E}_N L_j \Lambda_j f \Big\|_p^r\Big)^{1/r} \lesssim \begin{cases} \sup_{j>N} 2^{js}\|\Lambda_j f\|_p & \text{if } \frac{1}{p}-1 < s < \frac{1}{p}, \\ \sum_{j>N} 2^{js}\|\Lambda_j f\|_p & \text{if } s = \frac{1}{p}-1. \end{cases} \tag{34b}$$

The same inequalities hold with $\mathbb{E}_N$ *replaced by* $T_N[\cdot, \mathfrak{a}]$ *if* $\|\mathfrak{a}\|_\infty \le 1$.

Proposition 21 *Suppose* $1 \le p \le \infty$. *Then for all* $r > 0$

$$\Big(\sum_{k>N} 2^{ksr}\Big\| \sum_{j\le N} L_k \mathbb{E}_N^\perp L_j \Lambda_j f \Big\|_p^r\Big)^{1/r} \lesssim \sup_{j\le N} 2^{js}\|\Lambda_j f\|_p, \quad \text{if } s < \tfrac{1}{p}. \tag{35a}$$

Moreover, if $s = \frac{1}{p} < 1$ *then*

$$\sup_{k>N} 2^{ks}\Big\| \sum_{j\le N} L_k \mathbb{E}_N^\perp L_j \Lambda_j f \Big\|_p \lesssim \sup_{j\le N} 2^{js}\|\Lambda_j f\|_p, \tag{35b}$$

and if $s = p = 1$ *then*

$$\sup_{k>N} 2^{k}\Big\| \sum_{j\le N} L_k \mathbb{E}_N^\perp L_j \Lambda_j f \Big\|_1 \lesssim \sum_{j\le N} 2^{j}\|\Lambda_j f\|_1. \tag{35c}$$

The same inequalities hold with $\mathbb{E}_N^\perp$ *replaced by* $T_N[\cdot, \mathfrak{a}]$ *if* $\|\mathfrak{a}\|_\infty \le 1$.

Proposition 22 *Let $1 \le p \le \infty$ and $r > 0$. Then*

$$\Big(\sum_{k\le N} 2^{ksr}\Big\|\sum_{j\le N} L_k\mathbb{E}_N^{\perp}L_j\Lambda_j f\Big\|_p^r\Big)^{1/r} \lesssim \sup_{j\le N} 2^{js}\|\Lambda_j f\|_p \text{ if } -1 < s < 1. \tag{36a}$$

Moreover, for the case $s = -1$ we have

$$\sup_{k\le N} 2^{-k}\Big\|\sum_{j\le N} L_k\mathbb{E}_N^{\perp}L_j\Lambda_j f\Big\|_p \lesssim \sup_{j\le N} 2^{-j}\|\Lambda_j f\|_p, \tag{36b}$$

and for the case $s = 1$ we have

$$\Big(\sum_{k\le N} 2^{kr}\Big\|\sum_{j\le N} L_k\mathbb{E}_N^{\perp}L_j\Lambda_j f\Big\|_p^r\Big)^{1/r} \lesssim \sum_{j=0}^{N} 2^j\|\Lambda_j f\|_p. \tag{36c}$$

The same inequalities hold with $\mathbb{E}_N^{\perp}$ replaced by $T_N[\cdot, \mathfrak{a}]$ if $\|\mathfrak{a}\|_\infty \le 1$.

Proposition 23 *Let $1 \le p \le \infty$. Then for all $r > 0$,*

$$\Big(\sum_{k\le N} 2^{ksr}\Big\|\sum_{j>N} L_k\mathbb{E}_N L_j\Lambda_j f\Big\|_p^r\Big)^{1/r} \lesssim \sup_{j>N} 2^{js}\|\Lambda_j f\|_p \quad \text{if } s > \tfrac{1}{p} - 1 \tag{37a}$$

Moreover, for the case $s = \frac{1}{p} - 1$ and $1 \le p < \infty$,

$$\Big(\sum_{k\le N} 2^{k(\frac{1}{p}-1)r}\Big\|\sum_{j>N} L_k\mathbb{E}_N L_j\Lambda_j f\Big\|_p^r\Big)^{1/r} \lesssim \sum_{j=N+1}^{\infty} 2^{j(\frac{1}{p}-1)}\|\Lambda_j f\|_p. \tag{37b}$$

Finally, for the case $s = -1$ and $p = \infty$

$$\sup_{k\le N} 2^{-k}\Big\|\sum_{j>N} L_k\mathbb{E}_N L_j\Lambda_j f\Big\|_\infty \lesssim \sum_{j=N+1}^{\infty} 2^{-j}\|\Lambda_j f\|_\infty. \tag{37c}$$

The same inequalities hold when $\mathbb{E}_N$ is replaced by $T_N[\cdot, \mathfrak{a}]$, with $\|\mathfrak{a}\|_\infty \le 1$.

4.1 Proofs

The proofs of the four propositions involve Corollary 14 and an application of the triangle inequality for L^p when $p \ge 1$.

Proof of Proposition 20 Assume $s < 1/p$. By the triangle inequality and Corollary 14 the left hand side of (34b) is estimated by

$$\Big(\sum_{k>N} 2^{ksr}\Big[\sum_{j>N} \|L_k \mathbb{E}_N L_j \Lambda_j f\|_p\Big]^r\Big)^{\frac{1}{r}}$$
$$\lesssim \Big(\sum_{k>N}\Big[\sum_{j>N} U_{p,s}(j,k,N) 2^{js} \|\Lambda_j f\|_p\Big]^r\Big)^{\frac{1}{r}}$$
$$\lesssim \Big(\sum_{k>N} 2^{k(s-\frac{1}{p})r}\Big)^{\frac{1}{r}} \sum_{j>N} 2^{j(\frac{1}{p}-1-s)} 2^N 2^{js} \|\Lambda_j f\|_p.$$

When $s < 1/p$ the first factor is $c(p,s,r)2^{N(s-1/p)}$ and we see that the entire expression is dominated by a constant times

$$\sum_{j>N} 2^{(N-j)(s+1-\frac{1}{p})} 2^{js} \|\Lambda_j f\|_p$$

which proves (34b) and of course also (34a) when $s < 1/p$. Replacing the ℓ^r norm by a supremum in the above proof we see that (34a) is valid even for $s = 1/p$. □

Proof of Proposition 21 Let $s < 1/p$. The left hand side of (35a) is estimated by a constant times

$$\Big(\sum_{k>N}\Big[\sum_{j\le N} U_{p,s}(j,k,N) 2^{js} \|\Lambda_j f\|_p\Big]^r\Big)^{\frac{1}{r}}$$
$$\lesssim \Big(\sum_{k>N} 2^{k(s-\frac{1}{p})r}\Big)^{\frac{1}{r}} \sum_{j\le N} 2^{j(1-s)} 2^{N(\frac{1}{p}-1)} 2^{js} \|\Lambda_j f\|_p$$
$$\lesssim \sum_{j\le N} 2^{(j-N)(1-s)} 2^{js} \|\Lambda_j f\|_p.$$

This easily yields (35a). The proofs of (35b), (35c) are similar. □

Proof of Proposition 22 Assume $s > -1$. The left hand side of (36a) is estimated by a constant times

$$\Big(\sum_{k\le N}\Big[\sum_{j\le N} U_{p,s}(j,k,N) 2^{js} \|\Lambda_j f\|_p\Big]^r\Big)^{\frac{1}{r}}$$
$$\lesssim \Big(\sum_{k\le N} 2^{k(1+s)r}\Big)^{\frac{1}{r}} \sum_{j\le N} 2^{j(1-s)} 2^{-2N} 2^{js} \|\Lambda_j f\|_p,$$

and since the first factor is $\tilde{c}(p,q,r)2^{N(1+s)}$ we estimate the expression by a constant times

$$\sum_{j\le N} 2^{(j-N)(1-s)}2^{js}\|\Lambda_j f\|_p.$$

This easily yields (36a) and also (36c). The proof of (36b) which has a supremum in k for the case $s=-1$ is similar. □

Proof of Proposition 23 Let $s>-1$. The left hand side of (37a) is estimated by a constant times

$$\begin{aligned}
&\Big(\sum_{k\le N}\Big[\sum_{j>N} U_{p,s}(j,k,N)2^{js}\|\Lambda_j f\|_p\Big]^r\Big)^{\frac1r}\\
&\lesssim \Big(\sum_{k\le N}2^{k(1+s)r}\Big)^{\frac1r}\sum_{j>N}2^{j(\frac1p-1-s)}2^{-\frac{N}{p}}2^{js}\|\Lambda_j f\|_p\\
&\lesssim \sum_{j>N}2^{(j-N)(\frac1p-1-s)}2^{js}\|\Lambda_j f\|_p
\end{aligned}$$

which yields (37a) and also (37b). The proof of (37c) for the case $s=-1$ is similar. □

Remark 24 Similar reasonings as in Remark 19 justify the meaning of the extension formula for $\mathbb{E}_N$ in (32), for the ranges of indices in (i), (ii), (iii) in Theorem 8, and the cases (i), (ii) in Theorem 9. In the special case $s=1/p$, for $1<p\le\infty$, one has

$$\big\|\mathbb{E}_N\big(\textstyle\sum_{j=J_1}^{J_2} L_j\Lambda_j f\big)\big\|_{B^{1/p}_{p,\infty}} \lesssim_N 2^{-J_1}\|f\|_{B^s_{p,\infty}},$$

so the series $\sum_{j=0}^{\infty}\mathbb{E}_N(L_j\Lambda_j f)$ always converges in $B^{1/p}_{p,\infty}$, even though the series $\sum_{j=0}^{\infty} L_j\Lambda_j f$ only does if $f\in b^{1/p}_{p,\infty}$.

Proof of Theorems 8 and 9: Sufficiency for $1\le p\le\infty$ As before, one uses the previous four propositions combined with (18), (19) and trivial embeddings of Besov spaces. □

5 Necessary Conditions for Boundedness when $s=1/p$

It is well known that the characteristic function of a cube (and also the Haar functions) do not belong to $B^{1/p}_{p,q}$ for any $q<\infty$; see [15, 2.6.3 (18)]. In this section we elaborate a bit more on this result.

Recall that $b^s_{p,\infty}$ denotes the closure of the Schwartz space in the $B^s_{p,\infty}$ norm. Note also that $B^s_{p,q} \subset b^s_{p,\infty}$ for all $q < \infty$; see (15) above. Finally, span $\mathscr{H}_d$ denotes the vector space of all finite linear combinations of Haar functions.

Proposition 25 *Let* $0 < p \le \infty$. *Then*

$$b^{1/p}_{p,\infty} \cap \operatorname{span} \mathscr{H}_d = \{0\}.$$

Before proving this proposition we define, given $M \in \mathbb{N}$, certain test functions Ψ with vanishing moments of order up to $2M$ (which, along with their dilates $\Psi_N = 2^{Nd}\Psi(2^N\cdot)$, will be also be used in subsequent sections).

5.1 *Tensorized Test Functions*

Given $M \in \mathbb{N}$, consider a non-negative even function $\phi_0 \in C^\infty_c(-\frac{1}{8}, \frac{1}{8})$ such that $\phi_0^{(2M)}(t) > 0$ for all t in some interval $[-2\varepsilon, 2\varepsilon]$ (with $\varepsilon < 1/16$). Since $\widehat{\phi_0}(0) = \int \phi_0 > 0$, dilating if necessary we may also assume that $\widehat{\phi_0} \neq 0$ on $(-1, 1)$. Let $\varphi_0 \in C^\infty_c((-\frac{1}{8}, \frac{1}{8})^{d-1})$ be such that $\widehat{\varphi_0} \neq 0$ on $(-1, 1)^{d-1}$ and $\widehat{\varphi_0}(0) = 1$. For $M \ge 1$, let

$$\theta(t) = (\tfrac{d}{dt})^{2M}\phi_0(t), \qquad \vartheta(x_2, \dots, x_d) = \Big(\tfrac{\partial^2}{\partial x_2^2} + \cdots + \tfrac{\partial^2}{\partial x_d^2}\Big)^M \varphi_0(x').$$

In one dimension the function ϑ is obsolete and we just define $\Psi = \theta$. If $d \ge 2$ we define

$$\Psi(x) = \Delta^M[\phi_0 \otimes \varphi_0](x) = \theta(x_1)\varphi_0(x') + \phi_0(x_1)\vartheta(x'). \tag{38}$$

Clearly,

$$\int_{\mathbb{R}^d} \Psi(y) y_1^{m_1} \cdots y_d^{m_d}\, dy = 0, \quad \text{when } m_1 + \ldots + m_d < 2M.$$

If we choose $2M \gg |s| + d/p - d$ then for all $f \in B^s_{p,q}(\mathbb{R}^d)$,

$$\|f\|_{B^s_{p,q}} \gtrsim \Big\| \{2^{ks}\Psi_k * f\}_{k\in\mathbb{N}} \Big\|_{\ell^q(L^p)}. \tag{39}$$

5.2 Proof of Proposition 25

We argue by contradiction and assume that there is a nontrivial $f \in b^{1/p}_{p,\infty} \cap \operatorname{span} \mathscr{H}_d$. Then for some $N \in \mathbb{N}$ we can write f as

$$f = \sum_{\nu \in \Gamma} a_\nu \mathbb{1}_{I_{N,\nu}},$$

where $I_{N,\nu} = \prod_{i=1}^d [\nu_i 2^{-N}, (\nu_i + 1)2^{-N})$, Γ is a finite nonempty subset of $\mathbb{Z}^d$, and $a_\nu \in \mathbb{C}$ with $a_\nu \neq 0$ for $\nu \in \Gamma$.

Consider the usual partial order in $\mathbb{Z}^d$, that is for $\mu, \nu \in \mathbb{Z}^d$ we say that

$$\mu \leq \nu \quad \text{if} \quad \mu_i \leq \nu_i \quad \forall i = 1, \dots, d.$$

Pick a minimal element $\mu \in \Gamma$, meaning that that if $\nu \in \Gamma$ and $\nu \leq \mu$ then necessarily $\nu = \mu$. Now consider the function

$$g(x) = f(2^{-N}(x + \mu))/a_\mu,$$

which also belongs to $b^{1/p}_{p,\infty} \cap \operatorname{span} \mathscr{H}_d$. Note that g is now a linear combination of disjoint unit cubes and satisfies

$$g(x) = \begin{cases} 1 & \text{if } x \in [0,1)^d \\ 0 & \text{if } x \in (-1,1)^d \setminus [0,1)^d. \end{cases} \tag{40}$$

This last property is a consequence of the minimality of μ.

Consider now the function $\Psi \in C_c^\infty(\mathbb{R}^d)$ as in (38), with the pairs of functions ϕ_0, θ and φ_0, ϑ as in the paragraph preceding that definition. So, in particular,

$$\int_{\mathbb{R}^{d-1}} \varphi_0(x')\, dx' = 1 \quad \text{and} \quad \int_{\mathbb{R}^{d-1}} \vartheta(x')\, dx' = 0.$$

Observe further that for $t \in [-2\varepsilon, -\varepsilon]$

$$\int_0^\infty \theta(t-s)ds = \int_{-\infty}^t \theta(u)du = -\int_t^0 \theta(u)du \leq -\int_{-\varepsilon}^0 \theta(s)ds$$

since $\int_{-\infty}^0 \theta(s)ds = \phi_0^{(2M-1)}(0) = 0$ (because ϕ_0 is even) and $\theta(u) > 0$ for $u \in (-2\varepsilon, 0)$. Thus, if we set

$$c := \int_{-\varepsilon}^0 \theta(s)ds > 0$$

we obtain

$$\int_0^\infty \theta(t-s)ds \leq -c, \quad \forall\, t \in [-2\varepsilon, -\varepsilon]. \tag{41}$$

Next consider $\Psi_k(x) = 2^{kd}\Psi(2^k x)$, $k \geq 1$, and note that Ψ has enough vanishing moments so that

$$\|h\|_{B^{1/p}_{p,\infty}} \gtrsim \sup_{k\geq 1} 2^{k/p}\|\Psi_k * h\|_p, \quad h \in B^{1/p}_{p,\infty}.$$

Moreover, since $g \in b^{1/p}_{p,\infty}$ we also have

$$2^{k/p}\|\Psi_k * g\|_p \to 0, \quad \text{as} \quad k \to \infty. \tag{42}$$

We show that this leads to a contradiction. Indeed, if $x_1 \in [-\varepsilon 2^{1-k}, -\varepsilon 2^{-k}]$ and $x' \in [1/4, 3/4]^{d-1}$ then, using that $\operatorname{supp} \Psi_k(x - \cdot) \subset (-1,1) \times (1/8, 7/8)^{d-1}$, we may apply (40) and (41) to obtain

$$\begin{aligned}
g * \Psi_k(x) &= \int_{[0,1)^d} \Psi_k(x-y)\,dy \\
&= \int_0^1 \theta_k(x_1 - y_1)\,dy_1 \int_{(0,1)^{d-1}} \varphi_{0,k}(x'-y')\,dy' + \\
&\quad + \int_0^1 \phi_{0,k}(x_1-y_1)\,dy_1 \int_{(0,1)^{d-1}} \vartheta_k(x'-y')\,dy' \\
&= 2^k \int_0^1 \theta(2^k(x_1-y_1))\,dy_1 = \int_0^{2^k} \theta(2^k x_1 - u)\,du \\
&= \int_0^\infty \theta(2^k x_1 - u)\,du \leq -c/2.
\end{aligned}$$

Thus we must have

$$\|g * \Psi_k\|_p \geq (c/2)\,(\varepsilon 2^{-k})^{1/p}(1/2)^{\frac{d-1}{p}},$$

which contradicts (42). □

Remark 26 In the recent work [21] by Yuan, Sickel and Yang, the authors study regularity properties of the Haar system in other Besov-type spaces $B^{s,\tau}_{p,q}(\mathbb{R}^d)$ which serves as a first step to investigate its basis properties in these spaces.

6 Necessary Conditions for Boundedness when $s = 1$

We now consider the necessity of the condition $q \le p$ in part (iv) of Theorem 8. This restriction was also noticed in [8]. For $q > p$ we show that the operators $\mathbb{E}_N$ are bounded, but not uniformly bounded and determine the precise behavior of the operator norms as $N \to \infty$. The lower bounds will be obtained by testing with suitable functions with compact support; we refer to (10) for the notation in the next theorem.

Theorem 27 *Suppose that either*

(i) $\frac{d}{d+1} < p < 1$ *and* $p \le q \le \infty$, *or*
(ii) $p = 1$ *and* $q = \infty$.

Then for large N

$$\|\mathbb{E}_N\|_{B^1_{p,q} \to B^1_{p,q}} \approx N^{\frac{1}{p} - \frac{1}{q}}.$$

Moreover, for cubes Q *of side length* $\ge 1/2$,

$$\mathrm{Op}\left(\mathbb{E}_N, B^1_{p,q}, Q\right) \approx N^{\frac{1}{p} - \frac{1}{q}}.$$

6.1 *Proof of the Upper Bounds in Theorem 27*

Letting $s = 1$ in Propositions 15 and 18 (and noticing that $d(\frac{1}{p} - 1) < 1 < \frac{1}{p}$ when $\frac{d}{d+1} < p < 1$), we see that

$$\Big\| \{2^k \sum_{j>N} L_k \mathbb{E}_N L_j \Lambda_j f\}_{k=0}^{\infty} \Big\|_{\ell^q(L^p)} \lesssim \|f\|_{B^1_{p,\infty}} \le \|f\|_{B^1_{p,q}}.$$

On the other hand, letting $s = 1$ in Propositions 16 and 17, and using Hölder's inequality we obtain

$$\Big\| \{2^k \sum_{j \le N} L_k \mathbb{E}_N^{\perp} L_j \Lambda_j f\}_{k=0}^{\infty} \Big\|_{\ell^q(L^p)} \lesssim \Big(\sum_{j=0}^{N} 2^{jp} \|\Lambda_j f\|_p^p \Big)^{1/p} \lesssim N^{1/p - 1/q} \|f\|_{B^1_{p,q}}.$$

Combining this with (18) and (19) we obtain $\|\mathbb{E}_N\|_{B^1_{p,q} \to B^1_{p,q}} \lesssim N^{\frac{1}{p} - \frac{1}{q}}$. The above arguments also apply if $s = p = 1$, provided we let $q = \infty$. □

6.2 Proof of the Lower Bounds in Theorem 27

We shall actually prove a stronger result which gives a lower bound even for a $B^1_{p,q} \to B^1_{p,\infty}$ estimate and for functions supported in the open unit cube $Q_0 = (0,1)^d$.

Theorem 28 *If* $0 < p \le 1$ *and* $p \le q \le \infty$, *then there is* $c_{p,q} > 0$ *such that, for each* $N \ge 1$,

$$\sup\Big\{\|\mathbb{E}_N f\|_{B^1_{p,\infty}} : \|f\|_{B^1_{p,q}} \le 1,\ \operatorname{supp}(f) \subset Q_0\Big\} \ge c_{p,q}\, N^{\frac{1}{p}-\frac{1}{q}}. \tag{43}$$

Fix $u \in C_c^\infty(\mathbb{R})$ supported in $(1/8, 7/8)$ with $u(t) = 1$ for $t \in [1/4, 3/4]$, and $\chi \in C_c^\infty(\mathbb{R}^{d-1})$ supported in $(1/8, 7/8)^{d-1}$ with $\chi(x') = 1$ for $x' \in (1/4, 3/4)^{d-1}$; here $x' = (x_2, \dots, x_d)$. Define, for large N, functions of one variable

$$g_{N,j}(t) = e^{2\pi i 2^j t} u(Nt - 2j), \quad j \in \mathbb{N}, \tag{44}$$

and let

$$f_N(x) = \chi(x_2, \dots, x_d) \sum_{N/8<j<N/4} 2^{-j} g_{N,j}(x_1). \tag{45}$$

Lemma 29 *For* $p \le q \le \infty$ *we have*

$$\|f_N\|_{B^1_{p,q}} \lesssim N^{1/q-1/p}. \tag{46}$$

Proof We estimate $L_k f_N = \beta_k * f_N$. If $2^k \le N$, since $\beta_k * f_N$ is compactly supported and $\|\beta_k * f_N\|_\infty \lesssim \|f_N\|_\infty \lesssim 2^{-N/8}$, then

$$\Big(\sum_{k=0}^{\log_2 N} 2^{kq} \|\beta_k * f_N\|_p^q\Big)^{1/q} \lesssim N\, 2^{-N/8} \ll N^{1/q-1/p}.$$

Assume now that $2^k > N$. First notice that the sets

$$\operatorname{supp} \beta_k * g_{N,j} \subset \big\{\tfrac{2j}{N} + (-\tfrac{2}{N}, \tfrac{2}{N})\big\} \times (0,1)^{d-1}, \quad \tfrac{N}{8} < j < \tfrac{N}{4}, \tag{47}$$

are pairwise disjoint, and thus

$$\|\beta_k * f_N\|_p = \Big(\sum_{\frac{N}{8}<j<\frac{N}{4}} 2^{-jp} \|\beta_k * (g_{N,j} \otimes \chi)\|_p^p\Big)^{1/p}. \tag{48}$$

Next, we distinguish the cases $j \geq k$ and $j \leq k$. When $j \geq k$, if we integrate by parts M-times with respect to y_1 in the convolution we obtain

$$\beta_k * (g_{N,j} \otimes \chi)(x) = \int \frac{\partial^M}{\partial y_1}[\beta_k(x-y)u(Ny_1 - 2j)\chi(y')]\frac{e^{2\pi i 2^j y_1}}{(-2\pi i 2^j)^M} dy_1\, dy'$$

and thus, using that $N < 2^k$,

$$\|\beta_k * (g_{N,j} \otimes \chi)\|_p \lesssim 2^{-(j-k)M} N^{-1/p}, \quad j \geq k.$$

For $N/8 < j \leq k$ we use the cancellation of the β_k (with M vanishing moments) to obtain

$$\|\beta_k * (g_{N,j} \otimes \chi)\|_p \lesssim 2^{-kM} \|\partial^M g_{N,j}\|_\infty N^{-\frac{1}{p}} \lesssim 2^{-(k-j)M} N^{-1/p}, \quad j \leq k.$$

Thus

$$\Big(\sum_{2^k > N} 2^{kq} \|\beta_k * f_N\|_p^q\Big)^{\frac{1}{q}} = \Big(\sum_{2^k > N} 2^{kq}\Big[\sum_{\frac{N}{8} < j < \frac{N}{4}} 2^{-jp} \|\beta_k * (g_{N,j} \otimes \chi)\|_p^p\Big]^{\frac{q}{p}}\Big)^{\frac{1}{q}}$$
$$\lesssim \Big(\sum_{2^k > N}\Big[\sum_{\frac{N}{8} < j < \frac{N}{4}} 2^{(k-j)p} 2^{-|j-k|Mp}\Big]^{\frac{q}{p}}\Big)^{\frac{1}{q}} N^{-\frac{1}{p}} \lesssim N^{\frac{1}{q}-\frac{1}{p}}$$

and we are done. □

We now take $\Psi_N = 2^{Nd}\Psi(2^N\cdot)$ with Ψ as in Sect. 5.1, and we shall prove that

$$\|\mathbb{E}_N f_N\|_{B^1_{p,\infty}} \gtrsim 2^N \|\Psi_N * \mathbb{E}_N f_N\|_p \gtrsim 1. \tag{49}$$

Define $\Theta : \mathbb{R} \to \mathbb{R}$ by

$$\Theta(t) = \int_{-\infty}^{t} \theta(s)ds \tag{50}$$

with $\theta = \phi_0^{(2M)}$ as in Sect. 5.1 and observe that Θ is odd, supported in $(-\frac{1}{8}, \frac{1}{8})$ and $\int_{-\infty}^{\infty} \Theta(t)dt = 0$. In particular,

$$\int_0^{1/8} |\Theta(t)|^p dt \neq 0. \tag{51}$$

Let $\mathbb{E}_N^{(1)}$ and $\mathbb{E}_N^{(d-1)}$ be the dyadic averaging operators on $\mathbb{R}$ and $\mathbb{R}^{d-1}$, respectively. If we denote $\theta_N = 2^N\theta(2^N\cdot)$, $N \geq 1$, then we claim that

$$\Psi_N * (\mathbb{E}_N[g_{N,j} \otimes \chi])(x_1, x') = \theta_N * (\mathbb{E}_N^{(1)} g_{N,j})(x_1), \text{ for } x' \in (\tfrac{1}{3}, \tfrac{2}{3})^{d-1}. \tag{52}$$

Indeed, for $x' \in (\frac{1}{3}, \frac{2}{3})^{d-1}$ it is easily seen that

$$2^{N(d-1)}\varphi_0(2^N\cdot) * \mathbb{E}_N^{(d-1)}\chi(x') = \int \varphi_0(y')dy' = 1,$$

$$2^{N(d-1)}\vartheta(2^N\cdot) * \mathbb{E}_N^{(d-1)}\chi(x') = \int \vartheta(y')dy' = 0.$$

The proof of the lower bound in (49) will rely on the following lemma.

Lemma 30 *Let $\nu \in \mathbb{Z}$ and $\widetilde{I}_{N,\nu} = [\frac{\nu}{2^N}, \frac{\nu+1/8}{2^N}]$. Then, for every $t \in \widetilde{I}_{N,\nu}$ and $N/8 < j < N/4$ we have*

$$\theta_N * (\mathbb{E}_N^{(1)} g_{N,j})(t) = 2^{-N} g'_{N,j}\big(\tfrac{\nu}{2^N}\big)\, \Theta(2^N t - \nu) + O(2^{-2(N-j)}). \tag{53}$$

Moreover if $\frac{\nu}{2^N} \in [\frac{2j}{N} + \frac{1}{4N}, \frac{2j}{N} + \frac{3}{4N}]$ then $g'_{N,j}(2^{-N}\nu) = 2\pi i 2^j e^{2\pi i 2^{j-N}\nu}$.

Proof The last assertion is immediate by the definition of $g_{N,j}$ in (44). So we focus in proving (53). We split

$$\theta_N * \mathbb{E}_N^{(1)} g_{N,j} = \theta_N * (\mathbb{E}_N^{(1)} - I)g_{N,j} + \theta_N * g_{N,j}$$

and observe that from the cancellation properties of θ_N we have

$$\|\theta_N * g_{N,j}\|_\infty \lesssim 2^{-2N}\|g''_{N,j}\|_\infty \lesssim 2^{-2N}(2^{2j} + N^2)$$

which for $N/8 \leq j \leq N/4$ implies $\|\theta_N * g_{N,j}\|_\infty \lesssim 2^{-2(N-j)}$. Let $I_{N,\nu} = [2^{-N}\nu, 2^{-N}(\nu+1))$. For $t \in \widetilde{I}_{N,\nu}$ we have $\operatorname{supp}\theta_N(t-\cdot) \subset I_{N,\nu-1} \cup I_{N,\nu}$, so recalling (50) we obtain

$$\begin{aligned}
&\theta_N * (\mathbb{E}_N^{(1)} - I)g_{N,j}(t)\\
&= \int 2^N\Theta'(2^N(t-s)) \times \Big[\mathbb{1}_{I_{N,\nu}}(s)\Big(⨍_{I_{N,\nu}} g_{N,j}(w)dw - g_{N,j}(s)\Big)\\
&\qquad + \mathbb{1}_{I_{N,\nu-1}}(s)\Big(⨍_{I_{N,\nu-1}} g_{N,j}(w)dw - g_{N,j}(s)\Big)\Big]ds
\end{aligned}$$

For $s \in I_{N,\nu}$, using Taylor expansions one sees that

$$\begin{aligned}
&\fint_{I_{N,\nu}} g_{N,j}(w)dw - g_{N,j}(s)\\
&= \fint_{I_{N,\nu}} g'_{N,j}(s)(w-s)dw + \fint_{I_{N,\nu}} \int_0^1 (1-\sigma)g''_{N,j}(s+\sigma(w-s))d\sigma\,(w-s)^2dw\\
&= g'_{N,j}(\tfrac{\nu}{2^N})\frac{1}{2|I_{N,\nu}|}\Big[(\tfrac{\nu+1}{2^N}-s)^2-(\tfrac{\nu}{2^N}-s)^2\Big]+O(2^{2j-2N})\\
&= g'_{N,j}(\tfrac{\nu}{2^N})\Big[\tfrac{\nu+1/2}{2^N}-s\Big]+O(2^{2j-2N}).
\end{aligned}$$

Similarly, for $s \in I_{N,\nu-1}$,

$$\fint_{I_{N,\nu-1}} g_{N,j}(w)dw - g_{N,j}(s) = g'_{N,j}(\tfrac{\nu}{2^N})\Big[\tfrac{\nu-1/2}{2^N}-s\Big]+O(2^{2j-2N}).$$

Hence for $t \in \widetilde{I}_{N,\nu}$ we have

$$2^N\theta_N * (\mathbb{E}_N^{(1)} - I)g_{N,j}(t) \;=\; A_{1,j}(t) + A_{2,j}(t) + O(2^{-2(N-j)}) \tag{54}$$

where

$$A_{1,j}(t) = g'_{N,j}(\tfrac{\nu}{2^N})\int 2^N\Theta'(2^N(t-s))\frac{1}{2^{N+1}}\big(\mathbb{1}_{I_{N,\nu}}(s)-\mathbb{1}_{I_{N,\nu-1}}(s)\big)\,ds$$

and

$$A_{2,j}(t) = g'_{N,j}(\tfrac{\nu}{2^N})\int 2^N\Theta'(2^N(t-s))(\tfrac{\nu}{2^N}-s)\,ds\,.$$

Integration by parts yields (for $t \in \widetilde{I}_{N,\nu}$)

$$A_{2,j}(t) = g'_{N,j}(2^{-N}\nu)\int 2^N\Theta(2^N(s-t))\,ds = 0$$

since $\int\Theta(s)ds = 0$. To compute $A_{1,j}(t)$ we observe that

$$\begin{aligned}
&\int 2^N\Theta'(2^N(t-s))\big(\mathbb{1}_{I_{N,\nu}}(s)-\mathbb{1}_{I_{N,\nu-1}}(s)\big)\,ds\\
&= \Big[\int_{\frac{\nu}{2^N}}^{\frac{\nu+1}{2^N}} - \int_{\frac{\nu-1}{2^N}}^{\frac{\nu}{2^N}}\Big]\frac{d}{ds}\big(-\Theta(2^N(t-s)\big)\,ds\\
&= -\Theta\big(2^N(t-\tfrac{\nu+1}{2^N})\big)+2\Theta\big(2^N(t-\tfrac{\nu}{2^N})\big)-\Theta\big(2^N(t-\tfrac{\nu-1}{2^N})\big).
\end{aligned}$$

For $t \in \widetilde{I}_{N,\nu}$ we have $\Theta(2^N(t - 2^{-N}(\nu \pm 1))) = 0$ and thus

$$A_{1,j}(t) = 2^{-N} g_j'(\tfrac{\nu}{2^N})\Theta(2^N t - \nu), \quad t \in \widetilde{I}_{N,\nu}.$$

Inserting these expressions into (54) we are led to (53). □

We may now complete the proof of (49). Using (52), and the fact that, by (47), the functions $\theta_N * (\mathbb{E}_N^{(1)} g_{N,j})$ are supported in the disjoint intervals $J_{N,j} := \frac{2j}{N} + (-\frac{2}{N}, \frac{2}{N})$, we have

$$2^N \|\Psi_N * (\mathbb{E}_N f_N)\|_p \gtrsim 2^N \Big(\sum_{\frac{N}{8}<j<\frac{N}{4}} \|\theta_N * g_{N,j}\|_{L^p(\mathbb{R})}^p 2^{-jp} \Big)^{1/p}$$

$$\gtrsim \Big(\sum_{\frac{N}{8}<j<\frac{N}{4}} \sum_{\nu:\ \frac{\nu}{2^N} \in J_{N,j}} \Big[|2^{-j} g_{N,j}'(\tfrac{\nu}{2^N})|^p \int_{\widetilde{I}_{N,\nu}} |\Theta(2^N t - \nu)|^p dt - \frac{c 2^{(j-N)p}}{2^N} \Big] \Big)^{1/p}$$

using the previous lemma in the last step. Since by (51)

$$\int_{\widetilde{I}_{N,\nu}} |\Theta(2^N t - \nu)|^p dt = 2^{-N} \int_0^{1/8} |\Theta(t)|^p dt \gtrsim 2^{-N},$$

we obtain, for sufficiently large N,

$$2^N \|\Psi_N * (\mathbb{E}_N f_N)\|_p \gtrsim \Big(\sum_{\frac{N}{8}<j<\frac{N}{4}} \sum_{\nu:\ \frac{\nu}{2^N} \in J_{N,j}} 2^{-N}(1 - c' 2^{-pN/2}) \Big)^{1/p} \gtrsim 1.$$

This completes the proof of (49), which together with (46) establishes Theorem 28, and therefore also Theorem 27.

7 Necessary Conditions for Boundedness when $s \le 0$

7.1 *The Case* $1 < p \le \infty$, $s = 1/p - 1$, $q > 1$

In these cases the operator $\mathbb{E}_N$ is not bounded in $B^{1/p-1}_{p,q}$ because characteristic functions of cubes do not belong to the dual space $(B^{1/p-1}_{p,q})^* = B^{1/p'}_{p',q'}$; see Sect. 5. This also applies when $p = \infty$, since $(b^{-1}_{\infty,q})^* = B^1_{1,q'}$; see [10, §1.1.5].

7.2 The Case $p = \infty$, $s = -1$, $q \le 1$

We shall show

$$\|\mathbb{E}_N\|_{b^{-1}_{\infty,q}\to B^{-1}_{\infty,q}} \gtrsim N. \tag{55}$$

To prove this we argue by duality and first note that

$$\|\mathbb{E}_N\|_{b^{1}_{1,\infty}\to B^{1}_{1,\infty}} \approx N. \tag{56}$$

Indeed by Theorem 27 we have $\|\mathbb{E}_N\|_{B^{1}_{1,\infty}\to B^{1}_{1,\infty}} \approx N$ and the lower bound is obtained by testing $\mathbb{E}_N$ on the Schwartz-function f_N as in (45) satisfying $\|f_N\|_{b^1_{1,\infty}} = \|f_N\|_{B^1_{1,\infty}} \lesssim N^{-1}$ and $\|\mathbb{E}_N f_N\|_{B^1_{1,\infty}} \gtrsim 1$, *cf.* (46), (49).

To establish (55), since $\|\mathbb{E}_N\|_{b^{-1}_{\infty,q}\to B^{-1}_{\infty,q}} \ge \|\mathbb{E}_N\|_{b^{-1}_{\infty,q}\to B^{-1}_{\infty,1}}$, by (56) it suffices to prove that

$$\|\mathbb{E}_N\|_{b^{-1}_{\infty,q}\to B^{-1}_{\infty,1}} \gtrsim \|\mathbb{E}_N\|_{b^{1}_{1,\infty}\to B^{1}_{1,\infty}}. \tag{57}$$

We use that $(b^{-1}_{\infty,q})^* = B^1_{1,\infty}$ for $q \le 1$; see [15, 2.5.1/Remark 7]. Then for $f \in \mathcal{S}$

$$\|\mathbb{E}_N f\|_{B^1_{1,\infty}} = \|\mathbb{E}_N f\|_{(b^{-1}_{\infty,q})^*} = \sup_{\substack{g\in\mathcal{S} \\ \|g\|_{b^{-1}_{\infty,q}}\le 1}} \big|\langle \mathbb{E}_N f, g\rangle\big|. \tag{58}$$

Now, for each $g \in \mathcal{S}$, since $f = \sum_{j=0}^{\infty} L_j \Lambda_j f$ in $\mathcal{S}$, we have

$$\begin{aligned} |\langle \mathbb{E}_N f, g\rangle| = |\langle f, \mathbb{E}_N g\rangle| &\le \sum_{j=0}^{\infty} \|\Lambda_j f\|_1 \|L_j(\mathbb{E}_N g)\|_\infty \\ &\lesssim \|f\|_{b^1_{1,\infty}} \|\mathbb{E}_N g\|_{B^{-1}_{\infty,1}} \le \|f\|_{b^1_{1,\infty}} \|\mathbb{E}_N\|_{b^{-1}_{\infty,q}\to B^{-1}_{\infty,1}} \|g\|_{b^{-1}_{\infty,q}}. \end{aligned}$$

Inserting this into (58) we arrive at

$$\|\mathbb{E}_N f\|_{B^1_{1,\infty}} \le \|\mathbb{E}_N\|_{b^{-1}_{\infty,q}\to B^{-1}_{\infty,1}} \|f\|_{b^1_{1,\infty}}$$

and hence (57).

8 Density and Approximation

In this section we show two results regarding approximation by linear combinations of Haar functions. The main results in Sect. 8.1 are relevant to the formulation of Theorems 3 and 4 which rule out the case $s = 1$. We shall also obtain a positive result about approximation for the spaces $b^{1/p}_{p,\infty}$.

8.1 *The Case* $s = 1$

We shall show that no strongly admissible enumeration of the Haar system can form a Schauder basis on $B^1_{p,q}(\mathbb{R}^d)$ if $\frac{d}{d+1} \le p < 1$ and $q > 0$. Moreover if in addition $0 < q \le p$ we shall show that the Haar system is not dense in $B^1_{p,q}(\mathbb{R}^d)$. It seems plausible that this last assertion would continue to hold for all $0 < q \le \infty$, but we do not have a proof in this generality.

Let us start with an auxiliary result. It is well-known that

$$\|f\|_{B^s_{p,\infty}(\mathbb{R}^d)} \approx \|f\|_p + \sum_{j=1}^{d} \sup_{|h|\le 1} \frac{\|\Delta^2_{he_j} f\|_p}{|h|^s},$$

for all $p \le 1$ and $d(1/p - 1) < s < 2$, see [17, 2.6.1]. Below we show that a partial lower bound actually holds for all $0 < s < 2$, which allows to incorporate the endpoint $s = d(1/p - 1) = 1$ (i.e., $p = d/(d+1)$) to our later results.

Proposition 31 *Let* $0 < s < 2$ *and* $0 < p \le 1$. *Then*

$$\|g\|_p + \sum_{j=1}^{d} \sup_{|h|\le 1} \frac{\|\Delta^2_{he_j} g\|_p}{|h|^s} \lesssim \|g\|_{B^s_{p,\infty}} \tag{59}$$

holds for any function $g \in L^1(\mathbb{R}^d)$.

Proof Let $\widehat{\psi}_0 \in C^\infty_c(\mathbb{R}^d)$ supported in $\{|\xi| < 3/8\}$ and with $\widehat{\psi}_0(\xi) = 1$ if $|\xi| \le 1/4$, and let $\widehat{\psi}_k(\xi) = \widehat{\psi}_0(2^{-k}\xi) - \widehat{\psi}_0(2^{-k+1}\xi)$ if $k \ge 1$. Consider a standard dyadic frequency decomposition $g = \sum_{k=0}^{\infty} g_k$, with $g_k = \psi_k * g$, which converges in L^1 and also a.e. Since

$$\Big(\sum_{k=0}^{\infty} \|g_k\|_p^p\Big)^{1/p} \lesssim \|g\|_{B^s_{p,\infty}}$$

we also have $\|g\|_p \lesssim \|g\|_{B^s_{p,\infty}}$. In addition, for each $0 < |h| \le 1$, using the trivial estimate $\|\Delta^2_{he_j} g_k\|_p^p \le 4\|g_k\|_p^p$, we see that

$$\frac{\|\Delta^2_{he_j} \sum_{2^k \ge |h|^{-1}} g_k\|_p}{|h|^s} \le \Big(4 \sum_{2^k \ge |h|^{-1}} (2^k|h|)^{-sp} \|g_k\|_p^p 2^{ksp}\Big)^{1/p}$$
$$\lesssim \sup_{2^k|h| \ge 1} 2^{ks} \|g_k\|_p.$$

Let $\varphi \in \mathcal{S}$ be such that $\widehat{\varphi}(\xi) = 1$ if $|\xi| \le 1$. For every $0 < v < 1$, we let $K_{ve_j} = \Delta^2_{ve_j}\varphi$, so that

$$\widehat{K}_{ve_j}(\xi) = (e^{2\pi i \langle ve_j, \xi\rangle} - 1)^2 \widehat{\varphi}(\xi).$$

Then K_{ve_j} is a Schwartz function and we have the estimate

$$|K_{ve_j}(x)| \le C_N v^2 (1 + |x|)^{-2N},$$

for a large $N > d$. Hence, for each $k \ge 0$ such that $2^k|h| < 1$, we have

$$|\Delta^2_{he_j} g_k(x)| = 2^{kd} |K_{2^k he_j}(2^k \cdot) * g_k(x)|$$
$$\le C_N (2^k|h|)^2 \int 2^{kd} (1 + 2^k|y|)^{-2N} |g_k(x - y)| dy$$
$$\lesssim C_N (2^k|h|)^2 \sup_{y \in \mathbb{R}^d} (1 + 2^k|y|)^{-N} |g_k(x - y)|.$$

Choosing $N > d/p$ we can apply the Peetre maximal function estimate to obtain

$$\frac{\|\Delta^2_{he_j} \sum_{2^k < |h|^{-1}} g_k\|_p}{|h|^s} \le \Big(\sum_{2^k < |h|^{-1}} (2^k|h|)^{-sp} \|\Delta^2_{he_j} g_k\|_p^p 2^{ksp}\Big)^{1/p}$$
$$\lesssim \Big(\sum_{2^k < |h|^{-1}} (2^k|h|)^{(2-s)p} \|g_k\|_p^p 2^{ksp}\Big)^{1/p} \lesssim \sup_{k \ge 0} 2^{ks} \|g_k\|_p.$$

Combining the two estimates yields the result. □

Remark 32 The appropriate analogue for $B^s_{p,q}(\mathbb{R}^d)$-quasinorms for $q < \infty$, i.e.

$$\|g\|_p + \sum_{j=1}^d \Big(\int_{-1}^1 \frac{\|\Delta^2_{he_j} g\|_p^q}{|h|^{sq}} \frac{dh}{|h|}\Big)^{1/q} \lesssim \|g\|_{B^s_{p,q}}$$

remains valid (when $0 < s < 2$) but is not relevant in this section.

The following proposition is a modification of our argument in [5, Proposition 4.2]. It shows the necessity of the condition $s < 1$ in part (ii) of Theorem 3 and part (iii) of Theorem 4.

Proposition 33 *There exists a Schwartz function f supported in $(\frac{1}{16}, \frac{15}{16})^d$ such that for all $0 < p \le 1$ it holds*

$$\liminf_{N\to\infty} \|\mathbb{E}_N f - f\|_{B^1_{p,\infty}} > 0. \tag{60}$$

Proof Let $\eta \in C_c^\infty(\mathbb{R}^d)$ be supported in $(\frac{1}{16}, \frac{15}{16})^d$ with $\eta(x) = 1$ if $x \in (1/8, 7/8)^d$. Let $f(x) = x_1\, \eta(x)$. From (59) we have

$$\big\|\mathbb{E}_N f - f\big\|_{B^1_{p,\infty}} \gtrsim \sup_{0<h\le 1} \frac{\big\|\Delta^2_{he_1}\mathbb{E}_N f - \Delta^2_{he_1} f\big\|_p}{h},$$

Clearly, since f is a Schwartz function,

$$\|\Delta^2_{he_1} f\|_p \lesssim |h|^2, \quad 0 < h \le 1.$$

So, by an appropriate triangle inequality, it suffices to show that

$$\frac{\big\|\Delta^2_{2^{-N-2}e_1}\mathbb{E}_N f\big\|_p}{2^{-N-2}} \ge c > 0, \tag{61}$$

for large N. We now recall a calculation in [5, Prop 4.2]. Let $N > 10$ and let $h \in (0, 1/4)$. An explicit calculation shows that for $x \in (1/4, 3/4)^d$

$$\mathbb{E}_N f(x) = \sum_{2^{N-2}\le k<3\cdot 2^{N-2}} \tfrac{k+1/2}{2^N} \mathbb{1}_{[\frac{k}{2^N}, \frac{k+1}{2^N})\times[0,1)^{d-1}}(x).$$

Then, under the additional assumption $0 < h < 2^{-N-1}$,

$$\Delta_{he_1}\mathbb{E}_N f(x) = 2^{-N-1} \sum_{2^{N-2}\le k\le 3\cdot 2^{N-2}} \mathbb{1}_{[\frac{k}{2^N}-h, \frac{k}{2^N})\times[0,1)^{d-1}}(x),$$

and

$$\Delta^2_{he_1}\mathbb{E}_N f(x) =$$
$$2^{-N-1} \sum_{2^{N-2}<k<3\cdot 2^{N-2}} \big(\mathbb{1}_{[\frac{k}{2^N}-2h, \frac{k}{2^N}-h)\times[0,1)^{d-1}}(x) - \mathbb{1}_{[\frac{k}{2^N}-h, \frac{k}{2^N})\times[0,1)^{d-1}}(x)\big).$$

Therefore,

$$\|\Delta^2_{he_1}\mathbb{E}_N f\|_{L^p([0,1]^d)} \gtrsim 2^{N(1/p-1)}\, h^{1/p},$$

and in particular

$$\|\Delta^2_{2^{-N-2}e_1}\mathbb{E}_N f\|_{L^p([0,1]^d)} \gtrsim 2^{-N},$$

which implies (61). □

Finally, we conclude with the non-density result mentioned above.

Corollary 34 *Let* $\frac{d}{d+1} \le p < 1$, $0 < q \le p$. *Then* span $\mathscr{H}_d$ *is not dense in* $B^1_{p,q}(\mathbb{R}^d)$.

Proof By Proposition 33 and $B^1_{p,q} \hookrightarrow B^1_{p,\infty}$ we have, for some $f \in C_c^\infty$,

$$\liminf_{N\to\infty} \|\mathbb{E}_N f - f\|_{B^1_{p,q}} = c > 0, \tag{62}$$

By Theorem 8 the operators $\mathbb{E}_N$ are uniformly bounded on $B^1_{p,q}$. For $h \in$ span $\mathscr{H}_d$ we have $\mathbb{E}_N h = h$ for $N \ge N_0(h)$, with sufficiently large $N_0(h)$. Hence

$$\|\mathbb{E}_N f - f\|_{B^1_{p,q}} \lesssim \|\mathbb{E}_N[f-h]\|_{B^1_{p,q}} + \|f-h\|_{B^1_{p,q}} \lesssim \|f-h\|_{B^1_{p,q}}, \text{ for } N \ge N_0(h),$$

and the density of span $\mathscr{H}_d$ in $B^s_{p,q}$ would yield a contradiction to (62). □

Remark 35 When $d/(d+1) \le p < 1$, it follows from Theorem 8.iv (or vi), and from the results in Sect. 9 below, that each strongly admissible enumeration $\mathcal{U}$ of $\mathscr{H}_d$ is a *basic sequence* for $B^1_{p,p}(\mathbb{R}^d)$, that is, $\mathcal{U}$ is a Schauder basis for the subspace

$$\overline{\text{span}\, \mathscr{H}_d}^{B^1_{p,p}}.$$

It may be of interest to identify this subspace. By Oswald's result in [8], it contains the class $\mathscr{B}^1_{p,p,(1)}(\mathbb{R}^d)$ defined by first order differences.

8.2 An Approximation Result for $b^{1/p}_{p,\infty}$ when $1 < p < \infty$

In the limiting case $s = 1/p$, recall that $\mathscr{H}_d$ is contained in $B^{1/p}_{p,q}$ only if $q = \infty$. We show an approximation result in this case when $1 < p < \infty$. Recall that $B^s_{p,\infty}$ is not separable, and that $b^s_{p,\infty}$ denotes the closure of $\mathcal{S}$ in $B^s_{p,\infty}$. Recall also, from Proposition 25, that $b^{1/p}_{p,\infty} \cap$ span $\mathscr{H}_d = \{0\}$. However taking closures one obtains

Proposition 36 *Let* $1 < p \le \infty$. *Then*

$$b^{1/p}_{p,\infty}(\mathbb{R}^d) \subsetneq \overline{\text{span}(\mathscr{H}_d)}^{B^{1/p}_{p,\infty}}. \tag{63}$$

Proof Let $1 < p < \infty$. In view of [16, 2.5.12], we may use the equivalent norm

$$\|f\|_{B^{1/p}_{p,\infty}} = \|f\|_p + \sup_{h \ne 0} \frac{\|\Delta_h f\|_p}{|h|^{1/p}}.$$

By dimensional considerations it is clear that the characteristic function of any dyadic cube I of side length 2^{-k} can be written as a unique linear combination of Haar functions of frequency at most 2^{k-1} supported in the dyadic unit cube containing I. It therefore suffices to show that for every $f \in C^1_c(\mathbb{R}^d)$ we have $\|f - f_N\|_{B^{1/p}_{p,\infty}} \to 0$, where we choose

$$f_N = \sum_{I \in \mathscr{D}_N} f(c_I) \mathbb{1}_I,$$

and c_I denotes the center of I. Let $\mathcal{I} = \mathcal{I}(f)$ be the family of $I \in \mathscr{D}_N$ which intersect the support of f. Clearly for $f \in C^1_c$ we have

$$\|f - f_N\|_p \le \sqrt{d} 2^{-N} 2^{-Nd/p} \|f'\|_\infty (\#\mathcal{I}(f))^{1/p} \lesssim_f 2^{-N} \tag{64}$$

so that $\|f - f_N\|_p \to 0$ for $N \to \infty$. For the main term it suffices to show that

$$\sup_{h \ne 0} \frac{\|\Delta_h(f - f_N)\|_p}{|h|^{1/p}} \lesssim 2^{-N(1-\frac{1}{p})}, \tag{65}$$

and recall that we are assuming $p > 1$.

For $j > N$ we define the sets

$$\mathcal{U}_{N,j} = \Big\{(y_1, \ldots, y_d) \in \mathbb{R}^d : \min_{1 \le i \le d} \text{dist}(y_i, 2^{-N}\mathbb{Z}) \le 2^{-j-1}\Big\}. \tag{66}$$

Assume that $2^{-j-2} \le |h|_\infty < 2^{-j-1}$, for some $j > N$. If $I \in \mathscr{D}_N$ then

$$x \in I \setminus \mathcal{U}_{N,j} \quad \text{implies} \quad x + h \in I,$$

and thus $\Delta_h f_N(x) = 0$. So we have

$$\|\Delta_h(f - f_N)\|_p^p = A_N(h) + B_N(h)$$

where

$$A_N(h) = \int_{\mathcal{U}_{N,j}} |\Delta_h[f_N - f](x)|^p dx,$$

$$B_N(h) = \int_{\mathcal{U}_{N,j}^{\complement}} |f(x+h) - f(x)|^p \, dx \, .$$

In the second term we use $|\Delta_h f(x)| \le |h| \int_0^1 |\nabla f(x+sh)| ds$ to obtain $B_N(h) \le \|\nabla f\|_p^p \, |h|^p$ and thus

$$\sup_{|h|<2^{-N-2}} B_N(h)/|h| \lesssim_f 2^{-N(p-1)}.$$

For the term $A_N(h)$ we use that $\|f - f_N\|_\infty \le C_f 2^{-N}$, and also that f is compactly supported, and obtain the estimate

$$\begin{aligned} A_N(h) &\lesssim \sum_{I\in\mathscr{D}_N} \int_{I\cap\mathcal{U}_{N,j}} |f(x+h) - f_N(x+h)|^p + |f(x) - f_N(x)|^p \, dx \\ &\lesssim_f 2^{-Np} 2^N 2^{-j}, \end{aligned}$$

since $|I \cap \mathcal{U}_{N,j}| \approx 2^{-j} 2^{-(d-1)N}$. Hence $\sup_{|h|\approx 2^{-j}} A_N(h)/|h| \lesssim_f 2^{-N(p-1)}$. Putting the two estimates together we get

$$\sup_{|h|_\infty \le 2^{-N-2}} \frac{\|\Delta_h(f - f_N)\|_p}{|h|^{1/p}} \lesssim_f 2^{-N(1-\frac{1}{p})}.$$

Finally, if $|h| \gtrsim 2^{-N}$ we use (64) to have

$$\sup_{|h|_\infty \ge 2^{-N}} \frac{\|\Delta_h(f - f_N)\|_p}{|h|^{1/p}} \lesssim \frac{2\|f - f_N\|_p}{2^{-N/p}} \lesssim_f 2^{-N(1-\frac{1}{p})}.$$

This shows (65) and therefore $\|f - f_N\|_{B^{1/p}_{p,\infty}} \to 0$ for $p > 1$, completing the proof of the inclusion (63) when $p < \infty$. The case $p = \infty$ is immediate since for $f \in C_c^1$

$$\|f - f_N\|_{B^0_{\infty,\infty}} \lesssim \|f - f_N\|_\infty \lesssim 2^{-N}$$

by an elementary consideration. Finally, since $b^{1/p}_{p,\infty}(\mathbb{R}^d)$ is closed in $B^{1/p}_{p,\infty}(\mathbb{R}^d)$ Proposition 25 tells us that the inclusion (63) is proper. □

Remark 37 When $0 < p \le 1$, the same proof gives a version of (65), namely

$$\sup_{h \neq 0} \frac{\|\Delta_h(f - f_N)\|_p}{|h|^s} \lesssim 2^{-N(1-s)}, \quad \text{if } s < 1.$$

This can be used similarly to show that $\mathscr{H}_d$ is dense in the space $b^s_{p,\infty}$ when $d(1/p - 1) < s < 1$ and $d/(d+1) < p \le 1$.

Remark 38 Since $B^{1/p}_{p,\infty}$ is not separable, not every function $f \in B^{1/p}_{p,\infty}$ can be approximated by Haar expansions in the norm topology. However, (local) weak* convergence does hold, with norm-uniformly bounded partial sums. More precisely, if $1 < p \le \infty$ and $\chi \in C_c^\infty(\mathbb{R}^d)$, then

$$(f - S_R^{\mathcal{U}} f)\chi \xrightarrow{\text{w}^*} 0, \quad \text{and} \quad \sup_{R \ge 1} \|\chi\, S_R^{\mathcal{U}} f\|_{B^{1/p}_{p,\infty}} \lesssim \|f\|_{B^{1/p}_{p,\infty}}$$

for all $f \in B^{1/p}_{p,\infty}$ and any strongly admissible enumeration. This is a consequence of the duality relation $B^{1/p}_{p,\infty} = (B^{-1/p'}_{p',1})^*$ and the (local) norm convergence of $S_R^{\mathcal{U}} g \to g$ in the $B^{-1/p'}_{p',1}$ norm, when $g \in B^{-1/p'}_{p',1}$, see Theorem 4. We thank the referee for raising the question of weak* convergence.

9 Partial Sums and Localization

9.1 Partial Sums and Strongly Admissible Enumerations

We shall use a partition of unity to make statements on the structure of the partial sum operators $S_R^{\mathcal{U}}$ associated with a strongly admissible enumeration $\mathcal{U}$.

Let $\varsigma \in C_c^\infty$ be supported in a 10^{-2} neighborhood of $[0,1)^d$ and so that

$$\sum_{\nu \in \mathbb{Z}^d} \varsigma(\cdot - \nu) \equiv 1. \tag{67}$$

We shall denote $\varsigma_\nu = \varsigma(\cdot - \nu)$, $\nu \in \mathbb{Z}^d$.

In the sequel we will use the notation from Definition 2 and below. It is convenient to denote $\mathbb{E}_{-1}(g) \equiv 0$ and $T_{-1}[g, \mathfrak{a}] = \sum_{\mu \in \mathbb{Z}^d} a_\mu \langle g, h^{\mathbf{0}}_{0,\mu}\rangle h^{\mathbf{0}}_{0,\mu}$.

Lemma 39 *Let $\mathcal{U}$ be a strongly admissible enumeration of $\mathscr{H}_d$. Then, for every $R \in \mathbb{N}$ and $\nu \in \mathbb{Z}^d$ there is an integer $N_\nu = N_\nu(R) \ge -1$ and sequences $\mathfrak{a}^{\kappa,\nu}$, $0 \le \kappa \le b$, whose terms belong to $\{0, 1\}$, such that for all locally integrable functions g we have*

$$S_R^{\mathcal{U}}[g\varsigma_\nu] = \mathbb{E}_{N_\nu}[g\varsigma_\nu] + \sum_{\kappa=0}^{b} T_{N_\nu+\kappa}[g\varsigma_\nu, \mathfrak{a}^{\kappa,\nu}]. \tag{68}$$

Proof We write

$$S_R^{\mathcal{U}}[g\varsigma_\nu] = \sum_{n=1}^{R} u_n^*(g\varsigma_\nu)u_n = \sum_{n=1}^{R} 2^{k(n)d}\langle g\varsigma_\nu, h_{k(n),\mu(n)}^{\epsilon(n)}\rangle h_{k(n),\mu(n)}^{\epsilon(n)}. \tag{69}$$

Note that if $u_n^*(g\varsigma_\nu) \neq 0$ then necessarily u_n is supported in I_ν^{**}. Let

$$K_\nu = \max\{k(n) : \operatorname{supp}(h_{k(n),\mu(n)}^{\epsilon(n)}) \subset I_\nu^{**},\ n = 1, \dots, R\}.$$

If $K_\nu \leq b$ the asserted formula holds with $N_\nu = -1$. We therefore may assume $K_\nu > b$.

We let $n_\nu^* \in [1, R]$ such that $k(n_\nu^*) = K_\nu$. Now if $h_{k',\mu'}^{\epsilon'}$ is any other Haar function supported in I_ν^{**} there is a unique $n' \in \mathbb{N}$ such that $h_{k',\mu'}^{\epsilon'} = h_{k(n'),\mu(n')}^{\epsilon(n')}$. If in addition $k' \leq K_\nu - b$ (in other words if for $u_{n'} = h_{k(n'),\mu(n')}^{\epsilon(n')}$ we have that $|\operatorname{supp}(u_{n'})| \geq |\operatorname{supp}(u_{n_\nu^*})|2^b$) then by the admissibility condition we must have $n' \leq n_\nu^*$, in particular $n' \leq R$. That means that all Haar functions with frequency 2^k and $k \leq K_\nu - b$ which are supported in I_ν^{**} arise in the expansion (69). All other Haar functions that arise in this expansion have frequencies 2^k with $K_\nu - b + 1 \leq k \leq K_\nu$. This establishes the assertion with $N_\nu = K_\nu - b + 1$. The functions $\mathfrak{a}^{\kappa,\nu}$ defined on $\mathbb{Z}^d \times \Upsilon$ take values in $\{0, 1\}$. □

Remark 40 Formula (68) can be extended to all $g \in B_{p,q}^s$, when the indices (s, p, q) are as in Theorems 8 and 9. In that case, one must interpret

$$S_R^{\mathcal{U}}(g) = \sum_{j=0}^{\infty} S_R^{\mathcal{U}}(L_j\Lambda_j g);$$

see Remarks 19 and 24.

Proposition 41 *Suppose that*

$$\sup_{N\geq 0} \|\mathbb{E}_N\|_{B_{p,q}^s \to B_{p,q}^s} + \sup_{\substack{N\geq -1 \\ \|\mathfrak{a}\|_\infty \leq 1}} \|T_N[\cdot, \mathfrak{a}]\|_{B_{p,q}^s \to B_{p,q}^s} < \infty. \tag{70}$$

Then, for every strongly admissible enumeration $\mathcal{U}$ and every cube Q it holds

$$\sup_{R\in\mathbb{N}} \operatorname{Op}(S_R^{\mathcal{U}}, B_{p,q}^s, Q) < \infty. \tag{71}$$

Moreover, $\mathcal{U}$ is a local basic sequence of $B_{p,q}^s(\mathbb{R}^d)$, that is

$$\lim_{R\to\infty} \|\chi \cdot (S_R^{\mathcal{U}} f - f)\|_{B_{p,q}^s} = 0, \tag{72}$$

for all $\chi \in C_c^\infty(\mathbb{R}^d)$ and all $f \in \overline{\operatorname{span} \mathscr{H}_d}^{B_{p,q}^s}$.

Proof Using Lemma 39, the bound in (71) follows from (70). We now show the last assertion. Let $\chi \in C_c^\infty(\mathbb{R}^d)$ and $f \in \overline{\operatorname{span} \mathscr{H}_d}^{B^s_{p,q}}$. Suppose that $\operatorname{supp} \chi \subset (-N, N)^d$, and pick any $\tilde{\chi} \in C_c^\infty$ such that $\tilde{\chi} \equiv 1$ in $[-N, N]^d$ and $\operatorname{supp} \tilde{\chi}$ contained in $Q := (-2N, 2N)^d$. Observe that

$$u_n^*(g) = u_n^*(\tilde{\chi}\, g), \quad \text{if } g \in B^s_{p,q} \text{ and } \operatorname{supp} u_n \subset [-N, N]^d, \tag{73}$$

so we also have

$$\chi \cdot S_R^{\mathcal{U}}[g] = \chi \cdot S_R^{\mathcal{U}}[\tilde{\chi} g], \quad \forall\, g \in B^s_{p,q}. \tag{74}$$

Given $\varepsilon > 0$, let $h \in \operatorname{span} \mathscr{H}_d$ be such that $\|f - h\|_{B^s_{p,q}} < \varepsilon/(1 + A)$, with A the constant in (71). Let $R_0 = R_0(h)$ be such that $S_R^{\mathcal{U}}[h] = h$ for $R \geq R_0$. Then, for all such R we have

$$\begin{aligned}
\big\|\chi \cdot (S_R[f] - f)\big\|_{B^s_{p,q}} &= \big\|\chi \cdot \big(S_R[f - h] + h - f\big)\big\|_{B^s_{p,q}} \\
&\lesssim \big\|\chi \cdot S_R[\tilde{\chi}(f - h)]\big\|_{B^s_{p,q}} + \big\|\chi \cdot (h - f)\big\|_{B^s_{p,q}} \\
&\lesssim (A + 1)\, \big\|f - h\big\|_{B^s_{p,q}} < \varepsilon,
\end{aligned}$$

where in the second line we have used (74) with $g = f - h$. □

9.2 *Bourdaud Localizations of Besov Spaces*

In the unbounded setting of $\mathbb{R}^d$, the $B^s_{p,q}$-norms do not satisfy "localization properties" when $p \neq q$; see e.g. the discussion in [10, p. 66]. At the endpoint cases considered here, this creates a difficulty when trying to derive 'global' Schauder basis properties from the local ones in the previous subsection. This difficulty is not present in the case of $F^s_{p,q}$ spaces; see [5, 6].

To handle this problem one may consider the class of ℓ^p-local Besov spaces introduced by G. Bourdaud [2]

$$(B^s_{p,q})_{\ell^p} = \Big\{ f \in S' : \|f\|_{(B^s_{p,q})_{\ell^p}} = \Big[\sum_{\nu \in \mathbb{Z}^d} \|\varsigma(\cdot - \nu) \cdot f\|^p_{B^s_{p,q}}\Big]^{1/p} < \infty \Big\} \tag{75}$$

where $\varsigma \in C_c^\infty(\mathbb{R}^d)$ with $\sum_{\nu \in \mathbb{Z}^d} \varsigma(\cdot - \nu) \equiv 1$ as in (67). In [2] (see also [17, 2.4.7]) it is shown that this definition does not depend on the particular choice of ς, and that $(B^s_{p,q})_{\ell^p} = B^s_{p,q}$ if and only if $p = q$. Moreover one has the embeddings

$$B^s_{p,q} \hookrightarrow (B^s_{p,q})_{\ell^p} \text{ if } 0 < q \leq p, \tag{76}$$

$$(B^s_{p,q})_{\ell^p} \hookrightarrow B^s_{p,q} \text{ if } p \leq q \leq \infty. \tag{77}$$

Using this notation we can prove the following (Fig. 4).

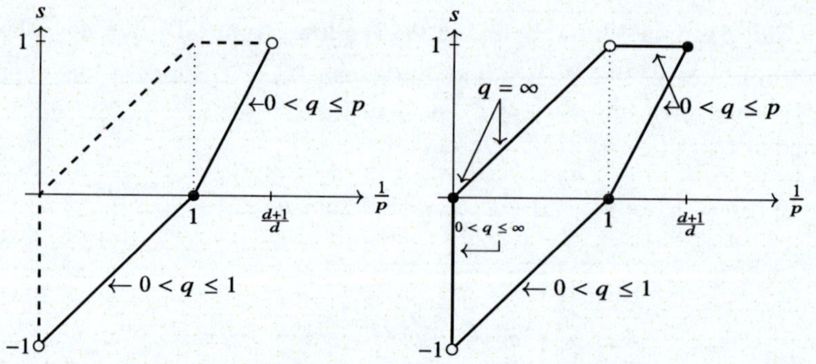

Fig. 4 The left caption shows the region in which strongly admissible enumerations form a Schauder basis of the Bourdaud localization $(B^s_{p,q})_{\ell^p}$; here always $q < \infty$. The right caption shows the corresponding region for the basic sequence property

Theorem 42 *Let $s \in \mathbb{R}$ and $0 < p, q \le \infty$. Suppose that (70) holds. Then, every strongly admissible enumeration $\mathcal{U}$ of $\mathscr{H}_d$ is a basic sequence of $\big(B^s_{p,q}\big)_{\ell^p}$. Moreover, $\mathcal{U}$ is a Schauder basis of $\big(B^s_{p,q}\big)_{\ell^p}$ in each of the cases (i) to (iv) in Theorem 4.*

Proof For the first assertion it suffices to show that the operator norms of $S_R \equiv S_R^{\mathcal{U}}$ in $(B^s_{p,q})_{\ell^p}$ are uniformly bounded in R. To do so we use the assumption (70), together with Lemma 39.

Observe first that $\varsigma_{\nu'} S_R(f\varsigma_\nu) = 0$ whenever $|\nu - \nu'|_\infty \ge 3$. Hence

$$\begin{aligned}\|S_R f\|_{(B^s_{p,q})_{\ell^p}} &= \Big(\sum_{\nu'} \big\|\varsigma_{\nu'} S_R\big(\sum_\nu \varsigma_\nu f\big)\big\|^p_{B^s_{p,q}}\Big)^{\frac{1}{p}} \\ &\lesssim \Big(\sum_{\nu'} \sum_{\nu:|\nu-\nu'|_\infty \le 2} \Big\|\varsigma_{\nu'} S_R(f\varsigma_\nu)\Big\|^p_{B^s_{p,q}}\Big)^{1/p} \\ &\lesssim \Big(\sum_\nu \|S_R(f\varsigma_\nu)\|^p_{B^s_{p,q}}\Big)^{1/p},\end{aligned}$$

using in the last step that $\varsigma_{\nu'}$ is a uniform multiplier in $B^s_{p,q}$; see [17, 4.2.2]. Then Lemma 39 and (70) give

$$\begin{aligned}\|S_R f\|_{(B^s_{p,q})_{\ell^p}} &\lesssim \Big(\sum_\nu \|\mathbb{E}_{N_\nu}(f\varsigma_\nu)\|^p_{B^s_{p,q}} + \big\|\sum_{\kappa=0}^{b} T_{N_\nu}[f\varsigma_\nu, \mathfrak{a}^{\kappa,\nu}]\big\|^p_{B^s_{p,q}}\Big)^{1/p} \\ &\lesssim_b \Big(\sum_\nu \|f\varsigma_\nu\|^p_{B^s_{p,q}}\Big)^{1/p} = \|f\|_{(B^s_{p,q})_{\ell^p}}.\end{aligned}$$

This shows the first part. Also, the Schauder basis property will hold if and only if span $\mathscr{H}_d$ is dense in $(B^s_{p,q})_{\ell^p}$.

We now show that density holds in the range of Theorem 4. Since $p < \infty$, for each $f \in (B^s_{p,q})_{\ell^p}$ and $\varepsilon > 0$ there is some $g \in B^s_{p,q}$ with compact support such that $\|f - g\|_{(B^s_{p,q})_{\ell^p}} < \varepsilon$. Moreover, in the asserted range span $\mathscr{H}_d$ is dense in $B^s_{p,q}$, so if supp $g \subset (-N, N)^d = Q$, then by Proposition 41 we may find a sufficiently large R such that $\|g - S_R g\|_{B^s_{p,q}} < \varepsilon / |Q|^{1/p}$. Since also $\operatorname{supp}(S_R g) \subset Q$ we deduce that

$$\|g - S_R g\|_{(B^s_{p,q})_{\ell^p}} \lesssim |Q|^{1/p} \|g - S_R g\|_{B^s_{p,q}} < \varepsilon,$$

which completes the proof. □

Finally, we gather as a corollary the positive Schauder results in the original scale of $B^s_{p,q}$ spaces.

Corollary 43 *Every strongly admissible enumeration $\mathcal{U}$ of $\mathscr{H}_d$ is a Schauder basis of $B^s_{p,q}(\mathbb{R}^d)$ in each of the cases (i), (ii), (iii) in Theorem 3.*

Proof When $q = p$ the result is a consequence of the identity $B^s_{p,p} = (B^s_{p,p})_{\ell^p}$ and the previous theorem. This covers the case (iii) in Theorem 3. For the other cases, in which $(1/p, s)$ lies in the interior of the pentagon $\mathfrak{P}$, one proceeds by real interpolation as follows. Pick two numbers s_0, s_1 such that $s_0 < s < s_1$ and $(1/p, s_i) \in \mathfrak{P}$, $i = 0, 1$. Then, for some $\theta \in (0, 1)$ we have

$$B^s_{p,q} = \big(B^{s_0}_{p,p}, B^{s_1}_{p,p}\big)_{\theta,q}, \quad 0 < q \le \infty.$$

Then the uniform boundedness of $S^{\mathcal{U}}_R$ on $B^s_{p,q}$ follows by interpolation from the diagonal cases. □

9.3 *Error Estimates for Compactly Supported Functions*

Here we include a technical result related to localization which will be used in the proof of Theorem 46 below.

Let f be supported in a dyadic cube Q with sidelength $\ell(Q) \ge 1$. Since the function $\Lambda_j f$ does not have compact support, the terms $L_k \mathbb{E}_N L_j \Lambda_j f(x)$ will contribute for x far away from the cube. We give a crude estimate which will suffice for our later application.

Let $\zeta \in C_c^\infty(\mathbb{R}^d)$ be supported on $(-2,2)^d$ and such that $\zeta \equiv 1$ on $[-\frac{3}{2}, \frac{3}{2}]^d$. If y_Q is the center of Q, we define

$$\zeta_Q(y) = \zeta\big((y - y_Q)/\ell(Q)\big).$$

Clearly $\zeta_Q f = f$ for every distribution f supported in Q. Moreover, this property continues to hold with ζ_Q replaced by $\tilde{\zeta}_Q$, where $\tilde{\zeta}(x) = \zeta(2x)$. For $n \geq 1$ we let

$$\zeta_{Q,n}(y) = \zeta\big(2^{-n}(y - y_Q)/\ell(Q)\big) - \zeta\big(2^{-n+1}(y - y_Q)/\ell(Q)\big).$$

Note that $\zeta_{Q,n}$ has support in $\big\{\frac{3}{4} \cdot 2^n \ell(Q) < |y - y_Q|_\infty < 2^{n+1}\ell(Q)\big\}$, and that $\sum_{n\geq 1} \zeta_{Q,n} \equiv 1$.

Lemma 44 *Let $s \leq 1$, $0 < p \leq 1$ and $0 < q \leq \infty$. Then, for every $M_1 > 1$ there exists a constant $C_{M_1} > 0$ such that, if $f \in B^s_{p,q}(\mathbb{R}^d)$ is supported in a cube Q with size $|Q| \geq 1$, then*

$$\|L_k \mathbb{E}_N L_j[\zeta_{Q,n} \Lambda_j f]\|_p \leq C_{M_1}\, 2^{-k/p} 2^{-jM_1} 2^{-nM_1} \|f\|_{B^s_{p,q}}, \tag{78}$$

for all $n \geq 1$, $k \geq 0$, $j \geq N$ and $N \geq 1$.

The same holds if $\mathbb{E}_N$ is replaced by $T_N[\cdot, \mathfrak{a}]$ with $\|\mathfrak{a}\|_\infty \leq 1$.

Proof Let $\phi_j(x) = 2^{jd}\phi(2^j x)$ be the convolution kernel of Λ_j, with $\phi \in \mathcal{S}$. Let

$$F_{j,n}(x) := \zeta_{Q,n}(x)\Lambda_j f(x) = \zeta_{Q,n}(x)\big\langle \phi_j(x - \cdot)\tilde{\zeta}_Q(\cdot), f\big\rangle$$

where we have used $f = f\tilde{\zeta}_Q$ for the second equation and the pairing $\langle\cdot,\cdot\rangle$ is in the sense of tempered distributions.

Pick a large $\gamma \in 2\mathbb{N}$ such that $B^s_{p,q} \subset B^{-\gamma}_{2,2}$ (e.g., $\gamma > d(\frac{1}{p} - \frac{1}{2}) - s$). Then by duality

$$\begin{aligned} |F_{j,n}(x)| &\lesssim |\zeta_{Q,n}(x)|\, \big\|(I - \Delta)^{\gamma/2}\big(\phi_j(x - \cdot)\tilde{\zeta}_Q(\cdot)\big)\big\|_2\, \|f\|_{B^{-\gamma}_{2,2}} \\ &\lesssim_{M_2} |Q|^{1/2}\, 2^{j(d+\gamma)}\, (1 + 2^{j+n}\ell(Q))^{-M_2}\, \|f\|_{B^s_{p,q}}. \end{aligned} \tag{79}$$

Observe that $F_{j,n}$, and hence $L_k \mathbb{E}_N L_j[F_{j,n}]$, are all supported in a set of diameter $C2^n\ell(Q)$. Then, if $k \leq N$ we have

$$\begin{aligned} \|L_k \mathbb{E}_N L_j(F_{j,n})\|_p &\lesssim (2^{nd}|Q|)^{1/p} \|L_k \mathbb{E}_N L_j(F_{j,n})\|_\infty \\ &\lesssim (2^{nd}|Q|)^{1/p} \|F_{j,n}\|_\infty. \end{aligned}$$

Inserting the bound (79) into this expression, with a sufficiently large M_2, and using that $k \leq N \leq j$, one easily obtains (78).

Assume now that $k > N$. We may use Proposition 12.i to obtain

$$\left\| L_k \mathbb{E}_N L_j(F_{j,n}) \right\|_p \lesssim 2^{-\frac{k}{p}} 2^{j(\frac{d}{p}-d)} 2^{N(d-\frac{d-1}{p})} \|\mathcal{M}_j F_{j,n}\|_p.$$

By the support properties of $F_{j,n}$ we have

$$\|\mathcal{M}_j F_{j,n}\|_p \lesssim (2^{nd}|Q|)^{1/p} \|F_{j,n}\|_\infty,$$

so again, using (79) with a sufficiently large M_2, and the assumption $N \le j$, one easily derives (78). □

10 The Case $s = d(\frac{1}{p} - 1)$ when $q > p$

In this section we restrict to the cases $q > p$ in the line $s = d/p - d$. We shall see that the individual operators $\mathbb{E}_N$ are not bounded, and hence positive results are not expected in this range.

Theorem 45 *Let $0 < p \le 1$. If $q > p$ then the operators $\mathbb{E}_N$ are unbounded on $B^{d/p-d}_{p,q}(\mathbb{R}^d)$.*

We shall actually prove something stronger, namely optimal estimates for the local version of the operator norms $\mathrm{Op}(\mathbb{E}_N, B^s_{p,q}, Q)$ defined in (10). This may be of interest in the context of Besov spaces in bounded domains; see Remark 6. We remark that Oswald [8] also proved some lower bounds in a local setting which grow with N. The following theorem provides optimal growth rates.

Theorem 46

(i) If $0 < p \le 1$ and $p \le q \le \infty$, then there is a constant $c_1 = c_1(p, q) > 0$ so that

$$\mathrm{Op}\left(\mathbb{E}_N, B^{d/p-d}_{p,q}, Q\right) \ge c_1 \, (2^{Nd}|Q|)^{\frac{1}{p}-\frac{1}{q}}.$$

(ii) If in addition $\frac{d}{d+1} \le p \le q \le 1$, then there is a constant $c_2 = c_2(p, q)$, such that for any dyadic cube Q with side length ≥ 1 and any $N > 10$

$$c_1 \le \frac{\mathrm{Op}\left(\mathbb{E}_N, B^{d/p-d}_{p,q}, Q\right)}{(2^{Nd}|Q|)^{\frac{1}{p}-\frac{1}{q}}} \le c_2. \tag{80}$$

Remark 47 From [16, 2.11.3] it is known that, when $0 < p \le 1$ and $1 < q < \infty$, it holds

$$\left(B^{d/p-d}_{p,q}\right)^* = B^0_{\infty,q'}.$$

As $q' < \infty$, this space does not contain the dual functionals u_n^*. In particular, the restriction in q in part (ii) of Theorem 46 is natural, since for $q > 1$ and $Q_0 = (0,1)^d$ we have

$$\mathrm{Op}\left(\mathbb{E}_0, B^{d/p-d}_{p,q}, Q_0\right) = \|\mathbb{1}_{Q_0}\|_{B^{d/p-d}_{p,q}} \|\mathbb{1}_{Q_0}\|_{(B^{d/p-d}_{p,q})^*} = \infty;$$

see also [8, Thm 2.ii.a].

10.1 Proof of Lower Bounds in Theorem 46

We fix $0 < p \le 1$ and choose an positive integer $M > d/p - d$.

Let $\eta \in C_c^\infty(\mathbb{R})$ be an odd function, supported on $(-1/2, 1/2)$, and such that $\int_0^{1/2} \eta(t)dt = 1$ and $\int_0^{1/2} t^n \eta(t)dt = 0$ for $n = 1, \ldots, M$. Let further

$$g_l(x_1, \ldots, x_d) = 2^{ld} \prod_{i=1}^{d} \eta(2^l x_i), \tag{81}$$

so that $\int g_l(x) P_M(x) dx = 0$ whenever P_M is a polynomial of degree $\le M$. By the properties of η, if $l \ge N$ we have

$$\mathbb{E}_N(g_l)(x) = 2^{Nd} \prod_{i=1}^{d} \left(\mathbb{1}_{[0,2^{-N})}(x_i) - \mathbb{1}_{[-2^{-N},0)}(x_i)\right) =: h_N(x). \tag{82}$$

Notice that h_N is not itself a Haar function, but up to a factor $(-2)^d$, it is a translate of a Haar function with Haar frequency 2^{N-1}. Moreover, we also have

$$\mathbb{E}_N[g_l(\cdot - \nu)] = h_N(\cdot - \nu), \quad \text{if } \nu \in 2^{-N}\mathbb{Z}^d, \quad l \ge N.$$

Let $\{\mathfrak{z}_m\}_{m=1}^\infty$ be an enumeration of $\mathbb{Z}^d$, and define

$$f_N(x) = \sum_{m=1}^{\infty} a_m g_{N+m}(x - 2^{-N+5}\mathfrak{z}_m). \tag{83}$$

Observe that the summands have disjoint supports. Also

$$\mathbb{E}_N f_N = \sum_{m=1}^{\infty} a_m h_N(\cdot - 2^{-N+5}\mathfrak{z}_m). \tag{84}$$

We claim that

$$\|f_N\|_{B^{d/p-d}_{p,q}} \lesssim \Big(\sum_{m=1}^{\infty} |a_m|^q\Big)^{1/q} \tag{85}$$

and

$$\|\mathbb{E}_N f_N\|_{B^{d/p-d}_{p,q}} \gtrsim \Big(\sum_{m=1}^{\infty} |a_m|^p\Big)^{1/p}. \tag{86}$$

This clearly implies that $\mathbb{E}_N$ cannot be a bounded operator on $B^{d/p-d}_{p,q}(\mathbb{R}^d)$ unless $q \le p$.

We first show (86). To do so we construct specific functions Ψ_n such that

$$\|g\|_{B^{d/p-d}_{p,q}} \ge \|g\|_{B^{d/p-d}_{p,\infty}} \gtrsim \sup_{n\ge1} 2^{n(\frac{d}{p}-d)} \|\Psi_n * f\|_p. \tag{87}$$

Let $\psi \in C_c^\infty(\mathbb{R})$ be supported in $(-1/2, 1/2)$, with

$$\int \psi(t) t^l \, dt = 0, \quad l = 0, \dots, M,$$

and such that, for some $\varepsilon > 0$,

$$\psi * (\mathbb{1}_{[0,\frac12)} - \mathbb{1}_{[-\frac12,0)})(t) \ge c > 0 \quad \text{when } t \in [\tfrac12, \tfrac12 + \varepsilon]. \tag{88}$$

We then define

$$\Psi_n(x) = 2^{nd} \prod_{i=1}^{d} \psi(2^n x_i), \tag{89}$$

which has enough vanishing moments to guarantee the validity of (87); see [17, 2.5.3]. In particular,

$$\|\mathbb{E}_N f_N\|_{B^{d(\frac1p-1)}_{p,q}} \gtrsim 2^{N(\frac{d}{p}-d)} \|\Psi_{N+1} * (\mathbb{E}_N f_N)\|_p.$$

Next, using (88) one shows that, for $x \in 2^{-N+5}\mathfrak{z}_m + 2^{-N-1}[\frac12, \frac12 + \varepsilon]^d$,

$$\Psi_{N+1} * h_N(x - 2^{-N+5}\mathfrak{z}_m) \ge 2^{Nd} c^d,$$

and therefore

$$\big\|\Psi_{N+1} * h_N(\cdot - 2^{-N+5}\mathfrak{z}_m)\big\|_p \gtrsim 2^{N(d-\frac{d}{p})}. \tag{90}$$

Also the functions $\Psi_{N+1} * h_N(\cdot - 2^{-N+5}\mathfrak{z}_m)$ have disjoint supports so that

$$\begin{aligned}\big\|\Psi_{N+1} * \mathbb{E}_N f_N\big\|_p &= \Big(\sum_{m=1}^{\infty} |a_m|^p \big\|\Psi_{N+1} * h_N(\cdot - 2^{-N+5}\mathfrak{z}_m)\big\|_p^p\Big)^{1/p} \\ &\gtrsim \Big(\sum_m |a_m|^p\Big)^{1/p} 2^{-Nd(\frac{1}{p}-1)}\end{aligned}$$

and (86) follows.

To prove (85) we examine $L_j g_l$ with $l = N + m$ and use the cancellation of the convolution kernel β_j of L_j when $j \geq l$, and the cancellation of g_l for $j < l$. Here cancellation refers to M vanishing moments. As a consequence we obtain the estimate

$$|L_j g_l(x)| \lesssim \begin{cases} 2^{ld}\mathbb{1}_{[-1,1]^d}(2^l x) 2^{-M|l-j|} & \text{for } j \geq l, \\ 2^{jd}\mathbb{1}_{[-1,1]^d}(2^j x) 2^{-M|l-j|} & \text{for } j \leq l; \end{cases} \tag{91}$$

see a similar argument in the proof of [5, Lemma 2.2]. From here one easily obtains

$$2^{jd(\frac{1}{p}-1)}\|L_j g_l\|_p \lesssim \left\{\begin{matrix} 2^{-(M-d(\frac{1}{p}-1))|l-j|} & \text{if } j \geq l \\ 2^{-M|l-j|} & \text{if } j \leq l \end{matrix}\right\} \leq 2^{-\delta|l-j|}, \tag{92}$$

if we set $\delta = M - d(\frac{1}{p} - 1) > 0$. This leads to

$$\begin{aligned}2^{jd(\frac{1}{p}-1)}\|L_j f_N\|_p &\leq 2^{jd(\frac{1}{p}-1)}\Big(\sum_{m=1}^{\infty} |a_m|^p \, \|L_j g_{N+m}(\cdot - 2^{-N+5}\mathfrak{z}_m)\|_p^p\Big)^{1/p} \\ &\lesssim \Big(\sum_{m=1}^{\infty} |a_m|^p 2^{-|N+m-j|\delta p}\Big)^{1/p},\end{aligned}$$

and consequently,

$$\Big(\sum_{j\geq 0}[2^{jd(\frac{1}{p}-1)}\|L_j f_N\|_p]^q\Big)^{1/q} \lesssim \Big(\sum_{j\geq 0}\Big(\sum_{m=1}^{\infty} |a_m|^p \, 2^{-|N+m-j|\,\delta p}\Big)^{q/p}\Big)^{1/q}.$$

Since $q \geq p$ we can apply the triangle inequality in $\ell^{q/p}$ to bound the previous expression, by

$$\Big(\sum_{j\geq 0}\Big(\sum_{n\in\mathbb{Z}}|a_{n+j-N}|^p 2^{-|n|\delta p}\Big)^{q/p}\Big)^{1/q}$$

$$\lesssim \Big(\sum_{j\geq 0}\sum_{n\in\mathbb{Z}}|a_{n+j-N}|^q 2^{-|n|q\frac{\delta}{2}}\Big)^{1/q} \lesssim \Big(\sum_{m=1}^{\infty}|a_m|^q\Big)^{1/q}.$$

This proves (85).

Finally, to establish the lower bound in Theorem 46, we simply chose

$$a_m = \begin{cases} 1 \text{ if } 2^{-N+5}\mathfrak{z}_m \in Q \\ 0 \text{ if } 2^{-N+5}\mathfrak{z}_m \notin Q. \end{cases}$$

Since $\{\mathfrak{z}_m\}$ enumerates $\mathbb{Z}^d$ and $\#(2^{-N+5}\mathbb{Z}^d \cap Q) \approx 2^{Nd}|Q|$ we obtain

$$\|f_N\|_{B^{d(\frac{1}{p}-1)}_{p,q}} \lesssim (2^{Nd}|Q|)^{1/q}$$

from (85), and

$$\|\mathbb{E}_N f_N\|_{B^{d(\frac{1}{p}-1)}_{p,q}} \gtrsim (2^{Nd}|Q|)^{1/p}$$

from (86). This establishes the desired lower bound for all $q \geq p$. □

10.2 *Proof of Upper Bounds in Theorem 46 (ii)*

In what follows let Q be a dyadic cube of side length ≥ 1. We assume $\frac{d}{d+1} \leq p \leq q \leq 1$.

We use the global estimates (18), (19) and examine the two expressions on the right hand side of (19) corresponding to the cases $j \leq N$ and $j \geq N$. The terms for $j \leq N$ cause no problem. Namely, by Propositions 16 and 17 we have (for $p \leq q$)

$$\Big(\sum_{k=0} 2^{k(\frac{d}{p}-d)r}\Big\|\sum_{j\leq N} L_k\mathbb{E}_N^{\perp} L_j\Lambda_j f\Big\|_p^r\Big)^{1/r} \lesssim \|f\|_{B^{d/p-d}_{p,\infty}} \text{ if } \frac{d}{d+1} < p \leq 1,$$

and in the endpoint $q \ge p = \frac{d}{d+1}$ (when $d/p - d = 1$) we have

$$\Big(\sum_{k=0} 2^{kr}\Big\|\sum_{j\le N} L_k\mathbb{E}_N^{\perp} L_j\Lambda_j f\Big\|_p^r\Big)^{1/r} \lesssim \Big(\sum_{j=0}^{N} 2^{jp}\|\Lambda_j f\|_p^p\Big)^{1/p} \lesssim N^{\frac1p-\frac1q}\|f\|_{B^1_{p,q}}, \quad p=\frac{d}{d+1}$$

where we have applied Hölder's inequality. This global bound is far better than what is need for the conclusion and this part satisfies the target upper bound in (80).

Hence it suffices to prove, for f supported in Q, the following bound

$$\Big(\sum_{k=0}^{\infty} 2^{kd(\frac1p-1)r}\Big\|\sum_{j\ge N+1} L_k\mathbb{E}_N L_j\Lambda_j f\Big\|_p^r\Big)^{1/r} \lesssim (2^{Nd}|Q|)^{\frac1p-\frac1q}\Big(\sum_{j=0}^{\infty} 2^{jd(\frac1p-1)q}\|\Lambda_j f\|_p^q\Big)^{1/q}, \tag{93}$$

for any $r > 0$. Notice that Lemma 44 reduces matters to show the following inequalities.

$$\Big(\sum_{k\ge N+1} 2^{kd(\frac1p-1)r}\Big\|\sum_{j\ge N+1} L_k\mathbb{E}_N L_j[\zeta_Q\Lambda_j f]\Big\|_p^r\Big)^{1/r} \lesssim (2^{Nd}|Q|)^{\frac1p-\frac1q}\Big(\sum_{j=0}^{\infty} 2^{jd(\frac1p-1)q}\|\Lambda_j f\|_p^q\Big)^{1/q} \tag{94}$$

and

$$\Big(\sum_{k\le N} 2^{kd(\frac1p-1)r}\Big\|\sum_{j\ge N+1} L_k\mathbb{E}_N L_j[\zeta_Q\Lambda_j f]\Big\|_p^r\Big)^{1/r} \lesssim (2^{Nd}|Q|)^{\frac1p-\frac1q}\Big(\sum_{j=0}^{\infty} 2^{jd(\frac1p-1)q}\|\Lambda_j f\|_p^q\Big)^{1/q}. \tag{95}$$

We first prove (94). Instead of using Proposition 15 directly we shall use a modification of its proof in [5, Proposition 2.1(i)]; we first recall some notation from that paper.

We let $\mathscr{D}_N$ be the collection of dyadic cubes of sidelength 2^{-N}. For $j > N$ we define $\mathcal{U}_{N,j}$ as in (66), that is a 2^{-j-1}-neighborhood of the set $\cup_{I\in\mathscr{D}_N}\partial I$. For $I \in \mathscr{D}_N$ and $l > N$ we denote by $\mathscr{D}_l[\partial I]$ the set of all $J \in \mathscr{D}_l$ such that $\bar{J}\cap\partial I \ne \emptyset$.

Likewise, $\mathscr{D}_N(I)$ denotes the collection of cubes $I' \in \mathscr{D}_N$ with $\bar{I} \cap \bar{I}' \neq \emptyset$, that is the collection of neighboring cubes of I.

We use the following result taken from [5, Lemma 2.3].

Lemma 48

(i) Let $k > N \geq 1$ and G be locally integrable. Then

$$L_k(\mathbb{E}_N G)(x) = 0, \quad \text{for all } x \in \mathcal{U}_{N,k}^{\complement} = \mathbb{R}^d \setminus \mathcal{U}_{N,k}\,. \tag{96}$$

(ii) Let $j > N \geq 1$, and F locally integrable.

$$\left|\mathbb{E}_N(L_j F)\right| \lesssim 2^{(N-j)d} \sum_{I \in \mathscr{D}_N} \sum_{J \in \mathscr{D}_{j+1}[\partial I]} \|F\|_{L^\infty(J)} \mathbb{1}_I. \tag{97}$$

Proof (of (94)) Observe that $F_j := \zeta_Q \Lambda_j f$ and the functions $L_k \mathbb{E}_N L_j[F_j]$ are all supported in a fixed C-dilate of Q (with say $C = 10$). By Lemma 48.i, $L_k \mathbb{E}_N[L_j F_j](x) = 0$ if $x \in \mathcal{U}_{N,k}^{\complement}$. We derive a pointwise estimate if $x \in \mathcal{U}_{N,k} \cap I$ for some $I \in \mathscr{D}_N$. From (97) and the fact that supp $\beta_k(x - \cdot)$ is contained in the union of all $I' \in \mathscr{D}_N(I)$ we have

$$\begin{aligned} |L_k \mathbb{E}_N[L_j F_j](x)| &\leq \int |\beta_k(x-y)| \left|\mathbb{E}_N(L_j F_j)(y)\right| dy \\ &\lesssim 2^{(N-j)d} \sum_{I' \in \mathscr{D}_N(I)} \sum_{J \in \mathscr{D}_{j+1}[\partial I']} \|F_j\|_{L^\infty(J)}. \end{aligned}$$

Let Q_* be the above C-dilate of Q. Then using $|\mathcal{U}_{N,k} \cap I| \approx 2^{-N(d-1)-k}$ we have

$$\begin{aligned} &\Big\| \sum_{j \geq N+1} L_k \mathbb{E}_N\big[L_j F_j\big] \Big\|_p \\ &\lesssim \Big(\sum_{\substack{I \in \mathscr{D}_N \\ I \cap Q_* \neq \emptyset}} |\mathcal{U}_{N,k} \cap I| \, |2^{Nd} \sum_{j \geq N+1} 2^{-jd} \sum_{I' \in \mathscr{D}_N(I)} \sum_{J \in \mathscr{D}_{j+1}[\partial I']} \|F_j\|_{L^\infty(J)}|^p \Big)^{1/p} \\ &\lesssim 2^{Nd} 2^{-(N(d-1)+k)/p} \Big(\sum_{\substack{I \in \mathscr{D}_N \\ I \cap Q_* \neq \emptyset}} \Big| \sum_{j \geq N+1} 2^{-jd} \sum_{J \in \mathscr{D}_{j+1}[\partial I]} \|\Lambda_j f\|_{L^\infty(J)}\Big|^p \Big)^{1/p}. \end{aligned}$$

The I-sum in the last display contains $\approx |Q|2^{Nd}$ terms. Let $p \le q \le 1$. Using Hölder's inequality in this sum we see that

$$\Big\| \sum_{j>N} L_k \mathbb{E}_N[L_j F_j] \Big\|_p \lesssim (2^{Nd}|Q|)^{\frac{1}{p}-\frac{1}{q}}$$

$$\times\, 2^{Nd} 2^{-(N(d-1)+k)/p} \Big(\sum_{I\in\mathscr{D}_N} \Big| \sum_{j>N} 2^{-jd} \sum_{J\in\mathscr{D}_{j+1}[\partial I]} \|\Lambda_j f\|_{L^\infty(J)} \Big|^q \Big)^{1/q}. \tag{98}$$

Consider the maximal function

$$\mathfrak{M}_j g(x) = \sup_{|h|_\infty \le 2^{-j+1}} |g(x+h)|.$$

Then $\|\mathfrak{M}_j \Lambda_j f\|_p \lesssim c_p \|\Lambda_j f\|_p$ for all $p > 0$. Moreover, as in [5, (22)], it holds

$$\sup_{x\in J} \mathfrak{M}_j g(x) \lesssim \Big[\fint_{J^*} |M_j g(x+h)|^p \, dh \Big]^{\frac{1}{p}},$$

where J^* is a C'-dilate of the cube $J \in \mathscr{D}_{j+1}$. Therefore,

$$\Big(\sum_{I\in\mathscr{D}_N} \Big| \sum_{j>N} 2^{-jd} \sum_{J\in\mathscr{D}_{j+1}[\partial I]} \|\Lambda_j f\|_{L^\infty(J)} \Big|^q \Big)^{1/q}$$

$$\lesssim \Big(\sum_{I\in\mathscr{D}_N} \Big| \sum_{j>N} 2^{-jd} \sum_{J\in\mathscr{D}_{j+1}[\partial I]} 2^{jd/p} \|M_j[\Lambda_j f]\|_{L^p(J^*)} \Big|^q \Big)^{1/q}.$$

Using the embeddings $\ell^p \hookrightarrow \ell^1$ (for the J-sum) and $\ell^q \hookrightarrow \ell^1$ (for the j-sum), and in the second step $\ell^{p/q} \hookrightarrow \ell^1$ (for the I-sum), the above quantity is further estimated by

$$\Big(\sum_{I\in\mathscr{D}_N} \sum_{j>N} 2^{jd(\frac{1}{p}-1)q} \Big(\sum_{J\in\mathscr{D}_{j+1}[\partial I]} \|M_j[\Lambda_j f]\|^p_{L^p(J^*)} \Big)^{q/p} \Big)^{1/q}$$

$$\lesssim \Big(\sum_{j>N} 2^{jd(\frac{1}{p}-1)q} \Big(\sum_{I\in\mathscr{D}_N} \sum_{J\in\mathscr{D}_{j+1}[\partial I]} \|M_j[\Lambda_j f]\|^p_{L^p(J^*)} \Big)^{q/p} \Big)^{1/q}$$

$$\lesssim \Big(\sum_{j>N} 2^{jd(\frac{1}{p}-1)q} \|M_j[\Lambda_j f]\|_p^q \Big)^{1/q} \lesssim \Big(\sum_{j>N} 2^{jd(\frac{1}{p}-1)q} \|\Lambda_j f\|_p^q \Big)^{1/q}.$$

Inserting this estimate into (98) we see that

$$\Big(\sum_{k>N} 2^{kd(\frac{1}{p}-1)r}\Big\|\sum_{j>N} L_k\mathbb{E}_N L_j[\zeta_Q\Lambda_j f]\Big\|_p^r\Big)^{1/r}$$

$$\lesssim (2^{Nd}|Q|)^{\frac{1}{p}-\frac{1}{q}}\Big(\sum_{k>N} 2^{(N-k)(d-\frac{d-1}{p})r}\Big)^{1/r}\Big(\sum_{j>N} 2^{jd(\frac{1}{p}-1)q}\|\Lambda_j f\|_p^q\Big)^{1/q},$$

and since the k-sum is $O(1)$ in the larger range $p > \frac{d-1}{d}$ we obtain (94) for $\frac{d}{d+1} \le p \le q \le 1$. □

Proof (of (95)) This case is simpler and can be obtained from the individual bounds of $\|L_k\mathbb{E}_N L_j[F_j]\|_p$ in Proposition 12. Recall that $F_j = \zeta_Q\Lambda_j f$ and $L_k\mathbb{E}_N L_j[F_j]$ are supported in a C-dilate of Q.

Let $k \le N$. Applying Hölder's inequality, the q-triangle inequality and Proposition 12.i we now obtain

$$2^{ks}\Big\|\sum_{j>N} L_k\mathbb{E}_N L_j[F_j]\Big\|_p \lesssim 2^{ks}\,|Q|^{\frac{1}{p}-\frac{1}{q}}\Big[\sum_{j>N}\|L_k\mathbb{E}_N L_j[F_j]\|_q^q\Big]^{\frac{1}{q}}$$

$$\lesssim |Q|^{\frac{1}{p}-\frac{1}{q}}2^{k(s+d+1-\frac{d}{q})}2^{-N}\Big(\sum_{j>N} 2^{j(\frac{d}{q}-d)q}\|\Lambda_j f\|_q^q\Big)^{1/q}.$$

We now use the extension of Young's inequality

$$\|\Lambda_j f\|_q \le 2^{j(\frac{d}{p}-\frac{d}{q})}\|\Lambda_j f\|_p; \tag{99}$$

see e.g. [16, 2.7.1/3]. As a result we obtain

$$2^{kd(\frac{1}{p}-1)}\Big\|\sum_{j>N} L_k\mathbb{E}_N L_j[\zeta_Q\Lambda_j f]\Big\|_p$$

$$\lesssim 2^{(k-N)(\frac{d}{p}-\frac{d}{q}+1)}(2^{Nd}|Q|)^{\frac{1}{p}-\frac{1}{q}}\Big(\sum_{j>N} 2^{j(\frac{d}{p}-d)q}\|\Lambda_j f\|_p^q\Big)^{1/q}.$$

Finally, we may sum over $k \le N$ using that $p \le q$, and therefore obtain (95). With this assertion, the proof of Theorem 46 is now complete. □

11 A Strongly Admissible Enumeration

We give explicit examples of strongly admissible enumerations for $\mathscr{H}_d$. We define the family of cubes

$$\mathfrak{Q}_5 = \Big\{ \prod_{i=1}^d [10\kappa_i - 5, 10\kappa_i + 5) : \kappa \in \mathbb{Z}^d \Big\}.$$

For $\ell = 0, 1, 2, \ldots$, let $\mathfrak{Q}_5(\ell)$ be a strictly increasing collection of finite families of cubes from $\mathfrak{Q}_5$ such that for each cube in $\mathfrak{Q}_5(\ell)$ all its neighboring cubes in $\mathfrak{Q}_5$ belong to $\mathfrak{Q}_5(\ell + 1)$, and such that $\mathfrak{Q}_5 = \cup_\ell \mathfrak{Q}_5(\ell)$.

For example, we may take $\mathfrak{Q}_5(\ell)$ to be family of all $Q \in \mathfrak{Q}_5$ such that $Q \subset [-10\ell - 5, 10\ell + 5)^d$.

Let $\mathcal{A}_0 = [-5, 5)^d$, and for $\ell \geq 1$ let $\mathcal{A}_\ell$ be the union of cubes in $\mathfrak{Q}_5$ which belong to $\mathfrak{Q}_5(\ell) \setminus \mathfrak{Q}_5(\ell - 1)$. For $\ell \geq 0$, let $\mathscr{H}_d(\ell, 0)$ be the family of characteristic functions of dyadic unit cubes contained in $\mathcal{A}_\ell$. For $k \geq 1, \ell \geq 0$, let $\mathscr{H}(\ell, k)$ be the family of Haar functions of mean value 0 and Haar frequency 2^{k-1} with the property that the interior of their support is contained in $\mathcal{A}_\ell$. Clearly, $\mathscr{H}_d = \cup_{\ell,k\geq 0} \mathscr{H}(\ell, k)$.

Let $N(\ell, k) = \#\mathscr{H}_d(\ell, k)$. We then have $N(0, 0) = 10^d$ and

$$N(\ell, k) = N(\ell, 0)2^{(k-1)d}(2^d - 1).$$

In the specific example above the sets $\mathcal{A}_\ell$, $\ell \geq 1$, are corridors of width 10, of the form $[-10\ell - 5, 10\ell + 5)^d \setminus [-10\ell + 5, 10\ell - 5)^d$ and we have $N(\ell, 0) = 10^d((2\ell + 1)^d - (2\ell - 1)^d)$.

We now define an admissible enumeration $\mathcal{U}$ associated with this collection. Let $P(m) = \sum_{i=0}^m N(m - i, i)$, for $m = 0, 1, 2, \ldots$, and let

$$R(m) = \sum_{j=0}^{m} P(j), \tag{100}$$

so that $R(m + 1) - R(m) = P(m + 1)$. First, for $n = 1, \ldots, R(0)$ we enumerate the functions in $\mathscr{H}(0, 0)$. Next, for $n = R(m)+1, \ldots, R(m+1)$ we enumerate the functions in $\cup_{i=0}^{m+1} \mathscr{H}(m + 1 - i, i)$ as follows: when

$$R(m) + 1 \leq n \leq R(m) + N(m + 1, 0)$$

we enumerate the functions in $\mathscr{H}_d(m+1, 0)$; subsequently, for each $\nu = 1, \ldots, m+1$, when

$$R(m) + \sum_{i=0}^{\nu-1} N(m + 1 - i, i) + 1 \leq n \leq R(m) + \sum_{i=0}^{\nu} N(m + 1 - i, i)$$

we enumerate the functions in $\mathscr{H}_d(m + 1 - \nu, \nu)$.

That is, the functions in $\mathscr{H}(\ell', k')$ occur earlier than those in $\mathscr{H}(\ell, k)$ if $\ell'+k' < \ell+k$. Moreover, $\mathscr{H}(\ell+1, k-1)$ also occurs earlier than $\mathscr{H}(\ell, k)$. Now, if u_n and $u_{n'}$ are both supported in I^{**}, the five-fold dilate of a fixed unit cube I, then their supports must be contained in cubes from $\mathcal{A}(\ell) \cup \mathcal{A}(\ell+1)$, for some smallest $\ell \geq 0$. Moreover, if $|\operatorname{supp} u_{n'}| \geq 2^d |\operatorname{supp} u_n|$, that is, $k(n') \leq k(n) - 1$, then the above observations imply that $u_{n'}$ must occur before u_n. Thus, the enumeration we just constructed for $\mathscr{H}_d$ is strongly admissible with $b = 1$.

In the next section it will be convenient to notice that, for the enumeration above, we have

$$S_{R(m)} f = \mathbb{E}_{m-\ell} f, \quad \text{if } \operatorname{supp}(f) \subset \mathcal{A}_\ell \text{ and } \ell \leq m. \tag{101}$$

In particular we have $S_{R(m)} f = \mathbb{E}_m f$ if f is supported in $(-5, 5)^d$.

12 Failure of Convergence for Strongly Admissible Enumerations

In this section we prove the remaining negative results for the Schauder basis property, as stated in Theorem 3; namely the cases

(a) $s = \frac{d}{p} - d$, $\frac{d}{d+1} \leq p \leq 1$ and $0 < q < p$
(b) $s = \frac{1}{p} - 1$, $1 < p < \infty$ and $0 < q \leq 1$.

We remark that in these cases the operators $\mathbb{E}_N$ are uniformly bounded, by Theorem 8 (iii) and (vi), and local positive results hold by Theorem 42. We disprove the possibility that the admissible enumerations in Sect. 11 may be *global* Schauder bases in $B^s_{p,q}(\mathbb{R}^d)$. It suffices to show that the corresponding partial sum operators S_R are not uniformly bounded.

12.1 The Case $0 < p \leq 1$

Proposition 49 *Let* $0 < q < p \leq 1$. *Then, for the strongly admissible enumerations defined in Sect. 11 we have*

$$\sup_{R \in \mathbb{N}} \sup \left\{ \|S_R f\|_{B^{d(\frac{1}{p}-1)}_{p,q}} : \|f\|_{B^{d(\frac{1}{p}-1)}_{p,q}} \leq 1 \right\} = \infty. \tag{102}$$

Proof We shall use a similar notation as in Sect. 10.1. Consider functions g_l as defined in (81). Fix $j \gg m$, and for $\ell \leq m$ pick $\mathfrak{z}_\ell \in \mathbb{Z}^d$ so that the threefold dilate of the cube $\mathfrak{z}_\ell + [0, 1)^d$ is contained in $\mathcal{A}_\ell$. Define

$$f_{m,j}(x) = \sum_{\ell=1}^{m} g_j(x - \mathfrak{z}_\ell). \tag{103}$$

Note that the summands $g_j(\cdot - \mathfrak{z}_\ell)$ have disjoint supports in $\mathscr{A}_\ell$. By (101)

$$S_{R(m)} f_{m,j} = \sum_{\ell=1}^{m} \mathbb{E}_{m-\ell}[g_j(\cdot - \mathfrak{z}_\ell)] = \sum_{\ell=1}^{m} h_{m-\ell}(\cdot - \mathfrak{z}_\ell),$$

where h_N was defined in (82).

Let Ψ_N be defined as in (89), so that by (90) we have

$$\|\Psi_{N+1} * h_N\|_p \gtrsim 2^{-Nd(\frac{1}{p}-1)}.$$

Then

$$\begin{aligned}
\|S_{R(m)} f_{m,j}\|_{B^{d(\frac{1}{p}-1)}_{p,q}} &\geq \Big(\sum_{N=1}^{\infty} 2^{N(\frac{d}{p}-d)q} \|\Psi_{N+1} * S_{R(m)} f_{m,j}\|_p^q\Big)^{1/q} \\
&= \Big(\sum_{N=1}^{\infty} 2^{N(\frac{d}{p}-d)q} \Big(\sum_{\ell=1}^{m} \|\Psi_{N+1} * h_{m-\ell}(\cdot - \mathfrak{z}_\ell)\|_p^p\Big)^{q/p}\Big)^{1/q} \\
&\geq \Big(\sum_{N=1}^{m-1} 2^{N(\frac{d}{p}-d)q} \|\Psi_{N+1} * h_N\|_p^q\Big)^{1/q} \gtrsim m^{1/q}.
\end{aligned}$$

Similarly, using the inequality in (92), that is

$$2^{kd(\frac{1}{p}-1)} \|L_k g_j\|_p \lesssim 2^{-|j-k|\delta},$$

for some $\delta > 0$, we may conclude that

$$\begin{aligned}
\|f_{m,j}\|_{B^{d(\frac{1}{p}-1)}_{p,q}} &\lesssim \Big(\sum_{k=0}^{\infty} 2^{kd(\frac{1}{p}-1)q} \|L_k(f_{m,j})\|_p^q\Big)^{\frac{1}{q}} \\
&= \Big(\sum_{k=0}^{\infty} 2^{kd(\frac{1}{p}-1)q} \Big(\sum_{\ell=1}^{m} \|L_k g_j(\cdot - \mathfrak{z}_\ell)\|_p^p\Big)^{\frac{q}{p}}\Big)^{\frac{1}{q}} \\
&\lesssim \Big(\sum_{k=0}^{\infty} 2^{-|j-k|\delta q}\Big)^{\frac{1}{q}} m^{1/p} \lesssim m^{1/p}.
\end{aligned}$$

Hence, the left hand side in (102) is $\gtrsim m^{1/q-1/p}$ which implies the assertion if $q < p$. □

12.2 *The Case* $1 < p < \infty$

We shall deduce this case from the previous one. First of all notice that Proposition 49 remains to be valid when $1 < p < \infty$. Indeed, the condition on p did not play any role in the proof. In particular, if the dimension $d = 1$ this implies

$$\sup_{R\in\mathbb{N}} \|S_R f\|_{B^{\frac{1}{p}-1}_{p,q} \to B^{\frac{1}{p}-1}_{p,q}} = \infty, \quad \text{when } 0 < q \le 1 < p. \tag{104}$$

To establish the same result for $d \ge 2$, we tensorize the previous example. Consider

$$F_{m,j}(x_1, x') = f_{m,j}(x_1)\,\chi(x'),$$

where $f_{m,j}$ is the 1-dimensional function in (103), and $\chi \in C_c^\infty((-2,2)^{d-1})$ with $\chi \equiv 1$ in $[-1,1]^{d-1}$. We claim that, for $s = 1/p - 1$ and $0 < q \le 1 < p$, we have

$$\|F_{m,j}\|_{B^s_{p,q}(\mathbb{R}^d)} \lesssim m^{1/p} \tag{105}$$

and

$$\|S_{R(m,d)}(F_{m,j})\|_{B^s_{p,q}(\mathbb{R}^d)} \gtrsim \|S_{R(m,1)}(f_{m,j})\|_{B^s_{p,q}(\mathbb{R})} \gtrsim m^{1/q}. \tag{106}$$

Here $R(m,d)$ are the numbers in (100), where we stress the dependence on the dimension. Notice that in either case they verify (101).

To justify these inequalities, we construct a function $\Psi \in C_c^\infty(\mathbb{R}^d)$ as in (38), that is

$$\Psi(x) = \Delta^M[\phi_0 \otimes \varphi_0](x) = \theta(x_1)\varphi_0(x') + \phi_0(x_1)\vartheta(x'), \tag{107}$$

for suitable $\phi_0, \varphi_0, \theta, \vartheta$ as in the paragraph preceding (38). We let

$$\Psi_0(x) = \phi_0(x_1)\varphi_0(x'), \quad \text{and} \quad \Psi_k(x) = 2^{kd}\Psi(2^k x), \quad k \ge 1.$$

These functions meet the required hypothesis to have

$$\|g\|_{B^s_{p,q}(\mathbb{R}^d)} \approx \Big(\sum_{k=0}^{\infty} 2^{ksq}\|\Psi_k * g\|^q_{L^p(\mathbb{R}^d)}\Big)^{\frac{1}{q}}.$$

Moreover, if we define, for $k \ge 1$,

$$\phi_k(x_1) = 2^k\theta(2^k x_1) \quad \text{and} \quad \varphi_k(x') = 2^{(d-1)k}\vartheta(2^k x'),$$

then the convolutions with ϕ_k (respectively φ_k), $k = 0, 1, 2, \ldots$, can be used to characterize the norms of $B^s_{p,q}$ in $\mathbb{R}$ (respectively in $\mathbb{R}^{d-1}$). Using this notation in (107) we can now write

$$\Psi_k = \phi_k \otimes \varphi_{0,k} + \phi_{0,k} \otimes \varphi_k, \tag{108}$$

with $\phi_{0,k}(x_1) = 2^k \phi_0(2^k x_1)$ and likewise for $\varphi_{0,k}$.

We now prove (106). First, using (101) one easily sees that

$$S_{R(m,d)}(F_{m,j}) = (S_{R(m,1)} f_{m,j}) \otimes (\mathbb{E}_m^{(d-1)}[\chi]).$$

Moreover, we claim that

$$\Psi_k * (S_{R(m,d)} F_{m,j})(x_1, x') = \phi_k * (S_{R(m,1)} f_{m,j})(x_1), \quad x' \in (\tfrac{1}{4}, \tfrac{3}{4})^{d-1}. \tag{109}$$

Indeed, this is a direct consequence of (108) and

$$\varphi_{0,k} * (\mathbb{E}_m^{(d-1)}\chi)(x') = \int \varphi_{0,k} = 1 \quad \text{and} \quad \varphi_k * (\mathbb{E}_m^{(d-1)}\chi)(x') = \int \varphi_k = 0.$$

Then (109) implies the first inequality in (106), and from the 1-dimensional result one obtains the second inequality.

We now prove (105). If $k \geq 1$ we can write

$$\begin{aligned} \Psi_k * F_{m,j} &= (\phi_k * f_{m,j}) \otimes (\varphi_{0,k} * \chi) + (\phi_{0,k} * f_{m,j}) \otimes (\varphi_k * \chi) \\ &= A_k + B_k \end{aligned} \tag{110}$$

(a similar formula holds for $k = 0$). Then

$$\|A_k\|_p \lesssim \|\phi_k * f_{m,j}\|_p \quad \text{and} \quad \|B_k\|_p \leq \|\phi_{0,k} * f_{m,j}\|_p \, \|\varphi_k * \chi\|_p. \tag{111}$$

From the previous calculation in one dimension we have

$$\|\phi_k * f_{m,j}\|_p \lesssim 2^{-|k-j|\delta} 2^{k(1-\frac{1}{p})} m^{1/p}.$$

We estimate the term

$$\|\phi_{0,k} * f_{m,j}\|_p = \Big(\sum_{\ell=1}^{m} \|\phi_{0,k} * g_j\|_p^p\Big)^{\frac{1}{p}} = m^{\frac{1}{p}} \|\phi_{0,k} * g_j\|_p.$$

Now, if $k \geq j$ then

$$\|\phi_{0,k} * g_j\|_p \leq \|\phi_{0,k}\|_1 \, \|g_j\|_p \lesssim 2^{(1-\frac{1}{p})j} \leq 2^{k(1-\frac{1}{p})}.$$

On the other hand, if $k < j$ then

$$\|\phi_{0,k} * g_j\|_p \leq \|\phi_{0,k}\|_p \, \|g_j\|_1 \lesssim 2^{k(1-\frac{1}{p})}.$$

Thus,

$$\|B_k\|_p \lesssim m^{1/p} \, 2^{k(1-\frac{1}{p})} \, \|\varphi_k * \chi\|_p,$$

which can be inserted into (111), and overall will imply

$$\|F_{m,j}\|_{B^{\frac{1}{p}-1}_{p,q}} \lesssim \|f_{m,j}\|_{B^{\frac{1}{p}-1}_{p,q}} + m^{\frac{1}{p}} \, \|\chi\|_{B^0_{p,q}} \lesssim m^{\frac{1}{p}}.$$

This completes the proof of (105), and hence of (104) for all $d > 1$.

13 Failure of Unconditionality when $s = d/p - d$

Theorem 3 states that strongly admissible enumerations of $\mathscr{H}_d$ form a Schauder basis of $B^{d/p-d}_{p,p}$ when $\frac{d}{d+1} < p \leq 1$. We show that the stronger conclusion of unconditionality fails. The argument will also apply to the Triebel-Lizorkin spaces $F^{d/p-d}_{p,q}$ and therefore we cover this case at the same time.

Theorem 50 *For every $N \geq 1$, there is a collection $A(N)$ of Haar functions, all supported in $[0, 1]^d$, with $\#(A(N)) \leq 2^d N$, and such that the orthogonal projection operators $P_{A(N)}$ satisfy the estimates*

$$\|P_{A(N)}\|_{B^{d(\frac{1}{p}-1)}_{p,q} \to B^{d(\frac{1}{p}-1)}_{p,q}} \gtrsim N^{1/q},$$

$$\|P_{A(N)}\|_{F^{d(\frac{1}{p}-1)}_{p,q} \to F^{d(\frac{1}{p}-1)}_{p,q}} \gtrsim N^{1/p}.$$

We shall use the following well-known identity.

Lemma 51 *For $N = 1, 2, \ldots,$ it holds*

$$2^{Nd} \mathbb{1}_{I_{N,0}} = \mathbb{1}_{I_{0,0}} + \sum_{k=0}^{N-1} 2^{kd} \sum_{\epsilon \in \Upsilon} h^{\epsilon}_{k,0}. \tag{112}$$

Proof The formula follows easily computing the Haar coefficients of the function on the left hand side of (112). □

Let $F_N(x) = 2^{Nd}\mathbb{1}_{[0,2^{-N})^d}(x)$, and let G_N be its odd extension $G_N(x) = F_N(x) - F_N(-x)$. Consider the finite dimensional subspace

$$A(N) = \operatorname{span}\Big(\{\mathbb{1}_{[0,1)^d}\} \cup \bigcup_{k=0}^{N-1}\{h_{k,0}^{\epsilon} : \epsilon \in \Upsilon\}\Big), \tag{113}$$

which has dimension $\dim A(N) = (2^d - 1)N + 1$. Let $P_{A(N)}$ be the orthogonal projection onto $A(N)$. Then, by Lemma 51, $F_N \in A(N)$ and

$$P_{A(N)}(G_N) = F_N.$$

The failure of unconditionality follows now from

Proposition 52 *Let $\frac{d-1}{d} < p < \infty$, $q > 0$. Then, for large N,*

$$\|G_N\|_{B^{\frac{d}{p}-d}_{p,q}} \lesssim 1\,, \tag{114a}$$

$$\|G_N\|_{F^{\frac{d}{p}-d}_{p,q}} \lesssim 1\,, \tag{114b}$$

and

$$\|P_{A(N)}G_N\|_{B^{\frac{d}{p}-d}_{p,q}} \gtrsim N^{1/q}\,, \tag{115a}$$

$$\|P_{A(N)}G_N\|_{F^{\frac{d}{p}-d}_{p,q}} \gtrsim N^{1/p}\,. \tag{115b}$$

Proof Since $\|\psi_0 * G_N\|_p \lesssim 1$ for any $\psi_0 \in \mathcal{S}$, we only need to estimate the terms involving $\psi_k * G_N$, with $k \geq 1$, in the $B^s_{p,q}$ or $F^s_{p,q}$ quasi-norms. Assume that $\psi_k(x) = 2^{kd}\psi(2^k x)$, where $\psi \in C_c^\infty((-1,1)^d)$ is such that $\psi(x) \geq 1$ for $x \in (-1/2, -1/8)^d$, and ψ has sufficient vanishing moments (to characterize the involved B and F norms). For $k \geq 1$, we analyze $\psi_k * G_N$. Note, that G_N is supported on $[-2^{-N}, 2^{-N}]^d$. Since $\int G_N(x)dx = 0$ we have

$$|\psi_k * G_N(x)| \lesssim 2^{kd}2^{k-N}\mathbb{1}_{[-2^{-k+1},2^{-k+1}]^d}, \qquad \text{for } k \leq N;$$

see (91). Hence

$$2^{k(\frac{d}{p}-d)}\|\psi_k * G_N\|_p \lesssim 2^{k-N}, \quad k \leq N. \tag{116}$$

For $k > N$ let D_N be the boundary of $I_N \cup -I_N$. Then $\psi_k * G_N$ is supported in a $C2^{-k}$ neighborhood $\mathcal{N}_{k,N}$ of D_N and $\psi_k * G_N = O(2^{Nd})$ on $\mathcal{N}_{k,N}$. The measure of $\mathcal{N}_{k,N}$ is $O(2^{-N(d-1)-k})$ and therefore we obtain

$$2^{k(\frac{d}{p}-d)}\|\psi_k * G_N\|_p \lesssim 2^{-(k-N)(d-\frac{d-1}{p})}, \quad k \geq N.$$

Since $p > \frac{d-1}{d}$ we can sum the estimates and obtain (114a).

Similarly

$$\Big\|\Big(\sum_{k=1}^{N}|\psi_k * G_N|^q 2^{k(\frac{d}{p}-d)q}\Big)^{1/q}\Big\|_p$$
$$\lesssim \Big\|\Big(\sum_{k=1}^{N}|2^{kd}2^{k-N}\mathbb{1}_{|x|\lesssim 2^{-k}}|^q 2^{k(\frac{d}{p}-d)q}\Big)^{1/q}\Big\|_p \lesssim 1$$

and

$$\Big\|\Big(\sum_{k=N+1}^{\infty}|\psi_k * G_N|^q 2^{k(\frac{d}{p}-d)q}\Big)^{1/q}\Big\|_p$$
$$\lesssim 2^{Nd}\Big\|\Big(\sum_{k=N+1}^{\infty}2^{k(\frac{d}{p}-d)q}\mathbb{1}_{\mathcal{N}_{k,N}}\Big)^{1/q}\Big\|_p$$
$$\lesssim 2^{Nd}\Big(\sum_{N\leq l<\infty}\operatorname{meas}(\mathcal{N}_{l,N})\Big(\sum_{N\leq k\leq l}2^{k(\frac{d}{p}-d)q}\Big)^{p/q}\Big)^{1/p}$$
$$\lesssim 2^{Nd}\Big(\sum_{N\leq l<\infty}2^{-l-N(d-1)}\Big(\sum_{N\leq k\leq l}2^{k(\frac{d}{p}-d)q}\Big)^{p/q}\Big)^{1/p} \lesssim 1.$$

Observe that the last inequality requires a slightly different argument in each of the cases $\frac{d-1}{d} < p < 1$, $p = 1$ and $p > 1$; we leave details to the reader. This proves (114b).

We now include the lower bound for $P_{\mathcal{A}(N)}G_N = F_N \equiv 2^{Nd}\mathbb{1}_{I_{N,0}}$. Let

$$\Omega_k = \Big(-\frac{3/8}{2^k}, -\frac{1/8}{2^k}\Big)^d.$$

Then, for $4 \leq k \leq N-4$,

$$\psi_k * F_N(x) = \int 2^{kd}\psi(2^k(x-y))2^{Nd}\mathbb{1}_{I_{N,0}}(y)dy \geq 2^{kd}, \qquad \text{for } x \in \Omega_k,$$

due to $2^k(x - [0, 2^{-N}]^d) \subset (-1/2, -1/8)^d$ and the assumptions on ψ. Hence

$$2^{k(\frac{d}{p}-d)}\|\psi_k * F_N\|_p \gtrsim 1, \quad 4 \le k \le N-4,$$

which implies (115a). Also

$$\|F_N\|_{F^{\frac{d}{p}-d}_{p,q}} \gtrsim \Big(\sum_{k=4}^{N-4} \int_{\Omega_k} 2^{k(\frac{d}{p}-d)p} 2^{kdp}\Big)^{1/p} \gtrsim N^{1/p}$$

and (115b) follows. □

14 Failure of Unconditionality when $s = 1/p - 1, 1 < p < \infty$

In dimension $d = 1$ the failure of unconditionality of $\mathscr{H}$ in $B^{\frac{1}{p}-1}_{p,q}(\mathbb{R})$ is already contained in Proposition 52. As happened in Sect. 12.2, the argument for $d \ge 2$ requires a slight variation of the above.

We consider the finite dimensional space

$$\mathcal{A}(N) := \mathrm{span}\Big\{h \otimes \mathbb{1}_{[0,1]^{d-1}} : h \in A^{(1)}(N)\Big\},$$

where $A^{(1)}(N)$ is the subspace defined in (113) (when $d = 1$). Note that $\dim \mathcal{A}(N) = \dim A^{(1)}(N) \approx N$. We now have

Theorem 53 *Let* $1 < p < \infty$. *Then*

$$\|P_{\mathcal{A}(N)}\|_{B^{\frac{1}{p}-1}_{p,q} \to B^{\frac{1}{p}-1}_{p,q}} \gtrsim N^{1/q}.$$

In particular, $\mathscr{H}_d$ *is not unconditional in* $B^{\frac{1}{p}-1}_{p,q}(\mathbb{R}^d)$ *for any* $q > 0$.

Proof We keep the notation F_N and G_N for the 1-dimensional functions in Proposition 52. We fix $\chi \in C_c^\infty((-1,2)^{d-1})$ with $\chi \equiv 1$ in $[0,1]^{d-1}$, and define

$$g_N(x) = G_N(x_1)\chi(x') \quad \text{and} \quad f_N(x) = F_N(x_1)\mathbb{1}_{[0,1]^{d-1}}(x').$$

Observe that $f_N \in \mathcal{A}(N)$ and

$$P_{\mathcal{A}(N)}(g_N) = f_N,$$

by our choice of χ. So, it suffices to show that, for large N,

$$\|g_N\|_{B^{\frac{1}{p}-1}_{p,q}} \lesssim 1 \quad \text{and} \quad \|f_N\|_{B^{\frac{1}{p}-1}_{p,q}} \gtrsim N^{1/q}. \tag{117}$$

The first assertion is proved as in Sect. 12.2; namely, one constructs functions Ψ_k as in (108) and observes that

$$\begin{aligned}\Psi_k * g_N &= (\phi_k * G_N) \otimes (\varphi_{0,k} * \chi) + (\phi_{0,k} * G_N) \otimes (\varphi_k * \chi)\\ &= A_k + B_k.\end{aligned}$$

A similar proof as the one following (111) gives

$$\|A_k\|_p \lesssim \|\phi_k * G_N\|_p \quad \text{and} \quad \|B_k\|_p \lesssim 2^{k(1-\frac{1}{p})} \|\varphi_k * \chi\|_p.$$

From here and the 1-dimensional results in (114a) it follows that

$$\|g_N\|_{B^{\frac{1}{p}-1}_{p,q}(\mathbb{R}^d)} \lesssim \|G_N\|_{B^{\frac{1}{p}-1}_{p,q}(\mathbb{R})} + \|\chi\|_{B^0_{p,q}(\mathbb{R}^{d-1})} \lesssim 1.$$

Likewise, to prove the second assertion in (117) one uses

$$\Psi_k * f_N(x_1, x') = \phi_k * F_N(x_1), \quad x' \in (\tfrac{1}{4}, \tfrac{3}{4})^{d-1}.$$

This identity, as before, follows from (108) and the facts

$$\varphi_{0,k} * \mathbb{1}_{[0,1]^{d-1}}(x') = \int \varphi_{0,k} = 1 \quad \text{and} \quad \varphi_k * \mathbb{1}_{[0,1]^{d-1}}(x') = \int \varphi_k = 0,$$

because the supports of $\varphi_{0,k}(x - \cdot)$ and $\varphi_k(x - \cdot)$ are contained in $[0, 1]^{d-1}$ for such values of x'. Thus,

$$\|f_N\|_{B^{\frac{1}{p}-1}_{p,q}(\mathbb{R}^d)} \gtrsim \|F_N\|_{B^{\frac{1}{p}-1}_{p,q}(\mathbb{R})} \gtrsim N^{1/q}.$$

This establishes (117) and completes the proof of Theorem 53. □

Acknowledgments The authors would like to thank the Isaac Newton Institute for Mathematical Sciences, Cambridge, for support and hospitality during the program *Approximation, Sampling and Compression in Data Science* where some work on this paper was undertaken. This work was supported by EPSRC grant no EP/K032208/1. G.G. was supported in part by grants MTM2016-76566-P, MTM2017-83262-C2-2-P and Programa Salvador de Madariaga PRX18/451 from Micinn (Spain), and grant 20906/PI/18 from Fundación Séneca (Región de Murcia, Spain). A.S. was supported in part by National Science Foundation grants DMS 1500162 and 1764295. T.U. was supported in part by Deutsche Forschungsgemeinschaft (DFG), grant 403/2-1.

References

1. Albiac, F., Ansorena, J.L., Berná, P., Wojtaszczyk, P.: Greedy approximation for biorthogonal systems in quasi-Banach spaces. Diss. Math. (to appear, 2021)
2. Bourdaud, G.: Localisations des espaces de Besov. Studia Math. **90**, 153–163 (1988)
3. Franke, J.: On the spaces $F^s_{p,q}$ of Triebel-Lizorkin type: Pointwise multipliers and spaces on domains. Math. Nachr. **125**, 29–68 (1986)
4. Garrigós, G., Seeger, A., Ullrich, T.: On uniform boundedness of dyadic averaging operators in spaces of Hardy-Sobolev type. Analysis Math. **43**(2), 267–278 (2017)
5. Garrigós, G., Seeger, A., Ullrich, T.: The Haar system as a Schauder basis in spaces of Hardy-Sobolev type. J. Fourier Anal. Appl. **24**(5), 1319–1339 (2018)
6. Garrigós, G., Seeger, A., Ullrich, T.: The Haar system in Triebel-Lizorkin spaces: Endpoint results. J. Geom. Anal. (2021). https://doi.org/10.1007/s12220-020-00577-x
7. Oswald, P.: On inequalities for spline approximation and spline systems in the space L_p, $(0 < p < 1)$. In: Ciesielski, Z. (ed.) Approximation and function spaces. Proceeding of the International Conference Gdansk 1979. PWN, Warszawa/North-Holland Amsterdam, pp. 531–552 (1981)
8. Oswald, P.: Haar system as Schauder basis in Besov spaces: the limiting cases for $0 < p \leq 1$, arXiv e-print: arXiv:1808.08156
9. Peetre, J.: On spaces of Triebel-Lizorkin type. Ark. Mat. **13**, 123–130 (1975)
10. Runst, T., Sickel, W.: Sobolev spaces of fractional order, Nemytskij operators, and nonlinear partial differential equations. In: de Gruyter series in nonlinear analysis and applications, vol. 3. Walter de Gruyter & Co., Berlin (1996)
11. Seeger, A., Ullrich, T.: Haar projection numbers and failure of unconditional convergence in Sobolev spaces. Math. Z. **285**, 91–119 (2017)
12. Seeger, A., Ullrich, T.: Lower bounds for Haar projections: deterministic Examples. Constr. Appr. **46**, 227–242 (2017)
13. Triebel, H.: Über die Existenz von Schauderbasen in Sobolev-Besov-Räumen. Isomorphiebeziehungen. Studia Math. **46**, 83–100 (1973)
14. Triebel, H.: On Haar bases in Besov spaces. Serdica **4**(4), 330–343 (1978)
15. Triebel, H.: Spaces of Besov-Hardy-Sobolev type. Teubner, Leipzig (1978)
16. Triebel, H.: Theory of function spaces. In: Monographs in Mathematics, vol. 78. Birkhäuser, Basel (1983)
17. Triebel, H.: Theory of function spaces II. In: Monographs in Mathematics, vol. 84. Birkhäuser, Basel (1992)
18. Triebel, H.: Function spaces in Lipschitz domains and on Lipschitz manifolds. Characteristic functions as pointwise multipliers. Rev. Mat. Complut. **15**(2), 475–524 (2002)
19. Triebel, H.: Bases in function spaces, sampling, discrepancy, numerical integration. In: EMS Tracts in Mathematics, vol. 11. European Mathematical Society (EMS), Zürich (2010)
20. Triebel, H.: Theory of Function Spaces IV, volume 107 of Springer Monogr. Math. Birkhäuser, Springer Nature Switzerland, Cham (2020)
21. Yuan, W., Sickel, W., Yang, D.: The Haar system in Besov-type spaces. Studia Math. **253**(2), 129–162. (2020)

Obstacle Problems Generated by the Estimates of Square Function

Irina Holmes and Alexander Volberg

Abstract In this note we give the formula for the Bellman function associated with the problem considered by B. Davis (Duke Math J 43:697–704, 1976) in 1976. In this article the estimates of the type $\|Sf\|_p \le C_p \|f\|_p$, $p \ge 2$, were considered for the dyadic square function operator S, and Davis found the sharp values of constants C_p. However, along with the sharp constants one can consider a more subtle characteristic of the above estimate. This quantity is called the Bellman function of the problem. It has never been proved that the confluent hypergeometric function from Davis' paper (second page) gives us this Bellman function. Here we fill out this gap by finding the exact Bellman function of the unweighted L^p estimate for the dyadic square function operator S. We cast the proofs in the language of obstacle problems. For the sake of comparison, we also find the Bellman function of weak $(1, 1)$ estimate of S. This formula was suggested by Bollobas (Math Proc Camb Phil Soc 87:377–382, 1980) and proved by Osekowski (Statist Probab Lett 79(13):1536–1538, 2009), so it is not new, but we like to emphasize the common approach to those two Bellman function descriptions.

Keywords Obstacle problems · Square function · Bellman function · Sharp constants

1 Obstacle Problems for Unweighted Square Function Operator: Burkholder–Gundy–Davis Function

In this note we find the exact Bellman function for the unweighted estimate of the dyadic square function operator in spaces L^p. The dyadic square function operator is one of the most basic and elementary singular operators. The sharp constants

I. Holmes · A. Volberg (✉)
Department of Mathematics, Michigan State University, East Lansing, MI, USA
e-mail: holmesir@math.msu.edu; volberg@math.msu.edu

P. Ciatti, A. Martini (eds.), *Geometric Aspects of Harmonic Analysis*,
Springer INdAM Series 45, https://doi.org/10.1007/978-3-030-72058-2_12

for its estimates from above and from below in L^p were found by Davis [16] and by Wang [30, 31] for various values of p's. However, some values of p's are still enigmatic even now.

Another strange thing is that the Bellman function for the problem was not exactly found. Notice that the dyadic problem has a Brownian motion counterpart (one can think that the discrete martingale problem has a close relative—a continuous martingale problem). In that continuous problem the part of dyadic square function is played by stopping time of Brownian motion.

Davis in [16] built the exact Bellman function for the continuous problem. In fact, this is how he found the sharp constants mentioned above. But the exact Bellman function for the dyadic problem was not constructed. The difference is not so simple: one should derive a certain finite difference inequality of elementary but quite complicated nature from its infinitesimal counterpart. This is what we are doing here. As a result we construct the exact Bellman function for L^p-estimate of dyadic square function.

In fact, the goal of this note is twofold: (1) to construct the exact Bellman function for L^p-estimate of dyadic square function as this has been done in [16] for stopping time of Brownian motion, (2) to present a small "theory" of reducing "all possible estimates" for dyadic square function to a class of obstacle problems for finite difference analogs of one special PDE.

This second aim is completely fulfilled and it allows us to write the Bellman equation and obstacle problem corresponding to "any" estimate of the square function. But notice that to write a PDE (in our case in its discrete form) is not the same as to solve it. However, in Davis case we solve it too.

Recall that h_J denotes the normalized in L^2 Haar function supported on interval J. Let now g be a test function on an interval I, then

$$g = \langle g \rangle_I \mathbf{1}_I + \sum_{J \in \mathcal{D}(I)} \Delta_J g$$

with $\Delta_J g = (g, h_J) h_J$. The square function of g is the following aggregate:

$$Sg(x) \stackrel{\text{def}}{=} \Big(\sum_{\substack{J \in \mathcal{D}(I) \\ x \in J}} (\Delta_J g)^2(x) \Big)^{1/2}.$$

Marcinkiewicz–Paley inequalities [20] relate the norms of $g - \langle g \rangle_I$ and Sg, claiming that for certain situations these norms can be equivalent.

Let $W(t)$ be the standard Brownian motion starting at zero, and T be any stopping time. Below $\|f\|_\alpha$ stands for L^α norm.

D. Burkholder [14], P. Millar [21], A. A. Novikov [23], D. Burkholder and R. Gundy [15], B. Davis [16], found the following norm estimates

$$c_\alpha \|T^{1/2}\|_\alpha \le \|W(T)\|_\alpha, \quad 1 < \alpha < \infty, \ \|T^{1/2}\|_\alpha < \infty; \tag{1}$$

$$\|W(T)\|_\alpha \le C_\alpha \|T^{1/2}\|_\alpha, \quad 0 < \alpha < \infty. \tag{2}$$

Davis [16] found the best possible values of constants above.

It was explained in [16] that the same sharp estimates (1) and (2) hold with $W(T)$ replaced by an integrable function g on [0, 1], and $T^{1/2}$ replaced by the dyadic square function of g.

More precisely, Davis proved that

$$c_\alpha \|Sg\|_\alpha \le \|g\|_\alpha, \quad 2 \le \alpha < \infty; \tag{3}$$

$$\|g\|_\alpha \le C_\alpha \|Sg\|_\alpha, \quad 0 < \alpha \le 2. \tag{4}$$

with the same constants as above, and these constants are sharp in those ranges of α. Inequality (4) with the same sharp constant as in (2) but for the range $\alpha \ge 3$ was proved by G. Wang [31]. In the range $\alpha \in (2, 3)$ the sharp constant in (4) is not known to the best of our knowledge. The same can be said about (3) in the range $\alpha \in (1, 2)$. Notice also that Wang's results are proved for square functions of conditionally symmetric martingales. So Wang's setting is more general than the dyadic setting presented here. For the weighted estimates see [3].

Our reasoning here first follows the original proof by B. Davis of estimates (1) and (3) based on the construction of a corresponding Bellman function. Davis considers two problems: (1) the continuous one, where stopping time serves as the replacement of the square function operator, (2) and a discrete one, concerning the dyadic square function operator S itself.

For the continuous problem he defines the Bellman function (on page 699 of [16] it is called $v(t, x)$). But he seems to be leaving the finding of the Bellman function for the estimate of S outside of the scope of his paper.

We just fill out this small gap in the present note. This is done by Theorem 8, the main part is Sect. 2.3.

But first we wish to cast the proofs in the language of obstacle problems. To prepare the ground we start with explanation what are obstacle problems related to square function estimates. The idea of using a specially designed function to find sharp constants is due to Burkholder [4–13] and Davis [16]. The reader can find various examples of this approach in [22, 25, 26], and [29].

1.1 Obstacle Problems Related to Square Function Estimates

We will always work with functions on some interval I, and $\mathcal{T} \stackrel{\text{def}}{=} \mathcal{T}(I)$ is the class of test functions. We say that $f \in \mathcal{T}$ if f is constant on each dyadic interval from $\mathcal{D}_N(I)$ for some finite N. By $\mathcal{D}_N(I)$ we denote the collection of dyadic subintervals of I of size $2^{-N}|I|$.

The main players will be an "arbitrary" function $O\colon \mathbb{R} \times \mathbb{R}_+ \to \mathbb{R}$ (an obstacle) and a function $U\colon \mathbb{R} \times \mathbb{R}_+ \to \mathbb{R}$, $U \ge O$, satisfying the following inequality

$$2U(p, q) \ge U(p + a, \sqrt{a^2 + q^2}) + U(p - a, \sqrt{a^2 + q^2}) . \tag{5}$$

We will call this *the main inequality*, functions U satisfying the main inequality will be precisely Bellman functions of various estimates concerning square function operator.

Of course the existence of U majorizing O and satisfying (5) is not at all ensured. Notice that (5) is invariant under taking infimum.

The reader may ask what is the future meaning of letters p, q, a in (5). Letter p is a "locum tenens" for the average of test function f: $\langle f\rangle_I$. Letter q is a locum tenens for a local version of square function of f, and a will serve as locum tenens for $\Delta_I f$.

Definition 1 We call the smallest U satisfying the main inequality and majorizing O *the heat envelope* of O.

We would like to find the heat envelope of some specific O.

Theorem 2 *Let U satisfy main inequality* (5). *Then for any $f \in \mathcal{T}(I)$*

$$\langle U(f, \sqrt{q^2 + (Sf)^2})\rangle_I \le U(\langle f\rangle_I, q)\,. \tag{6}$$

We denote by $\langle f\rangle_I := \frac{1}{|I|}\int_I f$, the average of f over the interval I. Here is a corollary relating the main inequality with square function estimates.

Corollary 3 *Let U satisfy main inequality* (5). *Then for any $f \in \mathcal{T}(I)$*

$$\langle U(f, Sf)\rangle_I \le U(\langle f\rangle_I, 0)\,. \tag{7}$$

Before proving Theorem 2, we wish to answer the question, when, given O, one can find a finite valued function majorizing O and satisfying the main inequality.

Theorem 4 *Let*

$$\mathbf{U}(p, q) \stackrel{\text{def}}{=} \sup_{\substack{f\in\mathcal{T}(I)\\ \langle f\rangle_I = p}} \langle O(f, \sqrt{q^2 + S^2 f})\rangle_I\,. \tag{8}$$

If this function is finite valued, then it satisfies the main inequality.

Now we wish to formulate results that can be considered as converse to Theorem 2. They concern the obstacle problem for (5).

As was already mentioned, by this we understand finding U satisfying (5) and majorizing a given function (obstacle) $O\colon \mathbb{R} \times \mathbb{R}_+ \to \mathbb{R}$. It turns out that one can give "simple" conditions necessary and sufficient for the solvability of the obstacle problem.

Theorem 5 *Let an obstacle function O, and a function $F\colon \mathbb{R} \to \mathbb{R}$ satisfying $F(p) \ge O(p, 0)$ be given. A finite valued function U satisfying*

- *main inequality* (5)
- $U \ge O$
- $U(p, 0) \le F(p)$

exists if and only if

$$\langle O(f, Sf)\rangle_I \le F(\langle f\rangle_I), \quad \forall f \in \mathcal{T}. \tag{9}$$

It will be especially important to use this result with one special F: $F = 0$.

Theorem 6 *Given an obstacle function O, to find U satisfying main inequality (5) and such that $U \ge O$ and $U(p, 0) \le 0$, it is necessary and sufficient to have*

$$\langle O(f, Sf)\rangle_I \le 0, \qquad \forall f \in \mathcal{T}. \tag{10}$$

Proof (of Theorem 2) Below by $\mathbb{E}_k$ we denote the expectation with respect to σ-algebra generated by dyadic intervals of family $\mathcal{D}_k$. We first prove Theorem 2. Let $f \in \mathcal{T}$, and let N be such that f is constant on each $J \in \mathcal{D}_N(I)$. Let us consider two siblings $\ell_+, \ell_- \in \mathcal{D}_N(I)$ with the same father $\ell \in \mathcal{D}_{N-1}(I)$.

Denote $p \stackrel{\text{def}}{=} \langle f\rangle_\ell$ and let $\langle f\rangle_{\ell_+} = p+a$, then $\langle f\rangle_{\ell_-} = p-a$, and $f(x) = p\pm a$ for all $x \in \ell_\pm$ correspondingly. Notice that for all $x \in \ell$, $|\Delta_\ell f(x)| = |a|$, and put $q_1 \stackrel{\text{def}}{=} \sqrt{S^2 f(x) - a^2}$, where $x \in \ell_\pm$ (the value $Sf(x)$ is the same for all $x \in \ell$). By the main inequality we have

$$\begin{aligned}
&\int_{\ell_+} U(f(x), \sqrt{q^2 + S^2 f(x)})\, dx + \int_{\ell_-} U(f(x), \sqrt{q^2 + S^2 f(x)})\, dx \\
&= |\ell| \left(\tfrac{1}{2} U(p+a, \sqrt{q^2 + a^2 + q_1^2}) + \tfrac{1}{2} U(p-a, \sqrt{q^2 + a^2 + q_1^2}) \right) \\
&\le |\ell| U(p, q) = \int_\ell U(f_1(x), \sqrt{q^2 + S^2 f_1(x)})\, dx,
\end{aligned}$$

where $f_1 \stackrel{\text{def}}{=} \mathbb{E}_{N-1} f$. We can continue now by recursion. We denote $f_k \stackrel{\text{def}}{=} \mathbb{E}_{N-k} f$, $k = 1 \dots N$. So $f_N(x) = \mathbb{E}_0 f = \langle f\rangle_I \mathbf{1}_I$. Notice that $Sf_N = 0$ identically, and after repeating the above recursion $N+1$ times we come to

$$\int_I U(f(x), \sqrt{q^2 + S^2 f(x)})\, dx \le |I| U(\langle f\rangle_I, q), \tag{11}$$

which is the claim of the theorem. □

Proof (of Theorem 4) It is clear by its definition and by rescaling, that $\mathbf{U}$ does not depend on the interval I, where test functions are defined. Therefore, given the data $(p + a, \sqrt{a^2 + q^2})$, we can find a function f_+ optimizing $\mathbf{U}(p + a, \sqrt{a^2 + q^2})$ up to ε, and we can think as well that it lives on I_+. Similarly, given the data $(p - a, \sqrt{a^2 + q^2})$, we can find a function f_- optimizing $\mathbf{U}(p - a, \sqrt{a^2 + q^2})$ up to ε, and we can think as well that it lives on I_-.

Concatenate functions $f_\pm$ on $I_\pm$ to the following function:

$$f(x) = \begin{cases} f_+(x), & x \in I_+ \\ f_-(x), & x \in I_- \end{cases}$$

Since $\langle f \rangle_I = p$, we have

$$\begin{aligned}
\mathbf{U}(p,q) &\geq \langle O(f, \sqrt{q^2 + S^2 f}) \rangle_I \\
&= \tfrac{1}{2}\langle O(f, \sqrt{q^2 + S^2 f}) \rangle_{I_+} + \tfrac{1}{2}\langle O(f, \sqrt{q^2 + S^2 f}) \rangle_{I_-} \\
&= \tfrac{1}{2}\langle O(f_+, \sqrt{q^2 + a^2 + S^2 f_+}) \rangle_{I_+} + \tfrac{1}{2}\langle O(f_-, \sqrt{q^2 + a^2 + S^2 f_+}) \rangle_{I_-} \\
&\geq \tfrac{1}{2}\mathbf{U}(p+a, \sqrt{a^2 + q^2}) - \varepsilon + \tfrac{1}{2}\mathbf{U}(p-a, \sqrt{a^2 + q^2}) - \varepsilon\,.
\end{aligned}$$

As ε is an arbitrary positive number we are done. □

Now we prove Theorem 5.

Proof First we prove the "if" part. We are given an obstacle O and a function F such that $F(p) \geq O(p, 0)$. We defined

$$\mathbf{U}(p,q) = \sup_{\substack{f \in \mathcal{T}(I) \\ \langle f \rangle_I = p}} \langle O(f, \sqrt{q^2 + S^2 f}) \rangle_I\,.$$

It is obvious that $\mathbf{U}(p,q) \geq O(p,q)$, one just plugs the constant function $f = p\mathbf{1}_I$.

It is also clear that $\mathbf{U}(p, 0) \leq F(p)$. Indeed,

$$\mathbf{U}(p,0) = \sup_{\substack{f \in \mathcal{T}(I) \\ \langle f \rangle_I = p}} \langle O(f, Sf) \rangle_I \leq F(\langle f \rangle_I) = F(p)$$

by assumption (9). Hence $\mathbf{U}(p, 0)$ is finite valued.

The fact that function $\mathbf{U}$ defined as above satisfies the main inequality (5) follows from Theorem 4. Then by (5) it is finite valued.

Now we prove the "only if" part. We need to prove that

$$\langle O(f, Sf) \rangle_I \leq F(\langle f \rangle_I)$$

if there exits a majorant U of O satisfying the main inequality and satisfying $U(p, 0) \leq F(p)$. This is easy:

$$\langle O(f, Sf) \rangle_I \leq \langle U(f, Sf) \rangle_I \leq U(\langle f \rangle_I, 0) \leq F(\langle f \rangle_I),$$

where the second inequality follows from Corollary 3. □

The following theorem sums up the results of this section.

Theorem 7 *There exists a finite valued function U majorizing O and satisfying the main inequality if and only if* **U** *from* (8) *is finite valued. Moreover, if* **U** *is finite valued, then the infimum of functions U majorizing O and satisfying the main inequality is equal to* **U**.

Proof We already saw in Theorem 4 that **U** from (8) (if finite valued) is one of those functions U that majorize O and satisfy the main inequality.

On the other hand, for any function U that majorize O and satisfy the main inequality we know from Theorem 2 that for any test function f and any non-negative q the following holds

$$U(\langle f\rangle_I, q) \geq \langle U(f, \sqrt{q^2 + S^2 f})\rangle_I \geq \langle O(f, \sqrt{q^2 + S^2 f})\rangle_I \,.$$

Take now the supremum over test functions in the right hand side. By definition we obtain $\mathbf{U}(\langle f\rangle_I, q)$. Theorem is proved. □

We will consider in detail examples 1, 2, and 3 below.

Example 1 Davis function that gives the proof of (3) for $\alpha \geq 2$. Here the obstacle function will be

$$O_0(p, q) = c_\alpha^\alpha |q|^\alpha - |p|^\alpha, \tag{12}$$

where the best value of c_α was found by Davis [16].

Example 2 Bollobás function. Here the obstacle function will be

$$O_1(p, q) = \mathbf{1}_{q\geq 1} - C|p|, \tag{13}$$

where the best value of C was suggested by B. Bollobás [2]. This was verified by A. Osękowski [24], see also [19].

Example 3 Bollobás function. Here the obstacle function will be

$$O_2(p, q) = \mathbf{1}_{p^2+q^2\geq 1} - C|p|, \tag{14}$$

where the best value of C was suggested by B. Bollobas [2] and also verified by A. Osękowski [24], see also [19].

Example 4 Bellman function associated with the Chang–Wilson–Wolff theorem.

$$O_3(p, q; \lambda) = \mathbf{1}_{[\lambda,\infty)}(p)\mathbf{1}_{[0,1]}(q) \,. \tag{15}$$

The function U is not fully known in the case. It is "almost" found in [22].

2 Davis Obstacle Problem

In this section we want to find the minimal value c_α for which there exists a function $\mathbf{U}\colon \mathbb{R}^2 \to \mathbb{R}$ that solves the problem with the obstacle function of Example 0, i. e.,

$$\mathbf{U}(p,q) \stackrel{\text{def}}{=} \sup\{\langle c_\alpha^\alpha (q^2 + (Sf)^2)^{\alpha/2} - |f|^\alpha\rangle_I : \langle f\rangle_I = p\}\,. \tag{16}$$

In other words, we want to find the heat envelope of O_0. Let $\alpha \geq 2$ and let $\beta = \frac{\alpha}{\alpha-1} \leq 2$ be the conjugate exponent of α. Let

$$\begin{aligned} N_\alpha(x) &\stackrel{\text{def}}{=} {}_1F_1\left(-\frac{\alpha}{2}, \frac{1}{2}, \frac{x^2}{2}\right) \\ &= \sum_{m=0}^{\infty} \frac{(-2x^2)^m}{(2m)!}\frac{\alpha}{2}\left(\frac{\alpha}{2}-1\right)\cdots\left(\frac{\alpha}{2}-m+1\right) \\ &= 1 - \frac{\alpha}{2}x^2 + \frac{\alpha}{12}\left(\frac{\alpha}{2}-1\right)x^4 \ldots \end{aligned} \tag{17}$$

be the confluent hypergeometric function. $N_\alpha(x)$ satisfies the Hermite differential equation

$$N_\alpha''(x) - xN_\alpha'(x) + \alpha N_\alpha(x) = 0 \quad \text{for} \quad x \in \mathbb{R} \tag{18}$$

with initial conditions $N_\alpha(0) = 1$ and $N_\alpha'(0) = 0$. Let c_α be the smallest positive zero of N_α.

Set

$$u_\alpha(x) \stackrel{\text{def}}{=} \begin{cases} -\dfrac{\alpha c_\alpha^{\alpha-1}}{N_\alpha'(c_\alpha)} N_\alpha(x), & 0 \leq |x| \leq c_\alpha; \\ c_\alpha^\alpha - |x|^\alpha, & c_\alpha \leq |x|. \end{cases} \tag{19}$$

Clearly $u_\alpha(x)$ is $C^1(\mathbb{R}) \cap C^2(\mathbb{R}\setminus\{c_\alpha\})$ smooth even concave function. The concavity follows from Lemma 9 on the page 434 and the fact that $N_\alpha'(c_\alpha) < 0$. Finally we define

$$U(p,q) \stackrel{\text{def}}{=} \begin{cases} |q|^\alpha u_\alpha\left(\frac{|p|}{|q|}\right), & q \neq 0, \\ -|p|^\alpha, & q = 0. \end{cases} \tag{20}$$

In this section we are going to prove the following result.

Theorem 8 *Function* $\mathbf{U}$ *from* (16) *is equal to* U *written above in* (20).

For the first time the function $U(p, q)$ appeared in [16]. Later it was also used in [30, 31] in the form $\widetilde{u}(p, t) = U(p, \sqrt{t}), t \geq 0$. Since we want to prove that

$$U = \mathbf{U},$$

at first we will verify the following properties:

$$U(p, q) \geq |q|^\alpha c_\alpha^\alpha - |p|^\alpha, \quad (p, q) \in \mathbb{R}^2, \tag{21}$$

$$2U(p, q) \geq U(p + a, \sqrt{a^2 + q^2}) + U(p - a, \sqrt{a^2 + q^2}), \ (p, q, a) \in \mathbb{R}^3. \tag{22}$$

When these two properties get proved, Theorem 7 ensures that

$$\mathbf{U} \leq U. \tag{23}$$

This inequality is the most difficult part of Theorem 8.

We called (21) the *obstacle condition*, and (22) the *main inequality*. The infinitesimal form of (22) is

$$\frac{1}{q} U_q + U_{pp} \leq 0, \tag{24}$$

which follows from the main inequality by expanding it into Taylor's series with respect to a near the origin and comparing the second order terms.

First we check (24). On domain $p/q \in (-c_\alpha, c_\alpha)$, $q > 0$, this follows from (20) and the first line of (19). Moreover, on this domain we have equality $U_q/q + U_{pp} = 0$, which easily follows from (18). On the complementary domain, where $|p| \geq c_\alpha q$, we have

$$\begin{aligned} \frac{1}{q} U_q + U_{pp} &= \alpha (c_\alpha^\alpha q^{\alpha - 2} - (\alpha - 1)|p|^{\alpha - 2}) \\ &= \alpha q^{\alpha - 2} c_\alpha^{\alpha - 2} \Big(c_\alpha^2 - (\alpha - 1) \big(\frac{|p|}{c_\alpha q}\big)^{\alpha - 2} \Big) < 0, \end{aligned}$$

because $\alpha \geq 2$ and, as we will see below in Lemma 9, $c_\alpha \leq 1$.

Inequality (24) guarantees that

$$X_t = U(W(t), \sqrt{t}) \quad \text{is a supermartingale for} \quad t \geq 0.$$

In fact, using Itô's formula, we get

$$dX(t) = \frac{1}{2\sqrt{t}} \frac{\partial U}{\partial q} dt + \frac{1}{2} \frac{\partial^2 U}{\partial p^2} dt + \frac{\partial U}{\partial p} dW(t),$$

and therefore (24) implies that $dX(t) - \frac{\partial U}{\partial p} dW(t) \leq 0$, so $X(t)$ is a supermartingale.

Finally, the supermartingale property gives us the second inequality below

$$\mathbb{E}(T^{\frac{\alpha}{2}} c_\alpha^\alpha - |B_T|^\alpha) \overset{(21)}{\leq} \mathbb{E}U(B_T, \sqrt{T}) \leq U(0,0) = 0,$$

which yields (3).

Now we are going to prove that $U(p, q)$ is the minimal function with properties (21) and (22).

The next step is to go from infinitesimal version (24) to finite difference inequality (22). For that we need several lemmas.

Lemma 9 *The minimal positive root c_α of N_α has the following properties.*

(1) *The estimate $0 < c_\alpha \leq 1$ is valid for $\alpha \geq 2$.*
(2) *c_α is decreasing in $\alpha > 0$.*
(3) *$N_\alpha'(t) \leq 0,\ N_\alpha''(t) \leq 0$ on $[0, c_\alpha]$ for $\alpha > 0$.*

Proof Consider $G_\alpha(t) \overset{\text{def}}{=} e^{-t^2/4} N_\alpha(t)$. Notice that the zeros of G_α and N_α are the same. It follows from (18) that

$$G_\alpha' + \left(\alpha + \frac{1}{2} - \frac{t^2}{4}\right) G_\alpha = 0, \quad G_\alpha(0) = 1 \quad \text{and} \quad G_\alpha'(0) = 0. \tag{25}$$

Besides we know that the solution is even. Consider the critical case $\alpha = 2$. In this case $G_2(t) = e^{-t^2/4}(1 - t^2)$ and the smallest positive zero is $s_2 = 1$. Therefore it follows from the Sturm comparison principle that $0 < c_\alpha < 1$ for $\alpha > 2$ (see below). Moreover, the same principle applied to G_{α_1} and G_{α_2} with $\alpha_1 > \alpha_2$ implies that G_{α_1} has a zero inside the interval $(-s_{\alpha_2}, s_{\alpha_2})$. Thus we conclude that c_α is decreasing in α.

To verify that $N_\alpha', N_\alpha'' \leq 0$ on $[0, c_\alpha]$, first we claim that

$$N_{\alpha_2} \geq N_{\alpha_1} \quad \text{on} \quad [0, s_{\alpha_1}]$$

for $\alpha_1 > \alpha_2 > 0$. Indeed the proof works in the same way as the proof of Sturm's comparison principle. For the convenience of the reader we decided to include the argument. As before, consider $G_{\alpha_j} = e^{-t^2/4} N_{\alpha_j}$. It is enough to show that $G_{\alpha_2} \geq G_{\alpha_1}$ on $[0, s_{\alpha_1}]$. It follows from (25) that $G_{\alpha_2}''(0) > G_{\alpha_1}''(0)$. Therefore, using the Taylor series expansion at the point 0, we see that the claim is true at some neighborhood of zero, say $[0, \varepsilon)$ with ε sufficiently small. Next we assume the contrary, i.e., that there is a point $a \in [\varepsilon, s_{\alpha_1}]$ such that $G_{\alpha_2} \geq G_{\alpha_1}$ on $[0, a]$, $G_{\alpha_2}(a) = G_{\alpha_1}(a)$ and $G_{\alpha_2}'(a) < G_{\alpha_1}'(a)$ (notice that the case $G_{\alpha_2}'(a) = G_{\alpha_1}'(a)$, by the uniqueness theorem for ordinary differential equations, would imply that $G_{\alpha_2} = G_{\alpha_1}$ everywhere, which is impossible). Consider the Wronskian

$$W = G_{\alpha_1}' G_{\alpha_2} - G_{\alpha_1} G_{\alpha_2}'.$$

We have $W(0) = 0$ and $W(a) = G_{\alpha_1}(a)(G'_{\alpha_1}(a) - G'_{\alpha_2}(a)) \geq 0$. On the other hand, we have

$$W' = (\alpha_2 - \alpha_1)G_{\alpha_1}G_{\alpha_2} < 0 \quad \text{on} \quad [0, a),$$

which is a clear contradiction, and this proves the claim.

It follows from (17) that

$$N''_\alpha = -\alpha N_{\alpha-2}, \tag{26}$$

and inequalities $N_{\alpha-2} \geq N_\alpha \geq 0$ on $[0, c_\alpha]$ imply that

$$N''_\alpha \leq 0 \text{ on } [0, c_\alpha].$$

Since $N'_\alpha(0) = 0$, and $N''_\alpha \leq 0$ on $[0, c_\alpha]$, we must have $N'_\alpha \leq 0$ on $[0, c_\alpha]$. □

Lemma 10 *For any $p \in \mathbb{R}$, the function*

$$t \mapsto U(p, \sqrt{t}) \quad \textit{is convex for} \quad t \geq 0. \tag{27}$$

Proof Without loss of generality, assume that $p \geq 0$. We recall that $U(p, \sqrt{t}) = t^{\alpha/2}u_\alpha(p/\sqrt{t})$. Since $\alpha \geq 2$, the only interesting case to consider is when $p/\sqrt{t} < c_\alpha$ (otherwise $t^{\alpha/2}$ is convex). In this case we have $U(p, \sqrt{t}) = \kappa_\alpha t^{\alpha/2} N_\alpha(p/\sqrt{t})$, where κ_α is a positive constant. In particular, by (18) we have $U(p, \sqrt{t})_t + \frac{1}{2}U(p, \sqrt{t})_{pp} = 0$. Using (26), we obtain

$$U(p, \sqrt{t})_t = -\frac{U(p, \sqrt{t})_{pp}}{2} = -\frac{\kappa_\alpha}{2} t^{\frac{\alpha}{2}-1} N''_\alpha(p/\sqrt{t}) = \frac{\alpha\kappa_\alpha}{2} t^{\frac{\alpha-2}{2}} N_{\alpha-2}(p/\sqrt{t}).$$

Therefore, it would be enough to show that for any $\gamma \geq 0$, the function $x^{-\gamma}N_\gamma(x)$ is decreasing for $x \in (0, s_{\gamma+2})$. Differentiating, and using (18) again, we obtain

$$\frac{\mathrm{d}}{\mathrm{d}x}\left(\frac{N_\gamma(x)}{x^\gamma}\right) = \frac{N''_\gamma(x)}{x^{\gamma+1}},$$

which is nonpositive by Lemma 9. □

The next lemma, together with Lemma 10 and (24), implies that $U(p, q)$ satisfies (22).

Lemma 11 (Barthe–Maurey [1]) *Let J be a convex subset of $\mathbb{R}$, and let*

$$V(p, q) \colon J \times \mathbb{R}_+ \to \mathbb{R}$$

be such that

$$V_{pp} + \frac{V_q}{q} \leq 0 \quad \text{for all} \quad (p, q) \in J \times \mathbb{R}_+; \tag{28}$$

$$t \mapsto V(p, \sqrt{t}) \quad \text{is convex for each fixed} \quad p \in J. \tag{29}$$

Then for all (p, q, a) *with* $p \pm a \in J$ *and* $q \geq 0$, *we have*

$$2V(p, q) \geq V(p + a, \sqrt{a^2 + q^2}) + V(p - a, \sqrt{a^2 + q^2}). \tag{30}$$

The lemma says that the global finite difference inequality (30) is in fact implied by its infinitesimal form (28) under the extra condition (29).

Proof The argument is borrowed from [1].

Without loss of generality assume $a \geq 0$. Consider the process

$$X_t = V(p + W(t), \sqrt{q^2 + t}), \quad t \geq 0.$$

Here $W(t)$ is the standard Brownian motion starting at zero. It follows from Itô's formula together with (28) that X_t is a supermartingale. Indeed, by Itô's formula we have

$$X_t = X_0 + \int_0^t V_p \, dW(t) + \frac{1}{2} \int_0^t \left(V_{pp} + \frac{V_q}{\sqrt{q^2 + t}} \right) dt$$

and notice that the drift term is negative. Let τ be the stopping time such that $W(\tau)$ hits a or $-a$, i. e.

$$\tau = \inf\{t \geq 0 \colon W(t) \notin (-a, a)\}.$$

The supermartingale property of X_t and concavity (29) yield the following chain of inequalities:

$$\begin{aligned}
V(p, q) = X_0 \geq \mathbb{E}X_\tau &= \mathbb{E}V(p + W(\tau), \sqrt{q^2 + \tau}) \\
&= P(W(\tau) = -a)\mathbb{E}(V(p - a, \sqrt{q^2 + \tau}) | W(\tau) = -a) \\
&\quad + P(W(\tau) = a)\mathbb{E}(V(p + a, \sqrt{q^2 + \tau}) | W(\tau) = a) \\
&= \tfrac{1}{2}\Big(\mathbb{E}(V(p - a, \sqrt{q^2 + \tau}) | W(\tau) = -a) \\
&\quad + \mathbb{E}(V(p + a, \sqrt{q^2 + \tau}) | W(\tau) = a)\Big)
\end{aligned}$$

$$\geq \tfrac{1}{2}\Big(V\big(p-a, \sqrt{q^2+\mathbb{E}(\tau|W(\tau)=-a)}\big)$$

$$+ V\big(p+a, \sqrt{q^2+\mathbb{E}(\tau|W(\tau)=a)}\big)\Big)$$

$$= \tfrac{1}{2}\Big(V\big(p-a, \sqrt{q^2+a^2}\big) + V\big(p+a, \sqrt{q^2+a^2}\big)\Big).$$

Notice that we have used $P(W(\tau)=a) = P(W(\tau)=-a) = 1/2, \mathbb{E}(\tau|W(\tau) = a) = \mathbb{E}(\tau|W(\tau)=-a) = a^2$, and the fact that the map $t \mapsto V(p, \sqrt{t})$ is convex together with Jensen's inequality. □

2.1 Majorization of the Obstacle Function

We have finished the proof of inequality (22). Now we are going to check (21) from page 433. Let

$$\kappa_\alpha = -\frac{\alpha c_\alpha^{\alpha-1}}{N'_\alpha(c_\alpha)}.$$

Function $\kappa_\alpha N_\alpha$ in the first line of (19) is equal to function $g \stackrel{\text{def}}{=} c_\alpha^\alpha - x^\alpha$ at $x = c_\alpha$. To prove that $\kappa_\alpha N_\alpha \geq g$ on $[0, c_\alpha]$, thus, it is enough to prove $\kappa_\alpha N'_\alpha \leq g'$ on this interval. At point c_α these derivatives coincide by the choice of κ_α. Notice that $\kappa_\alpha > 0$ and that N'_α and g' are negative. Therefore, to check that $-\kappa_\alpha N'_\alpha \geq -g'$ it is enough to show that function $-N'_\alpha/x^{\alpha-1}$ is decreasing on $[0, c_\alpha]$, i. e.

$$\left(\frac{-N'_\alpha}{x^{\alpha-1}}\right)' \leq 0. \tag{31}$$

But

$$\left(\frac{N'_\alpha}{x^{\alpha-1}}\right)' = \frac{xN''_\alpha - (\alpha-1)N'_\alpha}{x^\alpha} = \frac{N'''_\alpha}{x^\alpha},$$

where the last equality follows from (18).

On the other hand, from (17) it follows that $N'''_\alpha = -\alpha N'_{\alpha-2}$. This expression is positive by Lemma 9. Hence (31) is proved. This proves that

$$u_\alpha \geq c_\alpha^\alpha - |x|^\alpha, \qquad x \in [-c_\alpha, c_\alpha].$$

We conclude that the function U from page 432 majorizes the obstacle:

$$U(p, q) \geq c_\alpha^\alpha |q|^\alpha - |p|^\alpha. \tag{32}$$

2.2 Why Constant c_α is Sharp?

The example, which show that the value c_α given on page 432 cannot be replaced by larger value is based on results of A. Novikov [23] and L. Shepp [27]. Introduce the following stopping time

$$T_a \stackrel{\text{def}}{=} \inf\{t > 0\colon |W(t)| = a\sqrt{t+1}\}, \quad a > 0.$$

It was proved in [27] that $\mathbb{E}T_a^{\alpha/2} < \infty$ if $a < c_\alpha$ and that $\mathbb{E}t_{c_\alpha}^{\alpha/2} = \infty$, $\alpha > 0$. This gives us that $\mathbb{E}t_a^{\alpha/2} \to \infty$, when $a \to c_\alpha -$. From here we get

$$\lim_{a \to c_\alpha -} \frac{\mathbb{E}(T_a + 1)^{\alpha/2}}{\mathbb{E}T_a^{\alpha/2}} = 1\,.$$

By definition of T_a we have $|W(T_a)| = a\sqrt{T_a + 1}$, and hence

$$\lim_{a \to c_\alpha -} \frac{\mathbb{E}|W(T_a)|^\alpha}{\mathbb{E}T_a^{\alpha/2}} \to c_\alpha^\alpha\,.$$

Now it follows immediately that the best constant in (1) cannot be larger than c_α defined on page 432. Davis in [16] extended this estimate for the case of dyadic square function estimate (3).

2.3 Why U from Page 432 is the Smallest Function Satisfying (21) and (22)?

We know that on $\{(p, q)\colon q \ge 0,\ |p|^\alpha \le c_\alpha^\alpha q^\alpha\}$

$$|q|^\alpha c_\alpha^\alpha - |p|^\alpha \le \mathbf{U}(p, q) \le U(p, q)\,. \tag{33}$$

Indeed, we proved that U satisfies the main inequality and that it majorizes the obstacle $|q|^\alpha c_\alpha^\alpha - |p|^\alpha$. We also proved that $\mathbf{U}$ is the smallest such function (this is true for any obstacle whatsoever). Hence, (33) is verified.

But now we want to demonstrate that the Bellman function is already found: $\mathbf{U} = U$. To do that we need to work a little bit more.

By definition on page 432 $\mathbf{U}$ is homogeneous of degree α. We introduce $\mathbf{b}(p) \stackrel{\text{def}}{=} \mathbf{U}(p, 1)$, $b(p) \stackrel{\text{def}}{=} U(p, 1)$. Thus we need to prove that

$$b(p) = \mathbf{b}(p), \qquad p \in [-c_\alpha, c_\alpha]. \tag{34}$$

One can easily rewrite (22) in terms of $\mathbf{b}$: for all $x \pm \tau \in [-c_\alpha, c_\alpha]$ the following holds:

$$2\mathbf{b}(x) \geq (1+\tau^2)^{\alpha/2}\Big(\mathbf{b}\Big(\frac{x+\tau}{\sqrt{1+\tau^2}}\Big) + \mathbf{b}\Big(\frac{x-\tau}{\sqrt{1+\tau^2}}\Big)\Big). \tag{35}$$

Since by construction $U(p,q) = 0$ if $|q|^\alpha c_\alpha^\alpha - |p|^\alpha = 0$ we conclude that $b(\pm c_\alpha) = \mathbf{b}(\pm c_\alpha) = 0$.

Combining (22) with a simple observation that $\mathbf{U}$ by definition increases in q, we can conclude that function $\mathbf{U}$ is concave in p for every fixed q, $\mathbf{b}$ is concave.

Let us recall that for any concave function f the following holds (see e.g. [17]):

$$f(x+h) = f(x) + f'(x)h + \frac{1}{2}f''(x)h^2 + o(h^2), \quad h \to 0, \quad \text{for a.e. } x\,. \tag{36}$$

Then (36) and inequality (35) implies that $\mathbf{b}'' - x\mathbf{b}' + \alpha\mathbf{b} \leq 0$ a.e. But function $\mathbf{b}$ is concave. In particular, it is everywhere defined and continuous, and its derivative $\mathbf{b}'$ is also its distributional derivative, and it is everywhere defined decreasing function.

Let $(\mathbf{b})''$ denote the distributional derivative of decreasing function $\mathbf{b}'$. Thus it is a non-positive measure. We denote its singular part by symbol σ_s. Hence, in the sense of distributions

$$(\mathbf{b})'' - x\mathbf{b}'\,dx + \alpha\mathbf{b}\,dx = \big(\mathbf{b}'' - x\mathbf{b}' + \alpha\mathbf{b}\big)\,dx + d\sigma_s \leq 0\,. \tag{37}$$

Lemma 12 *Let $\alpha > 0$. Let even non-negative concave function v defined on $[-c_\alpha, c_\alpha]$ satisfy $v(\pm c_\alpha) = 0$. Let v satisfy $v'' - xv' + \alpha v \leq 0$ on $(-c_\alpha, c_\alpha)$ pointwise and in the sense of distributions. Assume also that v have finite derivative at c_α: $v'(c_\alpha) > -\infty$. Then $v'' - xv' + \alpha v = 0$ on $(-c_\alpha, c_\alpha)$ pointwisely and in the sense of distributions. Also $v = cu$ for some constant c.*

Proof Let $u \stackrel{\text{def}}{=} u_\alpha$ from (19). It is C^2 function and $u'' - xu' + \alpha u = 0$ on $[-c_\alpha, c_\alpha]$. Denote

$$g \stackrel{\text{def}}{=} v'' - xv' + \alpha v\,.$$

Function v is concave, so its second derivative is defined a.e., and we assumed that $g \leq 0$.

Consider everywhere defined function

$$w \stackrel{\text{def}}{=} v'u - u'v\,.$$

Its derivative is defined almost everywhere, and let us first calculate it a.e.:

$$w' = v''u - u''v = (g + xv' - \alpha v)u - (xu' - \alpha u)v = gu + xw\,.$$

Also in distributional sense

$$(w)' = (v)''u - u''v\,dx = (gu + xw)\,dx + u\,d\sigma_s\,.$$

Hence,

$$\frac{d}{dx}e^{-x^2/2}w = gue^{-x^2/2}, \quad \text{for almost every } x\,, \tag{38}$$

and

$$\left(e^{-x^2/2}w\right)' = gue^{-x^2/2}\,dx + ue^{-x^2/2}\,d\sigma_s, \text{ in distribution sense}\,. \tag{39}$$

Measure σ_s is non-positive, therefore, these two inequalities (38), (39) mean that for any two points $0 < a < b < 1$ we have

$$e^{-b^2/2}w(b) - e^{-a^2/2}w(a) \le \int_a^b gue^{-x^2/2}\,dx,$$

moreover, the inequality is strict, if $\sigma_s(a, b) \ne 0$.

Let us tend b to 1. Looking at the definition $w = v'u - u'v$ and using the assumptions of lemma, we conclude that $e^{-b^2/2}w(b) \to 0$. Hence,

$$e^{-a^2/2}w(a) \ge \int_a^1 (-g)ue^{-x^2/2}\,dx\,. \tag{40}$$

Again the inequality is strict if $\sigma_s(a, 1) \ne 0$.

Now let us tend $a \to 0$. By smoothness and evenness $u'(a) \to 0$. But $u(a) > 0$ and $v'(a) \le 0$ for a. e. $a > 0$. Therefore,

$$\limsup_{a \to 0+} e^{-a^2/2}w(a) \le 0\,.$$

Combining this with (40) we conclude that

$$\int_a^1 (-g)ue^{-x^2/2}\,dx \le 0$$

with the strict inequality if $\sigma_s(0, 1) \ne 0$. The strict inequality of course leads to contradiction (recall that $-g \ge 0$, $u > 0$), so we conclude that σ_s is a zero measure on $(0, 1)$. But also even a non-strict inequality implies that $g = 0$ a.e.

We conclude from (38), (39) that $e^{-x^2/2}w(x)$ is constant on $(0, 1)$. But we already saw that this function tends to 0 when x tends to 1. Thus, identically on $(0, 1)$

$$u'v - v'u = w = 0\,.$$

This means that $v/u = const$. Lemma is proved. □

Now it is easy to prove (34): $\mathbf{b} = b$. Choose $v = \mathbf{b}$, the assumptions on ordinary differential inequality is easy to verify, see (37). Of course this function vanishes at $\pm c_\alpha$. Also by the definition of b it is clear (see (19), (20)) that

$$b'(c_\alpha) = -\alpha c_\alpha^{\alpha-1} > -\infty\,.$$

We are left to see that the same is true for $\mathbf{b}'(c_\alpha)$.

Recall that $\mathbf{b}(\cdot) = \mathbf{U}(\cdot, 1)$, $b(\cdot) = U(\cdot, 1)$, then by (33) we definitely know that

$$c_\alpha^\alpha - |x|^\alpha \le \mathbf{b}(x) \le b(x), \quad x \in [-c_\alpha, c_\alpha]\,.$$

The functions on the left and on the right vanish at c_α and have the same derivative $-\alpha c_\alpha^{\alpha-1}$ at C_α. Hence, $\mathbf{b}$ is in fact differentiable at c_α (the left derivative exists), and its (left) derivative satisfies

$$\mathbf{b}'(c_\alpha) = b'(c_\alpha) = -\alpha c_\alpha^{\alpha-1} > -\infty\,.$$

But now Lemma 12 says that $\mathbf{b} = const \cdot b$. Since we have the above relationship on derivatives, the constant has to be 1. We proved (34). This gives

$$\mathbf{U} = U,$$

where U was defined in (19), (20). We found the Bellman function $\mathbf{U}$ for Burkholder–Gundy–Davis inequality, and we completely solved the obstacle problem with the obstacle $O(p, q) = c_\alpha^\alpha q^\alpha - |p|^\alpha$, $\alpha \ge 2$.

2.4 *When Obstacle Coincides with its Heat Envelope*

The next corollary immediately follows from the previous proposition, and it describes one possibility when the heat envelope coincides with its obstacle

Corollary 13 *Let $O(p, q)$ be $C^2(\mathbb{R} \times [0, \infty))$ obstacle such that*

$$O_{pp} + \frac{O_q}{q} \le 0 \quad \textit{and}$$

$$t \mapsto O(p, \sqrt{t}) \quad \textit{is convex.}$$

Then the heat envelope U of O satisfies $U(p, q) = O(p, q)$.

The next proposition says that if O satisfies "backward heat equation" then the convexity assumption $t \mapsto O(p, \sqrt{t})$ is necessary and sufficient for main inequality (5).

Proposition 14 *Let $O(p,q) \in C^4(\mathbb{R} \times [0,\infty))$ be such that*

$$O_{pp} + \frac{O_q}{q} = 0$$

for all $(p,q) \in \mathbb{R} \times (0,\infty)$. Then the following conditions are equivalent

(i) The map $t \mapsto O(p, \sqrt{t})$ is convex for $t \geq 0$.
(ii) $2O(p,q) \geq O(p+a, \sqrt{q^2+a^2}) + O(p-a, \sqrt{q^2+a^2})$ for all $p, a \in \mathbb{R}$ and all $q \geq 0$.

Proof The implication $(i) \Rightarrow (ii)$ follows from Lemma 11. It remains to show the implication $(ii) \Rightarrow (i)$. By Taylor's formula as $a \to 0$ we have

$$O(p+a, \sqrt{q^2+a^2}) + O(p-a, \sqrt{q^2+a^2})$$

$$= 2O(p,q) + \left(O_{pp} + \frac{O_q}{q}\right)a^2 + \left(O_{pppp} + 6\frac{O_{ppq}}{q} + 3\frac{O_{qq}}{q^2} - 3\frac{O_q}{q^3}\right)\frac{a^4}{12} + o(a^4)$$

Since $O_{pp} + \frac{O_q}{q} = 0$ we see that

$$O_{pppp} + 6\frac{O_{ppq}}{q} + 3\frac{O_{qq}}{q^2} - 3\frac{O_q}{q^3} = 2\left(\frac{O_q}{q^3} - \frac{O_{qq}}{q^2}\right).$$

Therefore,

$$0 \geq O(p+a, \sqrt{q^2+a^2}) + O(p-a, \sqrt{q^2+a^2}) - 2O(p,q) =$$

$$= \left(\frac{O_q}{q} - O_{qq}\right)\frac{a^4}{6q^2} + o(a^4).$$

Thus we obtain that $\frac{O_q}{q} - O_{qq} \leq 0$. On the other hand the latter inequality is equivalent to the fact that $t \to O(p, \sqrt{t})$ is convex. □

3 Bollobás Function

This part of the present article is taken from [19]. We put it here because the solution of the obstacle problem(s) in this section and the solution of the obstacle problem in the previous section have so much in common, and at the same time, they have essential differences. So we include the current section for the sake of comparison.

The classical Littlewood–Khintchine inequality states that

$$\Big(\sum_{k=1}^{n} a_k^2\Big)^{1/2} \le L \int_0^1 \Big|\sum_{k=1}^{n} a_k r_k(t) dt\Big|, \tag{41}$$

where $\{r_k(t)\}$ are Rademacher functions. It was one of Littlewood's problems to find the best value for constant L. The problem was solved by S. Szarek [28], see also [18]. The sharp constant is $L = \sqrt{2}$.

B. Bollobás [2] considered the following related problem, which we formulate in the form convenient for us. The problem of Bollobás was: what is the best value for the constant B for the following inequality

$$\lambda\big|\{t \in (0,1) \colon Sf(t) \ge \lambda\}\big| \le C\|f\|_1\,? \tag{42}$$

Consider $x_n \stackrel{\text{def}}{=} \sum_{k=1}^{n} a_k r_k(t)$. If we denote $\lambda \stackrel{\text{def}}{=} \big(\sum_{k=1}^{n} a_k^2\big)^{1/2} = Sx_n(x)$ (obviously Sx_n is a constant function), we get

$$\begin{aligned}\Big(\sum_{k=1}^{n} a_k^2\Big)^{1/2} &= \lambda\big|\{t \in (0,1) \colon Sx_n(t) \ge \lambda\}\big| \\ &\le C \int_0^1 \Big|\sum_{k=1}^{n} a_k r_k(t)\Big| dt\,.\end{aligned} \tag{43}$$

This says that $\sqrt{2} = L \le B$. On the other hand, D. Burkholder in [14] proved that $B \le 3$. B. Bollobás in [2] conjectured the best value of B, and in 2009 A. Osękowski [24] proved this conjecture. We will give a slightly different proof by solving the obstacle problem and finding the heat envelopes of two obstacles:

$$O_1(p,q) = \mathbf{1}_{q \ge 1} - C_1|p|, \tag{44}$$

$$O_2(p,q) = \mathbf{1}_{p^2+q^2 \ge 1} - C_2|p|. \tag{45}$$

We are interested in the smallest possible values of C_1 and C_2 such that these functions have (finite) heat envelopes. The reader will see, in particular, that $C_1 = C_2 = C$ and that the heat envelopes of these two functions coincide—see Theorem 19.

Define the following Bellman function:

$$\mathbf{B}(x,\lambda) \stackrel{\text{def}}{=} \inf\{\langle|\varphi|\rangle_J \colon \langle\varphi\rangle_J = x;\ S_J^2\varphi \ge \lambda \text{ a. e. on } J\}. \tag{46}$$

Some of the obvious properties of $\mathbf{B}$ are:

- Domain: $\Omega_{\mathbf{B}} \stackrel{\text{def}}{=} \{(x,\lambda) \colon x \in \mathbb{R};\ \lambda > 0\}$;
- $\mathbf{B}$ is increasing in λ and even in x;

- Homogeneity: $\mathbf{B}(tx, t^2\lambda) = |t|\mathbf{B}(x, \lambda)$;
- Range/Obstacle Condition: $|x| \le \mathbf{B}(x, \lambda) \le \max\{|x|, \sqrt{\lambda}\}$;
- Main Inequality:

$$2\mathbf{B}(x, \lambda) \le \mathbf{B}(x - a, \lambda - a^2) + \mathbf{B}(x + a, \lambda - a^2), \quad \forall\, |a| < \sqrt{\lambda}. \tag{47}$$

  Remark how this is going in the direction opposite to (5)—this is because the function $\mathbf{B}$ above is defined as an *infimum* and not the usual supremum. This function will mirror many of the properties of a supremum function, so convexity now becomes concavity.
- $\mathbf{B}$ is convex in x, and so it is easy to see that $\mathbf{B}$ is minimal at $x = 0$:

$$\mathbf{B}(0, \lambda) \le \mathbf{B}(x, \lambda), \quad \forall\, x, \tag{48}$$

  therefore we can use that $\mathbf{B}$ is increasing in λ and also use the minimality at $x = 0$ to obtain from (47) that $\mathbf{B}$ is non-decreasing in x for $x \ge 0$, and non-increasing in x for $x \le 0$;
- Greatest Subsolution: If $B(x, \lambda)$ is any continuous non-negative function on $\Omega_{\mathbf{B}}$ which satisfies the main inequality (47) and the range condition $B(x, \sqrt{\lambda}) \le \max\{|x|, \sqrt{\lambda}\}$, then $B \le \mathbf{B}$.

3.1 Bellman Induction

Theorem 15 *If B is any subsolution as defined above, then $B \le \mathbf{B}$.*

Proof We must prove that $B(x, \lambda) \le \langle|\varphi|\rangle_J$ for any function φ on J with $\langle\varphi\rangle_J = x$, $|J| = |\{x \in J : S_J^2\varphi(x) \ge \lambda\}|$. As before, we may assume that there is some dyadic level $N \ge 0$ below which the Haar coefficients of φ are zero.

If $\lambda \le (\Delta_J\varphi)^2$, then by the range/obstacle condition above

$$B(x, \lambda) \le \max\{|x|, \sqrt{\lambda}\} \le \max\{|x|, |\Delta_J\varphi|\} \le \langle|\varphi|\rangle_J,$$

and we are done. Otherwise, put $\lambda_{J_\pm} = \lambda - (\Delta_J\varphi)^2 > 0$, $x_{J_\pm} = \langle\varphi\rangle_{J_\pm}$. Then by the main inequality:

$$|J|B(x, \lambda) \le |J_-|B(x_{J_-}, \lambda_{J_-}) + |J_+|B(x_{J_+}, \lambda_{J_+}).$$

If $\lambda_{J_-} \le (\Delta_{J_-}\varphi)^2$, it follows as before that $|J_-|B(x_{J_-}, \lambda_{J_-}) \le \int_{J_-} |\varphi|$, and otherwise we iterate further on J_-.

Continuing this way down to the last level N and putting $\lambda_I \overset{\text{def}}{=} \lambda - (\Delta_{I^{(1)}}\varphi)^2 - \dots - (\Delta_J\varphi)^2$ for every $I \in \mathcal{D}_N(J)$, where $I^{(1)}$ denotes the dyadic father of I, the previous iterations have covered all cases where $\lambda_I \leq 0$, and we have (with $x_I I$)

$$|J|B(x,\lambda) \leq \sum_{\substack{I\in\mathcal{D}_N(J)\\ \lambda_I\leq 0}} \int_I |\varphi| + \sum_{\substack{I\in\mathcal{D}_N(J)\\ \lambda_I>0}} |I|B(x_I,\lambda_I). \tag{49}$$

Now note that for all $I \in \mathcal{D}_N(J)$ we must have $\lambda_I \leq (\Delta_I\varphi)^2$ just because $S_J^2\varphi(x) \geq \lambda$ everywhere on J, so we use the range/obstacle condition as before to obtain $B(x_I,\lambda_I) \leq \max\{|x_I|, |\Delta_I\varphi|\} \leq \langle|\varphi|\rangle_J$. Finally, (49) becomes:

$$|J|B(x,\lambda) \leq \sum_{I\in\mathcal{D}_N(J)} \int_I |\varphi| = \int_J |\varphi|.$$

Thus the proof of the claim

$$B \leq \mathbf{B}$$

is complete. □

3.2 *Finding the Candidate for* $\mathbf{B}(x,\lambda)$

We introduce

$$\mathbf{b}(\tau) \stackrel{\text{def}}{=} \mathbf{B}(\tau, 1)\,.$$

Using homogeneity, we write

$$\sqrt{\lambda}\mathbf{b}(\tau) = \sqrt{\lambda}\mathbf{B}\left(\frac{x}{\sqrt{\lambda}}, 1\right) = \mathbf{B}(x,\lambda)\,, \quad \text{where } \tau = \frac{x}{\sqrt{\lambda}}\,.$$

Then $\mathbf{b}\colon \mathbb{R} \to [0,\infty)$, $\mathbf{b}$ is even in τ, and from (48):

$$\mathbf{b}(0) \leq \mathbf{b}(\tau), \qquad \forall\,\tau. \tag{50}$$

Moreover, $\mathbf{b}$ satisfies

$$\mathbf{b}(\tau) = |\tau|, \qquad \forall\,|\tau| \geq 1\,. \tag{51}$$

We are looking for a candidate B for $\mathbf{B}$. We will assume now that $\mathbf{B}$ is smooth. We will find the candidate under this assumption, and later we will prove that thus found function is indeed **B**. Using again Taylor's formula, the infinitesimal version of (47) is

$$\mathbf{B}_{xx} - 2\mathbf{B}_\lambda \geq 0. \tag{52}$$

In terms of **b**, this becomes

$$\mathbf{b}''(\tau) + \tau \mathbf{b}'(\tau) - \mathbf{b}(\tau) \geq 0. \tag{53}$$

Since **b** is even, we focus next only on $\tau \geq 0$.

Let symbol Φ denote the following function:

$$\Phi(\tau) \stackrel{\text{def}}{=} \int_0^\tau e^{-y^2/2}\, dy\,.$$

Put

$$\Psi(\tau) = \tau\Phi(\tau) + e^{-\tau^2/2}, \qquad \forall\, \tau \geq 0.$$

The general solution of the differential equation

$$z''(\tau) + \tau z'(\tau) - z(\tau) = 0, \qquad \tau \geq 0$$

is

$$z(\tau) = C\Psi(\tau) + D\tau\,.$$

Note that

$$\Psi'(\tau) = \Phi(\tau), \qquad \Psi''(\tau) = e^{-\tau^2/2}\,. \tag{54}$$

Since $\mathbf{b}(\tau) = \tau$ for $\tau \geq 1$, see (51), a reasonable candidate for our function **b** is one already proposed by B. Bollobas [2]:

$$b(\tau) \stackrel{\text{def}}{=} \begin{cases} \frac{\Psi(\tau)}{\Psi(1)}, & 0 \leq \tau < 1 \\ \tau, & \tau \geq 1. \end{cases} \tag{55}$$

In other words, a candidate for **B** is

$$B(y, \lambda) = \begin{cases} \sqrt{\lambda}\frac{\Psi\left(\frac{|y|}{\sqrt{\lambda}}\right)}{\Psi(1)}, & \sqrt{\lambda} \geq |y|, \\ |y|, & \sqrt{\lambda} \leq |y|\,. \end{cases} \tag{56}$$

Our first goal will be to go from differential inequality (52) to its finite difference version (5).

Lemma 16 *The function B defined in* (56) *satisfies the finite difference main inequality* (*the analog of* (47)):

$$2B(y,\lambda) \le B(y-a,\lambda-a^2) + B(y+a,\lambda-a^2)\,. \tag{57}$$

We already saw in Lemma 11 that under some extra assumptions of convexity one can derive the finite difference inequalities from their differential form (infinitesimal form). Unfortunately, this approach will not work for function B defined in (56). This function exactly misses the extra property (29) of Lemma 11. In fact, we deal now with convexity paradigm rather than concavity conditions of Lemma 11, so the right analog of property (29) for B in the above formula would be

$$\lambda \to B(y,\lambda) \text{ is a concave function for every fixed } y\,.$$

But it is obvious that our candidate B does not have this property. This is why the proof of Lemma 16 requires direct calculations. This requires splitting the proof into several cases. One of them was considered in [2], but other cases were only mentioned there.

Proof (of Lemma 16) By symmetry we can think that $x \ge 0$. Case (1) will be when both points $x \pm t, \lambda - t^2)$ lie in Π (i. e. they lie over parabola $\lambda = x^2$).

Case (1). We follow [2]. Put

$$X(x,\tau) \stackrel{\text{def}}{=} \frac{x+\tau}{(1-\tau^2)^{1/2}},\quad \tau\in[0,x],\ x\in[0,1)\,. \tag{58}$$

In our case (57) can be rewritten as ($\tau \stackrel{\text{def}}{=} a/\sqrt{\lambda},\ \ x = y/\sqrt{\lambda}$):

$$2\Psi(x) \le \Psi(X(x,\tau)) + \Psi(X(x,-\tau)), \tag{59}$$

which is correct for $\tau = 0$. Let us check that

$$\frac{\partial}{\partial\tau}(\Psi(X(x,\tau)) + \Psi(X(x,-\tau))) \ge 0\,. \tag{60}$$

Using (54), we get the equality

$$\frac{\partial}{\partial\tau}(\Psi(X(x,\tau)) + \Psi(X(x,-\tau))) = \frac{1}{1-\tau^2}(\Phi(X(x,\tau)) - \Phi(X(x,-\tau)))$$
$$\frac{x\tau}{1-\tau^2}(\Phi(X(x,\tau)) + \Phi(X(x,-\tau))) - \frac{\tau}{(1-\tau^2)^{1/2}}(X(x,\tau))\Phi(X(x,\tau))$$
$$+ X(x,-\tau))\Phi(X(x,-\tau))) - \frac{\tau}{(1-\tau^2)^{1/2}}(e^{-X(x,\tau)^2/2} + e^{-X(x,-\tau)^2/2})\,.$$

After plugging (58) this simplifies to

$$\frac{\partial}{\partial\tau}(\Psi(X(x,\tau))+\Psi(X(x,-\tau)))=(\Phi(X(x,\tau))-\Phi(X(x,-\tau)))$$
$$-\frac{\tau}{(1-\tau^2)^{1/2}}(e^{-X(x,\tau)^2/2}+e^{-X(x,-\tau)^2/2}).$$

But $\frac{\tau}{(1-\tau^2)^{1/2}}=\frac{1}{2}(X(x,\tau)-X(x,-\tau))$, so to prove (60) one needs to check the following inequality.

$$⨍_{X(x,-\tau)}^{X(x,\tau)} e^{-s^2/2}ds\geq\frac{1}{2}(e^{-X(x,\tau)^2/2}+e^{-X(x,-\tau)^2/2}). \tag{61}$$

This inequality holds because in our case (1) we have $X(x,-\tau)\in[-1,1]$, $X(x,\tau)\in[-1,1]$, and function $s\to e^{-s^2/2}$ is concave on the interval $[-1,1]$. (It is easy that for every concave function on an interval, its average over the interval is at least its average over the ends of the interval.)

Case (2). Now suppose that the left point $(x-t,\lambda-t^2)$ lies on parabola. By homogeneity we can always think that $\lambda=1$. We continue to consider by symmetry $x\geq 0$ only. If $(x-t,1-t^2)$ is such that $(x-t)^2=1-t^2$ then we need to show that

$$2\frac{\Psi(x)}{\Psi(1)}\leq 2t \tag{62}$$

Clearly $0\leq t\leq 1, 0\leq x\leq t$. From $(x-t)^2=1-t^2$ we obtain that $t-\sqrt{1-t^2}\stackrel{\text{def}}{=}x(t)\geq 0$, so $t\geq\frac{1}{\sqrt{2}}$, and the inequality (62) simplifies to

$$x(t)\leq\Psi^{-1}(\Psi(1)t),\quad \frac{1}{\sqrt{2}}\leq t\leq 1.$$

The left hand side is convex and the right hand side is concave. Since at $t=1$ and $t=\frac{1}{\sqrt{2}}$ the inequality holds then it holds on the whole interval $[1/\sqrt{2},1]$.

So we proved that if the left point already left Π (and then automatically the right point also already left it), the desired inequality holds.

Case (3). It remains to show that if the right point already left Π but the left point is in Π, then (57) still holds. Again by homogeneity we can always think that $\lambda=1$. Then the required inequality amounts to

$$2\Psi(x)\leq\sqrt{1-t^2}\Psi\left(\frac{t-x}{\sqrt{1-t^2}}\right)+\Psi(1)(x+t)$$

where either $\sqrt{1-t^2}-t \le x \le t \le 1$ or $\sqrt{1-t^2}-t \le t \le x \le 1$. It is the same as to show

$$\Psi\left(\frac{t-x}{\sqrt{1-t^2}}\right)+\left(\frac{t-(\frac{2\Psi(x)}{\Psi(1)}-x)}{\sqrt{1-t^2}}\right)\Psi(1) \ge 0 \tag{63}$$

for all $0 \le x \le 1$ if $\frac{\sqrt{2-x^2}-x}{2} < t < \frac{x+\sqrt{2-x^2}}{2}$. The left inequality says that the right point already crossed parabola $\partial\Pi$ and the right inequality says that the left point is still inside Π.

Let as show that the derivative in t of the left hand side of (63) is nonnegative. If this is the case then we are done. Ψ is increasing (see (54)), and since $xt \le 1$ therefore $t \mapsto \Psi\left(\frac{t-x}{\sqrt{1-t^2}}\right)$ is increasing as a composition of two increasing functions. By the same logic, to check the monotonicity of the map $t \mapsto \frac{t-(\frac{2\Psi(x)}{\Psi(1)}-x)}{\sqrt{1-t^2}}$ it is enough to verify that $t(\frac{2\Psi(x)}{\Psi(1)}-x) \le 1$. The latter inequality follows from the following two simple inequalities

$$\Psi(x) \ge \Psi(1)x, \quad 0 \le x \le 1 \tag{64}$$

$$\left(\frac{x+\sqrt{2-x^2}}{2}\right)\left(\frac{2\Psi(x)}{\Psi(1)}-x\right) \le 1, \quad 0 \le x \le 1 \tag{65}$$

Indeed, to verify (64) notice that

$$\frac{d}{dx}\frac{\Psi(x)}{x} = \frac{x\Phi(x)-\Psi(x)}{x^2} = -\frac{e^{-\frac{x^2}{2}}}{x^2} < 0, \tag{66}$$

therefore $\frac{\Psi(x)}{x} \ge \Psi(1)$ when $0 \le x \le 1$.

To verify (65) it is enough to show that

$$\frac{\Psi(x)}{\Psi(1)x} \le \frac{1}{x^2+x\sqrt{2-x^2}}+\frac{1}{2}$$

If $x = 1$ we have equality. Taking derivative of the mapping

$$x \to \frac{\Psi(x)}{\Psi(1)x} - \frac{1}{x^2+x\sqrt{2-x^2}} - \frac{1}{2}$$

we obtain

$$\frac{2}{x^2}\left(-\frac{e^{-\frac{x^2}{2}}}{2\Psi(1)}+\frac{x+\frac{1-x^2}{\sqrt{2-x^2}}}{(x+\sqrt{2-x^2})^2}\right) \ge 0$$

To prove the last inequality it is the same as to show that

$$\frac{\sqrt{2-x^2}+x(2-x^2)}{x\sqrt{2-x^2}+1-x^2} \le \Psi(1)e^{\frac{x^2}{2}}.$$

For the exponential function we use the estimate $e^{\frac{x^2}{2}} \ge 1+\frac{x^2}{2}$. We estimate $\sqrt{2-x^2}$ from above in the numerator by $\sqrt{2}(1-\frac{x^2}{4})$, and we estimate $\sqrt{2-x^2}$ from below in the denominator by $(1-\sqrt{2})(x-1)+1$ (as $x \to \sqrt{2-x^2}$ is concave). Thus it would be enough to prove that

$$\frac{\sqrt{2}(1-\frac{x^2}{4})+x(2-x^2)}{\sqrt{2}x(1-x)+1} \le \Psi(1)\left(1+\frac{x^2}{2}\right), \quad 0 \le x \le 1$$

If we further use the estimates $\Psi(1) > \frac{29}{28}$, and $\frac{41}{29} < \sqrt{2} < \frac{17}{12}$ (for denominator and numerator correspondingly), then the last inequality would follow from

$$\frac{29}{240}\cdot\frac{246x^4-486x^3+233x^2-12x-8}{29+41x-41x^2} \le 0$$

The denominator has the positive sign. The negativity of $246x^4-486x^3+233x^2-12x-8 \le 0$ for $0 \le x \le 1$ follows from the Sturm's algorithm, which shows that the polynomial does not have roots on $[0,1]$. Since at point $x=0$ it is negative therefore it is negative on the whole interval. □

3.3 *Finding* B

Since it is easy to verify that B satisfies the range condition $B(x,\lambda) \le \max\{|f|, \sqrt{\lambda}\}$, we have then that B is a subsolution of (57), and so, by Theorem 15

$$B \le \mathbf{B}.$$

Now we want to prove the opposite inequality

$$\mathbf{B} \le B. \tag{67}$$

Lemma 17 *Let even functions* $\mathbf{b}$ *and* b *defined on* $[-1,1]$ *satisfy* $\mathbf{b}(1)=b(1)=1$, *and* $b''+xb'-b=0$, $b\in C^2$, $\mathbf{b}$ *being a convex function such that* $\mathbf{b}''+x\mathbf{b}'-\mathbf{b}\ge 0$ *on* $(-1,1)$ *in the sense of distributions. Then* $\mathbf{b}\le b$.

Proof If $\mathbf{b}$ were in C^2 as well, then this would be very easy. In fact, consider $a(x) \stackrel{\text{def}}{=} \mathbf{b}(x)-b(x)$. At end points it is zero, and $a''+xa'-a\ge 0$. Assume that function a

is strictly positive somewhere, then it should have a maximum, where it is positive. Let it be x_0. Then $a(x_0) > 0$, $a'(x_0) = 0$. So $a''(x_0) \geq a(x_0) > 0$. Then x_0 cannot be maximum, so we come to a contradiction.

If $\mathbf{b}$ is not C^2, we still consider $a(x) \stackrel{\text{def}}{=} \mathbf{b}(x) - b(x)$, which is still a continuous function on $[-1, 1]$ equal to 0 at the endpoints. If it is positive somewhere, it should have a positive maximum, let s_0 be a point of maximum.

Since $\mathbf{b}$ is assumed to be convex, function a' is of bounded variation, and as such it is the sum of f and g, where f is a continuous function and g is a jump function. Notice that (1) all jumps are positive, as they came only from $\mathbf{b}$, and (b) g is continuous everywhere except the countable set of jump points.

As a' is a function of bounded variation it has one-sided limits at any interior point. Let $a'(s_0\pm)$ be right and left limits correspondingly. Since all the jumps are positive we have

$$a'(s_0+) \geq a(s_0-).$$

But s_0 is a point of maximum of a, so $a'(s_0-) \geq 0$, $a'(s_0+) \leq 0$. All together may happen only if $a'(s_0+) = a'(s_0-) = 0$. But this means that s_0 is not a jump point.

By continuity at s_0, a' is small near s_0, but $a(s_0) > 0$, so we can choose a small neighborhood of s_0, where $|sa'(s)| < \frac{1}{2}a(s)$.

Since $a'' + sa' - a \geq 0$, in this neighborhood of s_0 we have

$$a'' \geq a - sa' > \tfrac{1}{2}a \geq 0$$

in the sense of distributions. But a convex function cannot have maximum strictly inside an interval. We come to a contradiction.

Lemma is proved. □

We found the Bellman function $\mathbf{B}$, the formula is given in the following theorem.

Theorem 18

$$\mathbf{B}(x, \lambda) = \begin{cases} \sqrt{\lambda}\dfrac{\Psi\left(\frac{|x|}{\sqrt{\lambda}}\right)}{\Psi(1)}, & x^2 \leq \lambda, \\ |x|, & x^2 \geq \lambda. \end{cases} \tag{68}$$

Let us introduce an obstacle function defined on $\mathbb{R}^2$.

$$O(x, \lambda) \stackrel{\text{def}}{=} \begin{cases} |x|, & x^2 \geq \lambda \\ \infty, & x^2 < \lambda. \end{cases} \tag{69}$$

Theorem 19 *Function $\mathbf{B}$ is the largest function satisfying the finite difference inequality such that it is majorized by the obstacle function $O(x, \lambda)$:*

$$\mathbf{B}(x, \lambda) \leq O(x, \lambda). \tag{70}$$

Moreover,

$$\mathbf{B}(x, \lambda) = \max\left(\sqrt{\lambda}\frac{\Psi(\frac{x}{\sqrt{\lambda}})}{\Psi(1)}, |x|\right). \tag{71}$$

Acknowledgments I. Holmes is supported by National Science Foundation as an NSF Postdoc under Award No.1606270, A. Volberg is partially supported by the NSF DMS-1600065. This paper is based upon work supported by the National Science Foundation under Grant No. DMS-1440140 A. Volberg was in residence at the Mathematical Sciences Research Institute in Berkeley, California, during the Spring and Fall 2017 semester.

References

1. Barthe, F., Maurey, B.: Some remarks on isoperimetry of Gaussian type. Annales de l'Institut Henri Poincare (B) Probab. Stat. **36**(4), 419–434 (2000)
2. Bollobás, B.: Martingale Inequalities. Math. Proc. Camb. Phil. Soc. **87**, 377–382 (1980)
3. Buckley, S.M.: Estimates for operator norms on weighted spaces and reverse Jensen inequalities. Trans. Am. Math. Soc. **340**(1), 253–272 (1993)
4. Burkholder, D.: A geometrical characterization of Banach spaces in which martingale difference sequences are unconditional. Ann. Probab. **9**, 997–1011 (1981)
5. Burkholder, D.: Boundary value problems and sharp estimates for the martingale transforms. Ann. Prob. **12**, 647–702 (1984)
6. Burkholder, D.: Martingales and Fourier analysis in Banach spaces. In: Probability and Analysis (Varenna, 1985) Lecture Notes in Mathematical, vol. 1206, pp. 61–108 (1986)
7. Burkholder, D.: An extension of classical martingale inequality. In: Chao, J.-A., Woyczyński, W.A. (eds.) Probability Theory and Harmonic Analysis. Marcel Dekker, New York (1986)
8. Burkholder, D.: Sharp inequalities for martingales and stochastic integrals. In: Colloque Paul Lévy sur les Processus Stochastiques (Palaiseau, 1987), Astérisque No. 157–158 (1988), pp. 75–94.
9. Burkholder, D.: A proof of the Pelczynski's conjecture for the Haar system. Studia Math. **91**, 79–83 (1988)
10. Burkholder, D.: Differential subordination of harmonic functions and martingales (El Escorial 1987). In: Lecture Notes in Mathematical, vol. 1384, pp. 1–23 (1989)
11. Burkholder, D.: Explorations of martingale theory and its applications. Lecture Notes in Math. **1464**, 1–66 (1991)
12. Burkholder, D.: Strong differential subordination and stochastic integration. Ann. Prob. **22**, 995–1025 (1994)
13. Burkholder, D.: Martingales and Singular Integrals in Banach spaces. In: Handbook of the Geometry of Banach Spaces, vol. 1(Ch. 6), pp. 233–269 (2001)
14. Burkholder, D.L.: Martingale transforms. Ann. Math. Statist. **37**, 1494–1504 (1966)
15. Burkholder, D.L., Gundy, R.F.: Extrapolation and interpolation of quasi-linear operators on martingales. Acta Math. **124**, 249–304 (1970)
16. Davis, B.: On the L^p norms of stochastic integrals and other martingales. Duke Math. J. **43**, 697–704 (1976)
17. Evans, L.C., Gariepy, R.F.: Measure Theory and Fine Properties of Functions (1992)
18. Hall, R.R.: On a conjecture of Littlewood. Math. Proc. Cambridge Philos. Soc. **78**, 443–445 (1975)
19. Holmes, I., Ivanisvili, P., Volberg, A.: The Sharp Constant in the Weak (1,1) Inequality for the Square Function: a new proof, Revista Mat. Iberoam. **36**(3), 741–770 (2020)

20. Marcinkiewicz, J.: Quelque théorèmes sur les séries orthogonales. Ann. Soc. Polon. Math. **16**, 84–96 (1937). (pages 307–318 of the Collected Papers)
21. Millar, P.W.: Martingale integrals. Trans. Am. Math. Soc. **133**, 145–166 (1968)
22. Nazarov, F., Vasyunin, V., Volberg, A.L.: On Bellman function associated with Chang–Wilson–Wolff theorem. Algebra & Analysis, vol. 33, No. 4, pp. 66–106 (2021)
23. Novikov, A.A.: On stopping times for Wiener processes. Theory Probab. Appl. **16**, 449–456 (1971)
24. Osekowski, A.: On the best constant in the weak type inequality for the square function of a conditionally symmetric martingale. Statist. Probab. Lett. **79**(13), 1536–1538 (2009)
25. Osekowski, A.: Sharp martingale and semimartingale inequalities. In: Monografie Matematyczne, vol. 72. Springer, Basel (2012)
26. Osekowski, A.: Weighted square function inequalities. Publ. Mat. **62**, 75–94 (2018)
27. Shepp, L.A.: A first passage problem for the Wiener process. Ann. Math. Statist. **38**, 1912–1914 (1967)
28. Szarek, S.J.: On the best constants in the Khintchine inequality. Studia Math. **18**, 197–208 (1976)
29. Vasyunin, V., Volberg, A.: *Bellman Function Technique in Harmonic Analysis* (Cambridge University Press, Cambridge, 2020), pp. 1–460. https://doi.org/10.1017/9781108764469
30. Wang, G.: Some sharp inequalities for conditionally symmetric martingales, Ph.D. Thesis. University of Illinois, Urbana-Champaign (1989)
31. Wang, G.: Sharp square function inequalities for conditionally symmetric martingales. Trans. Am. Math. Soc. **328**(1), 393–419 (1991)

Of Commutators and Jacobians

Tuomas P. Hytönen

Dedicated to Professor Fulvio Ricci

Abstract I discuss the prescribed Jacobian equation $Ju = \det \nabla u = f$ for an unknown vector-function u, and the connection of this problem to the boundedness of commutators of multiplication operators with singular integrals in general, and with the Beurling operator in particular. A conjecture of T. Iwaniec regarding the solvability for general datum $f \in L^p(\mathbb{R}^d)$ remains open, but recent partial results in this direction will be presented. These are based on a complete characterisation of the L^p-to-L^q boundedness of commutators, where the regime of exponents $p > q$, unexplored until recently, plays a key role. These results have been proved in general dimension $d \geq 2$ elsewhere, but I will here present a simplified approach to the important special case $d = 2$, using a framework suggested by S. Lindberg.

Keywords Commutator · Beurling transform · Jacobian determinant

1 The Prescribed Jacobian Problem

Given a vector-valued function $u = (u_j)_{j=1}^d \in \dot{W}^{1,pd}(\mathbb{R}^d)^d$ in the homogeneous Sobolev space

$$\dot{W}^{1,pd}(\mathbb{R}^d) = \{v \in L^1_{\text{loc}}(\mathbb{R}^d) : \partial_i v \in L^{pd}(\mathbb{R}^d)\ \forall i\},$$

it is clear that its Jacobian determinant—a linear combination of d-fold products of the various $\partial_i u_j$—satisfies $Ju := \det \nabla u := \det(\partial_i u_j)_{i,j=1}^d \in L^p(\mathbb{R}^d)$.

T. P. Hytönen (✉)
Department of Mathematics and Statistics, University of Helsinki, Helsinki, Finland
e-mail: tuomas.hytonen@helsinki.fi

P. Ciatti, A. Martini (eds.), *Geometric Aspects of Harmonic Analysis*,
Springer INdAM Series 45, https://doi.org/10.1007/978-3-030-72058-2_13

Our starting point is the reverse question: Given $f \in L^p(\mathbb{R}^d)$, is there $u \in \dot{W}^{1,pd}(\mathbb{R}^d)^d$ such that $Ju = f$? This is a nonlinear PDE, known as the "prescribed Jacobian equation". It has been mostly studied for *smooth* functions f on *bounded* domains Ω [4, 12], in which case there are signifcant geometric applications (e.g. [1]). In the global L^p case that we discuss, there is:

Conjecture 1 ([6]) For $p \in (1,\infty)$, there exists a continuous $E : L^p(\mathbb{R}^d) \to \dot{W}^{1,pd}(\mathbb{R}^d)^d$ such that $J \circ E = I$.

As suggested in [6], such an E could be interpreted as a "fundamental solution of the Jacobian equation".

The case $p = 1$ had already been addressed a little earlier. In this case, a simple integration by parts confirms that

$$u \in \dot{W}^{1,d}(\mathbb{R}^d)^d \quad \Rightarrow \quad \int Ju = 0 \quad \Rightarrow \quad J(\dot{W}^{1,d}(\mathbb{R}^d)^d) \subsetneq L^1(\mathbb{R}^d).$$

A somewhat more careful argument yields:

Theorem 2 ([3]) *For $u \in \dot{W}^{1,d}(\mathbb{R}^d)^d$, $d \geq 2$, we have*

$$\|Ju\|_{H^1(\mathbb{R}^d)} \lesssim \|u\|^d_{\dot{W}^{1,d}(\mathbb{R}^d)^d}$$

where $H^1(\mathbb{R}^d)$ denotes the Hardy space.

Again in the reverse direction, [3] asked: Given $f \in H^1(\mathbb{R}^d)$, is there $u \in \dot{W}^{1,d}(\mathbb{R}^d)^d$ such that $Ju = f$? As a partial positive evidence, they proved:

Theorem 3 ([3]) *For every $f \in H^1(\mathbb{R}^d)$, there are $u^i \in \dot{W}^{1,d}(\mathbb{R}^d)^d$ and $\alpha_i \geq 0$ such that*

$$f = \sum_{i=1}^{\infty} \alpha_i J(u^i), \qquad \|u^i\|_{\dot{W}^{1,d}(\mathbb{R}^d)^d} \leq 1, \qquad \sum_{i=1}^{\infty} \alpha_i \lesssim \|f\|_{H^1(\mathbb{R}^d)}.$$

What about the (perhaps more usual) non-homogeneous Sobolev space

$$\begin{aligned} W^{1,p}(\mathbb{R}^d) &:= \{v \in L^p(\mathbb{R}^d) : \nabla v \in L^p(\mathbb{R}^d)^d\}, \\ \subsetneq \dot{W}^{1,p}(\mathbb{R}^d) &:= \{v \in L^1_{\text{loc}}(\mathbb{R}^d) : \nabla v \in L^p(\mathbb{R}^d)^d\}. \end{aligned}$$

Given $f \in L^p(\mathbb{R}^d)$ (resp. $H^1(\mathbb{R}^d)$ if $p = 1$), could we even hope to find $u \in W^{1,pd}(\mathbb{R}^d)^d$ with $Ju = f$? It was only fairly recently that this was shown to fail, and in fact quite miserably:

Theorem 4 ([10]) *The set*

$$\Big\{\sum_{i=1}^{\infty} \alpha_i J(u^i) : \|u^i\|_{W^{1,pd}(\mathbb{R}^d)^d} \leq 1, \sum_{i=1}^{\infty} |\alpha_i| < \infty\Big\},$$

which obviously contains the image $JW^{1,pd}(\mathbb{R}^d)^d$, *has* first category *in* $L^p(\mathbb{R}^d)$ *if* $p \in (1,\infty)$ *resp. in* $H^1(\mathbb{R}^d)$ *if* $p = 1$.

Very roughly speaking, the reason for this negative result is the incompatibility of scaling in $W^{1,pd}(\mathbb{R}^d)^d$ on the one hand, and in $L^p(\mathbb{R}^d)$ if $p \in (1,\infty)$ resp. in $H^1(\mathbb{R}^d)$ if $p = 1$ on the other hand, but the precise argument is more delicate.

2 Functional Analysis Behind the Results

Both the existence (in Theorem 3) and the non-existence (in Theorem 4) of the representation $f = \sum \alpha_i J(u^i)$ are based on the following functional analytic lemma from [3] and its elaboration from [10]:

Lemma 5 ([3]) *Let* $V \subset X$ *be a symmetric bounded subset of a Banach space* X. *Then the following are equivalent:*

1. *Every* $x \in X$ *can be written as* $x = \sum_{k=1}^{\infty} \alpha_k v_k$, *where* $v_k \in V$, $\alpha_k \geq 0$ *and* $\sum_{k=1}^{\infty} \alpha_k < \infty$.
2. V *is* norming *for* X^*, *i.e.,* $\|\lambda\|_{X^*} \eqsim \sup_{v \in V} |\langle \lambda, v \rangle| \quad \forall \lambda \in X^*$.

Lemma 6 ([10]) (1) *either holds for all* $x \in X$, *or in a subset of first category.*

For the mentioned theorems, these lemmas are applied with the symmetric set $V = J(B)$, where B = unit ball of $\dot{W}^{1,pd}(\mathbb{R}^d)^d$ or $W^{1,pd}(\mathbb{R}^d)^d$, which is a bounded subset of the Banach space $X = L^p(\mathbb{R}^d)$ or $X = H^1(\mathbb{R}^d)$. Via the equivalent condition (2), the well-known dual spaces $X^* = L^{p'}(\mathbb{R}^d)$ or $X^* = \mathrm{BMO}(\mathbb{R}^d)$ enter the considerations.

In order to obtain Theorem 3, [3] proved that

Proposition 7 ([3]) *Let* $d \geq 2$. *For every* $b \in \mathrm{BMO}(\mathbb{R}^d)$, *we have*

$$\|b\|_{\mathrm{BMO}(\mathbb{R}^d)} \eqsim \sup\Big\{\Big|\int bJ(u)\Big| : \|\nabla u\|_d \leq 1\Big\}.$$

The analogous result for $p \in (1,\infty)$ read as follows:

Theorem 8 ([5]) *Let* $d \geq 2$ *and* $p \in (1,\infty)$. *For every* $f \in L^p(\mathbb{R}^d)$, *there are* $u^i \in \dot{W}^{1,dp}(\mathbb{R}^d)^d$ *and* $\alpha_i \geq 0$ *such that*

$$f = \sum_{i=1}^{\infty} \alpha_i J(u^i), \quad \|u^i\|_{\dot{W}^{1,dp}(\mathbb{R}^d)^d} \leq 1, \quad \sum_{i=1}^{\infty} \alpha_i \lesssim \|f\|_{L^p(\mathbb{R}^d)}.$$

Proposition 9 ([5]) *Let* $d \geq 2$ *and* $p \in (1,\infty)$. *For every* $b \in L^{p'}(\mathbb{R}^d)$, *we have*

$$\|b\|_{L^{p'}(\mathbb{R}^d)} \eqsim \sup\Big\{\Big|\int bJ(u)\Big| : \|\nabla u\|_{dp} \leq 1\Big\}.$$

3 Complex Reformulation and Connection to Commutators for $d = 2$

The various results formulated above are valid, as stated, in all dimensions $d \geq 2$, and their proofs in this generality can be found in the quoted references. We now restrict ourselves to dimension $d = 2$ in order to discuss an alternative complex-variable approach that is available in this situation, as suggested by Lindberg [10].

For $u = (u_1, u_2) \in \dot{W}^{1,2p}(\mathbb{R}^2; \mathbb{R}^2)$, let us denote

$$h := u_1 + iu_2 \in \dot{W}^{1,2p}(\mathbb{C}; \mathbb{C}), \quad \partial := \frac{1}{2}(\partial_1 - i\partial_2), \; \bar{\partial} := \frac{1}{2}(\partial_1 + i\partial_2).$$

Then, after some algebra, we find that

$$Ju = \det\begin{pmatrix} \partial_1 u_1 & \partial_2 u_1 \\ \partial_1 u_2 & \partial_2 u_2 \end{pmatrix} = |\partial h|^2 - |\bar{\partial} h|^2 =: |S(v)|^2 - |v|^2,$$

where $v := \bar{\partial} h \in L^{2p}(\mathbb{C})$ is in isomorphic correspondence with $h \in \dot{W}^{1,2p}(\mathbb{C}; \mathbb{C})$, and S is the (Ahlfors–)Beurling (or 2D Hilbert) transform

$$Sv(z) = -\frac{1}{\pi}\text{p.v.}\int_{\mathbb{C}} \frac{v(y)\,\mathrm{d}y_1\,\mathrm{d}y_2}{(z-y)^2},$$

which satisfied the fundamental relation $S \circ \bar{\partial} = \partial$ and maps $S : L^p(\mathbb{C}) \to L^p(\mathbb{C})$ bijectively and isometrically for $p = 2$ and isomorphically for all $p \in (1, \infty)$.

Let us now see how Propositions 7 and 9 are connected to commutators when $d = 2$. By the reformulations just discussed, we have

$$\sup\Big\{\Big|\int bJ(u)\Big| : \|u\|_{\dot{W}^{1,2p}(\mathbb{R}^2;\mathbb{R}^2)} \leq 1\Big\} \eqsim \sup\Big\{\Big|\int b(|Sv|^2 - |v|^2)\Big| : \|v\|_{L^{2p}(\mathbb{C})} \leq 1\Big\}$$

denoting $v = \bar{\partial}(u_1 + iu_2)$. We claim that the right side can be further written as

$$\eqsim \sup\Big\{\Big|\int b(Sv\overline{Sw} - v\bar{w})\Big| : \|v\|_{L^{2p}(\mathbb{C})}, \|w\|_{L^{2p}(\mathbb{C})} \leq 1\Big\}. \tag{1}$$

In fact, "$\lesssim$" is obvious, while "$\gtrsim$" follows from the elementary polarisation identity

$$a\bar{b} = \frac{1}{4}\sum_{\varepsilon = \pm 1, \pm i} \varepsilon|a + \varepsilon b|^2, \qquad a, b \in \mathbb{C},$$

applied pointwise to both $(a, b) = (Sv, Sw)$ and $(a, b) = (v, w)$, which implies that

$$\begin{aligned} Sv\overline{Sw} - v\overline{w} &= \frac{1}{4}\sum_{\varepsilon=\pm 1,\pm i} \varepsilon|Sv - \varepsilon Sw|^2 - \frac{1}{4}\sum_{\varepsilon=\pm 1,\pm i} \varepsilon|v - \varepsilon w|^2 \\ &= \frac{1}{4}\sum_{\varepsilon=\pm 1,\pm i} \varepsilon\Big(|S(v - \varepsilon w)|^2 - |v - \varepsilon w|^2\Big), \end{aligned}$$

where $\|v - \varepsilon w\|_{2p} \leq \|v\|_{2p} + \|w\|_{2p} \leq 2$ if $\|v\|_{2p}, \|w\|_{2p} \leq 1$.

Denoting $g := \overline{Sw}$, we have $\overline{g} = Sw$ and hence $S^*\overline{g} = S^*Sw = w$, where we denoted by S^* the conjugate-linear adjoint of S and used the fact that S^*S is the identity. With this substitution, $g \in L^{2p}(\mathbb{C})$ and $w \in L^{2p}(\mathbb{C})$ are in isomorphic correspondence, and we have

$$(1) \eqsim \sup\left\{\left|\int b(Sv \cdot g - v\overline{S^*\overline{g}})\right| : \|v\|_{L^{2p}(\mathbb{C})}, \|g\|_{L^{2p}(\mathbb{C})} \leq 1\right\}$$

Finally, using the duality $\int \phi\overline{S^*\psi} = \int S\phi \cdot \overline{\psi}$ with $\phi = bv$ and $\psi = \overline{g}$, we have

$$\int b(Sv \cdot g - v\overline{S^*\overline{g}}) = \int (b \cdot Sv \cdot g - S(bv) \cdot \overline{\overline{g}}) = \int g \cdot [b, S]v, \qquad (2)$$

where we finally introduced the commutator

$$[b, S]v = bSv - S(bv).$$

Now the supremum of (the absolute value of) (2) over $\|g\|_{2p} \leq 1$ is the dual norm $\|[b, S]v\|_{(2p)'}$, and the supremum of this over $\|v\|_{2p} \leq 1$ is the operator norm

$$\|[b, S]\|_{L^{2p}(\mathbb{C}) \to L^{(2p)'}(\mathbb{C})}.$$

Summarising the discussion, we have proved:

Lemma 10 *Let* $p \in [1, \infty)$. *Then*

$$\sup\left\{\left|\int bJ(u)\right| : \|u\|_{\dot{W}^{1,2p}(\mathbb{R}^2;\mathbb{R}^2)} \leq 1\right\} \eqsim \|[b, S]\|_{L^{2p}(\mathbb{C}) \to L^{(2p)'}(\mathbb{C})}.$$

Thus Propositions 7 and 9, for $d = 2$, are reduced to understanding the norm of the Beurling commutator $[b, S] : L^{2p}(\mathbb{C}) \to L^{(2p)'}(\mathbb{C})$. When $p = 1$, we have $2p = (2p)' = 2$, and we are talking about L^2-boundedness of commutators, which is a well-studied topic since the pioneering work of [2]. When $p \in (1, \infty)$, we have $2p > 2 > (2p)'$, and we are talking about the boundedness of commutators between different L^p spaces. This, too, has been well studied in the case that the target space exponent is larger (cf. [7]), but we are now precisely in the complementary regime. In this case, the result was only achieved very recently.

4 The Commutator Theorem

Complementing various classical results starting with [2], the following result was recently completed in [5]:

Theorem 11 *Let $T = S$ with $d = 2$, or more generally, let T be any "uniformly non-degenerate" Calderón–Zygmund operator on $\mathbb{R}^d$, $d \geq 1$. Let $1 < p, q < \infty$ and $b \in L^1_{\mathrm{loc}}(\mathbb{R}^d)$. Then*

$$[b, T] : L^p(\mathbb{R}^d) \to L^q(\mathbb{R}^d) \quad \textit{boundedly}$$

if and only if

1. *$p = q$ and $b \in \mathrm{BMO}$ [2], or*
2. *$p < q \leq p^*$, where $\frac{1}{p^*} := (\frac{1}{p} - \frac{1}{d})_+$, and $b \in C^{0,\alpha}$ with $\alpha = d(\frac{1}{p} - \frac{1}{q})$, or*
3. *$q > p^*$ and b is constant (this and the previous case are due to [7]), or*
4. *$p > q$ and $b = a + c$, where c is constant and $a \in L^r$ for $\frac{1}{r} = \frac{1}{q} - \frac{1}{p}$ [5].*

Aside from the new regime of exponents $p > q$, another novelty of [5] (also when $p \leq q$) is the validity of the "only if" implication under the fairly general "uniform non-degeneracy" assumption on T. Recall that [2] proved this direction only for the Riesz transfroms, and [7, 11] for "smooth enough" kernels, which has been gradually relaxed in subsequent contributions.

The usual Calderón–Zygmund size condition requires the upper bound

$$|K(x, y)| \leq \frac{c_K}{|x - y|^d}.$$

on the kernel K of T. "Uniform non-degeneracy" means that we have a matching lower bound essentially over all positions and length-scales, more precisely: For every $y \in \mathbb{R}^d$ and $r > 0$, there is x such that $|x - y| \eqsim r$ and

$$|K(x, y)| \geq \frac{c_0}{|x - y|^d}.$$

This is manifestly the case for the Beurling operator, whose kernel $K(x, y) = -\pi^{-1}/(x - y)^2$ satisfies both bounds with an equality.

More generally, Theorem 11 holds for both

1. two-variable kernels $K(x, y)$ (with very little continuity), and
2. rough homogeneous kernels

$$K(x, y) = K(x - y) = \frac{\Omega((x - y)/|x - y|)}{|x - y|^d}$$

as soon as Ω is not identically zero; this was conjectured by Lerner et al. [9], who came very close for $p = q$.

We refer the reader to [5] for the proof of Theorem 11 in the stated generality; below we give a much simpler argument in the particular case of the Beurling operator $T = S$, which is relevant for the two-dimensional Jacobian problem, as discussed above.

Indeed, for $d = 2$, Theorems 3 and 8 are direct corollaries of Theorem 11 (via the earlier discussion). For $d > 2$, they are not direct consequences of Theorem 11 itself, but they can nevertheless be proved by adapting the ideas of the proof of Theorem 11; see again [5] for details.

5 The Classical Implications

We begin with a brief discussion of the "if" implications in Theorem 11:

1. The case $p = q$ and $b \in$ BMO is the only non-trivial "if" statement in Theorem 11. There are many excellent discussions of this bound (including two entirely different proofs already in [2]), so we skip it here.
2. If $p < q$ and $b \in C^{0,\alpha}$, we only need the size bound $|K(x, y)| \lesssim |x - y|^{-d}$ to see that

$$\begin{aligned}|[b, T]f(x)| &= \left|\int (b(x) - b(y))K(x, y)f(y)\,\mathrm{d}y\right| \\ &\leq \int |b(x) - b(y)||K(x, y)||f(y)|\,\mathrm{d}y \\ &\lesssim \int |x - y|^{\alpha}|x - y|^{-d}|f(y)|\,\mathrm{d}y.\end{aligned}$$

   This is a fractional integral with well-known $L^p \to L^q$ bounds!
3. If $b = c =$ constant, then $[b, T] = 0$ is trivially bounded.
4. If $p > q$ and $b \in L^r$, we use the boundedness of $T : L^p \to L^p$ and $T : L^q \to L^q$ together with Hölder's inequality

$$\|bf\|_q \leq \|b\|_r \|f\|_p, \qquad \frac{1}{q} = \frac{1}{r} + \frac{1}{p}$$

   to see that both bT and Tb individually are $L^p \to L^q$ bounded.

We then turn to the "only if" part, starting with the beautiful classical argument of [2] for $p = q$. Given a function $b \in L^1_{\mathrm{loc}}(\mathbb{C})$ and a ball (disc) $B = B(z, r) \subset \mathbb{C}$, we can pick an auxiliary function σ with $|\sigma(x)| = 1_B(x)$ so that

$$\begin{aligned}\int_B |b(x) - \langle b\rangle_B|\,\mathrm{d}x &= \int_B (b(x) - \langle b\rangle_B)\sigma(x)\,\mathrm{d}x,\\ &= \frac{1}{|B|}\int_B\int_B (b(x) - b(y))\sigma(x)\,\mathrm{d}x\,\mathrm{d}y\\ &= \int_B\int_B \frac{b(x) - b(y)}{(x-y)^2}\frac{(x-z)^2 - 2(x-z)(y-z) + (y-z)^2}{\pi r^2}\sigma(x)\,\mathrm{d}x\,\mathrm{d}y\\ &= \sum_{i=1}^{3}\int g_i(x)\Big(\int \frac{b(x)-b(y)}{(x-y)^2} f_i(y)\,\mathrm{d}y\Big)\,\mathrm{d}x = \sum_{i=1}^{3}\int g_i[b, S]f_i,\end{aligned}$$

for suitable functions f_i, g_i with $|f_i(x)| + |g_i(x)| \lesssim 1_B(x)$, whose explicit formulae can be easily deduced from above. Thus

$$\int_B |b - \langle b\rangle_B| \leq \sum_{i=1}^{3} \|[b, S]\|_{L^p\to L^p}\|f_i\|_p\|g_i\|_{p'} \lesssim \|[b, S]\|_{L^p\to L^p}|B|^{1/p}|B|^{1/p'}.$$

Dividing by $|B|^{1/p}|B|^{1/p'} = |B|$ and taking the supremum over all B proves that $\|b\|_{\mathrm{BMO}} \lesssim \|[b, S]\|_{L^p\to L^p}$.

With a simple modification of the previous display observed by Janson [7], we also find that

$$\int_B |b - \langle b\rangle_B| \leq \sum_{i=1}^{3} \|[b, S]\|_{L^p\to L^q}\|f_i\|_p\|g_i\|_{q'} \lesssim \|[b, S]\|_{L^p\to L^q}|B|^{1/p}|B|^{1/q'},$$

where

$$|B|^{1/p+1/q'} = |B|^{(1/p-1/q)+1} \approx |B|\cdot r_B^{d(1/p-1/q)} = |B|\cdot r_B^{\alpha}.$$

Thus

$$\fint_B |b - \langle b\rangle_B| \lesssim r_B^{\alpha},$$

which a well-known characterisation of $b \in C^{0,\alpha}$. For $\alpha > 1$, this space has nothing but the constant functions, completing the sketch of the proof of all the classical "only if" statements of Theorem 11.

6 The New Case $p > q$

We finally discuss the proof of the "only if" implication of Theorem 11 in the case $p > q$ that was only recently discovered in [5]. The above estimate

$$\int_B |b - \langle b \rangle_B| \lesssim |B|^{1/p+1/q'} = |B|^{(1/p-1/q)+1} = |B|^{-1/r+1} = |B|^{1/r'}$$

is still true but seems to be useless in this range. How do we even check that a given function is in L^r + constants?

A convenient tool is as follows:

Lemma 12 ([5], Lemma 3.6) *If we have the following bound uniformly for cubes $Q \subset \mathbb{R}^d$:*

$$\|b - \langle b \rangle_Q\|_{L^r(Q)} \leq C,$$

then there is a constant c ($= \lim_{Q \to \mathbb{R}^d} \langle b \rangle_Q$) such that

$$\|b - c\|_{L^r(\mathbb{R}^d)} \leq C.$$

To estimate the local L^r norm, the following result is useful. Depending on one's background, one may like to call it an iterated Calderón–Zygmund or atomic decomposition; one can also view it as a toy version of an influential formula of [8], featuring merely measurable functions in place $L^1(Q_0)$, the median of b in place of the mean $\langle b \rangle_{Q_0}$, etc. "Sparse bounds" of this type have been extensively used in the last few years; the version below is very elementary compared to several recent highlights, but quite sufficient for the present purposes.

Lemma 13 *Given a cube $Q_0 \subset \mathbb{R}^d$ and $b \in L^1(Q_0)$, there is a* sparse *collection $\mathbb{S}$ of the family $\mathbb{D}(Q_0)$ of dyadic subcubes of Q_0 such that*

$$1_{Q_0}(x)|b(x) - \langle b \rangle_{Q_0}| \lesssim \sum_{Q \in \mathbb{S}} 1_Q(x) \fint_Q |b - \langle b \rangle_Q|.$$

A collection of cubes $\mathbb{S}$ is called *sparse* (or almost disjoint) if there are pairwise disjoint *major subsets* $E(Q) \subset Q$ for each $Q \in \mathbb{S}$, meaning that

$$E(Q) \cap E(Q') = \varnothing \quad (\forall Q \neq Q'), \qquad |E(Q)| \geq \frac{1}{2}|Q|.$$

For L^p estimates, sparse is almost as good as disjoint; namely,

$$\Big\| \sum_{Q \in \mathbb{S}} \lambda_Q 1_Q \Big\|_p \eqsim \Big(\sum_{Q \in \mathbb{S}} \lambda_Q^p |Q| \Big)^{1/p}, \qquad \forall \lambda_Q \geq 0, \tag{3}$$

where equality would hold for a disjoint collection.

With these tools at hand, we are ready to prove that $[b, S] : L^p \to L^q$ for $1 < q < p < \infty$ only if $b = a + c$, where $a \in L^r$ with $\frac{1}{r} = \frac{1}{q} - \frac{1}{p}$ and c is constant. For any cube $Q_0 \subset \mathbb{R}^d$, we estimate

$$\begin{aligned}
\|b - \langle b \rangle_{Q_0}\|_{L^r(Q_0)} &\lesssim \Big\| \sum_{Q \in \mathbb{S}} 1_Q \fint_Q |b - \langle b \rangle_Q| \Big\|_{L^r(Q_0)} && \text{(by Lemma 13)} \\
&\eqsim \Big(\sum_{Q \in \mathbb{S}} |Q| \Big[\fint_Q |b - \langle b \rangle_Q| \Big]^r \Big)^{1/r} && \text{(by (3))} \\
&= \sum_{Q \in \mathbb{S}} |Q| \lambda_Q \fint_Q |b - \langle b \rangle_Q| = \sum_{Q \in \mathbb{S}} \lambda_Q \int_Q |b - \langle b \rangle_Q|,
\end{aligned}$$

with a suitable dualising sequence λ_Q such that

$$\sum_{Q \in \mathbb{S}} |Q| \lambda_Q^{r'} = 1. \tag{4}$$

By the same considerations as in Sect. 5 in the case of just one ball B, for each of the cubes $Q \in \mathbb{S}$ above we find functions f_Q^i, g_Q^i with

$$|f_Q^i| + |g_Q^i| \lesssim 1_Q \tag{5}$$

such that

$$\int_Q |b - \langle b \rangle_Q| = \sum_{i=1}^{3} \int g_Q^i [b, S] f_Q^i.$$

Summarising the discussion so far, we have

$$\|b - \langle b \rangle_{Q_0}\|_{L^r(Q_0)} \lesssim \sum_{i=1}^{3} \sum_{Q \in \mathbb{S}} \lambda_Q \int g_Q^i [b, S] f_Q^i, \tag{6}$$

where the coefficient λ_Q and the functions f_Q^i, g_Q^i satisfy (4) and (5).

We now enter independent random signs ε_Q on some probability space, and denote by $\mathbb{E}$ the expectation. (For the Jacobian theorem in $d > 2$: we need to use random dth roots of unity at the analogous step, see [5].) With the basic orthogonality $\mathbb{E}(\varepsilon_Q \varepsilon_{Q'}) = \delta_{Q,Q'}$ and Hölder's inequality after observing that

$$\frac{1}{r} = \frac{1}{q} - \frac{1}{p} \quad \Rightarrow \quad \frac{1}{r'} = \frac{1}{q'} + \frac{1}{p} \quad \Rightarrow \quad 1 = \frac{r'}{q'} + \frac{r'}{p},$$

we have

$$
\begin{aligned}
RHS(6) &= \sum_{i=1}^{3}\mathbb{E}\int\Big(\sum_{Q\in\mathbb{S}}\varepsilon_Q\lambda_Q^{r'/q'}g_Q^i\Big)[b,S]\Big(\sum_{Q'\in\mathbb{S}}\varepsilon_{Q'}\lambda_{Q'}^{r'/p}f_{Q'}^i\Big)\\
&\lesssim \|[b,S]\|_{L^p\to L^q}\Big\|\sum_{Q\in\mathbb{S}}\lambda_Q^{r'/q'}1_Q\Big\|_{q'}\Big\|\sum_{Q\in\mathbb{S}}\lambda_Q^{r'/p}1_Q\Big\|_p \qquad \text{(by (5))}\\
&\lesssim \|[b,S]\|_{L^p\to L^q}\Big(\sum_{Q\in\mathbb{S}}\lambda_Q^{r'}|Q|\Big)^{1/q'}\Big(\sum_{Q\in\mathbb{S}}\lambda_Q^{r'}|Q|\Big)^{1/p} \qquad \text{(by (3))}\\
&= \|[b,S]\|_{L^p\to L^q} \qquad \text{(by (4))}.
\end{aligned}
$$

This shows that

$$\|b-\langle b\rangle_{Q_0}\|_{L^r(Q_0)}\lesssim\|[b,S]\|_{L^p\to L^q}$$

for every cube Q_0, and hence

$$\|b-c\|_{L^r(\mathbb{C})}\lesssim\|[b,S]\|_{L^p\to L^q}$$

for some constant c by Lemma 12. If we a priori know that $b\in L^r(\mathbb{C})$ (as in Proposition 9), then necessarily $c=0$, and we obtain the desired quantitative bound for $\|b\|_{L^r(\mathbb{C})}$.

Acknowledgments The author is supported by the Academy of Finland via project Nos. 307333 (Centre of Excellence in Analysis and Dynamics Research) and 314829 (Frontiers of singular integrals).

References

1. Avila, A.: On the regularization of conservative maps. Acta Math. **205**(1), 5–18 (2010)
2. Coifman, R.R., Rochberg, R., Weiss, G.: Factorization theorems for Hardy spaces in several variables. Ann. Math. (2) **103**(3), 611–635 (1976)
3. Coifman, R., Lions, P.-L., Meyer, Y., Semmes, S.: Compensated compactness and Hardy spaces. J. Math. Pures Appl. (9) **72**(3), 247–286 (1993)
4. Dacorogna, B., Moser, J.: On a partial differential equation involving the Jacobian determinant. Ann. Inst. H. Poincaré Anal. Non Linéaire **7**(1), 1–26 (1990)
5. Hytönen, T.P.: The L^p-to-L^q boundedness of commutators with applications to the Jacobian operator. Preprint, arXiv:1804.11167 (2018)
6. Iwaniec, T.: Nonlinear commutators and Jacobians. J. Fourier Anal. Appl. **3**, 775–796 (1997)
7. Janson, S.: Mean oscillation and commutators of singular integral operators. Ark. Mat. **16**(2), 263–270 (1978)
8. Lerner, A.K.: A pointwise estimate for the local sharp maximal function with applications to singular integrals. Bull. Lond. Math. Soc. **42**(5), 843–856 (2010)

9. Lerner, A.K., Ombrosi, S., Rivera-Ríos, I.P.: Commutators of singular integrals revisited. Bull. Lond. Math. Soc. **51**(1), 107–119 (2019)
10. Lindberg, S.: On the Hardy space theory of compensated compactness quantities. Arch. Ration. Mech. Anal. **224**(2), 709–742 (2017)
11. Uchiyama, A.: On the compactness of operators of Hankel type. Tôhoku Math. J. (2) **30**(1), 163–171 (1978)
12. Ye, D.: Prescribing the Jacobian determinant in Sobolev spaces. Ann. Inst. H. Poincaré Anal. Non Linéaire **11**(3), 275–296 (1994)

On Regularity and Irregularity of Certain Holomorphic Singular Integral Operators

Loredana Lanzani and Elias M. Stein

Dedicated to Fulvio Ricci, on the occasion of his 70th birthday

Abstract We survey recent work and announce new results concerning two singular integral operators whose kernels are holomorphic functions of the output variable, specifically the Cauchy–Leray integral and the Cauchy–Szegő projection associated to various classes of bounded domains in $\mathbb{C}^n$ with $n \geq 2$.

Keywords Hardy space · Cauchy integral · Diederich-Fornæss worm domain

1 Introduction

This is a review of recent and forthcoming work concerning a menagerie of singular integral operators in several complex variables whose kernels are holomorphic functions of the output variable. (All proofs have appeared or will appear elsewhere.) Our family of operators consists of the *Cauchy–Szegő Projection*, namely the orthogonal projection of $L^2(bD, \mu)$ onto the holomorphic Hardy space $\mathcal{H}^2(bD, \mu)$, as well as various higher-dimension analogs of the Cauchy integral for a planar curve that are collectively known as *Cauchy–Fantappiè integrals* and include the *Cauchy–Leray*

The author Loredana Lanzanis was supported in part by the National Science Foundation, award DMS-1503612.

The author Elias M. Stein was supported in part by the National Science Foundation, award DMS-1700180.

L. Lanzani (✉)
Department of Mathematics, Syracuse University, Syracuse, NY, USA
e-mail: llanzani@syr.edu

E. M. Stein
Department of Mathematics, Princeton University, Princeton, NJ, USA

P. Ciatti, A. Martini (eds.), *Geometric Aspects of Harmonic Analysis*,
Springer INdAM Series 45, https://doi.org/10.1007/978-3-030-72058-2_14

integral as a particularly relevant example. We will henceforth denote such operators $\mathcal{S}$ and C, respectively. Here D is a bounded domain in complex Euclidean space $\mathbb{C}^n$ with $n \geq 2$; bD is the topological boundary of D, while μ is an appropriate measure supported on bD, and we will pay particular attention to two such measures, namely *induced Lebesgue measure* Σ, and the *Leray–Levi measure* λ.

To be precise, we are interested in *the L^p-regularity problem for C and for $\mathcal{S}$*, that is:

Determine regularity and geometric conditions on the ambient domain D that grant that

- $C : L^p(bD, \lambda) \to L^p(bD, \lambda)$ is bounded for all $p \in (1, \infty)$.
- $\mathcal{S} : L^p(bD, \Sigma) \to L^p(bD, \Sigma)$ is bounded for p in an interval of maximal size about $p_0 = 2$.

In complex dimension 1 (that is, for $D \Subset \mathbb{C}$) these problems are well understood; see [42] and references therein. Here we focus on dimension $n \geq 2$.

We point out that the Cauchy–Leray integral is a Calderón–Zygmund operator, thus L^p-regularity for $p = 2$ is equivalent to regularity in L^p for $1 < p < \infty$. On the other hand, the Cauchy–Szegő projection is automatically bounded in $L^2(bD, \mu)$ but L^2-regularity does not guarantee L^p-regularity for $p \neq 2$: indeed, establishing L^p-regularity of $\mathcal{S}$ for $p \neq 2$ is, in general, a very difficult problem. The main difficulty stems from the fact that the Schwartz kernel for $\mathcal{S}$ (that is the *Cauchy–Szegő kernel*) is almost never explicitly available, even in the favorable setting when D is smooth and strongly pseudoconvex, so direct estimates cannot be performed and one has to rely on other methods, such as asymptotic formulas analogous to those obtained by C. Fefferman [22] (for the Bergman kernel) and Boutet de Monvel–Sjöstrand [10], or a paradigm discovered by N. Kerzman and E. M. Stein [29] that relates $\mathcal{S}$ to a certain Cauchy–Fantappiè integral associated to D (the *Kerzman–Stein identity*).

About 30 years later, a surge of interest in singular integral operators in a variety of "non-smooth" settings led us to a new examination of these problems from the following point of view:

> to what extent is the L^p-boundedness of the aforementioned operators reliant upon the boundary regularity and (natural to this context) upon the amount of convexity of the ambient domain D?

Stripping away the smoothness assumptions brings to the fore the geometric interplay between the operators and the domains on which they act: it soon became apparent that new ideas and techniques were needed, even to deal with rather tame singularities such as the class $C^{2,\alpha}$. The following results were proved in [48] and [45].

(I) $\mathcal{S}$: $L^p(bD, \Sigma) \to L^p(bD, \Sigma)$ is bounded for $1 < p < \infty$ if

(i.) $D \Subset \mathbb{C}^n$ is strongly pseudoconvex, and
(ii.) bD is of class C^2.

(II) $\mathcal{C}$: $L^p(bD, \lambda) \to L^p(bD, \lambda)$ is bounded for $1 < p < \infty$ if

(i.) $D \Subset \mathbb{C}^n$ is strongly $\mathbb{C}$-linearly convex, and
(ii.) bD is of class $C^{1,1}$.

The purpose of this note is to summarize the main points in the proof of (I) and (II), and to announce new results that will appear in forthcoming papers [46] and [50] pertaining the optimality of the assumptions made in (I) and (II). Related results in the extensive literature can be found in e.g., [3, 5, 7, 8, 12, 14, 16, 18, 20, 21, 23–25, 27, 28, 30–36, 39–41, 43, 44, 52, 55, 56, 58–60, 63, 66, 70–73, 76].

2 The L^p-Regularity of the Cauchy–Leray Integral

Here we take as our model the seminal one-dimensional theory of Calderòn [11], Coifman et al. [17], and David [19] for the Cauchy integral of a planar curve, and in particular its key theorem on the *Lipschitz* case. As is well known, the initial result was the classical theorem of M. Riesz for the Cauchy integral on the unit disc (i.e. the Hilbert transform on the circle); the standard proofs which developed from this then allowed an extension to a corresponding result where the disc is replaced by a domain $D \subset \mathbb{C}$ whose boundary is relatively smooth, i.e. of class $C^{1,\alpha}$, for $\alpha > 0$. However, going beyond that to the limiting case of regularity, namely C^1 and other variants "near C^1", required further ideas. The techniques introduced in this connection led to significant developments in harmonic analysis such as the "$T(1)$ theorem" and various aspects of multilinear analysis and analytic capacity, [15, 54, 74, 75]. The importance of those advances suggests the following fundamental question: *what might be the corresponding results for the Cauchy integral in several variables*. However, in the context of higher dimension geometric obstructions arise (pseudoconvexity or, equivalently, lack of conformal mapping) which in the one-dimensional setting are irrelevant. As a consequence, there is no canonical notion of *holomorphic* Cauchy kernel: all such kernels must be domain-specific. Indeed, the only kernel that can be deemed "canonical"[1] is the Bochner–Martinelli kernel [38], but such kernel is nowhere holomorphic and thus of no use in the applications described below. One is therefore charged with the further task of *constructing* a holomorphic kernel that is fitted to the specific geometry of the domain and, after that, with supplying proof of regularity of the resulting singular integral operator. As in the one-dimensional setting, this theory was first conceived within the context of *smooth* ambient domains; if the domain is not sufficiently smooth (of class $C^{2,\alpha}$ or better) the original kernel constructions by Henkin and

[1]"Canonical" in the sense that it is the restriction to bD of a *universal* kernel defined in $\mathbb{C}^n \times \mathbb{C}^n \setminus \{z = w\}$.

Ramirez, [26] and [67], and the "osculation by the Heisenberg group" technique in Kerzman–Stein [29] are no longer applicable. In [45] it is shown that the $T(1)$-theorem technique for a space of homogeneous type fitted to the geometry and regularity of the ambient domain can be applied to prove $L^p(bD, \mu)$-regularity, for $1 < p < \infty$ of the *Cauchy–Leray integral*:

$$\mathcal{C}f(z) = \frac{1}{(2\pi i)^n} \int\limits_{w \in bD} f(w) \frac{\Omega(z,w) \wedge (\mathrm{d}_w \Omega(z,w))^{n-1}}{\langle \Omega(z,w), w - z \rangle^n}, \quad z \in D \tag{1}$$

whenever $D \subset \mathbb{C}^n$ is a bounded, strongly $\mathbb{C}$-linearly convex domain whose boundary satisfies the minimal regularity condition given by the class $C^{1,1}$ (that is, the domain admits a defining function ρ of class $C^{1,1}$). Here the *generating 1-form* $\Omega(z, w)$ is the complex gradient of the domain's defining function (and we should point out that the definition of $\mathcal{C}$ is independent of the choice of defining function ρ). More precisely: $\Omega(z, w) = j^*\partial\rho(w)$, where j^* denotes the pull-back under the inclusion map $j : bD \hookrightarrow \mathbb{C}^n$. The boundary measure μ belongs to a family that includes induced Lebesgue measure Σ, as well as the Leray–Levi measure

$$\mathrm{d}\lambda(w) := (2\pi i)^{-2n} j^* \left(\partial\rho \wedge (\overline{\partial}\partial\rho)^{n-1} \right)(w), \quad w \in bD. \tag{2}$$

We remark that under our assumptions (class $C^{1,1}$) the factor $\overline{\partial}\partial\rho$ in the definition of the Leray–Levi measure λ, as well as the factor $\mathrm{d}_w\Omega(z, w)$ in the Schwartz kernel for $\mathcal{C}$, are only in $L^\infty(\mathbb{C}^n)$ and therefore may be undefined on bD because the latter is a zero-measure subset of $\mathbb{C}^n$, however it turns out that the *tangential component* of each of $\overline{\partial}\partial\rho(w)$ and $\mathrm{d}_w\Omega(z, w)$, namely $j^*\overline{\partial}\partial\rho(w)$ and $j^*\mathrm{d}_w\Omega(z, w)$, are in fact meaningful, leading to a kernel that is well-defined even in our singular context.

3 Counter-Examples to the L^p-Theory for the Cauchy–Leray Integral

In [46] we construct two examples that establish the optimality of the assumptions made on the ambient domain for the Cauchy–Leray integral $\mathcal{C}$. Both examples are real ellipsoids of the form

$$D_{r,q} := \left\{ (z_1, z_2) \in \mathbb{C}^2 \;\middle|\; |\mathrm{Re}\, z_1|^r + |\mathrm{Im}\, z_1|^q + |z_2|^2 < 1 \right\}. \tag{3}$$

For the first example, $(r, q) = (2, 4)$; in this case the domain is smooth, strongly pseudoconvex and strictly convex, but it is not strongly $\mathbb{C}$-linearly convex. In the second example, $(r, q) = (m, 2)$ for any $1 < m < 2$; this domain is strongly $\mathbb{C}$-linearly convex but is only of class $C^{1,m-1}$ (and no better). In both cases we show that the associated Cauchy–Leray integral C is well-defined on a dense subset of $L^p(bD_{r,q}, \mu)$ but *does not extend to a bounded operator*: $L^p(bD_{r,q}, \mu) \mapsto L^p(bD_{r,q}, \mu)$ for any $1 < p < \infty$. Specifically, we prove that there is a function $f \in C^1(bD_{r,q})$ supported in a proper subset of $bD_{r,q}$ such that

(i.) $Cf(z)$ can be defined as an absolutely convergent integral whenever $z \in bD_{r,q}$ is at positive distance from the support of f.

(ii.) The inequality: $\|C(f)\|_{L^p(S,d\mu)} \leq A_p\|f\|_{L^p(bD_{r,q},d\mu)}$ (with A_p independent of S) *fails* whenever $S \subset D_{r,q}$ is disjoint from the support of f.

Here μ is a boundary measure that belongs to a family that includes induced-Lebesgue measure Σ and the Leray–Levi measure λ, as well as Fefferman's measure μ^F, see [22]; for the first example (the smooth domain $D_{2,4}$) all such measures are mutually absolutely continuous. For the second example (the non-smooth domain $D_{m,2}$) these measures are essentially different, yet the counter-example holds in all cases.

The main tool for proving (i.) and (ii.) is a scaling and limiting process that transfers the problem to specific, unbounded smooth domains, namely $\{2\,\mathrm{Im}\, z_2 > (\mathrm{Re}\, z_1)^2\}$ in the first case, and $\{2\,\mathrm{Im}\, z_2 > |\mathrm{Re}\, z_1|^m\}$ in the second. On the unbounded domains, explicit computations are carried out to prove failure of the L^p-boundedness of the transported operator.

There is also the matter of showing that the Cauchy–Leray integral for $D_{r,q}$ maps L^p into the holomorphic Hardy space $\mathcal{H}^p$: this question is addressed in [47] where it is shown that

(iii.) $Cf(z)$, for $z \in bD_{r,q}$ as in item (i.) above, arises as "boundary value" of a function F holomorphic in $D_{r,q}$.

The proof of (iii.) requires three different approaches, each tailored to the particular type of singularity displayed by the example under consideration: in dealing with the non-smooth domain $D_{m,2}$ one has to distinguish the case when $1 < m \leq 3/2$ from the case $3/2 < m < 2$: in the second case, a global integration by parts gives that $Cf(z)$ is the restriction to $bD_{m,2}$ of a holomorphic $F \in C^1(\overline{D}_{m,2})$. On the other hand, when $1 < m \leq 3/2$ such method is no longer viable but we show nonetheless, that Cf extends to a holomorphic F that is continuous everywhere on $\overline{D}_{m,2}$ except for a 0-measure subset of the boundary (namely the sphere $\{|\mathrm{Re}\, z_1|^2 + |z_2|^2 = 1\}$).

Finally, the lack of strong $\mathbb{C}$-linear convexity in the first example (the domain $D_{2,4}$) prevents us from carrying a global integration by parts: instead, one shows that Cf extends to a holomorphic $F \in C(\overline{D}_{2,4})$ by using a local integration by parts which depends on the location of the coordinate patch with respect to the "flat" part of the boundary. It should be noted that an earlier result in Barrett–Lanzani [6] already gave an example with irregularity in $L^2(bD, \mu)$, however the less explicit and more complex nature of the construction did not provide insight for $L^p(bD, \mu)$ when $p \neq 2$.

4 The L^p-Regularity of the Cauchy–Szegő Projection

4.1 Discussion of the Problem

We recall that the Cauchy–Szegő projection $\mathcal{S}$ is the unique, *orthogonal* (equivalently, *selfadjoint*) projection operator of $L^2(bD, \Sigma)$ onto the Hardy space of holomorphic functions; here Σ is the induced Lebesgue measure on bD. As mentioned earlier, one must come to terms with the fact that, in general, orthogonal projections are *not* Calderón–Zygmund operators, thus L^p-regularity for $p \neq 2$ does not follow from L^2-regularity; also, one may have L^p-regularity only for p in a proper *sub-interval* of $(1, \infty)$, see e.g. [42]. (By contrast, for a Calderón–Zygmund operator boundedness in L^2 *implies* boundedness in L^p for $1 < p < \infty$.) Regularity properties of the Cauchy–Szegő projection, in particular L^p-regularity, have been the object of considerable interest for more than 40 years. When the boundary of the domain D is sufficiently smooth, decisive results were obtained in the following settings: *(a)*, when D is strongly pseudoconvex [10, 64]; *(b)*, when $D \subset \mathbb{C}^2$ and its boundary is of finite type [51, 62]; *(c)*, when $D \subset \mathbb{C}^n$ is convex and its boundary is of finite type [51, 53]; *(d)*, when $D \subset \mathbb{C}^n$ is of finite type and its Levi form is diagonalizable [13]. Related results include [1, 2, 4, 9, 37, 61, 65, 68, 69, 77]. The main difference when dealing with the situation when D has lower (in fact minimal) regularity than the setting of the more regular domains treated in *(a)*–*(d)*, is that in each of those cases known formulas for the Cauchy–Szegő kernel, or at least size estimates, played a decisive role. In our general situation such estimates are unavailable and one must proceed by a different analysis that relies upon `(i.)`, the $T(1)$-theorem technique of [45] and `(ii.)`, a new, tricky variant of the original Kerzman–Stein paradigm [29] described below.

4.2 L^p-Regularity of the Cauchy–Szegő Projection

Strong $\mathbb{C}$-linear convexity *implies* strong pseudoconvexity whenever the domain enjoys enough regularity for the latter to be meaningful. In [48] some of the techniques from [45] are adapted to study the L^p-regularity problem for the Cauchy–Szegő projection of strongly Levi-pseudoconvex domains $D \Subset \mathbb{C}^n$ with minimal boundary regularity, namely the class C^2 (which is the minimal regularity for strong Levi-pseudoconvexity to hold), leading to the conclusion that L^p-boundedness of $\mathcal{S}$ holds in the full range $1 < p < \infty$. As mentioned above, in this general setting a direct analysis of the Cauchy–Szegő kernel does not lead to the desired result. Instead, our starting point is the original Kerzman–Stein paradigm [29] for domains that are sufficiently smooth: this proceeded by constructing a holomorphic Cauchy–Fantappiè integral C in the same spirit of (1) but for a different choice of generating form Ω. The analysis of $\mathcal{S}$ begins with the representation: $C = \mathcal{S}(I - \mathcal{A})$ on $L^2(bD, \Sigma)$, where I is the identity and $\mathcal{A}$ denotes the difference of C and its formal L^2-adjoint, that is: $\mathcal{A} = C^* - C$. This identity follows from the fact that, just like the Cauchy–Szegő projection, the Cauchy–Fantappiè integral C is also a projection of $L^2(bD, \Sigma)$ onto the holomorphic Hardy space[2] (albeit not the orthogonal projection!). In particular, since $\mathcal{A}^* = -\mathcal{A}$ it follows that the operator $(I - \mathcal{A})$ is invertible in $L^2(bD, \Sigma)$ with bounded inverse, and we obtain:

$$\mathcal{S} = C(I - \mathcal{A})^{-1} \quad \text{on } L^2(bD, \Sigma). \tag{4}$$

Kerzman and Stein [29] proved that if the (strongly pseudoconvex) domain is sufficiently smooth (e.g. of class C^3) the singularities of C and C^* cancel out and as a result $\mathcal{A}$ is "small" in the sense that it is compact in $L^2(bD, \Sigma)$ (indeed smoothing); from this it follows that the righthand side of the above identity is bounded in $L^p(bD, \Sigma)$ for all $1 < p < \infty$ and therefore so is $\mathcal{S}$, giving the solution to the L^p-regularity problem for $\mathcal{S}$ in the full range $1 < p < \infty$.

If the domain is only of class C^2 this argument is no longer applicable because $\mathcal{A}$ in general fails to be compact on $L^2(bD, \Sigma)$, see [6]. Instead, in [48] we work with a *family* of holomorphic Cauchy–Fantappiè integrals $\{C_\epsilon\}_\epsilon$ whose kernels are constructed via a first-order perturbation of the Cauchy–Leray kernel (1) that makes use of a smooth approximation $\{\tau_\epsilon\}_\epsilon$ of certain second-order derivatives of the defining function of the domain. As in the case of the Cauchy–Leray integral C, here there are two boundary measures at play: the induced Lebesgue measure Σ, and the Leray–Levi measure λ, see (2), which in this new context is absolutely continuous with respect to Σ because of the relation

$$d\lambda(w) \approx |\varphi(w)| d\Sigma(w)\,, \quad w \in bD \tag{5}$$

where $\varphi(w)$ is the determinant of the Levi matrix. The operators $\{C_\epsilon\}_\epsilon$ are then seen to be bounded in $L^p(bD, \lambda)$ *and* $L^p(bD, \Sigma)$ for all $1 < p < \infty$ by an application of the $T(1)$-theorem. On the other hand, in defining the Cauchy–Szegő projection it is imperative to specify the underlying measure for bD that arises in the notion of orthogonality that is being used. Correspondingly, we now have two distinct Cauchy–Szegő projections $\mathcal{S}_\Sigma$ and $\mathcal{S}_\lambda$ but these, in our general setting, are not directly related to one another. It turns out that the Leray–Levi measure λ has a "mitigating" effect that leads to a new smallness argument for the difference $C_\epsilon^\dagger - C_\epsilon$ that occurs *when the adjoint $C_\epsilon^\dagger$ is computed with respect to* λ. While the $\{C_\epsilon\}_\epsilon$ do not approximate $\mathcal{S}_\lambda$ (in fact the norms of the C_ϵ are in general unbounded as $\epsilon \to 0$), we show that for each fixed $1 < p < \infty$ (in fact for $p < 2$) there is $\epsilon = \epsilon(p)$ such that $C_\epsilon^\dagger - C_\epsilon$ splits as the sum $\mathcal{B}_\epsilon + \mathcal{A}_\epsilon$, where $\mathcal{B}_\epsilon : L^p(bD, \lambda) \to C(bD)$, and $\|\mathcal{A}_\epsilon\|_{L^p \to L^p} \leq \epsilon$: this is the new, "tricky" variant of the original Kerzman–Stein paradigm that was alluded to earlier, and it gives us the identity

$$\mathcal{S}_\lambda = (\mathcal{S}_\lambda \mathcal{B}_\epsilon + C_\epsilon)(I - \mathcal{A}_\epsilon)^{-1} \quad \text{in} \quad L^2(bD, \lambda). \tag{6}$$

[2] It is failure of this property that renders the Bochner–Martinelli integral unsuitable for the analysis of $\mathcal{S}$.

Then one proves that the righthand side is bounded on $L^p(bD, \lambda)$ (here we also use that $p < 2$ and that D is bounded) and we conclude that $\mathcal{S}_\lambda$ is bounded in $L^p(bD, \lambda)$ whenever $1 < p < 2$; the result for $p > 2$ follows by duality. A similar argument is needed to treat $\mathcal{S}_\Sigma$, but there is no direct way to show smallness for $C_\epsilon^* - C_\epsilon$ when the adjoint C_ϵ^* is computed with respect to the induced Lebesgue surface measure Σ. Instead, one recovers such smallness from the corresponding result for $C_\epsilon^\dagger - C_\epsilon$, by observing that $C_\epsilon - C_\epsilon^* = C_\epsilon - C_\epsilon^\dagger + |\varphi|^{-1}\big[|\varphi|, C_\epsilon^\dagger\big]$, where φ is as in (5), and by controlling the size of the operator norm of the commutator $\big[|\varphi|, C_\epsilon^\dagger\big]$.

To complete the proof one also needs the requisite representation formulae and density results for the holomorphic Hardy spaces of the domains that satisfy the minimal boundary regularity conditions stated in (I) and (II): these are obtained in [49].

5 A Counter-Example to the L^p-Theory for the Cauchy–Szegő Projection

The forthcoming work [50] investigates a long-standing open question concerning *L^p-irregularity* of the Cauchy–Szegő projection for the *Diederich-Fornæss worm domains*:

$$W_{k,h} := \left\{ (z_1, z_2) \in \mathbb{C}^2\,,\ \left| z_2 - ie^{ih(|z_1|)} \right|^2 < 1 - k(|z_1|) \right\}\,. \tag{7}$$

Appropriate choices of the functions h and k produce domains that are smooth and pseudoconvex but only *weakly* pseudoconvex along a 2-dim subset of their topological boundary. (The nick-name "worm" is meant to illustrate winding caused by the argument $h(|z_1|)$.)

Developed by Diederich and Fornæss in 1977 as examples of smooth, weakly pseudoconvex domains with non-trivial Nebenhülle,[3] the class (7) has since proved to be a reliable source of counter-examples to a variety of phenomena in complex function theory. Of special relevance here are the seminal paper [4] and the related work [37] that prove *irregularity* of the Bergman projection[4] for the worm domain in the Sobolev- and Lebesgue-space scales, respectively, when the following choices are made for h and k:

$$h(|z_1|) := \log|z_1|^2\,; \qquad k(z_1) := \phi(h(|z_1|)) \tag{8}$$

[3]The domain is pseudoconvex but cannot be "exhausted" by smooth pseudoconvex "super-domains".

[4]That is, the orthogonal projection of $L^2(D, dV)$ onto the Bergman space $A^2(D) := \vartheta(D) \cap L^2(D, dV)$.

with ϕ a smooth, non-negative even function chosen so that $W_{h,k}$ is smooth, bounded, connected and pseudoconvex, and moreover $\phi^{-1}(0) = \{\, |t| \leq \beta - \pi/2 \}$ for fixed, given $\beta > \pi/2$.

In contrast with the situation for the Cauchy–Leray integral, the Cauchy–Szegő and Bergman projections are *always* bounded in L^2 (that is for $p = 2$) so in this context "L^p-irregularity" should be interpreted as "failure of L^p-regularity in the *full range* $1 < p < \infty$".

The results described henceforth will appear in [50].

Theorem 1 (Main Result) *For any $p \neq 2$ there is $\beta = \beta(p) > \pi/2$ such that for $W = W_{h,k}$ with h, k as in (8), the Cauchy–Szegő projection associated to W is not bounded: $L^p(bW, \Sigma) \to L^p(bW, \Sigma)$.*

Here Σ is induced Lebesgue measure for bW. The strategy of proof is similar in spirit to the original arguments [4] and [37] for the Bergman projection (which also inspired the strategy of proof for the examples for the Cauchy–Leray integral described in the previous section): one starts with a (biholomorphic) scaling of the original domain W leading to a family of smooth domains $\{W_\lambda\}_\lambda$; then a limiting process transfers the L^p-regularity problem to a specific, unbounded limiting domain W_∞. On the latter, explicit computations are carried out that prove failure of L^p-regularity of the relevant operator for W_∞. The scaling and limiting arguments then allow to percolate failure of L^p-regularity back to W via a suitable transformation law under the scaling map.

When carrying out this scheme for the Cauchy–Szegő projection several new obstacles arise that were nonexistent in the analysis of the Bergman projection and of the Cauchy–Leray integral: here we focus on just one, namely the fact that the limiting domain W_∞ is unbounded and *non-smooth* (it is a Lipschitz domain), thus for W_∞ there is no canonical notion of holomorphic Hardy space nor of Cauchy–Szegő projection (by contrast, the definition of the Bergman space $A^2(W_\infty, dV)$ is standard, and so is the associated Bergman projection). It is not hard to see that the topological boundary of W_∞ splits into three distinct parts: two of these, denoted $\dot{W}_\infty$ and $\ddot{W}_\infty$ have full induced-Lebesgue measure, while the third part is the distinguished boundary $d_b W_\infty$. In [57] the authors prove irregularity of the Cauchy–Szegő projection associated to $d_b W_\infty$ (defined with respect to induced Lebesgue measure for $d_b W_\infty$). However the small size of the distinguished boundary (it is a codimension-1 subset of the topological boundary) makes it impossible to percolate the result for $d_b W_\infty$ back to the Cauchy–Szegő projection for the full boundary of the original worm W. Here we focus instead on the full-measured part of the boundary denoted $\dot{W}_\infty$ because this particular piece of the boundary supports a natural notion of "quasi-product measure" μ_∞ that captures the main features of the full boundary of W_∞, as indicated by the following key observation:

Proposition 2 *Suppose that* $F \in C_0\left(\overline{\bigcup_{\lambda>0} W_\lambda}\right)$. *Then*

$$\lim_{\lambda\to\infty} \int\limits_{bW_\lambda} F\, d\mu_\lambda = \int\limits_{\dot{W}_\infty} F\, d\mu_\infty .$$

Here μ_λ is the transported induced Lebesgue measure for W via the scaling map. (In fact a more sophisticated version of the above proposition is needed, one that is valid for F in a larger function space that is dense in the Hardy space for W, but the above already provides the required "supporting evidence".)

It turns out that the quasi-product measure μ_∞ leads to a meaningful notion of Hardy space for $\dot{W}_\infty$ and furthermore, that the topological boundary of the original (smooth) worm W also supports a "quasi-product" measure μ_0 that is mutually absolutely continuous with respect to induced Lebesgue measure Σ and enjoys a certain stability under the scaling maps, leading us to the following result:

Theorem 3 *Let $\mathcal{S}_\infty$ denote the Cauchy–Szegő projection for $H^2(\dot{W}_\infty, \mu_\infty)$, and let $\mathcal{S}_{bW}$ denote the Cauchy–Szegő projection for $H^2(bW, \Sigma)$. If $\mathcal{S}_{bW} : L^p(bW, \Sigma) \to L^p(bW, \Sigma)$ is bounded, then $\mathcal{S}_\infty : L^p(\dot{W}_\infty, \mu_\infty) \to L^p(\dot{W}_\infty, \mu_\infty)$ is bounded and*

$$\|\mathcal{S}_\infty\|_{L^p(\dot{W}_\infty,\, \mu_\infty)\circlearrowleft} \leq \|\mathcal{S}_{bW}\|_{L^p(bW,\, \Sigma)\circlearrowleft} .$$

Finally, a direct examination shows that $\mathcal{S}_\infty$ is *unbounded* on $L^p(\dot{W}_\infty, \mu_\infty)$, giving us the proof of Theorem 1.

References

1. Ahern, P., Schneider, R.: A smoothing property of the Henkin and Cauchy–Szegő projections. Duke Math. J. **47**, 135–143 (1980)
2. Ahern, P., Schneider, R.: The boundary behavior of Henkin's kernel. Pac. J. Math. **66**, 9–14 (1976)
3. Andersson, M., Passare, M., Sigurdsson, R.: Complex Convexity and Analytic Functionals Birkhäuser, Boston (2004)
4. Barrett, D.: Behavior of the Bergman projection on the Diederich-Fornæss worm. Acta Math. **168**, 1–10 (1992)
5. Barrett, D.E., Ehsani, D., Peloso, M.: Regularity of projection operators attached to worm domains. Doc. Math. **20**, 1207–1225 (2015)
6. Barrett, D., Lanzani, L.: The spectrum of the Leray transform for convex Reinhardt domains in $\mathbb{C}^2$. J. Funct. Anal. **257**, 2780–2819 (2009)
7. Barrett, D.E., Vassiliadou, S.: The Bergman kernel on the intersection of two balls in $\mathbb{C}^2$. Duke Math. J. **120**, 441–467 (2003)

8. Bonami, A., Charpentier, P.: Comparing the Bergman and Cauchy–Szegő projections. Math. Z. **204**, 225–233 (1990)
9. Bonami, A., Lohoué, N.: Projecteurs de Bergman et Cauchy–Szegő pour une classe de domaines faiblement pseudo-convexes et estimations L^p. Compos. Math. **46**(2), 159–226 (1982)
10. Boutet de Monvel, L., Sjöstrand J.: Sur la singularité des noyaux de Bergman et the Szegő, Journées équations aux dérivées partialles de Rennes (1975), pp. 123–164. Astérisque, no. 34–35. Soc. Math. France, Paris (1976)
11. Calderón, A.: Cauchy integrals on Lipschitz curves and related operators. Proc. Nat. Acad. Sci. U.S.A. **74**, 1324–1327 (1977)
12. Chakrabarti, D., Zeytuncu, Y.: L^p mapping properties of the Bergman projection on the Hartogs triangle. Proc. AMS **144**(4), 1643–1653 (2016)
13. Charpentier, P., Dupain, Y.: Estimates for the Bergman and Cauchy–Szegő projections for pseudoconvex domains of finite type with locally diagonalizable Levi forms. Publ. Mat. **50**, 413–446 (2006)
14. Chen, L.W., Zeytuncu, Y.: Weighted Bergman projections on the Hartogs triangle: exponential decay. New York J. Math. **22**(16), 1271–1282 (2016)
15. Christ, M.: A $T(b)$ theorem with remarks on analytic capacity and the Cauchy integral. Colloq. Math. **60/61**(2), 601–628 (1990)
16. Christ, M.: Lectures on Singular Integral Operators, vol. 77. AMS-CBMS (1990)
17. Coifman, R.R., McIntosh, A., Meyer, Y.: L'intégrale de Cauchy définit un opérateur borné sur L^2 pour les courbes Lipschitziennes. Ann. Math. **116**, 361–387 (1982)
18. Cumenge, A.: Comparaison des projecteurs de Bergman et Cauchy–Szegő et applications. Ark. Mat. **28**, 23–47 (1990)
19. David, G.: Opérateurs intégraux singuliers sur certain courbes du plan complexe. Ann. Sci. École Norm. Sup. **17**, 157–189 (1984)
20. Duong, X.-T., Lacey, M., Li, J., Wick, B., Wu, Q.: Commutators of Cauchy-type integrals for domains in $\mathbb{C}^n$ with minimal smoothness. Preprint (2018) (ArXiv: 1809.08335)
21. Duren, P.L.: Theory of H^p Spaces. Dover, New York (2000)
22. Fefferman, C.: The Bergman kernel and biholomorphic mappings of pseudoconvex domains. Invent. Math. **26**, 1–65 (1974)
23. Fefferman, C.: Parabolic invariant theory in complex analysis. Adv. Math. **31**, 131–162 (1979)
24. Gupta, P.: Lower-dimensional Fefferman measures via the Bergman kernel. Contemp. Math. **681**, 137–151 (2017)
25. Hansson, T.: On Hardy spaces in complex ellipsoids. Ann. Inst. Fourier (Grenoble) **49**, 1477–1501 (1999)
26. Henkin, G.M.: Integral representation of functions which are holomorphic in strictly pseudo-convex regions, and some applications (Russian). Mat. Sb. (N.S.) **78**, 611–632 (1969)
27. Hörmander, L.: Notions of Convexity. Progress in Mathematics, vol. 127. Birkhäuser, Boston (1994)
28. Kenig, C.: Weighted H^p spaces on Lipschitz domains. Am. J. Math. **102**, 129–163 (1980)
29. Kerzman, N., Stein, E.M.: The Cauchy–Szegő kernel in terms of Cauchy–Fantappié kernels. Duke Math. J. **45**, 197–224 (1978)
30. Kerzman, N., Stein, E.M.: The Cauchy kernel, the Cauchy–Szegő kernel and the Riemann mapping function. Math. Ann. **236**, 85–93 (1978)
31. Kiselman, C.: A study of the Bergman projection in certain Hartogs domains, in *Several Complex Variables and Complex Geometry, Part 3*, Proc. Sympos. Pure Math, vol. 52, Part 3, Amer. Math. Soc., Providence (1991)
32. Kœnig, K.D.: Comparing the Bergman and Cauchy–Szegő projections on domains with subelliptic boundary Laplacian. Math. Ann. **339**, 667–693 (2007)
33. Kœnig, K.D.: An analogue of the Kerzman–Stein formula for the Bergman and Cauchy–Szegő projections. J. Geom. Anal. **14**, 63–86 (2004)
34. Kœnig, K.D., Lanzani, L.: Bergman vs. Szegö via conformal mapping. Indiana U. Math. J. **58**, 969–997 (2009)

35. Krantz, S.G.: Integral formulas in complex analysis, in Bejing Lectures in Harmonic Analysis, Ann. of Math. Stud., vol. 112, pp. 185–240 (1987)
36. Krantz, S.G.: Function Theory of Several Complex Variables. Wiley, London (1982)
37. Krantz, S., Peloso, M.: The Bergman kernel and projection on non-smooth worm domains. Houst. J. Math. **34**, 873–950 (2008)
38. Kytmanov, A.M.: The Bochner–Martinelli Integral and Its Applications. Birkhäuser, Basel (1992)
39. Lanzani, L.: Cauchy–Szegő projection versus potential theory for non-smooth planar domains. Indiana Univ. Math. J. **48**, 537–556 (1999)
40. Lanzani, L.: Cauchy transform and Hardy spaces for rough planar domains. Contemp. Math. **251**, 409–428 (2000)
41. Lanzani, L.: Harmonic analysis techniques in several complex variables. Bruno Pini Math. Anal. Seminar Ser. **1**, 83–110 (2014). ISSN: 2240-2829
42. Lanzani, L., Stein, E.M.: Cauchy–Szegő and Bergman projections on non-smooth planar domains. J. Geom. Anal. **14**, 63–86 (2004)
43. Lanzani, L., Stein, E.M.: The Bergman projection in L^p for domains with minimal smoothness. Illinois J. Math. **56**(1), 127–154 (2013)
44. Lanzani, L., Stein, E.M.: Cauchy-type integrals in several complex variables. Bull. Math. Sci. **3**(2), 241–285 (2013)
45. Lanzani, L., Stein, E.M.: The Cauchy integral in $\mathbb{C}^n$ for domains with minimal smoothness. Adv. Math. **264**, 776–830 (2014)
46. Lanzani, L., Stein, E.M.: The Cauchy–Leray integral: counter-examples to the L^p-theory. Indiana U. Math. J. **68**(5), 1609–1621 (2019)
47. Lanzani, L., Stein, E.M.: The role of an integration identity in the analysis of the Cauchy–Leray transform. Sci. China Math. **60**, 1923–1936 (2017)
48. Lanzani, L., Stein, E.M.: The Cauchy–Szegő projection for domains with minimal smoothness. Duke Math. J. **166**(1), 125–176 (2017)
49. Lanzani, L., Stein, E.M.: Hardy Spaces of Holomorphic functions for domains in $\mathbb{C}^n$ with minimal smoothness. In: Harmonic Analysis, Partial Differential Equations, Complex Analysis, and Operator Theory: Celebrating Cora Sadosky's Life, AWM-Springer vol. 1, pp. 179–200 (2016). ISBN-10:3319309595.
50. Lanzani, L., Stein, E.M.: On irregularity of the Cauchy–Szegő projection for the Diederich-Fornæss worm domain. Manuscript in preparation
51. McNeal, J.: Boundary behavior of the Bergman kernel function in $\mathbb{C}^2$. Duke Math. J. **58**(2), 499–512 (1989)
52. McNeal, J.: Estimates on the Bergman kernel of convex domains. Adv. Math. **109**, 108–139 (1994)
53. McNeal, J., Stein, E.M.: Mapping properties of the Bergman projection on convex domains of finite type. Duke Math. J. **73**(1), 177–199 (1994)
54. Meyer, Y., Coifman, R.: Ondelettes et Opérateurs III, Opérateurs multilinéaires, Actualités Mathématiques, Hermann (Paris) (1991), pp. i–xii and 383–538. ISBN: 2-7056-6127-1
55. Monguzzi, A.: Hardy spaces and the Cauchy–Szegő projection of the non-smooth worm domain D'_β. J. Math. Anal. Appl. **436**, 439–466 (2016)
56. Monguzzi, A.: On Hardy spaces on worm domains. Concr. Oper. **3**, 29–42 (2016)
57. Monguzzi, A., Peloso, M.: Sharp estimates for the Cauchy–Szegő projection on the distinguished boundary of model worm domains. Integr. Equ. Oper. Theory **89**, 315–344 (2017)
58. Monguzzi, A., Peloso, M.: Regularity of the Cauchy–Szegő projection on model worm domains. Complex Var. Elliptic Equ. **62**(9), 1287–1313 (2017)
59. Munasinghe, S., Zeytuncu, Y.: Irregularity of the Szegő projection on bounded pseudoconvex domains in $\mathbb{C}^2$. Integr. Equ. Oper. Theory **82**, 417–422 (2015)
60. Munasinghe, S., Zeytuncu, Y.: L^p-regularity of weighted Cauchy–Szegő projections on the unit disc. Pac. J. Math. **276**, 449–458 (2015)
61. Nagel, A., Pramanik, M.: Diagonal estimates for the Bergman kernel on certain domains in $\mathbb{C}^n$ (2011). Preprint

62. Nagel, A., Rosay, J.-P., Stein, E.M., Wainger, S.: Estimates for the Bergman and Cauchy–Szegő kernels in $\mathbb{C}^2$. Ann. Math. **129**(2), 113–149 (1989)
63. Nazarov, F., Treil, S., Volberg, A.: Cauchy integral and Calderón–Zygmund operators on nonhomogeneous spaces. Int. Math. Res. Not. **15**, 703–726 (1997)
64. Phong, D., Stein, E.M.: Estimates for the Bergman and Cauchy–Szegő projections on strongly pseudoconvex domains. Duke Math. J. **44**(3), 695–704 (1977)
65. Poletsky, E., Stessin, M.: Hardy and Bergman spaces on hyperconvex domains and their composition operators. Indiana Univ. Math. J. **57**, 2153–2201 (2008)
66. Range, R.M.: Holomorphic Functions and Integral Representations in Several Complex Variables. Graduate Texts in Mathematics, vol. 108. Springer, Berlin (1986)
67. Ramírez de Arellano, E.: Ein Divisionsproblem und Randintegraldarstellungen in der komplexen analysis. Math. Ann. **184**, 172–187 (1969/1970)
68. Rotkevich, A.S.: Cauchy–Leray-Fantappiè integral in linearly convex domains. J. of Math. Sci. **194**, 693–702 (2013)
69. Rotkevich, A.S.: The Aizenberg formula in non convex domains and some of its applications. Zap. Nauchn. Semin. POMI **389**, 206–231 (2011)
70. Semmes, S.: The Cauchy integral and related operators on smooth curves. Thesis, Washington University (1983)
71. Stein, E.M.: Boundary Behavior of Holomorphic Functions of Several Complex Variables. Princeton University Press, Princeton (1972)
72. Stein, E.M.: Harmonic Analysis Princeton University Press, Princeton (1993)
73. Stout, E.L.: H^p functions on strictly pseudoconvex domains. Am. J. Math. **98**, 821–852 (1976)
74. Tolsa, X.: Analytic capacity, rectifiability, and the Cauchy integral, International Congress of Mathematicians, vol. II, 1505–1527. Eur. Math. Soc., Zürich (2006)
75. Tolsa, X.: The semiadditivity of continuous analytic capacity and the inner boundary conjecture. Am. J. Math. **126**, 523–567 (2004)
76. Verdera, J.: L^2 boundedness of the Cauchy integral and Menger curvature. In: Harmonic Analysis and Boundary Value Problems. Contemp. Math., vol. 277, pp. 139–158. Amer. Math. Soc., Providence (2001)
77. Zeytuncu, Y.: L^p-regularity of weighted Bergman projections. Trans. Am. Math. Soc. **365**, 2959–2976 (2013)